Advanced Statistical Methods in Life Science

This book introduces the principles and foundations of advanced statistical methods for designing experiments and testing hypotheses in life sciences. Advanced statistical methods, such as testing of hypotheses, recent methods of sample size determination/imputation, estimation techniques, probability distributions, and univariate analysis demonstrated with real data, and their integration into life sciences are included in this book. Advanced topics are presented with sufficient conceptual depth and examples to explain the use of recent statistical techniques and to demonstrate what conclusions can be drawn at the right time using modeling in life science research.

Key features:

- Explains the derivation of statistical models to prove disease transmission using massive real-world datasets to explore practical applicability
- Incorporates the application of innovative advanced statistical and epidemiological models and demonstrates the possible solutions for public health policy intervention
- Helps to understand the process of hypothesis testing in small or larger observations by using weighted parameters
- Presents suitable examples and real-life research datasets, and all models can easily be followed in formulating statistical and mathematical derivations and key points
- Includes machine learning (ML), statistical methods for meta-analysis, testing of hypotheses, methods of imputation, estimation techniques, probability distributions, univariate analysis, and recent nonparametric methods, all illustrated through actual data.

This textbook is for students and scholars of various courses in life sciences, medicine, mathematics, and statistical science. It will also help academicians and researchers to understand the foundation of this topic.

Advanced Statistical Methods in Life Science

Basavarajaiah D.M and Narasimhamurthy B

CRC Press
Taylor & Francis Group
Boca Raton London New York

CRC Press is an imprint of the
Taylor & Francis Group, an **informa** business
A CHAPMAN & HALL BOOK

Designed cover image: ShutterStock

First edition published 2026
by CRC Press
2385 NW Executive Center Drive, Suite 320, Boca Raton FL 33431

and by CRC Press
4 Park Square, Milton Park, Abingdon, Oxon, OX14 4RN

CRC Press is an imprint of Taylor & Francis Group, LLC

ISBN: 9781032711539 (hbk)
ISBN: 9781032713823 (pbk)
ISBN: 9781032713816 (ebk)

DOI: 10.1201/9781032713816

Typeset in Times
by codeMantra

Heartily dedicated to

Highness of his holy feet God

"Avadoot Datta Guru"

Sri. Sadanandaswamiji

Contents

Preface

Data in life science are derived from investigations, observations, and experiments with living things. They include ecological observations, biochemical measurements, genetic sequences, and clinical trial outcomes. Students at the undergraduate level need to be exposed to and tutored in basic statistical methods to understand elementary data from life sciences. These methods include an overview of descriptive statistics, including the coefficient of variation; theory of probability; statistical inference, including estimation; correlation and various regression methods; types of study designs; sampling techniques; sample size determination; and nonparametric tests. For postgraduate students with exposure to basic statistical methods, advanced statistical techniques are essential to understand and comprehend the complexity of data analysis, including inference and preparation of a report. These include statistical tools for clinical trials; statistical models using real-life data for life science research; strategies to study molecular genetics, including artificial neural network modeling using real-life data; and meta and systematic analysis. Research scholars and investigators need exposure to the latest developments in statistical methodology, for example, statistical ethics and future research. Various statistical techniques entail methodically examining and analyzing data to extract insightful information. Beyond simple calculations, they reveal latent patterns and approximate both observed and unobserved components. Empirically based conclusions for future research are produced by robust statistical techniques. In the field of life science data analysis, a variety of statistical tests are employed to extract meaningful information and draw trustworthy conclusions from complex datasets. The most appropriate statistical test will depend on the kind of data and the current research question. To meet the needs of undergraduate, postgraduate, and research scholars/investigators in statistics, this book covers the fundamental and advanced statistical tests and their applications in life science, medical, and biological research. These methods assist in overcoming challenges related to data analysis in many disciplines. The steps for important tests using SAS and R software packages are provided. This book provides a comprehensive set of statistical tools for analyzing the complex data structures of high- and micro-throughput datasets. The entire book is divided into 14 chapters, each containing visually appealing graphs and figures that are used to explain various real-world examples. We have aimed to reconnect the basic ideas of statistics with biology in each chapter. To aid in the understanding of fundamental ideas, it offers scholars, researchers, and inexperienced researchers fresh viewpoints. The diagrammatic representations of life science research are discussed in Chapter 1. Chapter 2 includes fundamental statistical methods and its applications to biological and life sciences research with various real-world examples and software codes. Chapter 3 addresses the applications of correlation and regression analysis. Readers will find it easy to understand the practical applications of this technique and its wide range of uses in theoretical and biological science research. The law of probability and various distributions are extensively used in real-world applications discussed in Chapters 4 and 5. Various theoretical and generalized biologically related illustrations are described in an exhaustive manner, assisting postgraduate and doctoral students in deriving important applications of the law of probability from various research problems. An introduction to statistical hypothesis testing for life sciences is covered in Chapter 6. It also covers various hypothesis testing methods, how to use ANOVA and ANCOVA in research, and how to construct a data frame for analysis using R software. Chapter 7 and 8 explored the statistical performance and different strategies used to study the molecular genetics and potential of proteomics/data analysis with various statistical measures. Chapter 9 deals with statistical tools for clinical trials in medical research. Several clinical trials are conducted worldwide, and clinical researchers find that in-depth knowledge of statistical methods, design settings, and data compilation is crucial in the early stages of experiments. Medical researchers and clinical experts might find this chapter useful. Many scientists have made arbitrary assumptions about the size of the sample, chosen from the target or parental population, because there was a dearth of statistical literature. The sample size for a study needs to be estimated at the time the study is proposed; too large a sample is unnecessary and unethical, while too small a sample is unscientific and also unethical. If no assumptions can be made, then an arbitrary sample size is set for a pilot study. In keeping with this, the basic concepts of sampling techniques and sample size determination in biological science research are covered in Chapter 10. The comprehensive overview of machine learning and artificial neural network modeling is covered in Chapter 11. We address the state-of-the-art machine learning applications in the Cloud Force and AI Generative eras in this chapter. Twenty thousand scientific journals publish over one hundred thousand research articles annually. A human being cannot possibly read all the published articles (metadata) in any field. A concise and trustworthy literature summary is needed to understand new insights from genuine research after the untrustworthy and unnecessary articles are removed from the global metadata. A purposeful discussion of the entire research findings summary will be conducted using statistical techniques. Regarding the research's value and visibility, numerous researchers have published meta-data studies with a wide scope and minimal use of statistical tools. Applying meta-analysis in the new trend of scientific scenarios is described in Chapter 12, based on the practical approach of metadata and systematic analysis. Readers can easily understand the flowchart and analytical steps for systematic review and meta-analysis. Researchers and non-researchers start using ambient analytical techniques to perform large-scale meta-analyses after reading this chapter. In some biological experiments, skewed data series prevented scientists from testing hypotheses and drawing conclusions about the study material or subjects. Viewed from this perspective, Chapter 13 gives readers a summary of various nonparametric fundamental and recent tools. Readers will find it easy to use these tools for gathering and grouping variables. New and advanced tools like Fried–Dehling (FD), Beier, Buning's adaptive tests, Terpstra–Magel, and New tests were developed and demonstrated using real datasets. Readers will find the analytical procedures very easy to comprehend. Chapter 14 concludes with a discussion of future line of research work and the challenges of statistical ethics. This book can be used as a textbook for undergraduate students in schools and colleges. This textbook can also be prescribed to postgraduate students in schools and universities. Researchers and academicians will find this book useful as a reference tool for further research in compiling life science datasets as it examines specific applications of the statistical approach to life sciences research. In the new era of global virtual vetted massive data (GVVMD), all students (UG, PG, and Ph.D.), regardless of discipline, should use this opportunity to learn new insights into analytical applications and how to integrate statistics into life science, medical, and biological research. Above all, this book will hopefully equip users to demystify the art of statistics in its application to life sciences and assist policymakers, planners, and administrators in furthering global health. Jai Hind!

Authors

Basavarajaiah D.M is currently the Head of the Department of Statistics and Computer Science at Karnataka Veterinary, Animal, and Fisheries Sciences University (Bidar), Hebbal, Bangalore. He obtained his Ph.D. from the National Institute of Epidemiology (NIE), ICMR, Chennai (affiliated with the University of Madras). His research focuses on statistical theory and statistical modeling of high-dimensional datasets in agriculture, medicine, veterinary medicine, and animal sciences. He has authored 100 research articles and six academic books with reputed publishers, including works on HIV transmission statistical modeling, the design of experiments and advanced statistical techniques in clinical research, and COVID transmission modeling (published by CRC Press, UK). He serves as an editorial board member and scientific board advisor for various international indexed journals. He is a life member of numerous academic organizations and has received several accolades for his academic and research excellence. These include the Chartered Scientist award stalwart by the Science Council, the United Kingdom, in collaboration with the Royal Statistical Society; the Outstanding International Scientist award from VDGOOD organization; the Best Reviewer award (2016) from 'TRANSSTELLAR' Journal Publications and Research Consultancy, TJPRC Ltd. (NAAS-rated journals); and fellowships from the Royal Statistical Society, UK (London) and the Mathematical Society, UK (London). He has been honored with the Bharathshikshratana and Indo-Dubai Achiever's Pacific awards by the Global Society of Health and Educational Growth, New Delhi. Other awards include Best Reviewer (2015), Best Scientific Board Advisor (2016), and Best Editorial and Researcher awards conferred by the International Academy of Engineering Science and Technology, USA. He is an active member of the Board of Studies and the Academic Council of various universities.

Narasimhamurthy B is a former Scientist 'G' (Director Grade cadre) at the National Institute of Epidemiology, ICMR, Chennai. Currently, he is an independent National Level Monitor for the National Health Mission, MOH&FW, Government of India. He obtained his Ph.D. from the International Institute of Population Science (IIPS), affiliated with Bombay University, in 1985. His areas of research include mathematical statistics, operational research, clinical research, public health, demography and vital statistics, and mathematical modeling in medical science and public health. He has authored more than 100 research articles published in reputed national and international journals and co-authored four books, including HIV transmission statistical modeling, design of experiments and advanced statistical techniques in clinical research, and COVID transmission modeling (published by CRC Press, UK). He has also guided research scholars for their Ph.D. He has served as an editorial board member and scientific board advisor for various international MCI and UGC-recognized journals and is a life member of various academic organizations. He has been honored with several accolades for his academic and research excellence, including the Prof. R. N. Srivastava Gold Medal Award, the Bharat Jyothi Award, the Bharat Ratna Indira Gandhi Gold Medal Award, and the Best Citizens of India award. Recently, he was recognized as one of the Asian Admirable Achievers. He serves as a member of the Board of Studies for various national universities, including the University of Madras, Acharya Nagarjuna University, and the University of Kerala. He is also a visiting emeritus professor at Johns Hopkins University, Bloomberg School of Public Health, Baltimore, Maryland, and North Carolina University, USA, among others.

1

Diagrammatic Representation of Life Science Research: A Comprehensive Overview

1.1 Introduction

The datum is singular, data is plural. A datum can be derived from numerical and alphabetical expressions such as numbers, sequences, or arrays. Any statistical figures collected from the experimental area are called data, which can include both textual and non-textual data series. Examples of textual data include photos of experimental plots, videos, and images of glass ware (Zelterman & Holmes, 1986). Non-textual data include measurements or numerical expressions such as the efficacy of a drug, the number of animals considered for each replication, the number of days spent in hospitals, BMI, and the age of the respondents. Both data series are required for researchers to compile further investigations and estimate the relatedness of variables to obtain conclusive results. After obtaining and editing the data, the next step for the researcher is to classify the data series (Sab et al., 2017). Classification is a meaningful scientific method for arranging data series into groups or categories based on common characteristics. In other words, classification is the process of arranging the data into sequential order according to their common characteristics and separating them into different homogeneous numerical expressions. The common essential characteristics of data classification are clarity, stability, flexibility, and level of utilization (Scaife & Rogers, 1996). The important objectives of classification are to reduce flaws in collected data by dividing them into classes and statistical measures, to unravel the significance of hidden information in the data, to make the data easily understandable, and to facilitate the relative comparison of related variables. Classification can provide scientific data presentation through tabulation, diagrams, and graphs. Statistical classification requires certain rules for properly measuring variables (Rahmasari & Kuswanto, 2023). These rules include exhaustibility, exclusiveness, homogeneity, consistency, flexibility, and appropriability.

Exhaustibility: This means that the classification should be made in an exhaustive manner so that each item of the data series belongs to one of the classes without leaving any item to be shown under any other class.

Exclusiveness: The different classes should be mutually exclusive from each other, without any overlapping.

Homogeneity: All the data included in a particular class should be of a homogeneous or similar nature, and each different class should include data of a homogeneous nature.

Consistency: Basis of classification should be maintained not only for the same variable but also for all related variables.

Flexibility: Manner of classification should allow for changes in times and situations. Outdated classes should be removed and new classes should be added for suitable analysis.

Appropriability: The basis of classification should be appropriate to the nature of the data series; otherwise, the purpose of the classification may be defeated (Table 1.1).

TABLE 1.1

Types of Biological Research Data Classification

Classification Types	Descriptions	Examples
Geographical	Classification based on the area or place is known as spatial classification	*Tectonagrandis* (Teak) tree species are distributed across different geographical areas. The prevalence of lumpy skin disease in different disease affected areas
Chronological	The data are classified based on the time of their occurrence (purely known as time series)	The growth of Candida species inoculated in petri plates was estimated at different time intervals
Qualitative	Classified based on certain descriptive characteristics or qualitative aspects of a phenomenon, e.g. gender-wise distribution with respect to beauty, honesty, IQ, religion, eyesight, or clothing preferences, etc.	Selection of pea plants based on its color of the flowers, shape, and the size of the seeds etc.
Quantitative	Under this type of classification, the collected data are classified based on certain variables	Classification of the students based on marks, income, age height, economic holdings, etc

DOI: 10.1201/9781032713816-1

1.2 Research Data Tabulation

The research data has been arranged in rows and columns. It is done to enable meaningful analysis and interpretation of the data. The objective of data collection procedures is to obtain numerical measures of some sort, to be used in administration, evaluation, and research. These measures may represent the results of laboratory experiments, clinical trials, or observations of characteristics of a population. These are usually presented in a tabular form. Such a form makes it possible for us to grasp the overall picture at a glance. Diagrams help us to see the pattern and shape of any complex data. They register a meaningful impression before we think. However, both tables and diagrams sacrifice details and accuracy of data in order to improve communication. The sacrifice is justifiable under the following circumstances: the audience is not technically trained and will therefore not gain as much from raw data as from a table or diagram, or when it is required to get a point across quickly.

Independent and dependent variables in a table: Suppose we are creating a table to show the number of patients based on their age. In this case, the number of patients is the dependent variable, while age is the independent variable (Table 1.2). A typical table classifies data based on a single variable (the independent variable), and the table displays the number of observations in each class of the independent variable. This relationship is known as a frequency distribution, where frequency refers to the number of observations in a particular class. The table below is an example of a frequency distribution.

Table 1.3 is called a one-way table or a one-dimensional table. The dimension of a table is equal to the number of independent variables used in the table. The following table, which gives the number of children by age and sex, is a two-dimensional table.

Table 1.4 describes age and gender as two independent variables, while the number of children is the dependent variable. An example of a three-dimensional table is shown below:

TABLE 1.2

Age Wise Distribution of TB Cases

Age Class (Years)	Male ($n = 556$)	Females ($n = 386$)	Total
<18	125	89	214
19–25	44	32	76
26–30	78	45	123
31–35	69	55	124
36–40	32	24	56
41–45	25	26	51
46–50	84	62	146
51–55	36	20	56
56–60	38	23	61
>60	25	10	35

TABLE 1.3

Income Distribution of Families

Monthly Income	No. of Families ($n = 253$)
Under Rs.1,000	200
Rs. 1,000 – Rs.4,999	50
Rs. 5,000 or more	0.3

TABLE 1.4

Age Wise Distribution of Males and Females

Age (Years)	Males ($n = 150$)	Females ($n = 140$)
0–4	32	24
5–9	48	36
10–14	70	80

TABLE 1.5

Distribution of Disease Status

Vaccination Status	Disease Developed	Disease Not Developed
Males		
Vaccinated	45	36
Not vaccinated	25	18
Females		
Vaccinated	55	32
Not vaccinated	28	35
Total	153	121

From Table 1.5 sex, vaccination status, and subsequent status are the three independent variables, while the number of persons is the dependent variable. If a table has more than three independent variables, it becomes challenging to prepare and understand. According to Tuttle, a statistical table is a logical listing of related quantitative data in vertical columns and horizontal rows of numbers, with different explanatory and qualifying words, phrases, and statements in the form of titles, headings, and notes to make the full meaning of the data and their origin clear.

1.3 Characteristics

- It is an art and science for the arrangement of quantitative data with a symmetrical nature.
- The data must be related to each other on some logical basis.
- The data are arranged in particular cells without any jargon.
- It must have explanatory notes and qualifying words to make the meaning of the data clear.
- This is done after the data are classified.

1.4 Reasons for Tabulations

- To simplify the complexity of the data.
- To facilitate comparative analysis of the data.
- To ensure the economy of space and time.
- To indicate the trends and patterns of the data.
- To facilitate references.
- To facilitate the computation of various factors.
- To identify errors in research data.
- To know the source of data in the appropriate reference heading.

1.5 Components of the Research Data Tabulation Aid of a Table

The aim of a table is to clearly communicate the information contained in the collected data in a condensed form. It should be simple and easy to understand.

1.5.1 Identifying the Numbers

A number is assigned to each table to identify and refer to it when necessary.

1.5.2 Title, Footnote, and Source

The title is an essential part of a table. It should indicate the nature of the data in the table (dependent variable), the variables used to classify the data (independent variables), and the place and time period to which the data refer. A helpful way to remember the title rule is "What, how classified, where, and when." Any remarks about a specific row, column, or cell are usually given in a footnote. In the case of secondary data, the source of the data is provided in a footnote. The source is typically the primary source, such as the publication where the data originally appeared. This allows the reader to check the data at the source for further details. An example of a typical table is provided (see Table 1.1).

1.5.3 Trial-and-Error Methods in Tabulation

Experimenting with different styles of table presentation can be challenging, even for an experienced investigator. Decisions need to be made regarding whether to present actual numbers, percentages, or both; the number of classes to use; the dimensions assigned to the rows and columns; and the various headings, including a "total" line.

1.5.4 Subs

A stub is a heading given to a row of a table for identification and easy reference to the presented data. It should be prominently mentioned at the left end of each row.

1.5.5 Body of the Table

This is the main part of the table where the numerical figures are presented, coordinating the respective columns and rows. The arrangement of the body is typically from left to right in the rows and from top to bottom in the columns. If a figure is not available for a coordinated cell of the table, it should be indicated by an abbreviation.

1.5.6 Total

Figures of each column and sub-column, as well as each row, should be totaled. To clearly display the totals of different columns, a separate "row of totals" is provided at the bottom of the table.

Diagrammatic representation is a visually appealing way to present data series through diagrams and graphs. It is often used by statisticians to present data to a general audience who may not understand numerical figures. Most experimental datasets are expressed in unstructured form, making it difficult for reviewers and readers to understand the data. Diagrammatic representation uses geometric figures such as points, lines, bars, squares, rectangles, circles, pie charts, cubes, pictures, cartograms, maps, charts, and advanced AI contour and heat maps to represent the data (Table 1.6).

TABLE 1.6
Rules of Tabulations

Steps	Rules	Description
1	Numbering	Assigning numbers to all tables, overlapping numbers should be avoided.
2	Headings of the table	The heading of the tables should be presented in the simplest way to understand the overall information for the readers.
3	Abbreviation	The longest words should be abbreviated and presented in the footnote of the table.
4	Ditto marks	It should be used, as at times it creates confusion on the part of the readers or viewers.
5	Clarity	Every table should give meaningful information pertinent to your research results.
6	Units	The unit must be presented as a prefatory heading of the table, immediately below the title line of the table. If the different data have different units, they should be stated at the top of the respective columns (kgs, micro/dL, grams, etc.).
7	Fixing the number of columns and rows	Keeping in view the nature and types of data to be presented, the number of rows and columns should be carefully fixed. Any mistake in this respect will cause overlapping.
8	Size	The length and width should always be kept in mind.
9	Marking overcrowding	Every column and row of the table should be marked with numbers in serial order.
10	Minimization of main headings	The number of main headings should be kept few so that the main points of the table can be easily understood by the readers.
11	Self explanatory	The captions (column headings) and stubs (row headings) should be self-explanatory without leaving any room for further clarification.

(Continued)

TABLE 1.6 (*Continued*)

Rules of Tabulations

Steps	Rules	Description
12	Vicinity	Columns to be compared with each other should always be kept close to each other. Similarly, columns for percentages, averages, etc., should be mention closer to the columns of the datasets.
13	Approximation	Figures should always be mentioned in the body of the table and should also be approximate to the real values.
14	Totality	The sums of the rows ought to be displayed in the far-right column (while last row of the column).
15	Arrangement	The items should be arranged in a logical order (alphabetical, geographical, chronological, by size, importance, or causal relationship) to facilitate comparison and analysis of the datasets.
16	Indicating the emphasis	When certain figures need more emphasis, they should be shown in boxes, circles, or between two thick bars.
17	Logical sequence	In preparing the table, a logical sequence must be maintained without any ambiguity.
18	Suitability	The table must be prepared to suit the purpose of the inquiry.
19	Overall beauty of the table	The table should be attractive and appealing to the reader's eyes and minds; it should be easy to understand without much effort.

TABLE 1.7

Rules of Diagrammatic Representation of Datasets

Rule No	Rules	Description
1	Selection of proper diagram	The proper diagram should be selected for construction to give appropriate information. If the diagram selection is not suitable for the data, it may cause confusion and appear distorted to the observer.
2	Simplicity	The primary purpose of resorting to diagrammatic representation of data is to eliminate complexity and make them easily understandable for the reader or viewer.
3	Attractiveness	One of the chief objectives of a diagram is that it should be attractive and convey the pith and substance of the subject matter under review.
4	Size	The size of the diagram should neither be too big nor too small. If it is too big, it may look unattractive and inelegant, and it may not be possible to accommodate all the data in a standard size.
5	Proportion of length and breadth	Though there is no such hard and fast rule regarding the proportion of the length and breadth of a diagram, it should be in proportion $\sqrt{2}$:1, i.e. 1.41 to 1. This means the longer side (vertical or horizontal) should be 1.414 times the length of the shorter side.
6	Scaling	Proper scaling of a diagram is important for the representation of the data and for deriving conclusion from them. Basically, there are two axes, namely horizontal and vertical, of a diagram that need to be scaled properly.
7	Note of reference	For ready reference and comparative study, it is required that each diagram be assigned a number or alphabetical letter.
8	Note of source	To provide the veracity of the information represented through a diagram, a note should be made of the source from which the data have been obtained. This will augment the authenticity and reliability of the information to the satisfaction of learned observers.
9	Foot note	To clarify certain ambiguities, some notes should be added at the foot of the diagram. For instance, if a particular year is a financial year and all the rest are calendar years, it should be mentioned in the footnote to prevent any confusion that may arise in the reader's mind.
10	Index	The index should be provided on the left or right side of the corner of all diagrams, depicting different components used in the construction of the diagram.

1.6 Characteristics of Diagrammatic Representation

1.6.1 Pictorial Representation

Diagrams provide a visual view of quantitative data, making them more attractive and easier to understand.

1.6.2 Concerned Only with Homogeneous Data

Diagrams are suitable for representing data that is comparable and related to each other. Heterogeneous data, which is unrelated and incomparable, is not suitable for diagrammatic representation.

1.6.3 Overexpression of Constant Shape

Diagrams do not have a constant shape or size like numerical figures. Their size and shape change with the scale and manner of presentation.

1.6.4 Bird's-Eye View of Research Data

Diagrams provide a high-level overview of the collected data related to a problem. They do not reveal the entire story of the problem and cannot be further analyzed or compiled using statistical software (Table 1.7).

1.7 Types of Diagram

Diagram

Pictogram	Cartogram	Dimensional diagram
a.One dimensional diagram 1.Simple (Absolute, broken, deviation, flow) 2.Multiple Spheres 3. Subdivided (Absolute, relative)	b. Two dimensional 1. Rectangle 2. Square 3. Pie (Simple, angular)	c. Three 1. Cube 2. Spheres 3. Cylinder
a. Pictogram	b. Cartograms	c. Dimensional diagram
Appropriate pictures are drawn to represent the quantitative data relating to a phenomenon, e.g. doctor-to-patient ratio in hospitals or public health centers, and the proportion of males and females (gender distribution).	A cartogram refers to a map through which statistical information is represented in different manners, viz. shades, data, pictogram, and columns, e.g. the distribution of rainfall in various parts of a country or a pandemic outbreak's worldwide distribution with respect to symptomatic and asymptomatic cases.	A diagram in which certain are displayed prominently and proportionally is called a dimensional diagram. It is classified into three types: one-dimensional, two-dimensional, and three-dimensional diagrams.

1.7.1 One-Dimensional Diagram

A one-dimensional diagram is a diagram in which only one dimension, typically length, is fixed in proportion to the value of the data. These diagrams are also commonly referred to as bar diagrams. They can be drawn vertically or horizontally, the only difference being the length dimension.

Advantages

- Simple to draw and read
- Can represent a large amount of data on a single piece of paper
- Can be drawn both vertically and horizontally
- Provides a better appearance and facilitates comparison

Disadvantages

- Cannot exhibit a large number of aspects of the data
- The width of the bars is arbitrarily fixed

1.7.2 Classification of One-Dimensional Diagrams

Simple bar diagram: The length of this type of diagram is fixed in proportion to the magnitude of the data, while the width is fixed arbitrarily based on the number of diagrams to be accommodated on the given piece of information. The diagram can be scaled vertically or horizontally based on the viewer's convenience and paper size. The height of a bar is proportional to the frequency of the variable. Each bar has the same width. A bar chart represents a dataset using vertical or horizontal bars (Figure 1.1). The greater the height of the bar, the higher the value for the individual category.

1.7.3 Steps to Draw a Bar Chart

A diagram that prominently features specific elements.

Step 1: Prepare a pair of axes. Usually, the horizontal axis (x-axis) is labeled with the categories of the dataset and the vertical axis (y-axis) is the frequency. Your axes must be labeled.

Step 2: Mark the frequencies on the vertical axis in equal intervals.

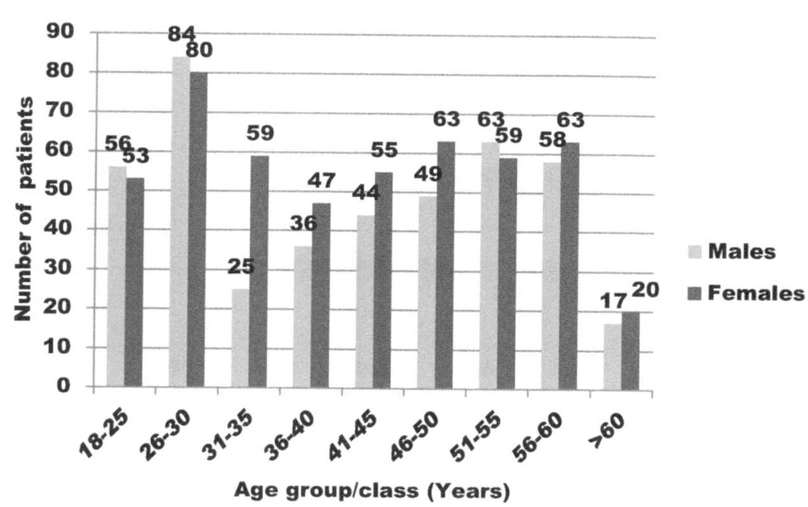

FIGURE 1.1 Bar chart representing the distribution of males and females.

Step 3: Ensure the bars have equal gaps between them, as they represent discrete data.

Step 4: Ensure the bars are of equal width.

Step 5: Add a title to the chart.

When bar charts are represented vertically, they are sometimes called column charts. Bar charts can also be represented horizontally, where the vertical axis represents each class and the horizontal axis represents the frequency.

1.8 Deviation Bar Diagram

When a simple bar diagram is drawn to represent only one difference or deviation of certain related values, it is called a deviation bar diagram. Such diagrams depict only the net quantities, such as surplus or deficit, net profit, net loss, balance of payment for employees, trade business, drug potency, etc. As such, these diagrams can have both positive and negative directions (Figure 1.2).

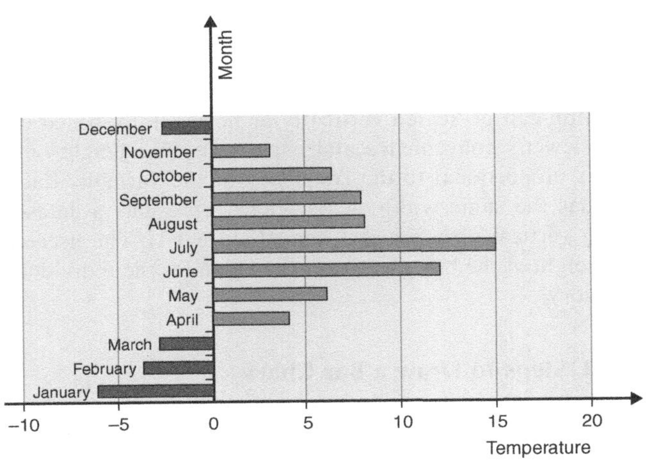

FIGURE 1.2 Bar chart that depicts temperature variations in Bangalore.

1.9 Multiple-Bar Diagrams

Multiple-bar diagrams are created when different sets or groups of data are represented using bars to show two or more aspects of the data. These diagrams, also referred to as compound bar diagrams, are used when comparing two or more aspects of a dataset, such as the male, female, and children components of a population or the arts, science, and commerce components (Figure 1.3).

1.10 Pie Chart

A pie chart is a circular diagram divided into different sectors, with the size of each sector proportional to the frequency it represents (Figure 1.2). A pie diagram is a circular graph where the different sectors represent different frequencies, and the areas of the sectors are proportional to the frequencies. To create a pie diagram, the central angle for each component is calculated, and a circle is drawn. Radii are then drawn for each component, with the sectors shaded and labeled accordingly (Table 1.8).

1.10.1 To Create a Pie Chart

Step 1: Calculate the central angle for each component using the formula provided (Figure 1.4).

Step 2: Draw a circle with a convenient radius.

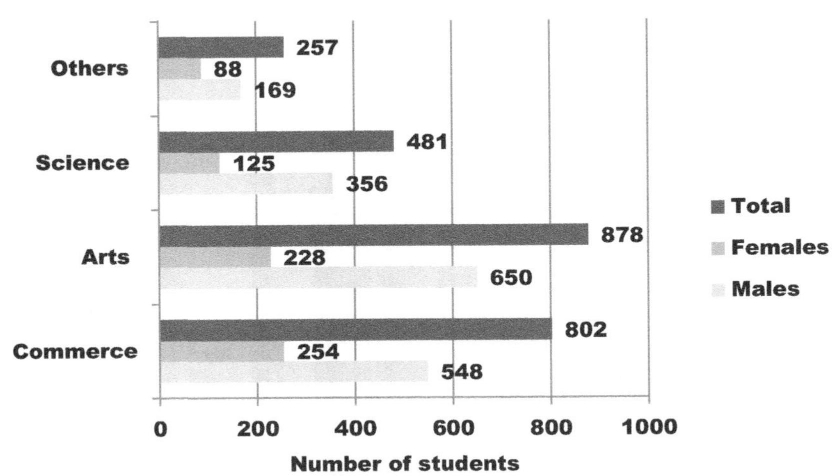

FIGURE 1.3 Multiple-bar chart showing the distribution of students across various disciplines.

Step 3: Draw a horizontal radius within the circle.

Step 4: Draw a radius for each component, starting with the first component and making its central angle with the horizontal radius. Continue drawing radii for each component until all components are represented.

TABLE 1.8

Construction of Pie Chart with Different Components

Particulars	Expenses	Central Angle
Food	4,800	$\frac{4,800}{12,000} \times 360° = 140°$
House rent	2,400	$\frac{2,400}{12,000} \times 360° = 72°$
Clothing	1,600	$\frac{1,600}{12,000} \times 360° = 48°$
Education	800	$\frac{800}{12,000} \times 360° = 24°$
Savings	1,000	$\frac{1,000}{12,000} \times 360° = 30°$
Miscellaneous	1,400	$\frac{1,400}{12,000} \times 360° = 42°$
Total	12,000	360°

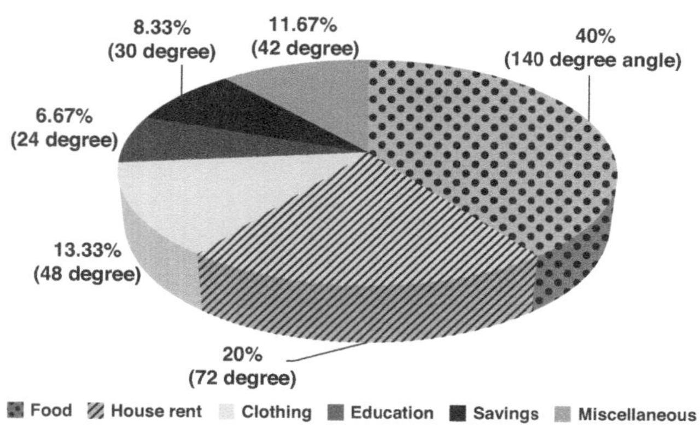

FIGURE 1.4 Pie chart of expenditures incurred for purchasing various items to enhance the quality of life.

Step 5: Shade each sector differently and label it with the component it represents.

Step 6: Provide a heading for each component.

The central angle of a component is calculated using the formula provided in Step 1.

$$= [\text{Value of the component} / 100] \cdot 360°$$

We may calculate the central angles for various components as follows:

1.11 Line Diagram

A line diagram shows time (e.g. years) on the x-axis and the value of a quantity or index (e.g. couple protection rate) on the y-axis (Figure 1.5). It should include a number, title, footnote, and source. The independent variable is marked on the x-axis, and the dependent variable on the y-axis. The description should be clear and easy to understand.

1.11.1 Precautions

1. When classes have different widths, divide the frequency by the number of standard classes represented in the non-standard classes.
2. For open-ended classes, find the average and divide the original frequency by this average to get the modified frequency. If the average cannot be computed, break the x-axis and mark the frequency at some point.
3. The choice of origin is important. If the origin cannot be started for one or both axes, use a scale break to warn the reader.
4. The choice of scale for the x- and y-axes is important. Changing the scale can give a false impression of a steep rise or fall in frequency. Choose scales that result in a rectangular diagram.

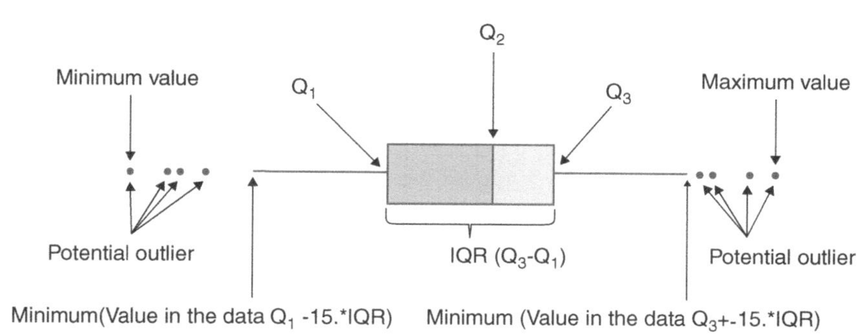

FIGURE 1.5 Box plot (Q_1, Q_2, and Q_3).

1.12 Stem-and-Leaf Diagram

A stem-and-leaf diagram is a histogram where the rectangles are built using the data itself. Each data value is split into a stem (the first digit) and a leaf (the second digit). For example, 36 has a stem of 3 and a leaf of 6, while 9.4 has a stem of 9 and a leaf of 4 (Table 1.9).

1.13 Box Plot

A box plot is drawn using the minimum value, first quartile, median, third quartile, and maximum value of the data. It is used to explore the skewness in a set of data and compare the average and dispersion of different datasets. The median is denoted by a '+' sign in the box plot (Zelterman & Holmes, 1986). The boxes indicate the middle 50% of the data, which represents the middle two quartiles of the data's distribution. Lines, called whiskers, can be considered (Figure 1.6) to display all points within 1.5 times the interquartile range or all points at the maximum extent of the data (Figure 1.6).

1.13.1 Steps to Create a Box Plot

A box plot for the data set can be created by initially determining five key values: the minimum, maximum, median (Q2), first quartile (Q1), and third quartile (Q3). (Figure 1.5):

Step 1: Connect to the Sample–Superstore data source.

TABLE 1.9

Examples Given Below Two 'Stem-and-Leaf' Diagrams

Example 1		Example 2		
9	04			
9				
10	0022244	10	24	
10	666		10	
11	000222444		11	022444
11	6688		11	6666888
12	0244		12	
000022224444				
12	6688		12	66668888
13	02		13	024
13			13	68
14	24			
14				
15	4			
15	6			
16	4			
Hb values of 40 pregnant women			Hb values of 40 pregnant women	
During AN period, in Sample 1			During AN period in Sample 2.	

Step 2: Drag the Segment dimension to the Columns shelf.

Step 3: Drag the Discount measure to the Rows shelf.

Step 4: Tableau will create a vertical axis and display a bar chart, which is the default chart type when there is a dimension on the Columns shelf and a measure on the Rows shelf.

Step 5: Drag the Region dimension to the Columns shelf and drop it to the right of the Segment dimension (Table 1.10).

1.13.2 Illustrative Table

Table 1.11 explains how individuals were counted in a study on tuberculosis treatment in South India. Adult males were counted as one unit, adult females as 0.8 units, and children below 15 as 0.6 units. The study excluded certain individuals, such as families of deceased patients and a nun with no family or income. The information is sourced from a publication by the Tuberculosis Chemotherapy Centre in Madras, 1959.

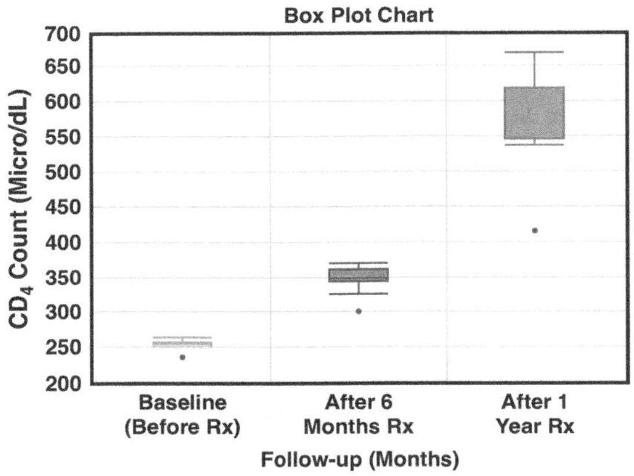

FIGURE 1.6 Box plot techniques applied to CD_4 counts in HIV-infected cases.

TABLE 1.10

CD_4 Distribution of HIV Infected Patients

CD_4 Counts Follow Up of HIV Cases		
Baseline	**@ 6 Months**	**@ 12 Months**
255	325	415
263	300	598
257	345	566
250	365	547
236	347	538
234	344	625
251	352	669
256	350	563
258	365	633
256	370	547

TABLE 1.11

Total Family Income Before Treatment, Expressed in Standard Units* for Home and Sanatorium Patients with Pulmonary Tuberculosis in South India, (1956–1957)

Income per Standard Unit (in rupees per month)	Male Patient in Study Families of Families of Home Patients San.Patients	Female Patient in Study Families of Families of Home Patients San.Patients
0–9.9	1 3	6 4
10–19.9	11 13	12 15+
20–20.0	15 18	9 6
30–39.9	9+ 9	5 1
40–49.9	5 1	1 1
50–59.9	2 1	0 2
60 or more	2 4	0 0
Total families	45 49	33 29

* Vials.

1.14 Advanced Diagram

1.14.1 Sankey Diagram

A Sankey diagram is a visualization that shows the flow of values from one set to another. The connected elements are referred to as nodes, while the connections themselves are called links. A Sankey diagram is a specific type of flow diagram commonly used to represent the flow of energy, materials, or costs between various stages or processes (Nadiyah & Dewi, 2022). The diagram is composed of nodes, which represent the stages or processes, and flow arrows, which depict the flow of energy, materials, or costs between the nodes. The width of the flow arrows is proportional to the magnitude of the flow, allowing for easy comparison of different flows (Figure 1.7). Sankey diagrams are especially helpful in illustrating energy or material balances in industrial processes, but they can also be used to represent other types of flows, such as financial transactions, water consumption, or web traffic.

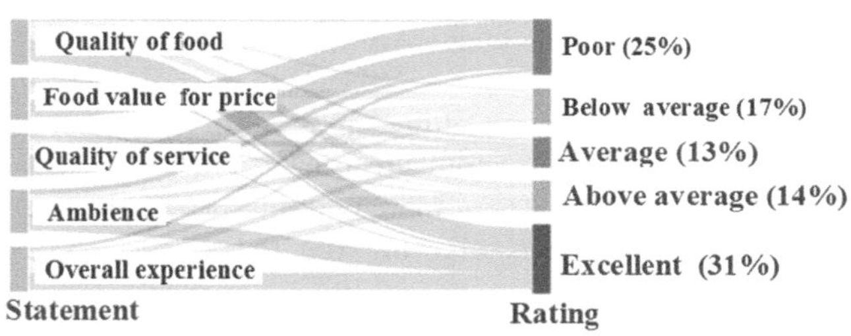

FIGURE 1.7 Sankey diagram representing the acceptance levels of products.

1.15 Bland–Altman Plot

A Bland-Altman plot, also known as a difference plot, is a graphical method used to compare two measurement techniques or assess the agreement between two sets of data. The plot displays the difference between the two measurements on the y-axis and the average of the two measurements on the x-axis. It consists of a scatter plot of the differences against the averages, with a horizontal line representing the mean difference and lines representing the standard deviation from the mean difference (Makarov et al., 2021). The plot allows for visualizing the agreement between the methods, detecting systematic bias or random error, and identifying outliers. It is widely utilized in medical research, industrial quality control, and other fields that require comparing multiple measurement methods. Additionally, it is useful for interpreting the agreement between a new measurement technique and a gold standard (Figure 1.8).

1.16 Pareto Chart

A Pareto chart is a type of chart that combines a bar graph and a line graph to display the relative importance of different factors. It is used to show the relative frequency of different categories of data and is commonly used in quality control and process improvement. The chart consists of a bar graph

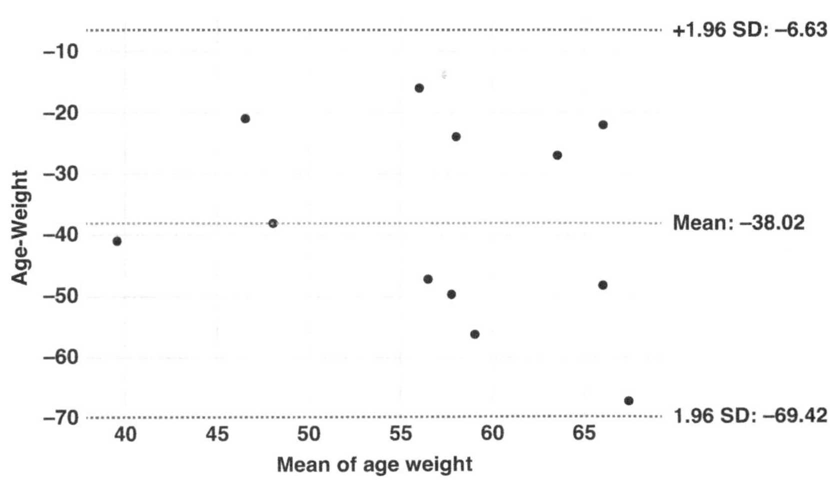

FIGURE 1.8 Scattered Bland-Altman plot correlating age and weight.

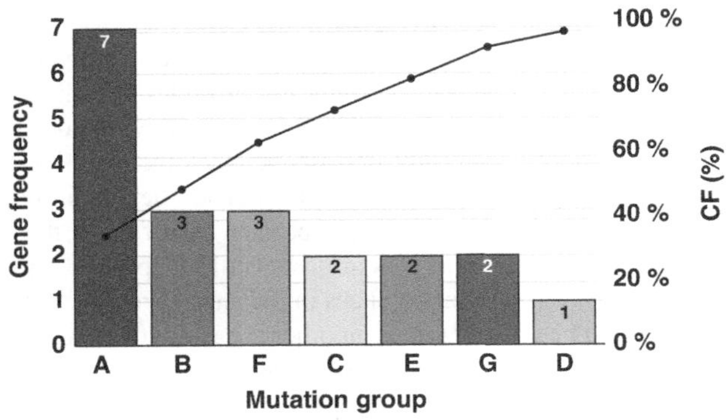

FIGURE 1.9 Pareto chart illustrating the failure rates of chromosomal aberrations across different groups.

that shows the frequency of each category and a line graph that indicates the cumulative percentage of the total frequency (Makarov et al., 2021). The bars are typically arranged in descending order of frequency, with the most frequent category on the left and the least frequent on the right. The Pareto chart is valuable for identifying the most significant factors or causes of a specific problem and for directing improvement efforts toward the areas with the greatest impact. It can be applied in various fields, including industry, healthcare, finance, and more (Figure 1.9).

1.17 Scatter Plot

A scatter plot, also known as a scatter chart or scatter diagram, is a type of plot used to display the relationship between two quantitative variables. The plot consists of points represented by *x*- and *y*-coordinates, where the *x*-coordinate represents one variable's value and the *y*-coordinate represents the other variable's value (Liu et al., 2023). These points are plotted on a Cartesian coordinate system, and their pattern can reveal the relationship between the variables. Scatter plots are useful for identifying trends, patterns, and outliers in data. They can also explore the strength and direction of the relationship between two variables. Scatter plots find applications in various fields such as economics, statistics, social sciences, engineering, and more. Additional information can be added to a scatter plot by including trendlines, regression lines, and confidence intervals. Furthermore, the plot can be colored or shaped based on a third variable to provide more information (Figure 1.10).

1.18 Histogram

The histogram maker divides your data into equal-sized sections, known as bins. These bins are then plotted on a graph, with the height of each bin representing the number of values it contains. If your data is ordinal or nominally scaled, a histogram is not the appropriate choice; instead, you should create a bar chart (Figure 1.11).

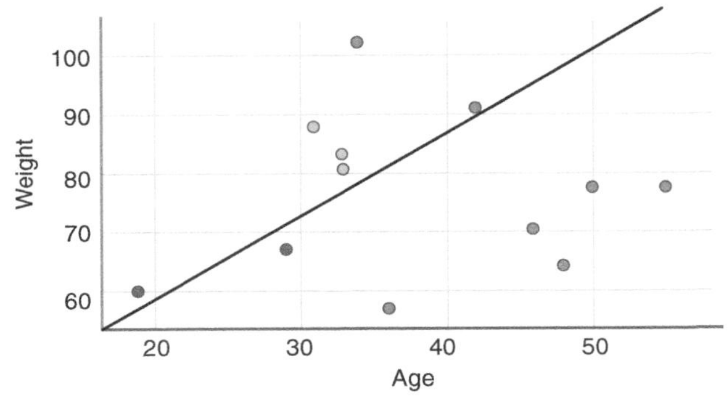

FIGURE 1.10 Scatter plot depicting the relationship between age and weight.

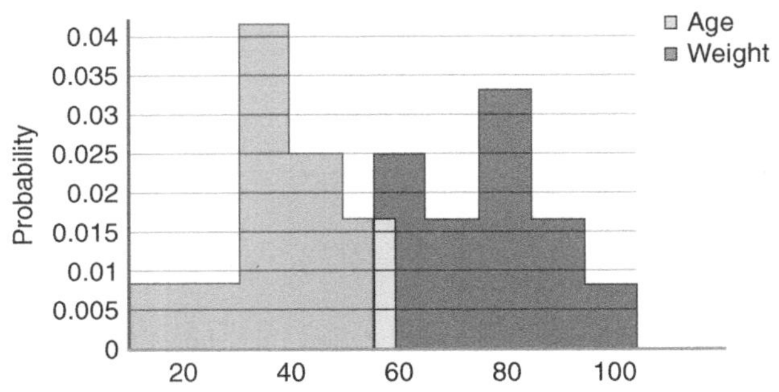

FIGURE 1.11 Histogram of age and body weight.

1. Convenience: Gone are the days of manually plotting data. Online tools make the process swift and effortless.
2. Accuracy: Manual plotting can result in errors, but histogram makers ensure precision.
3. Customization: These tools typically offer options for customization, allowing you to select colors, labels, and more.
4. Accessibility: Create a histogram from anywhere, at any time, without the need for specialized software.

1.19 Practical Application of Diagrammatic Representation of Data

A diagrammatic representation of data is defined as an overall bird's-eye view of the data, aided by suitable diagrams to boost the simplicity of the statistics surrounding the concerned data series (Cavilli et al., 2013). These diagrams are just geometrical figures used to enhance the overall conceptual reading of data and provide visual aids to the researcher and reader in an easily understandable manner. A properly constructed diagrams appeal to the human eye and the mind since they are practical, clear, and easy to review, presenting the entire series of data without any complexity. The researcher should construct the diagrams more attractively with effective means of presentation; all lines should be very clear and precise to give full information. Sketching with different colors and signs attracts the eye and does not strain the observer's mind. A common person who does not typically engage in figures can grasp a message from a well-prepared diagram. Because of this, diagrams are used to present statistical data in different areas (Bajo-Benito et al., 2023). It is a widely used technique in research, economic analysis, business administration, communication technology, engineering science, etc. A diagram depicts more information than the data shown in a table. Information concerning data for the general public becomes easier to understand through a diagram and can be absorbed by a person with ordinary knowledge.

1.20 Youden Plot

A Youden plot is a scatter plot representation of results obtained from *N* laboratories on two similar materials (Figure 1.12). The main objective of the plot is to find out which of the *N* labs provided unreliable or faulty results.

As shown in Figure 1.12, the following elements are present in a Youden plot:

x-axis: readings of sample I from *N* labs.
y-axis: readings of sample II from *N* labs.

1.20.1 Steps for Construction of the Youden Plot

Two parallel lines are drawn parallel to the *x*- and *y*-axes in such a way that the data points on both sides of each line are the same.

1.20.2 Manhattan Median

The point where the two parallel lines intersect is known as the Manhattan median. A 45° reference line is drawn passing through this point.

There is a **95% coverage circle** around the Manhattan median.

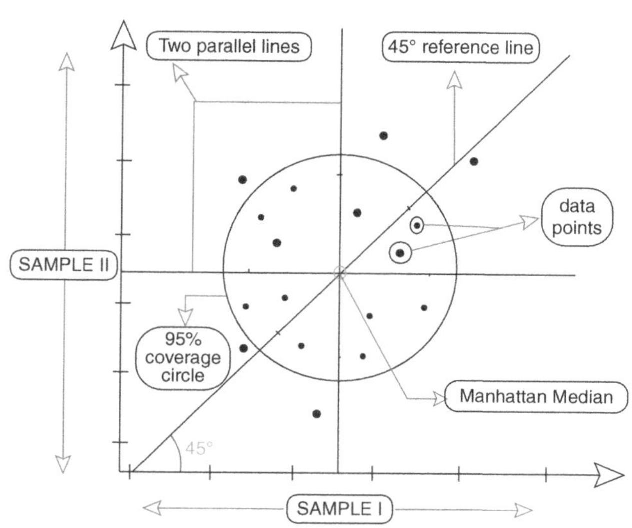

FIGURE 1.12 Structure of a Youden's lot.

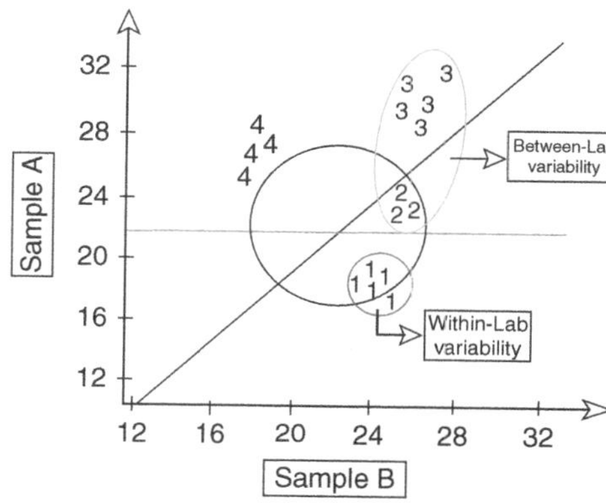

FIGURE 1.13 Variability represented in the Youden's plot.

Youden plots help in understanding the variations that occur in the measurements of two similar samples. They provide insights into two types of variability:

1.20.3 Within-Lab Variability

The variations in the results when multiple tests are performed on the two samples (*A* and *B*) within the same lab lead to repeatability problems.

1.20.4 Between-Lab Variability

The variations in the results among different labs performing tests on the same two samples (*A* and *B*) lead to reproducibility issues (Figure 1.13).

The plot also helps in detecting the presence of various types of errors, such as systematic and random errors.

1.20.5 Systematic Error

These errors occur due to inconsistencies in the experiment, such as faulty or uncalibrated equipment. They are fixable, as we can identify the source of the error by replicating the experiments. These errors are usually found within a single lab performing the experiment.

1.20.6 Random Error

These errors are unsystematic, as they occur at random. They cannot be fixed due to their unpredictable nature.

If random error >> systematic error, then the data points will cluster in the circle around the Manhattan median. They will lie further away from the 45° reference line.

If systematic error >> random error, then the data points will cluster around the 45° reference line in an elliptical pattern.

1.20.7 R-Code for Construction of the Youden Plot

```
YoudenPlot <- function(A, B){
 plot(A,B,asp = 1, xlab = "A", ylab = "B",
pch=".")
 # manhattan median of sample A
```

```
  MMofA <- median(A)
  # manhattan median of sample B
  MMofB <- median(B)
  # in-built function to create horizontal and
vertical lines in the plot
  abline(h = MMofB, v = MMofA)
  # in-built function to create the circle
around the manhattan median.
  curve(x-(MMofA-MMofB),add=TRUE)
  d <- mean(A-B)
  d_prime <- A-B-d
  r <- 2.45*mean(abs(d_prime))*sqrt(pi)/2
  t <- seq(0,2*pi,by=0.01)
  x <- r*cos(t)+MMofA
  y <- r*sin(t)+MMofB
  lines(x,y)
}
  # rnorm (data_points, mean, variation)
A <- rnorm(500,10,100)
B <- rnorm(500,10,100)
  # Function call to create a sample youden
plot.
YoudenPlot(A,B)
```

The Youden plot is a very powerful graphical tool used to analyze inter-laboratory data and help in finding both within-laboratory and between-laboratory errors. Moreover, the Youden plot is considered to be useful for analyzing the performance of laboratories. It is prominently used in the field of medical research for the purpose of quality assessment.

1.21 Polar Plot

It is known that plotting the frequency response signifies sketching the variations in magnitude and phase angle with respect to the input frequency. These plots are known as the magnitude plot (gain plot) and phase plot, respectively. In the Bode plot, the frequency response is sketched using a logarithmic scale. Similarly, in a polar plot, a sketch between the magnitude and phase angle of the transfer function $G(j\omega)$ is formed for different values of ω. So, to construct a polar plot, the different values of magnitude and phase angle is tabulated, and then the sketch is formed (Figure 1.14). The table is given below:

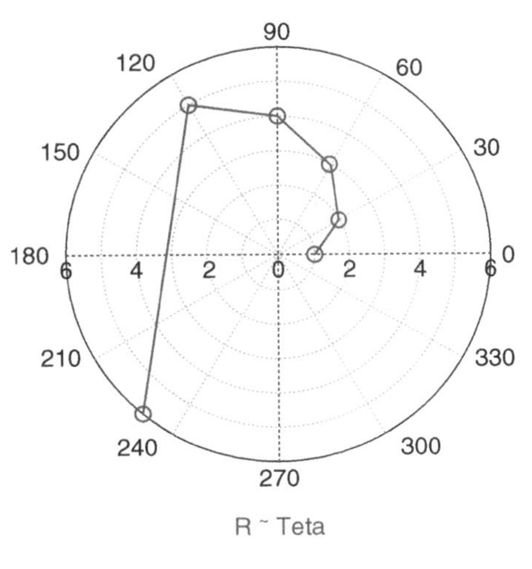

FIGURE 1.14 Structure of a polar plot.

Frequency	Magnitude	Phase Angle
0	M_0	φ_0
ω_1	M_1	φ_1
ω_2	M_2	φ_2
|	|	|
|	|	|
∞	M_∞	φ_∞

Basically, each point on the polar plot is significantly plotted for each specific value of magnitude and phase angle for particular frequency ω. For example, from the above table, for $\omega=\omega_1$, $M=M_1$, and $\varphi=\varphi_1$, a point in the polar coordinate system is decided that represents $M_1\angle\varphi_1$. Hence, the point on the plot corresponds to the tip of the phase or of magnitude M_1 plotted at an angle φ_1.

1.22 Waterfall Chart

The waterfall chart is frequently used in financial analysis to understand the positive and negative effects of multiple factors on a particular asset. The chart can show the effect based on either time or category. Category-based charts represent gains or losses over expenses or sales, or any other variable with sequentially positive and negative values. Time-based charts represent the gain or loss over a time period. The waterfall chart is mostly horizontal (Zelterman & Holmes, 1986). They start from the horizontal axis and are connected by a series of floating columns that represent negative or positive changes. Sometimes, the bars are connected with lines in the charts. The waterfall chart is invaluable in understanding the intricate dynamics of incremental changes. Its ability to visually represent positive and negative shifts over time or steps offers clarity in various scenarios. Whether tracking financial performance or analyzing project progress, the waterfall chart brings insights to the forefront (Figure 1.15, Table 1.12).

1.23 Forest Plot

The forest plot is a mainstay figure in systematic reviews that demonstrates the results from any meta-analyses that have been undertaken. Getting comfortable with forest plots will allow for easy and efficient interpretation of these

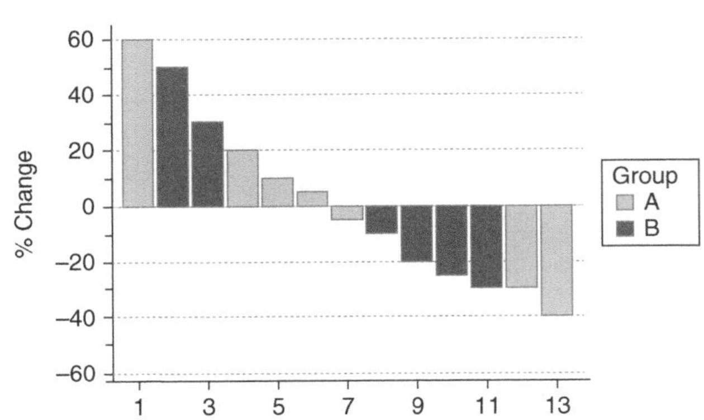

FIGURE 1.15 Structure of a waterfall chart.

TABLE 1.12

Percentage Variation among the Groups

Group	A	B	A	B	A	A	A	B	B	A	A	A	B
%change	60	−10	−30	30	10	5	−5	−25	−20	−40	−30	20	50

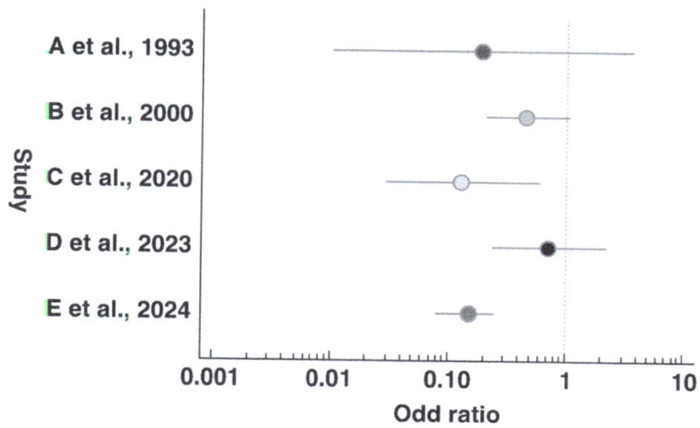

FIGURE 1.16 Waterfall chart structure.

TABLE 1.13

Data View of Construction of Forest Plot

Study	Odd Ratio	Low	High
A et al. (1993)	0.19	0.01	3.60
B et al. (2000)	0.46	0.21	1.07
C et al. (2020)	0.13	0.03	0.60
D et al. (2023)	0.72	0.24	2.18
E et al. (2024)	0.15	0.08	0.25

results and could save you from spending a lot of time pouring over the text and tables (Zhang et al., 2021). In a standard meta-analysis, results for two treatments (e.g. drug *A* vs placebo) from multiple studies have been compared for a single outcome in each analysis (e.g. a dichotomous outcome such as total mortality). This comparison is represented in the form of a ratio describing the chances or 'risk' of the outcome occurring in patients treated with one treatment (drug *A*) versus the other (placebo) (Figure 1.16).

From Figure 1.16, the forest plots show the ratio and confidence interval from each individual study using a box and horizontal line plot. The location of the box on the x-axis represents the odds ratio value for that outcome in that particular study, and the 95% confidence interval extends out as lines from the sides of this box. The *x*-axis of this plot is important, as it indicates which side of the 'no effect' line favors the treatment versus the placebo. Note as well that the *x*-axis scale is often logarithmic rather than linear for dichotomous variables to allow for symmetrical confidence intervals (Table 1.13).

When studies are compared in a meta-analysis, each study is not given equal weighting in the comparison. There are several reasons for this, the most obvious being that studies can vary widely in size. Meta-analyses also adjust the weight based on the estimate's uncertainty; studies with wider confidence intervals will therefore be given a lower weighting than those with narrow CIs. Some analyses, such as those conducted by the Cochrane review, adjust the weighting further based on the risk of bias assessment. Visually, the weighting of each study in the meta-analysis is indicated by the size of the square, with a larger square showing a higher weighting and vice versa. In addition to the visual content, forest plots typically show the numerical data supporting Figure 1.16. This can include the number of events in each group from each study (n/N), the actual percentage weighting value assigned to each study, the numerical value of the ratio and its confidence interval, and statistical test results for overall effect and heterogeneity. Often, you may want to take a closer look at these numbers as they can explain additional details that are not immediately apparent in Figure 1.15. The main results of the analysis, however, can always be ascertained from a glance at the forest plot.

1.24 Heat Map

Heat maps are used to show relationships between two variables, one plotted on each axis. By observing how cell colors change across each axis, you can observe if there are any patterns in value for one or both variables. The variables plotted on each axis can be of any type, whether they take on categorical labels or numeric values. In the latter case, the numeric value must be binned, like in a histogram, to form the grid cells where colors associated with the main variable of interest will be plotted. Cell coloring can correspond to all manner of metrics, like a frequency count of points in each bin, or summary statistics like the mean or median for a third variable. One way of thinking about the construction of a heat map is as a table or matrix, with color encoding on top of the cells. In certain applications, it is also possible for cells to be colored based on non-numeric values (e.g. general qualitative levels of low, medium, high). Different visualization applications can have different ways of accepting data for plotting as a heat map. In one major form, data can be supplied in the same way it would be naturally displayed as a table. The first column will hold values for one axis of the heat map, while the names of the remaining columns will correspond with bins for the remaining axis (Figure 1.17). Values in those columns will be encoded into the heat map itself.

Figure 1.17 shows each square representing the correlation between the variables on each axis. Correlation ranges from −1 to +1. Values closer to zero mean there is no linear trend between the two variables. The closer the correlation is to 1, the more positively correlated they are; that is, as one increases, so does the other, and the closer to 1, the stronger this relationship is. A correlation closer to −1 is similar, but instead of both increasing, one variable will decrease as the other increases (Yoo et al., 2023). The diagonals are all 1/black because those squares are correlating each variable to itself (so it's a perfect correlation). For the rest, the larger the number and darker the color, the higher the correlation between the two variables. The plot is also symmetrical about

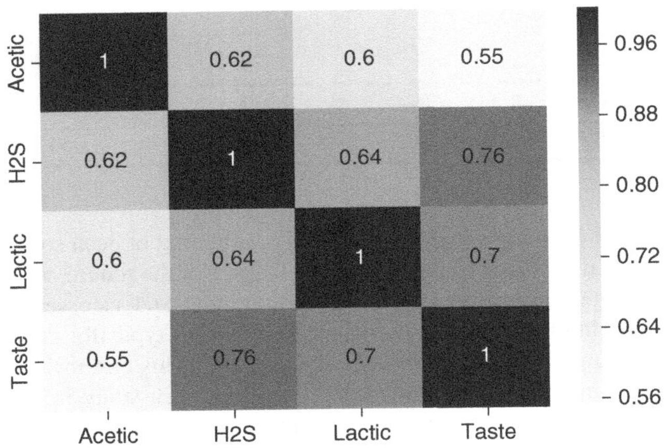

FIGURE 1.17 Structure of a heat map, clearly showing the correlation.

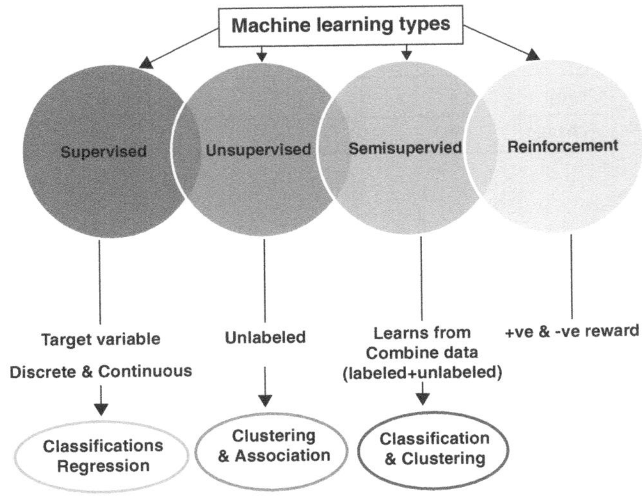

FIGURE 1.18 Types and classifications of machine learning.

the diagonal since the same two variables are being paired together in those squares.

1.25 Machine Learning Graphs

Machine learning (ML) is the field of study that gives computers the capability to learn without being explicitly programmed. ML is one of the most exciting technologies that one could ever come across (Xia et al., 2021). As it is evident from the name, it makes computers more similar to humans. Machine learning is a sub-branch of artificial intelligence that involves the development of algorithms and statistical models that enable computers to improve their performance in tasks through experience. These algorithms and models are designed to learn from data and make predictions or decisions without explicit instructions. There are several types of machine learning, including supervised learning, unsupervised learning, and reinforcement learning (Li et al., 2023; Lu & Uddin, 2023). Supervised learning involves training a model on labeled data, while unsupervised learning involves training a model on unlabeled data. Reinforcement learning involves training a model through trial and error. Machine learning is used in a wide variety of applications, including image and speech recognition, natural language processing, and recommender systems. A computer program is said to *learn* from experience **'E'** with respect to some class of tasks **'T'** and performance measure **'P'**, if its performance at tasks **'T'**, as measured by **'P'** improves with experience **'E'**.

1.26 Examples

1.26.1 Handwriting Recognition Learning Problem

Task **T**: Recognizing and classifying handwritten words within images

Performance **P**: Percentage of words correctly classified

Training experience **E**: A dataset of handwritten words with given classifications

1.26.2 A Robot Driving Learning Problem

Task **T**: Driving on highways using vision sensors

Performance **P**: Average distance traveled before an error

Training experience **E**: A sequence of images and steering commands recorded while observing a human driver

Machine learning implementations are classified into four major categories, depending on the nature of the learning 'signal' or 'response' available to a learning system (as shown in Figure 1.18)

1.27 Types of Machine Learning

1.27.1 Supervised Learning

Supervised learning is the machine learning task of learning a function that maps an input to an output based on example input-output pairs (Xia et al., 2021). The given data is labeled. Both *classification* and *regression* problems are supervised learning problems. Consider the following data regarding patients entering a clinic: the data consists of the gender and age of the patients, and each patient is labeled as 'healthy' or 'sick'.

Gender	M	M	F	M	F	M	M
Age	48	67	53	49	32	34	21
Label	Sick	Sick	Healthy	Sick	Healthy	Healthy	Healthy

1.27.2 Unsupervised Learning

Unsupervised learning is a type of machine learning algorithm used to draw inferences from datasets consisting of input data without labeled responses. In unsupervised learning algorithms, classification or categorization is not included in the observations. Example: Consider the following data regarding patients entering a clinic. The data consists of the gender and age of the patients.

Gender	M	M	F	M	F	M	M
Age	48	67	53	49	32	34	21

As a kind of learning, it resembles the methods humans use to figure out that certain objects or events are from the same class, such as by observing the degree of similarity between objects. Some recommendation systems found on the web, in the form of marketing automation, are based on this type of learning (Yoo et al., 2023).

1.27.3 Reinforcement Learning

Reinforcement learning is the problem of getting an agent to act in the world to maximize its rewards. A learner is not told what actions to take, as in most forms of machine learning, but instead must discover which actions yield the most reward by trying them. For example, consider teaching a dog a new trick: we cannot tell it what to do or what not to do, but we can reward or punish it if it does the right or wrong thing.

1.27.4 Semi-Supervised Learning

Where an incomplete training signal is given, a training set with some (often many) of the target outputs missing, there is a special case of this principle known as transduction, where the entire set of problem instances is known at learning time, except that part of the targets is missing. Semi-supervised learning is an approach to machine learning that combines a small amount of labeled data with a large amount of unlabeled data during training. Semi-supervised learning falls between unsupervised learning and supervised learning.

1.27.5 Categorizing Based on Required Output

Another categorization of machine-learning tasks arises when one considers the desired output of a machine-learned system:

1.28 Classification of Machine Learning

When inputs are divided into two or more classes, the learner must produce a model that assigns unseen inputs to one or more (multi-label classification) of these classes. This is typically tackled in a supervised way. Spam filtering is an example of classification, where the inputs are email (or other) messages, and the classes are 'spam' and 'not spam'.

1.28.1 Regression

Regression in machine learning is a method for predicting outcomes by analyzing the relationship between variables. Here are some examples of regression in machine learning.

1.28.2 Clustering

Clustering is when a set of inputs is to be divided into groups. Unlike in classification, the groups are not known beforehand, making this typically an unsupervised task.

1.29 Graph of Machine Learning (GML)

At its core, graph machine learning (GML) is the application of machine learning to graphs, specifically for predictive and prescriptive tasks. GML has a variety of use cases across supply chain, fraud detection, recommendations, customer 360, drug discovery, and more (Figure 1.19).

1.29.1 Node Property Prediction

Predicting a discrete or continuous node property. One can think of node property prediction as predicting an adjective about a thing, such as whether an account on a financial services platform should be classified as fraud or how to categorize a product on an online retail store.

1.29.2 Link Prediction

Link prediction is predicting whether or not a relationship should exist between two nodes and potentially some properties about the relationship. It is helpful for applications like entity resolution, where we want to predict whether two nodes reflect the same underlying entity; recommendation systems, where we want to predict what a user will want to purchase or interact with next; and bioinformatics, for predicting things like protein and drug interactions. In each case, we care about predicting an association, similarity, or potential action or interaction between entities.

1.29.3 Graph Property Prediction

Graph property prediction is predicting a discrete or continuous property of a graph or subgraph and is useful in domains where you want to model each entity as an individual graph for prediction rather than modeling entities as nodes within a larger graph representing a complete dataset (Scaife & Rogers,

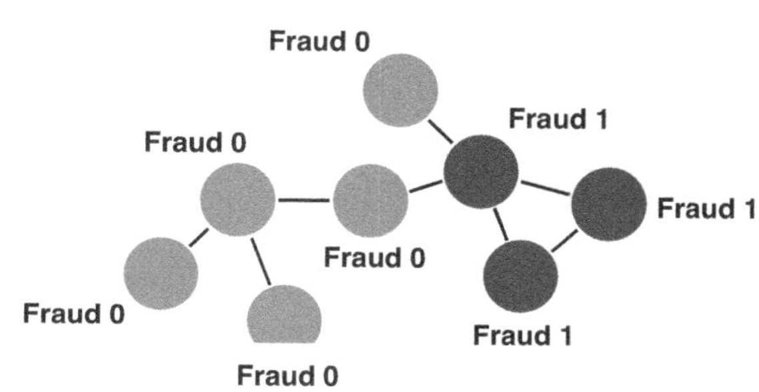

FIGURE 1.19 Node property prediction-agglomerative classification techniques.

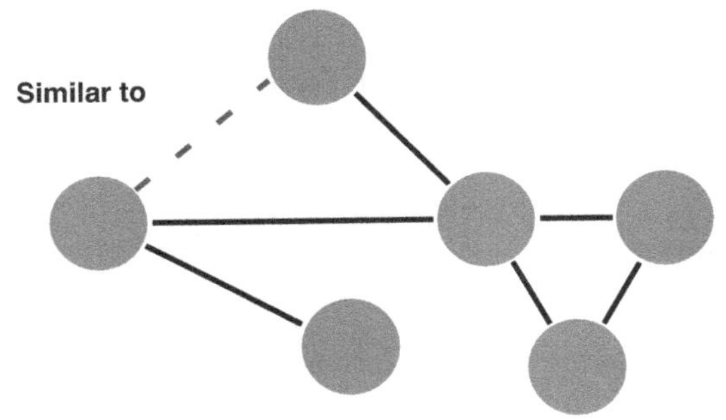

FIGURE 1.20 Link prediction on similarities of events occurring over different time periods.

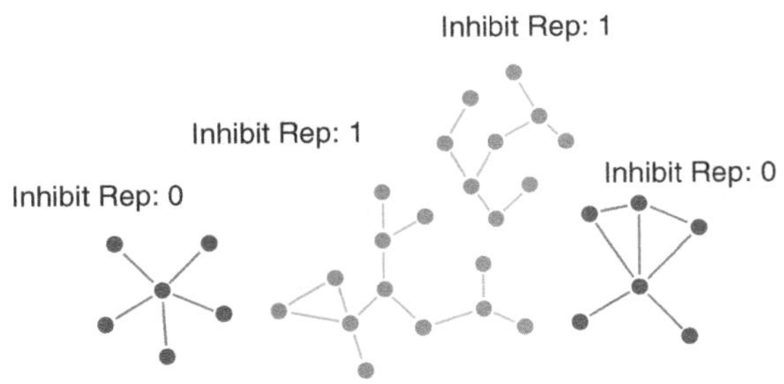

FIGURE 1.21 Prediction of inhibitory factors of various replications connected to internodes.

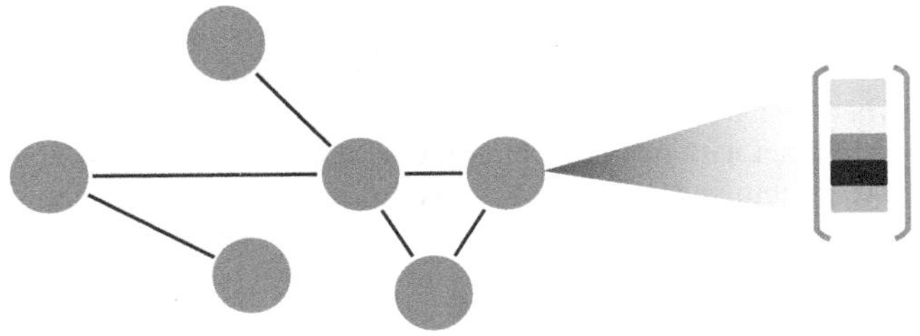

FIGURE 1.22 Primary nodes through child nodes derived from training datasets.

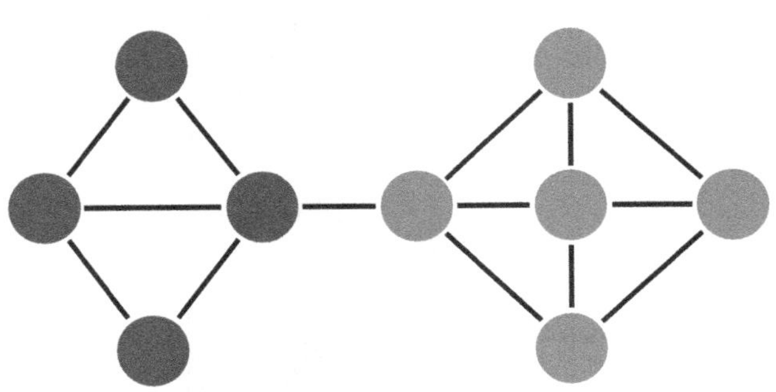

FIGURE 1.23 Centrality and path finding.

1996) Use cases include material sciences, bioinformatics, and drug discovery, where individual graphs can represent molecules or proteins that you want to make predictions about (Figures 1.20–1.23).

1.30 Unsupervised GML Tasks

1.30.1 Representation Learning

Representation learning involves reducing dimensionality while maintaining important signals, which is a central theme for GML applications. Graph representation learning does this explicitly by generating low-dimensional features from graph structures, usually to use them for downstream exploratory data analysis (EDA) and ML.

1.30.2 Community Detection (Clustering for Relationships)

Community detection is a clustering technique for identifying groups of densely interconnected nodes within a graph. It has various practical applications in anomaly detection, fraud and investigative analytics, social network analysis, and biology.

1.30.3 Similarity

Similarity in GML refers to finding and measuring similar pairs of nodes in a graph. It is applicable to many use cases, including recommendation, entity resolution, and anomaly and fraud detection. Common similarity techniques include node similarity algorithms, topological link prediction, and K-Nearest Neighbor (KNN).

1.31 Centrality and Path Finding

I'm grouping these together, as they tend to be less associated with ML tasks and more with analytical measures. However, they still technically fit here, so I will cover them for completeness. Centrality finds important or influential nodes in a graph. Centrality is ubiquitous throughout many use cases, including fraud and anomaly detection, recommendations, supply chain, logistics, and infrastructure problems. Pathfinding is used to find the lowest-cost paths in a graph or evaluate the quality and availability of paths. Path finding can benefit many use cases dealing with physical systems, such as logistics, supply chain, transportation, and infrastructure.

1.32 New Advanced Statistical Software for Diagrammatic Analysis

With the advancement of science and technology, many software are available for data analysis on a paid and trial bases. The following table and matrices present the important software for data analysis in a diagrammatic representation (Table 1.14).

TABLE 1.14

Different Artificial Intelligence (AI)-Based Software Used for Diagrammatic Representation and Flow Chart

Software	Application	URL
Lucid chart	Advanced functionality	https://zapier.com/blog/flowchart-diagramming-software/#lucidchart
Microsoft vision	For Microsoft 365 and Window users	https://www.microsoft.com/en-us/microsoft-365/visio/flowchart-software
Textgrafo	For text to flow chart	https://textografo.com/
draw.in	For the best free diagram and flow chart app	https://www.drawio.com/
Smart draw	For the fastest diagramming	https://www.smartdraw.com/\
Zapier Canvas	For diagrammatic business process	https://zapier.com/canvas\
Whimsical	AI chart generation	https://whimsical.com/
Flow map	Web designers and agencies	https://www.flowmapp.com/
Venngage	Business templates	https://venngage.com/\
Creately	Two-way data synchronization	https://creately.com/
Figjam	For remote team management	https://www.figma.com/
Mural	For the best templates	https://www.mural.co/

Practical Exercise

Illustration 1

From the following data relating to cement production of a factory (2018–2023), prepare a simple bar diagram.

Year	2018	2019	2020	2021	2022	2023
Production (in M. tonnes)	125	175	260	310	450	520

Illustration 2

Present the following data by multiple-bar diagram.

	Group 1	Group 2	Group3
Quarter	CD_4 Counts (micro/dL)	CD_4 Counts (micro/dL)	CD_4 Counts (micro/dL)
I	420	200	140
II	320	240	300
III	380	360	480

Illustration 3

Represent the data relating to the cost of production in a factory by means of percentage diagram:

Element of Cost (in '000 tonnes)	Items		
	A	B	C
Raw materials	75	55	60
Wages	45	30	25
Factory overheads	15	10	30
Office overheads	5	15	5

Illustration 4

From the following data of imports of a firm, prepare a suitable diagram.

Year	2015	2016	2017	2018	2019
Imports (â,¹ in lakhs)	16	22	34	170	25

Illustration 5

Represent the following data relating to net exports of a company by deviation bar diagram.

Year	2014	2015	2016	2017	2018	2019
Net exports (â,¹ in crores)	250	160	(−)80	50	(−)110	70

Illustration 6

The estimated percentage break-up of the cost of construction of a house (excluding the cost of land) is as follows:

Labor	Bricks	Cement	Steel & Timber	Marble	Miscellaneous
20%	12%	25%	15%	13%	15%

Construct a pie diagram to represent the given data.

Illustration 7

The following table gives the monthly expenditure of two families A and B. Represent them by means of sub-divided bar diagram and percentage diagram.

Items of Expenditure	Family A (income: â,¹ 50,000)	Family B (Income: 80,000)
Food	15,000	23,000
Clothing	9,000	17,000
House Rent	11,000	11,000
Education	5,000	9,000
Fuel and Lighting	4,000	13,000
Miscellaneous	6,000	7,000

REFERENCES

Bajo-Benito, J. M., Gavilán-Izquierdo, J. M. & Sánchez-Matamoros García, G. The concept of number sequence in graphical representations for secondary school students. *European Journal of Educational Research* **12,** 159–172 (2023).

Cavilli, S., Coates, R., Doherty, M. & Maycock, B. Beyond survivor: Women's identity & recovery after domestic violence. *Journal of Sexual Medicine* **10,** 307 (2013).

Li, H., Wang, X. & Zhu, W. Curriculum graph machine learning: A survey, in *IJCAI International Joint Conference on Artificial Intelligence* 2023-August, 6674–6682 (2023).

Liu, L., Zhang, L., Tan, X. & Deng, Y. Graphical representations and worm algorithms for the O(N) spin model. *Communications in Theoretical Physics* **75,** 115702 (2023).

Lu, H. & Uddin, S. Disease prediction using graph machine learning based on electronic health data: A review of approaches and trends. *Healthcare* **11,** 1031 (2023).

Makarov, I., Kiselev, D., Nikitinsky, N. & Subelj, L. Survey on graph embeddings and their applications to machine learning problems on graphs. *PeerJ Computer Science* **7,** 1–62 (2021).

Nadiyah, K. & Dewi, G. S. Quality control analysis using flowchart, check sheet, p-chart, Pareto diagram and fishbone diagram. *OPSI* **15,** 183 (2022).

Rahmasari, A. & Kuswanto, H. The effectiveness of problem-based learning physics pocketbook integrating augmented reality with the local wisdom of catapults in improving mathematical and graphical representation abilities. *Journal of Technology and Science Education* **13,** 886–900 (2023).

Sab, S., Schütte, K. & Lindner, M. A. Test-takers' eye movements: Effects of integration aids and types of graphical representations. *Computers and Education* **109,** 85–97 (2017).

Scaife, M. & Rogers, Y. External cognition: How do graphical representations work? *International Journal of Human Computer Studies* **45,** 185–213 (1996).

Xia, F. et al. Graph learning: A survey. *IEEE Transactions on Artificial Intelligence* **2,** 109–127 (2021).

Yoo, Y., Shin, J. & Kyeong, S. Medicare fraud detection using graph analysis: A comparative study of machine learning and graph neural networks. *IEEE Access* **11,** 88278–88294 (2023).

Zelterman, D. & Holmes, N. Designer's guide to creating charts and diagrams. *Journal of the American Statistical Association* **81,** 875 (1986).

Zhang, Z., Wang, X. & Zhu, W. Automated machine learning on graphs: A survey, in *IJCAI International Joint Conference on Artificial Intelligence*, 4704–4712 (2021). https://doi.org/10.24963/ijcai.2021/637

2

A Comprehensive Guide to Descriptive Statistics on Measures of Central Tendency and Dispersion

2.1 Introduction

It is often useful and important to summarize a large volume of data in a single figure. Such a figure may be called a 'measure of central tendency' or a 'measure of location'. These measures are used for making comparisons between groups of values (Ali & Bhaskar, 2016). There are three measures in common use: the mean, the median, and the mode. A single value can describe the characteristics of an unwinding data series. The measure of central tendency deals with both population and sample. The population refers to the universe or group of individuals, such as a group of animals, the number of hens in a pen, the number of flowers in plants, or the number of amino acids responsible for binding proteins. The population is classified into two distinct categories: finite population and infinite population. A group of individuals with distinct categories and a known sample size (*n*) is a finite population. An infinite population is a group without distinct categories and an unknown sample size (Dimitriadis et al., 2019). In research, two types of populations are widely used for measuring variability and statistical significance. The target population refers to the entire group of individuals or objects to which researchers are interested in generalizing effective conclusions based on the objective of interest. The target population usually has varying characteristics and is also known as the theoretical population. The accessible population is the population in research to which researchers can apply their conclusions (Anilkumar & Samiyya, 2013). This population is a subset of the target population and is also known as the study population. Researchers draw their samples from the accessible population, such as the infected and uninfected populations for a vaccination drive based on its characteristics (Figure 2.1).

The study population, also known as the accessible population, is the actual sampling frame from which we randomly select our sample. This group is the one to which we can generalize our findings because we sampled from them.

In simpler terms:

- The target population is the population you are interested in studying.
- The study population is a subset of the target population that you select for your study.
- The theoretical population is the same as the target population, which is the population you want your

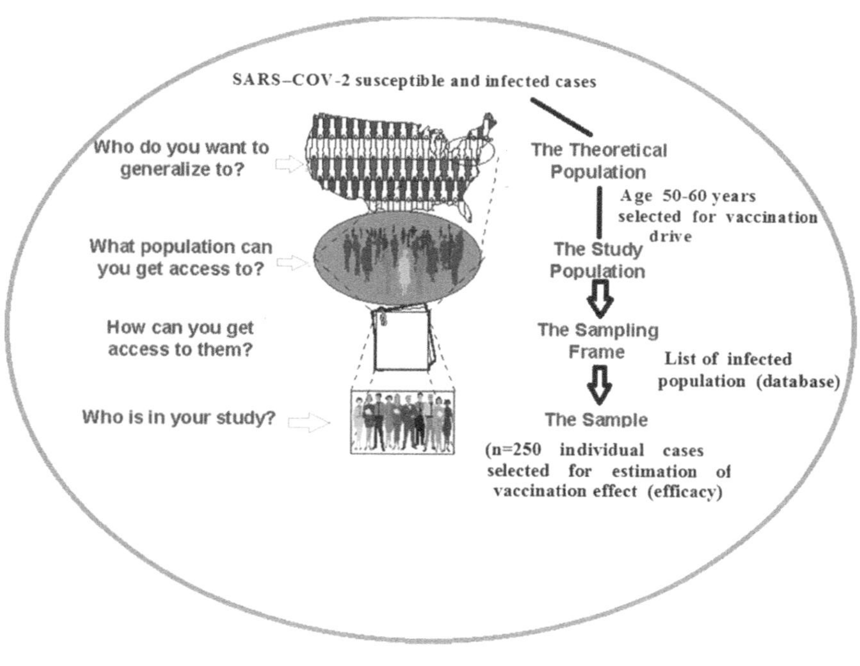

FIGURE 2.1 Characteristics of population in biological research.

DOI: 10.1201/9781032713816-2

TABLE 2.1

Flowchart Categorizing Averages into Three Distinct Types

Mathematical Average	Positional Average	Miscellaneous Average
• Arithmatic mean	• Median	• Moving average (ARMIA)
• Geometric mean	• Mode	• Progressive average
• Harmonic mean	• QD	
	• Deciles	
	• Percentiles	
	• Octiles	
	• Quantiles	
	• Hexiles	

study to generalize. For example, if you want to study the average effect of SARS-CoV-2 vaccination on men between the ages of 50 and 60 in Indian states, the target and theoretical population would be all men between the ages of 50 and 60. Your study population would be men between the ages of 50 and 60 in the selected states.

It is not possible to remember every single fact related to a field of inquiry, as the human mind is not capable of storing all the facts. Trying to remember everything may lead to confusion. After collecting data, we apply different statistical measures to understand the descriptive characteristics of the target population. The measure of central tendency is a study of averages. An average value is a single value within the range of the data and is sometimes called a measure of central tendency. In other words, an average is a typical value that represents all the individual values in a series of a variable expressed in some quantitative form. It lies between the extreme values of the series and represents all the values in the series. The measure of central tendency can be classified in the following ways (Table 2.1).

2.2 Arithmetic Mean or Average $(\overline{X})$

In biological research, the arithmetic average is the most widely used and popular. It can be applied in real-life situations and is commonly referred to as the mean. This average is popular because it satisfies most of the requirements of an ideal average, such as simplicity and rigidity. Mathematically, the arithmetic average is defined as the quotient obtained by dividing the total of the values of a variable by the total number of observations or items (Cazorla et al., 2023).

$$\text{Mathematical formula}: \left(\overline{X} = \frac{\sum_{i=1}^{n} X_i}{N}\right) = \frac{9{,}005.0}{23}$$

$$= 39.50\% \text{ population vaccinated}$$

From Table 2.2, we calculated the sum of the vaccinated $\sum_{i=\text{HP}}^{n=\text{Bihar}} X_i = 63.5+44.9+47.3...+23.9 = 9{,}005.0$ and the number of observations $(N) = 23$.

$$\text{Similarly, fully unvaccinated} \left(\overline{X} = \frac{\sum_{i=1}^{n} X_i}{N}\right) = \frac{233.34}{23} =$$

10.50% population vacciànted (fully)

2.2.1 Case I: Direct Method

TABLE 2.2

Descriptive Statistics of SARS-CoV-2 Vaccination Coverage on the Target Population in Indian States

	% of over 18 Years Population Covered	
State/UT	Vaccinated (X_i)	Direct Method Formula
HP	63.5	
Kerala	44.9	$\left(\overline{X} = \frac{\sum_{i=1}^{n} x_i}{n}\right)$
Delhi	47.3	
Uttarakhand	50.1	
Gujarat	46.1	$\sum_{i=\text{HP}}^{n=\text{Bihar}} X_i = 63.5+44.9+47.3....23.9$
NE minus Assam	53.4	
Karnataka	44.9	$= 9{,}005.0$
Rajasthan	44.3	$n = 23$
AP	36.7	
Jammu & Kashmir	48.0	$\left(\overline{X} = \frac{9{,}005.0}{23}\right) = 39.15\%$
WB	25.5	
Chhattisgarh	45.1	
Maharashtra	33.2	
Haryana	41.4	
Odisha	34.4	
Telangana	39.1	
MP	38.5	
Punjab	31.9	
Assam	30.6	
TN	27.5	
Jharkhand	27.3	
UP	22.9	
Bihar	23.9	

2.2.2 Case II: Indirect Method

$$\left(\overline{X} = A + \frac{\sum_{i=1}^{n} d}{N}\right) d = |X_i - A|$$

where A is the assumed or arbitrary value taken from the given series of data, from Table 2.3.

2.2.3 Ungrouped Series

TABLE 2.3

SARS-CoV-2 Vaccination Data

State/UT	Vaccinated (X_i)	Assumed $A = 22.9$, $d = \|X_i - A\|$	Average Calculation Formula (Indirect Method)
HP	63.5	(63.5−22.9)=40.6	
Kerala	44.9	(44.9−22.90)=22	$\left(\bar{X} = A + \frac{\sum_{i=1}^{n} d}{N}\right)$
Delhi	47.3	(47.30−22.90)=24.4	
Uttarakhand	50.1	(50.10−22.90)=27.2	
Gujarat	46.1	(46.10−22.90)=23.2	
NE minus Assam	53.4	(53.40−22.90)=30.5	$A = 22.90$
Karnataka	44.9	(44.90−22.90)=22	
Rajasthan	44.3	(44.30−22.90)=21.4	$\sum_{i=1}^{23} d = \|X_i - A\| = 373.8$
AP	36.7	(36.70−22.90)=13.8	
Jammu Kashmir	48	(48.00−22.90)=25.1	$N = 23$
WB	25.5	(25.50 −22.90)=2.6	$\bar{X} = 22.9 + \frac{\sum_{i=1}^{n} 373.80}{23}$
Chhattisgarh	45.1	(45.10−22.90)=22.2	
Maharashtra	33.2	(33.20−22.90)=10.3	$\bar{X} = 39.15$
Haryana	41.4	(41.40−22.90)=18.5	
Odisha	34.4	(34.40−22.90)=11.5	
Telangana	39.1	(39.10−22.90)=16.2	
MP	38.5	(38.50−22.90)=15.6	
Punjab	31.9	(31.90−22.90)=9	
Assam	30.6	(30.60−22.90)=7.7	
TN	27.5	(27.50−22.90)=4.6	
Jharkhand	27.3	(27.30−22.90)=4.4	
UP	22.9	(22.90−22.90)=0	
Bihar	23.9	(23.90−22.90)=1	
$\sum_{i=1}^{23} d = \|X_i - A\| = 373.8$			

2.2.4 Working Rules

When performing calculations by hand, it is possible to simplify the process using 'working units'. For example, let's calculate the mean of 107, 102, 108, 105, and 108 using this method. We can choose 100 as our working origin. The next step is to subtract 100 from each value, resulting in 7, 2, 8, 5, and 8. The mean of these values is (7+2+8+5+8)/5=30/5=6. To obtain the mean of the original values, we add the working origin (100) to this mean, giving us (6+100=106). It should be noted that adding or subtracting a constant value to each observation will increase or decrease the mean by the same constant, respectively. Similarly, multiplying or dividing each observation by a constant value will result in the mean being multiplied or divided by the same value (Table 2.4).

2.2.5 Mean of Several Groups Combined

Suppose we want to calculate the overall mean by combining the means of multiple groups. To do this, we need to obtain the total values for each group (as shown in the last column below), add them together, and then divide by the total number of observations across all groups (Table 2.5).

Mean of all groups = 2,000/50 = 40.

$$\overline{X}_{\text{Groups }(1,2,3)} = \left(\frac{N_1\overline{X_1} + N_2\overline{X_2} + N_3\overline{X_3}}{N_1 + N_2 + N_3}\right)$$

$$\bar{X}_{\text{Groups }(1,2,3)} = \frac{(10*41)+(15*36)+(25*42)}{10+15+25} = \frac{2,000}{50} = 40$$

2.3 Applications of Pooled Mean

The pooled mean, or combined mean, serves as a basis for interpretation, and a unified figure is employed across various fields. In the realm of biological sciences, the area-combined mean is applied in the following contexts: the assessment of average body weight among different mice utilized for experimental purposes; the average impact of growth promoters on plant development; and the collective effect of various food formulations on enhancing body weight. In the field of medical sciences, the influence of vaccination in relation to body weight and the current state of disease is examined

TABLE 2.4

Immunization Details of SARS-CoV-2

Covid Vaccination Efficacy (%)@ First Dose (X_i)	Number of Infected Population Immunized (F_i)	X_iF_i	Formula
86	10	(86*10)=860	Direct method
88	14	(88*14)=1,232	$\bar{X} = \frac{\sum_{i=1}^{n} X_iF_i}{\sum_{i=1}^{n} F_i}$
92	6	(92*6)=552	
93	9	(93*9)=837	
94	13	(94*13)=1,222	$\bar{X} = \frac{20,654}{226} = 91.38\%$
93	18	(93*18)=1,674	
88	13	(88*13)=1,144	Indirect method
88	16	(88*16)=1,408	$\bar{X} = A + \frac{\sum_{i=1}^{n} d_iF_i}{\sum_{i=1}^{n} F_i}$
84	20	(84*20)=1,680	
92	23	(92*23)=2,116	
92	27	(92*27)=2,484	$\sum_{i=1}^{226} d = \lvert X_i - A \rvert$
94	28	(94*28)=2,632	
97	29	(97*29)=2,813	
	$\sum_{i=1}^{n} F_i = N = 226$	$\sum_{i=1}^{n} X_iF_i = 20,654$	

TABLE 2.5

Combined Mean Sample Size

Group (i)	Size (i_n)	Mean (X_i)	Total (n_i X_i)
X_1	10	41	(10*41)= 410
X_2	15	36	(15*36)=540
X_3	25	42	(25*42)=1,050
Total	50		2,000

(Chakrabarty, 2021). Additionally, in numerous economic and social sciences, the combined mean is utilized for informed decision-making.

2.4 Geometric Mean (GM)

This is extensively used in microbiological and serological research. Observations are sometimes expressed as 'titers', which are dilutions of certain suspensions or reagents at which a specified phenomenon, such as the agglutination of red cells, first takes place (Darling, 2022). If repeated observations are made during the same investigation, the possible values of a titer will usually be multiples of the same dilution factor, e.g. 2, 4, 8, 16, etc. for a two-fold dilution. It is commonly found that a series of titers obtained, for example, from different sera, are not symmetrically distributed. However, the logarithms of these values are nearly symmetrically distributed. Consequently, the arithmetic mean of the logarithmic values is calculated, and its antilogarithm is taken to obtain the mean in the original units. Such a mean is referred to as the geometric mean (GM) and is used when dealing with concentrations of drugs estimated in biological fluids, minimal inhibitory concentrations of drugs in sensitivity tests, etc. To take an example, let us calculate the GM of 5, 10, 20, 25, and 40 µg/mL. The logarithm of these values is 0.70, 1.00, 1.30, 1.40, and 1.60, respectively. The arithmetic mean of the logarithms is 6.00/5 = 1.20, and the GM is the antilog of 1.20, i.e., 15.85 µg/mL. If we multiply X_g, the GM of n numbers, by itself n times, we should get the product of the n numbers X_1, X_2, …, X_n. In this example, the product of the five numbers 5, 10, 20, 25, and 40 is 1,000,000. If we multiply the GM 15.85 five times, *i.e.*, 15.85 × 15.85 × 15.85 × 15.85 × 15.85, we obtain 1,000,337. The difference observed is due to rounding-off errors in taking logarithms (Table 2.6).

TABLE 2.6

Drug Efficacy of Novel Regimen

Drug Efficacy (%) (X_i)	Log (X_i)
90	1.95
95	1.98
85	1.93
84	1.92
82	1.91
96	1.98
93	1.97
98	1.99
93	1.97
94	1.97
$\sum_{i=1}^{n} \text{Log } X_i$	**19.58**
GM	2.41

$$\text{GM} = n\sqrt{(X_1)(X_2)(X_3)..(X_n)}$$

$$\text{GM} = [X_1, X_2, X_3 \ldots . X_n]^{\frac{1}{n}}$$

Take log on both sides:

$$\text{GM} = \left[\text{Log }(X_1) + \text{Log }(X_2) \ldots . \text{Log } X_n\right]^{\frac{1}{n}}$$

$$\text{GM} = \frac{1}{n}\sum_{i=1}^{n}\text{Log } X_i$$

Take antilog for the right-hand side:

$$\text{GM} = \text{Antilog } \frac{1}{n}\sum_{i=1}^{n}\text{Log } X_i$$

$$\text{GM} = \text{Antilog } \frac{1}{n}\sum_{i=1}^{n}\text{Log } X_i = \frac{1}{10}\sum_{i=1}^{n}19.58 \ = 2.41$$

2.4.1 GM of Several Groups Combined

If we have several groups, each with a known GM and number of values, we can obtain the overall GM without needing the individual values. For example, let's consider three groups of patients (Groups A, B, and C) with the following geometric mean isoniazid concentrations (Table 2.7).

First, we write down the logarithms of the three GMs: 8.5, 10.2, and 9.4. Then, we calculate $n_i \times \log$ GM and add the three values (see last column). Next, we divide by the total number of observations, which is 48.4=0.9684. Finally, we take the antilogarithm to obtain the overall geometric mean as 9.68 µg/mL.

$$\text{GM} = \text{Antilog}\left[\frac{N_1 \text{ Log } X_1 + N_2 \ Log \ X_2 + \ldots X_k \text{ Log } X_k}{N_1 + N_2 + \ldots N_3 \ldots N_k}\right] \tag{2.1}$$

$$\text{GM} = \text{Antilog}\left[\frac{20{*}0.93 + 18{*}1.01 + \ldots 12{*}0.97}{20 + 18 + 12}\right]$$

$$= \text{Antilog}\sum_{i=1}^{n} N_i * \text{Log } X_i$$

$$\text{GM} = \text{Antilog}\left[\frac{48.42}{50}\right] = 0.968(9.68\%) \text{ concentration/m L}$$

2.5 Harmonic Mean (HM)

This measure is rarely used in statistics (Jankowski & Flannelly, 2015). The harmonic mean of a set of n numbers is defined as the reciprocal of the arithmetic mean of the reciprocals.

TABLE 2.7

Geometric Mean Calculation with Different Groups

Group (i)	Number of Patients (n_i)	GM (µg/mL) of Isoniazid Concentrations	Log GM	N_i * log GM
A	20	8.5	0.93	18.60
B	18	10.2	1.01	18.18
C	12	9.4	0.97	11.64
Total	50	–	–	48.42

TABLE 2.8

Harmonic Data Distribution for Average Calculation

Speed (X) mph	$\left(\frac{1}{X_i}\right)$
20	0.0500
25	0.0400
30	0.0333

This is best explained by a practical example. If one drives equal distances at speeds of 20, 25, and 30 mph, the average speed is not 25 mph, which is the arithmetic mean of the three speeds. A drive of 300 miles at 20 mph will take 15 hours, at 25 mph will take 12 hours, and at 30 mph will take 10 hours. Thus, a distance of 900 miles will be covered in 37 (15+12+10) hours, giving an average speed of 900/37=24.3 mph. This can be calculated using the formula for the harmonic mean (Table 2.8).

$$\text{Average of reciprocals} = 0.1233 = 0.0411\ 3$$

$$\text{Reciprocal of average} = 1 = 24.3 \text{ mph}$$

For a set of 'n' observations, the AM > GM > HM. These three measures will be equal if all the 'n' observations are identical.

$$\text{RMS} \geq \text{AM} \geq \text{GM} \geq \text{HM} \tag{2.2}$$

$$\sqrt{\frac{a^2 + b^2}{2}} \geq \frac{a+b}{2} \geq \sqrt{ab} \geq \frac{2}{\frac{1}{a} + \frac{1}{b}}, \tag{2.3}$$

For positive numbers a and b:

For any positive numbers $\text{GM} = \sqrt{\text{AM} * \text{HM}}$

Let a and b be the two positive numbers.

$$\text{AM} = \frac{a+b}{2}, \text{GM} = \sqrt{ab} \ \ \text{HM} = \frac{2ab}{a+b} \tag{2.4}$$

$$\text{AM} * \text{HM} = \frac{a+b}{2} * \frac{2ab}{a+b} = ab = (\text{GM})^2 \tag{2.5}$$

$$\text{GM} = \sqrt{\text{AM} * \text{HM}} \tag{2.6}$$

$$\text{AM} - \text{GM} = \frac{a+b}{2} - \sqrt{ab} \tag{2.7}$$

$$\frac{\left(\sqrt{a} - \sqrt{b}\right)^2}{2} \geq 0 \tag{2.8}$$

2.6 Quartiles, Deciles, and Percentiles

Quartiles divide a series into four equal parts, resulting in three quartiles: Q_1 (lower quartile), Q_3 (upper quartile), and Q_2 (median value). There are two methods used for determining quartiles: the tabular method and the graphical method.

In the tabular method, the following steps are involved:

1. Arrange the series in ascending order.
2. Calculate the cumulative frequencies of the values in the case of frequency distribution.
3. Determine the required factor for the formula represented in the table.

Factor	Individual and Discrete Series	Continuous Series
Q_1=Value of q_1	Value of $\left(\frac{N+1}{4}\right)^{\text{th}}$ item.	Value of $\left(\frac{N}{4}\right)^{\text{th}}$ item.
Q_2=Value of q_2	Value of $\left(\frac{2\,(N+1)}{4}\right)^{\text{th}}$ item.	Value of $\left(\frac{2N}{4}\right)^{\text{th}}$ item
Q_3=Value of q_3	Value of $\left(\frac{3\,(N+1)}{4}\right)^{\text{th}}$ item	Value of $\left(\frac{3N}{4}\right)^{\text{th}}$ item

Illustration

The protein sequences have expressed regression slopes rendered by a chromatograph. The following observations are recorded during the relative expression of proteins:

Protein Sequences (b^+)									
S_1	S_2	S_3	S_4	S_5	S_6	S_7	S_8	S_9	S_{10}
2,131	2,365	2,614	3,244	2,032	3,214	3,265	1,984	1,968	3,621

Steps

1. Arrange the data series in ascending order of magnitude

S_9	S_8	S_5	S_1	S_2	S_3	S_6	S_4	S_7	S_{10}
1,968	1,984	2,032	2,131	2,365	2,614	3,214	3,244	3,265	3,621

2. Determine the Quartile for the seq

$$Q_1 = \text{Value of } \left(\frac{N+1}{4}\right)^{\text{th}} \text{item. Value of } \left(\frac{10+1}{4}\right)^{\text{th}} \text{item}$$

$$= 11/4 = 2.75$$

Value of second item+0.7 (Value of third item – Value of second item)

$$1{,}984 + 0.7(2{,}131 - 2{,}032)$$

$$1{,}984 + 99 = 2{,}712$$

$$Q_2 = \text{Value of } \left(\frac{2\,(N+1)}{4}\right)^{\text{th}} \text{item}$$

$$Q_3 = \text{Value of } \left(\frac{3\,(N+1)}{4}\right)^{\text{th}} \text{item}$$

2.7 Median

When the distribution is not symmetrical (e.g. income distribution), it is often better to calculate a different type of average, called the median (Khorana et al., 2023). Another situation where the median would be more appropriate than the mean is if there are unusually high or unusually low values (e.g. duration of hospital stay), which may be genuine or may be due to measurement errors. Finally, when there are indeterminate values at either extreme, it will not be possible to compute the mean, and hence the median should be computed. Calculation of the median presents an entirely different approach to the problem of selecting a single figure to represent a group of data. If the observations are arranged in increasing or decreasing order, the median is the middle value. If the number of observations, n, is odd, there will be a unique median – the $(n+1)/2$th observation from either end in the ordered sequence. If 'n' is even, there is strictly no middle observation, but the median is defined by convention as the mean of the two middle observations – the nth and the $(n+1)/2$th from either end.

Suppose we wish to find the median of the following values.

$$(n = 10,\ 20,\ 12,\ 3,\ 18,\ 16,\ 14,\ 25,\ 2)$$

Arranging the numbers in increasing order, we have

$$(n = 2,\ 3,\ 10,\ 12,\ 14,\ 16,\ 18,\ 20,\ 25)$$

Hence $n=9$ is an odd number; hence the median is $9+1$th/2=5th value, that is, 14.

Suppose one more observation, 8, is included in the previous example. Then, arranging in increasing order, we have

$$(n = 2,\ 3,\ 8,\ 10,\ 12,\ 14,\ 16,\ 18,\ 20,\ 25)$$

Here $n=10$ is an even number, and the median is therefore the mean of the 10th/2 and 10/2+1th values—the mean of the 5th and 6th values, i.e. $(12+14)/2=13$.

Consider another example. The duration (in days) of absence from work of 21 laborers owing to sickness is given below.

$$\left(\begin{array}{c} n = 1,\ 1,\ 2,\ 2,\ 3,\ 3,\ 4,\ 4,\ 4,\ 4,\ 5,\ 6, \\ 6,\ 6,\ 6,\ 7,\ 8,\ 9,\ 10,\ 10,\ 59,\ 80 \end{array}\right)$$

The arithmetic mean (11 days) of absence is not really typical of the series as a whole since 19 of the 21 laborers were absent for less than 11 days. The median of 5 days would be a better measure.

However, the median has some disadvantages in comparison with the mean:

1. It takes no account of the precise magnitude of most of the observations and is, therefore, less efficient than the mean.
2. If two groups of observations are pooled, the median of the pooled group cannot be estimated from the two individual group medians.

3. The median is much less amenable than the mean to mathematical treatment and is not much used in the more elaborate statistical techniques.

Case 1: if n is odd

$$\text{Me} = \left(\frac{n+1}{2}\right)^{\text{th}} \quad (2.9)$$

Case 2: if n is even

$$\text{Me} = \frac{\left(\frac{n}{2}\right)^{\text{th}} + \left(\frac{n}{2}+1\right)^{\text{th}}}{2} \quad (2.10)$$

2.7.1 Steps for Determination of Median

1. First, arrange the series in a table in either ascending or descending order of their magnitude.
2. Determine cumulative frequencies in a separate column headed by CF.
3. Find the median item by applying the formula:

$$\text{Me} = \left(\frac{N+1}{2}\right)^{\text{th}} \text{ item for individual series} \quad (2.11)$$

$$\text{Me} = \left(\frac{N}{2}\right)^{\text{th}} \text{ item for continuous series} \quad (2.12)$$

4. Locate the median in the value column with reference to the median item in the CF column.
5. In the case of a continuous series, from the median class so located, determine the value of the median by applying any of the interpolation formulas cited as follows:

$$\text{Me} = L_1 + \frac{L_2 - L_1}{f_1}(m-c), \text{Me} = L_2 + \frac{L_2 - L_1}{f_1}(m-c) \quad (2.13)$$

Illustration

1. Determine the value of the median from the following series

Microorganisms' growth (days)	0–10	10–15	15–20	20–25	25–30
Number of organisms' growth	7	5	8	38	42
CF	7	12	20	58	100

Median = value of '*m*'

$$= \text{Value of } \left(\frac{N}{2}\right)^{\text{th}} \text{ item}$$

$= \text{Value of } \left(\frac{100}{2}\right)^{\text{th}}$ item or 50th item that falls within the CF of 58. The value of this item lies in the class (20–25).

Apply the interpolation formula:

$$\text{Me} = L_1 + \frac{L_2 - L_1}{f_1}(m-c)$$

where Me = Median to be determined

L_1 = Lower limit of the median class = 20

L_2 = Upper limit of the median class = 25

f_1 = Frequency of the median class = 38

m = median item is 50

c = cumulative frequency of the class preceding the median class = 20

$$\text{Me} = 20 + \frac{25-20}{38}(50-20)$$

$$= 23.95 \text{ or } 24 \text{ approx}$$

2.8 The Mode

The mode may be defined simply as 'the value that occurs most frequently'. It is not widely used in analytical statistics; however, it is easy to understand. It has the advantage over other averages in that it can be obtained when some characteristic cannot be directly measured, such as the preferred color of cars for people (Kuna-Broniowska & Smal, 2017).

For example, suppose in a survey we get the following data:

Color Preference	Number of Persons
Green	354
Blue	852
Grey	310
Black	474

From this, we conclude that the mode color is blue.

Forming a frequency distribution and calculating the mean, median, and mode is a useful first step in summarizing a fairly large quantitative data collection. A frequency distribution is a table that shows the number of observations (frequency) at different values or within certain ranges of values of the variable. If we are dealing with a discrete variable (i.e. a variable with distinct and separated possible values), the frequency can be tabulated for each value of the variable (Tableau) or for groups of values of the variable.

Number of X-rays Taken	Number of Patients (n=239)	Number of Doses of Drugs Received	Number of patients (n=123)
0	10	0–4	01
1	18	5–9	05
2	128	10–14	07
3	51	15–19	24
4	21	20–24	56
5	11	25–29	30

In the case of a continuous variable (i.e. a variable that can assume a continuous, uninterrupted range of values) class intervals must be formed and the corresponding frequencies tabulated:

Weight (kg)	Number of Patients ($n = 100$)
20–24.9	02
25–29.9	04
30–34.9	20
35–39.9	33
40–44.9	33
45–49.9	05
50–54.9	03

Grouped data of 'Z'

$$Z = L + h\left(\frac{f_m - f_1}{2f_m - f_1 - f_2}\right)$$

where L = Lower boundry of modal class

h = Size of modal class

f_m = Frequency corresponding to modal class

f_1 = Frequency preceding to modal class

f_2 = Frequency succeeding to modal class

2.9 Forming a Frequency Distribution

Given a large dataset, let us consider the task of creating a frequency distribution. If the variation is continuous (refer to Table 2.3) or occurs in discrete steps that are small compared to the entire range of variation (refer to Table 2.2), it is necessary to group the variable under study into classes (Marshall & Jonker, 2010). For convenience, the width of each class should be the same (i.e., the class interval must be constant) to ensure that the number of observations in different classes is comparable. Generally, it is advisable to have no more than 20 classes. Often, because the total number of observations is not large, it may not be necessary to have more than five classes (Manikandan, 2011). The limits of the intervals must be clearly stated, and the choice should be made to minimize the effects of any recognized tendencies for bias (e.g., rounding off age to the nearest 5 years). Finally, the count should be made by systematically going through the list of measurements and using the method of tally marks.

Illustration/Example

From the following data, determine the values of the mode by the Interpolation method

CD_4:CD_8 %	5–10	10–15	15–20	20–25	25–30	30–35	35–40	40–45
Number of students	5	6	15	10	5	4	2	2

The formula of interpolation of mode is

$$Z = L_1 + \frac{f_1 - f_0}{2f_1 - f_0 - f_2} \times (L_2 - L_1) \tag{2.14}$$

$$Z = 15 + \frac{15 - 6}{2(15) - 6 - 10} \times (20 - 15) = Z(\text{Mode}) = 18.21$$

where Z = Modal value

L_1 = Lower limit of the modal class = 15

L_2 = Upper limit of the model class = 20

f_1 = Frequency of the modal class = 15

f_0 = Frequncy of the class preceding the modal class = 6

f_2 = Frequncy of the class suceeding the modal class = 10

2.10 Applications and Limitations of Measures of Central Tendency

Before compiling research data, it is important for researchers to understand how the experimental data series are distributed. Generally, the central tendency of a population is characterized by typical numerical values that describe the raw data. In life science research, descriptive statistical tools such as mathematical, positional, and other averages are used to measure the central tendency of a population. These tools are utilized to determine a numerical value that reflects the distribution of the population. The main goal of these tools is to offer a single numerical value that effectively summarizes the collected data from the experimental material (Rosidah & Ikram, 2021). This value aims to capture the core or typical aspects of the data, providing a concise overall summary of the data series. However, measures of central tendency have limitations because they cannot undergo any mathematical algebraic treatment. Researchers can only use them for the summarization of raw data and they have not been tested for statistical hypothesis central values.

2.11 Measures of Dispersion

Measurement of variation is often important for descriptive reasons (Saidi & Siew, 2018). For instance, it is not sufficient to be told that the mean systolic blood pressure of a group of subjects is 110 mm Hg. One is also interested in knowing how much the blood pressure varies from subject to subject. As we shall see later, the measurement of variation also plays a central part in methods of statistical inference. To take an example, the reliability of the mean of 100 values of some variable depends on the extent to which the 100 values differ among themselves. We can appreciate intuitively that the smaller the variation, the greater the reliability, and in the extreme case where there is no variation (that is, all the values are equal), the mean is determined with certainty and is equal to the common value.

TABLE 2.9

Flowchart for the Classification of Measures of Dispersion

Method of Limits	Method of Compilation	Method of Graphs
• Range	• Mean deviation	• Lorenz curve
• IQR	• SD	• Progressive average
• Semi IQR	• QD	
• Decile range	• CV	
• Percentile range	• Variance	

One often finds in medical journals a bland setting out of mean values, with no indication of the variability of the observations. This is a bad practice, for a mean of 200 could have been obtained from three values of 100, 200, and 300, or equally well from three values of 199, 200, and 201, and the lack of information regarding the variability could be misleading. Indeed, there can be instances where the differences in variation are more important than the differences in mean values. The frequency distribution is one obvious way of indicating the extent of variation. However, there may not always be sufficient numbers to obtain a frequency distribution. In any case, it is more convenient to have a single quantity to measure variation because this would result in economy in presentation and also because statistical methods require such an index (Table 2.9).

2.12 Different Types of Measures of Dispersion

2.12.1 Range (*R*)

The range is defined as the difference between the two extreme values of a series; it is denoted as '*R*'

$$\text{Range } (R) = \text{Largets value } (L) - \text{Smallest value}(S)$$

$$\text{Co eff. } R = \frac{L - S}{L + S} \tag{2.15}$$

It is determined purely on the basis of the two extreme values.

In the case of a continuous distribution, the range is determined by taking the difference between the upper limit of the largest class interval and the lower limit of the smallest class intervals. In such case range is given by

$$R = L_2(L) - L_1(S) \tag{2.16}$$

$$\text{Co eff. } R = \frac{L_2(L) - L_1(S)}{L_2(L) + L_1(S)} \tag{2.17}$$

This is defined as the difference between the maximum and minimum values. For example, if the lowest weight in a group of 100 patients is 30kg and the highest is 65kg, the range is calculated as 65 – 30=35kg. The range is determined by only two of the many observations made and does not provide any information about the distribution of the remaining values. These values could be evenly spread between 30 and 65kg, unevenly spread, or clustered around a specific value. The range has two additional disadvantages. Firstly, it may increase with the size of the sample. In the previous example, if we include ten more patients, the range will remain the same (35kg) only if all the newly added patients have weights between 30 and 65kg. If any of these ten patients have a weight of less than 30kg or more than 65kg, the range will be larger. Therefore, ranges based on different sample sizes should not be compared. Secondly, the range can be distorted by outliers, which may be genuine or due to experimental errors (Table 2.10).

TABLE 2.10

Distribution of CD_4 Counts of HIV Infected 100 Cases (Continuous Data)

CD_4 Counts (micro/dL) Class Limits	No. of Cases '*f*'
<50–100	2
100–150	6
151–200	8
201–250	19
251–300	22
301–350	43
Total	$\sum f =$

CD_4 Counts (micro/dL) of HIV-infected: 325, 360, 345, 310, 333, 328, 369, 425

2.13 Practical Illustration of Calculation of Range Value

$$\text{Range}(R) = 369 - 310 = 59 \text{ micro/dL}$$

$$\text{Coeff. } R = \frac{369 - 310}{369 + 310} = \frac{59}{679}$$
$$= \mathbf{0.086} \text{ micro/dL Discrete data series}$$

$$R = L_2(L) - L_1(S)$$

$$R = 350 - 50 = 300 \text{ micro/dL}$$

$$\text{Co eff. } R = \frac{L_2(L) - L_1(S)}{L_2(L) + L_1(S)}$$

$$\text{Co eff. } R = \frac{350 - 50}{350 + 50} = 0.75 \text{micro/dL}$$

2.14 Inter-Quartile Distance

Just as the median breaks the distribution into two equal parts, we can define a quantity Q_1, which divides the distribution in the ratio of ¼: ¾, and another quantity Q_3, which divides the distribution in the ratio of ¾: ¼. $Q_3 - Q_1$ is then defined as the inter-quartile distance and can be used as a measure of variation. Unlike the range, it is not dependent on the number of observations; however, it can be used only when the number of observations is large enough to form a frequency distribution.

2.15 Absolute Deviation

If x.... is the mean of n observations $X_1, X_2, \ldots X_n$, $\Sigma X_i = X n$, $\sum_{i=1}^{n} X_i = n\bar{X}$. The mean deviation, i.e. the mean of the deviations of each observation from the mean, is always zero. For example, let us consider the values 3, 5, 4, and 8. The mean is 20/4=5. The deviations from the mean are −2, 0, −1, and 3; their sum is zero and hence the mean deviation is zero. In general, $\Sigma (X_i - X\ldots) = \Sigma X_i - \Sigma X = nX - nX = 0$. If the sign of the deviations from the mean is ignored, we have values of 2, 0, 1, and 3, the mean of which is 1.5. This value is called the absolute deviation. In general, the absolute deviation $= \Sigma d_i$ where d_i is the positive value of the deviation of the ith observation from its mean. The absolute deviation is not usually employed because it does not lend itself well to statistical treatment (Table 2.11).

TABLE 2.11

ATP (Adinosine Triphosphate) Molecules Data are Obtained from C3 Pants, Data Distribution Were Taken as per the SOP

Number of ATP	$\delta_X = \mathrm{MD}_x = \lvert X - \bar{X} \rvert$	Working Rule
556	$(556 - 670.08) = 114.08$	$\bar{X} = \frac{\sum_{i=1}^{n} X_i}{N}$
587	$(587 - 670.08) = 83.08$	
478	$(478 - 670.08) = 192.08$	$\bar{X} = \left(\frac{8041}{12}\right)$
496	$(496 - 670.08) = 174.08$	
458	$(458 - 670.08) = 212.08$	$N = 12$
475	$(475 - 670.08) = 195.08$	δ_X
655	$(655 - 670.08) = 15.08$	$\delta_X = \frac{\lvert \sum_{i=1}^{n} X_i - \bar{X} \rvert}{N}$
647	$(647 - 670.08) = 23.08$	
875	$(875 - 670.08) = 204.92$	
897	$(897 - 670.08) = 226.92$	$\mathrm{MD}_X = \left(\frac{2{,}013}{12}\right)$
954	$(954 - 670.08) = 283.92$	
963	$(963 - 670.08) = 292.92$	
$\sum X = 8{,}041$	$\sum \lvert X_i - \bar{X}_i \rvert = 2{,}013$	

2.16 Mean Deviation

2.16.1 R-Code

```
x <- c(556,587,478,496,458,475,655,647,875,897,954,963)
  n=length(x)
  sum(abs(x-mean(x)))/n
  mean(x)
  length(x)
```

2.17 Standard Deviation

For obtaining the absolute deviation, the sign of deviations was ignored. An alternative way is to square the deviations. Consider the following five values, which denote the number of X-rays taken in a year for five patients: The mean is 45/5=9 X-rays. The variance is defined as the sum of squares of the deviations from the mean, divided by one less than the number of observations (this latter is called degrees of freedom) (Wilcox & Keselman, 2003).

Discrete series:

$$\sigma = \sqrt{\frac{\sum_{i=1}^{N} \left(X_i - \bar{X}\right)^2}{N}} \text{ or}$$

$$\sigma = \frac{1}{N}\sqrt{\sum_{i=1}^{n} \left(X_i - \bar{X}\right)^2} \text{ Population}$$

$$\mathrm{SD} = \sqrt{\frac{\sum_{i=1}^{n} (x_i - \bar{x})^2}{n}} \text{ or}$$

$$\frac{1}{n}\sqrt{\sum_{i=1}^{n} (x_i - \bar{x})^2} \text{ Sample}$$

Continuous series:

$$\sigma = \sqrt{\frac{\sum_{i=1}^{N} F_i \left(X_i - \bar{X}\right)^2}{N}} \text{ or}$$

$$\sigma = \frac{1}{N}\sqrt{\sum_{i=1}^{n} \left(X_i - \bar{X}\right)^2} \text{ Population}$$

$$N = \sum F_i$$

$$\mathrm{SD} = \sqrt{\frac{\sum_{i=1}^{n} f_i (x_i - \bar{x})^2}{n}} \text{ or}$$

$$\frac{1}{n}\sqrt{\sum_{i=1}^{n} (x_i - \bar{x})^2} \text{ Sample}$$

$$n = \sum f_i$$

2.17.1 Illustration

The agricultural entomologist recorded the number of insects per plot during his study intervention to control insects in experimental plots. The following insect counts (per plot) were recorded: 10, 12, 23, 23, 16, 23, 21, 16. Find the standard deviation of the above discrete series.

$$\sigma = \sqrt{\frac{1}{N}\sum_{i=1}^{n} (x_i - \mu)^2}$$

$$\sigma^2 = \frac{\sum (10-18)^2 + \sum (12-18)^2 + \ldots \sum (16-18)^2}{8} = 24$$

$$\sigma = \sqrt{24}$$

$$\sigma = 4.89$$

2.18 Margin of Error (Confidence Interval)

The sampling mean most likely follows a normal distribution. In this case, the standard error of the mean (SEM) can be calculated using the following equation. $\sigma_{\bar{x}} = \frac{\sigma}{\sqrt{N}}$. Based on the SEM, the following are the margins of error (or confidence intervals) at different confidence levels. Depending on the field of study, a confidence level of 95% (or statistical significance of 5%) is typically used for data representation (Tables 2.12 and 2.13).

CI 95% margin error =

$$\left[\bar{x} - \mu * 1.96 \text{ SEM} \leq \text{CI } 95\% \quad \bar{x} + \mu * 1.96 \text{ SEM}\right]$$

Solution

$$\bar{X} = \frac{1}{N}\sum_{i=1}^{n} f_i X_i = \frac{665.50}{60} = 11.07\ \%$$

$$S_x^2 = \frac{1}{N-1}\left(\sum_{i=1}^{n} f_i X_i - \frac{\left(\sum_{i=1}^{n} f_i X_i\right)^2}{N}\right)$$

$$S_x^2 = \frac{1}{59}\left(665.5 - \frac{7{,}413.75}{60}\right)$$

$$S_x^2 = 0.5465$$

$$\text{HB}(\%) = 8.77,\ 8.9,\ 9.84,\ 9.85,\ 11.24,\ 12.33,\ 12.5,\ 13.26$$

$$S_x^2 = \frac{\sum_{i=1}^{n}\left(X_i - \bar{X}\right)^2}{n-1} = \frac{SS_x}{n-1} = \frac{21.02}{7} = 3.00$$

TABLE 2.13

In Case of Continuous Datasets

Class Interval HB (%)	Class Boundaries	Mid-Value (x_i)	Freq f_i No. of Patients	$f_i * x_i$	$f_i * x_i^2$
9.25–9.75	9.25–9.75	9.5	2	19	180.5
9.75–10.25	9.75–10.25	10	5	50	500
10.25–10.75	10.25–10.75	10.5	12	126	1,323
10.75–11.25	10.75–11.25	11	17	187	2,057
11.25–11.75	11.25–11.75	11.5	14	161	1,851.5
11.75–12.25	11.75–12.25	12	6	72	864
12.25–12.75	12.25–12.75	12.5	3	37.5	468.75
12.75–13.25	12.75–13.25	13	1	13	169
Total			60	665.5	7,413.75

2.19 Coefficient of Variation (CV %)

The mean and standard deviation of a set of measurements are expressed in the same units as the measurements. Sometimes, it is useful to describe the variability by expressing the standard deviation as a percentage of the mean. For instance, in several laboratory tests, the errors, as measured by the standard deviation, are proportional to the mean, leading to assertions that the error of a test is, say, ±5% (Table 2.14).

$$S_x^2 = \frac{SS}{N-1} = \frac{964.47}{16} = 60.27$$

$$\text{SD} = \sqrt{S_x^2} = \sqrt{60.27} = 7.76$$

$$\text{CV}(\%) = \frac{\text{SD}}{\bar{x}} * 100, \text{CV}(\%) = \left(\frac{7.76}{59.18}\right) * 100 = 13.12\%$$

TABLE 2.12

Distribution of Confidence Level and Calculation of Experimental Error

Confidence Level	Margin of Error	$\% E = \left(\frac{E-T}{T}\right)$	Experimental Value Error $E = T\left(1 \pm \frac{\% \text{error}}{100}\right)$
68.30%, $\sigma_{\bar{x}}$	18±1.732 (±9.62%)	4.56	$E = 47.07\left(1 \pm \frac{\% 4.55}{100}\right)$
90%, 1.645 $\sigma_{\bar{x}}$	18±2.849 (±15.83%)	4.59	
95%, 1.960 $\sigma_{\bar{x}}$	18±3.395 (±18.86%)	4.56	$E = [44.93 - 49.22]$
99%, 2.576 $\sigma_{\bar{x}}$	18±4.462 (±24.79%)	4.55	
99.9%, 3.291 $\sigma_{\bar{x}}$	18±5.7 (±31.67%)	4.55	
99.99%, 3.891 $\sigma_{\bar{x}}$	18±6.739 (±37.44%)	4.56	
99.999%, 4.417 $\sigma_{\bar{x}}$	18±7.65 (±42.50%)	4.55	
999.9999%, 4.892 $\sigma_{\bar{x}}$	18±8.473 (±47.07%)	4.55	

Working steps % error 4.55, theoretical error 47.07 (*T*).

TABLE 2.14
Distribution of Disability Score

Disability Score for Children	Deviation from the Mean ($X-d$)	Squared Deviation
65	5.82	33.91
70	10.82	117.15
55	–4.18	17.44
63	3.82	14.62
64	4.82	23.27
48	–11.18	124.91
54	–5.18	26.8
52	–7.18	51.5
63	3.82	14.62
69	9.82	96.5
68	8.82	77.85
67	7.82	61.21
48	–11.18	124.91
49	–10.18	103.56
63	3.82	14.62
52	–7.18	51.5
56	–3.18	10.09
$\bar{x}$ = 59.18		SS:964.47

2.20 Descriptive Statistics Exploratory Data Analysis by *R*-Statistical Software

data <- iris # load the iris dataset and rename it data

## Sepal.	LengthSepal.	WidthPetal.	LengthPetal.	WidthSpecies
##	1	5.1	3.5	1.40.2 setosa
##	2	4.9	3.0	1.40.2 setosa
##	3	4.7	3.2	1.30.2 setosa
##	4	4.6	3.1	1.50.2 setosa
##	5	5.0	3.6	1.40.2 setosa
##	6	5.4	3.9	1.70.4 setosa

Minimum and maximum values can be found using the `min()` and `max()` functions:

```
  Min(dat$Sepal.Length)
  ## [1] 4.3
  Max(dat$Sepal.Length)
  ## [1] 7.9
```

Alternatively the `range()` function:

```
  rng<- range(dat$Sepal.Length)
  rng
  ## [1] 4.3 7.9
  rng[1] # rng=name of the object specified
above
  ## [1] 4.3
  max(dat$Sepal.Length) - min(dat$Sepal.Length)
  range2 <- function(x){
  range<- max(x) - min(x)
  return(range)
  }
  Range2(dat$Sepal.Length)
  Median(dat$Sepal.Length)
  Quantile(dat$Sepal.Length, 0.5)
  Quantile(dat$Sepal.Length, 0.25) # first
quartile
  IQR(dat$Sepal.Length)
  Quantile(dat$Sepal.Length, 0.75) -
quantile(dat$Sepal.Length, 0.25)
  Sd(dat$Sepal.Length)# standard deviation
  lapply(dat[, 1:4], sd)
  ## $Sepal.Length
  ## [1] 0.8280661
  ## ## $Sepal.Width
  ## [1] 0.4358663
  ## ## $Petal.Length
  ## [1] 1.765298
  ## ## $Petal.Width
  ## [1] 0.7622377
```

Output summary

```
  ## Sepal.Length Sepal.Width Petal.Length
Petal.Width
  ## Min. :4.300 Min. :2.000 Min. :1.000Min.
:0.100
  ## 1st Qu.:5.100 1st Qu.:2.800 1st Qu.:1.600
1st Qu.:0.300
  ## Median :5.800 Median :3.000 Median :4.350
Median:1.300
  ## Mean :5.843 Mean :3.057 Mean :3.758
Mean:1.199
  ## 3rd Qu.:6.400 3rd Qu.:3.300 3rd Qu.:5.100
3rd Qu.:1.800
  ## Max. :7.900 Max. :4.400 Max. :6.900
Max.:2.500
  ## Species
  ## setosa:50
  ## versicolor:50
  ## virginica:50
  ##
```

2.21 Applications of MC

The measure of MC is a statistical method that represents the numerical value of a dataset (Walesiak et al., 2024). It is highly useful in many real-life applications, including a brief review of the following areas where it is widely applied.

2.21.1 Agricultural Research

Estimation of mean yield with respect to different concentrations of NPK chemical fertilizers, potency levels of bioassay studies, nutritional formulations of different crops, breeding and agronomical trials for evaluating different varieties of crops, understanding cropping patterns, and soil texture across different geographical areas, etc.

2.21.2 Business

Calculating sales revenue, profit, or employee salaries
 Setting benchmarks
 Assessing financial performance

2.21.3 Social Sciences

Summarizing and understanding data
 Assessing central tendencies in surveys or psychological studies

2.21.4 Psychology

Understanding what is normal or 'average' for a set of data

Condensing large amounts of data into a single representative value

2.21.5 Sociology

Studying the social structure of the population and demographic characteristics at the community level.

2.21.6 Economics

Conducting econometric analysis (micro and macro), assessing GDP, inflation, socioeconomic evaluations, and studying poverty levels, among other applications.

2.21.7 Medical Science

Conducting drug trial experiments to calculate the potency of drugs, estimatng the mean survivability of patients, analyzing the relationship between treatment length and complications, and including biomedical assays in public health studies.

2.21.8 Genetics

Conducting studies on heritability, analyzing quantitative and qualitative traits, exploring molecular genetics, and examining the genetic complexity of species or organisms.

2.21.9 Animal Husbandry

Study of breeding trials, evolution of new germplasm, cross-breeding, and inbreeding coefficients, etc.

2.21.10 Actuarial Science

Study of the average life expectancy of insurers, default analysis associated with loan borrowers, age-specific mortality after adjusting life expectancy, etc.

Practical Exercise

Illustration 1

A number of people invested $1,000 each in the Canara Bank Mutual Funds. The frequency table below shows the current values of those investments. Compute the mean, median, and mode.

Value	0	50	75	100	150	200	250
Frequency	51	56	31	28	23	38	39

Illustration 2

Find the median of the data: 5, 7, 4, 9, 5, 4, 4, 3

Illustration 3

Find the Upper and Lower Quartiles for 17, 10, 9, 14, 13, 17, 12, 20, 14

Illustration 4

Find the range for the dataset: 17, 10, 9, 14, 13, 17, 12, 20, 14

Illustration 5

A petrol station owner records the number of cars that visit his premises in 10 days.

The numbers are: 204, 310, 279, 314, 257, 302, 232, 261, 308, 217

a. Find the mean number of cars per day.
b. The owner hopes the mean will increase if he includes the number of cars the next day. If 252 cars use the petrol station the next day, will the mean increase or decrease?

Illustration 6

A research station kept records of the number of microorganisms growing in Petri dishes per day in the thrust area of Microbiology over 100 days. The figures below give the number of microorganisms growing per day.

1	4	3	5	5	2	5	4	3	2	0	3	1	2	2	3	0	5	2	1
3	3	2	6	2	1	6	1	2	2	3	2	2	2	2	5	4	4	2	3
3	1	4	1	7	3	3	0	2	5	4	3	3	4	3	4	5	3	5	2
4	4	6	5	2	4	5	5	3	2	0	3	3	4	5	2	3	3	4	4
1	3	5	1	1	2	2	5	6	6	4	6	5	8	2	5	3	3	5	4

Find the mean growth of organisms per day.

Illustration 7

The mean of a sample of 6 numbers is 3.2. An extra value of 3.9 is included in the sample. What is the new mean?

Illustration 8

The mean number of a set of 5 numbers is 12.7. What extra number must be added to bring the mean up to 13.1?

Illustration 9

The mean height of a class of 28 students is 162 cm. A new student of height 149 cm joins the class. What is the mean height of the class now?

Illustration 10

A salesman sells a mean of 4.6 solar power systems per day for 5 days. How many must he sell on the sixth day to increase his mean to 5 sales per day?

Illustration 11

Five numbers have a mean of 12. When one number is removed, the mean is 11. What is the value of the number removed?

Illustration 12

The ages of students in a small primary school were recorded in the table below.

Age	5–6	7–8	9–10
Frequency	29	40	38

(a) Estimate the mean. (b) Estimate the median. (c) Find the modal class.

Illustration 13

The number of days that students were absent from school due to sickness in 1 year was recorded:

Number of days off sick	1–5	6–10	11–15	16–20	21–25
Frequency	12	11	10	4	3

(a) Estimate the mean. (b) Find the median class. (c) Find the modal class.

Illustration 14

The table shows the distribution of scores for 40 students in a Mathematics test.

Score	10–12	13–15	16–18	19–21	22–24
Frequency	4	6	13	9	8

a. Estimate the mean score obtained in the test.
b. Estimate the probability that a student selected at random will score at most 15 marks in the test.
c. Using a scale of 1 cm to represent 1 unit on the frequency axis and 2 cm to represent 5 units on the scores axis, use graph paper to draw a frequency polygon to represent the distribution of scores shown in the table.

REFERENCES

Ali, Z. & Bhaskar, S. B. Basic statistical tools in research and data analysis. *Indian Journal of Anaesthesia* **60,** 662–669 (2016).

Anilkumar, D. R. P. & Samiyya, N. Refining measure of central tendency and dispersion. *IOSR Journal of Mathematics* **6,** 1–4 (2013).

Cazorla, I. M., Utsumi, M. C. & Magina, S. M. The conceptual field of measures of central tendency: A first approximation. *International Electronic Journal of Mathematics E*ducation **18,** em0748 (2023).

Chakrabarty, D. Measuremental data: Seven measures of central tendency. *International Journal of Electronics and Applied Research* **8,** 15–24 (2021).

Deshpande, S., Gogtay, N. J. & Thatte, U. M. Measures of central tendency and dispersion. *Journal of Association of Physicians of India* **64,** 64–66 (2016).

Dimitriadis, T., Patton, A. J. & Schmidt, P. Testing forecast rationality for measures of central tendency. *SSRN Electronic Journal* (2019). https://doi.org/10.2139/ssrn.3466442

Jankowski, K. R. B. & Flannelly, K. J. Measures of central tendency in chaplaincy, health care, and related research. *Journal of Health Care Chaplaincy* **21,** 39–49 (2015).

Khorana, A. et al. Choosing the appropriate measure of central tendency: Mean, median, or mode? *Knee Surgery, Sports Traumatology, Arthroscopy* **31,** 12–15 (2023).

Kuna-Broniowska, I. & Smal, H. Statistical measures of the central tendency for H+ activity and pH. *Soil Science Annual* **68,** 174–181 (2017).

Manikandan, S. Measures of central tendency: The mean. *Journal of Pharmacology and Pharmacotherapeutics* **2,** 140–142 (2011).

Marshall, G. & Jonker, L. An introduction to descriptive statistics: A review and practical guide. *Radiography* **16,** e1–e7 (2010).

Rosidah, R. & Ikram, F. Z. Measure of central tendency: Undergraduate students' error in decision-making perspective. *International Journal of Education* **14,** 39–47 (2021).

Saidi, S. S. & Siew, N. M. Assessing students' understanding of the measures of central tendency and attitude towards statistics in rural secondary schools. *International Electronic Journal of Mathematics Education* **14,** 73–86 (2018).

Walesiak, M., Dehnel, G. & Dudek, A. A dynamic approach to relative taxonomy and robust measures of central tendency. *Communications in Statistics: Simulation and Computation* **53,** 2645–2661 (2024).

Wilcox, R. R. & Keselman, H. J. Modem robust data analysis methods: measures of central tendency. *Psychological Methods* **8,** 254–274 (2003).

3

Correlation and Regression Analysis: A Beginner's Guide

3.1 Introduction

In biological studies, it is common practice to measure several quantitative variables on each individual and study the data for associations between pairs of variables. For example, the measurements could be height and weight, age and vital capacity, or circulating protein and surface area (Bhutto et al., 2016). To assess whether there is any relationship between two characteristics, the simplest method is to plot the two characteristics on a graph sheet – for example, age on the *X*-axis and systolic BP on the *Y*-axis. Such a diagram is called a scatter diagram and can often be very instructive. From the above diagram, we see that as age increases, there is a general tendency to increase systolic BP. However, the scatter diagram does not give a quantitative estimate of the degree of the relationship. Such an estimate or index would be very valuable as a summary measure, and also as a tool for comparing the degree of association in different groups. For example, one may be interested in knowing whether the degree of association between age and systolic BP is the same (or different) in males and females or sedentary workers and laborers. An index, known as the 'correlation coefficient', is usually employed for these purposes. This coefficient is denoted by the letter '*r*' and takes values between –1 and +1. It must be stressed that '*r*' is purely a measure of the degree of association and does not provide evidence of a cause-effect relationship. The following important points must be noted.

3.2 Types of Correlation

The scatter plot explains the correlation between the two attributes or variables (Boller, 2024). It represents how closely the two variables are connected (Figure 3.1). There can be three such situations to observe the relationship between the two variables.

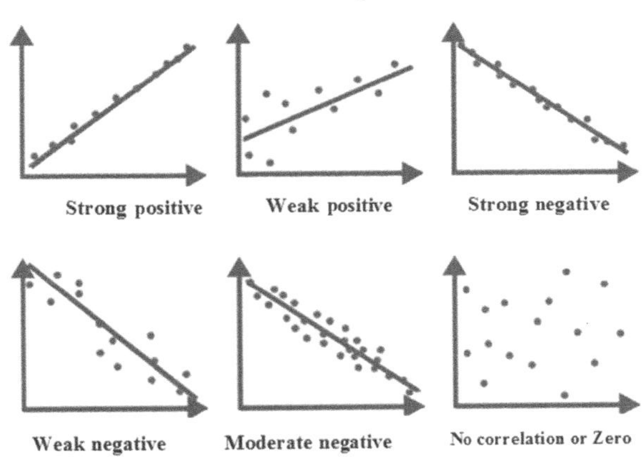

FIGURE 3.1 Graphical representation of various types of correlation.

3.2.1 Positive Correlation

When the values of two variables move in the same direction, an increase or decrease in the value of one variable is followed by an increase or decrease in the value of the other variable.

3.2.2 Negative Correlation

When the values of two variables move in the opposite direction, an increase or decrease in the value of one variable is followed by a decrease or increase in the value of the other variable.

3.2.3 Zero Correlation

Zero correlation is a statistical term that means there is no linear relationship between two variables. When two variables have zero correlation, changes in one variable do not predict changes in the other. (Table 3.1).

i. If high values of one variable tend to occur with high values of the other variable (and low with low), we say that the correlation between the two variables is positive. When high values of one variable occur with low values of the other variable (and vice versa), the correlation between the two variables is said to be negative. Some relationships are mathematically precise. For example, the distance covered by a train will be directly proportional to the hours of travel. In this case, we say that the correlation is positive and perfect, equal to +1 (Figure 3.2)

   To take another example, if the distance is fixed, the time taken by the train will decrease as the speed increases; in this case, the correlation is negative and

TABLE 3.1

Age-Wise Distribution of Respondents

Age (years)	33	33	34	42	29	19	50	55	31	46	36	48
Weight (kg)	80	82.5	101	90	67	60	77	77	87	70	57	64

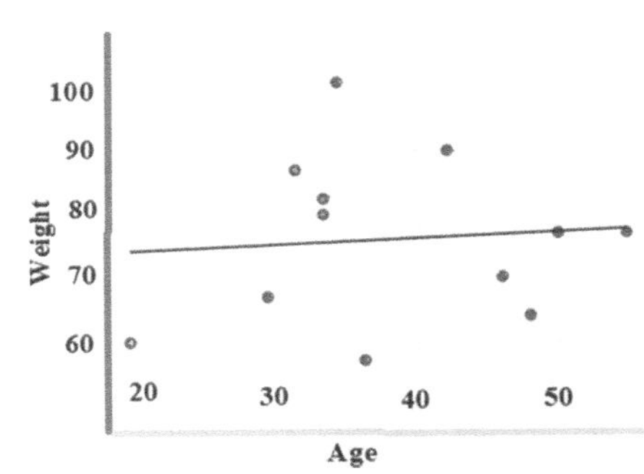

FIGURE 3.2 Scatter diagrams on the relationship between age and BMI.

DOI: 10.1201/9781032713816-3

perfect, equal to −1. However, in the natural sciences, the relationship between variables is seldom perfect. For example, consider the height and weight of a group of individuals (Cui et al., 2024). In general, it is known that taller people have larger weights than shorter people. But this need not be true in every case. For example, it is possible that a person with a height of 160 cm can have a weight less than that of another person with a smaller height, say 150 cm. In this case, we say there is a positive correlation between height and weight, but not a perfect one; the correlation coefficient will be between 0 and 1 (Durgadevi et al., 2018). This sets out the case of a negative non-perfect correlation, i.e. r lies between −1 and 0. An example of this is the association between the regularity of clinic attendance of patients and the distance of their homes from the health center. Finally, this illustrates a case of zero correlation, i.e. no obvious relationship between the values of 'X' and the values of 'Y'.

ii. The interpretation is valid only in the range of values studied. In other words, extrapolation of the association may not always be valid. For example, consider the association between age and grip strength. Grip strength increases with age up to a certain point (say, up to 35 years), then remains static up to a certain age (say, up to 55 years), and thereafter declines as age advances. Thus, for the same pair of variables (age and grip strength), the association is positive up to a certain age, and no association exists after a certain age. This is a good example to illustrate that extrapolation of the association beyond the range of values studied is not valid.

iii. The 'r' measures the degree of linear relationship; hence, $r=0$ does not necessarily mean that there is no relationship between the two characteristics under study. For example, the relationship could be curvilinear.

iv. The correlation coefficient does not give the 'rate of change' in one variable for changes in the other variable. Thus, if the correlation coefficient between age and BP is 0.7 in males and 0.5 in females, one should not conclude that as age increases, BP increases at a higher rate among males than in females. For studying the rate of change, one has to calculate the regression coefficient.

v. The correlation coefficient is unaffected by the addition or subtraction of a constant value to all the values of X and another (or same) constant value to all the values of Y. For example, if the correlation coefficient between X and Y is 0.7, then the correlation coefficient between $(X+10)$ and $(Y-6)$ will also be 0.7.

vi. The correlation coefficient is also unaffected by multiplication or division by a constant of all the values of one variable or both. For example, if 0.8 is the correlation coefficient between the variables X and Y, the correlation coefficient between $5X$ and $2Y$ will also be equal to 0.8. If 0.6 is the correlation coefficient between height in inches and weight in pounds, the correlation coefficient between height in centimeters and weight in pounds or kilograms will also be equal to 0.6 (Table 3.2).

Illustration

TABLE 3.2

Data for the Age and BMI Variables to Find the Sample Correlation Coefficient

Age (years)	25	26	32	28	34	40	36	41	43	45
BMI (kg/mm^3)	22	19	23	25	24	26	28	32	35	38

3.3 Working Rule of Simple Correlation

Step 1: Define the dependent (y) and independence t(x) variables. From the above illustration, we consider the independent variable as age and the dependent variablel as BMI.

Step 2: Compute the correlation coefficient from the following table matrices:

Age (X)	BMI (Y)	$X*Y$	X^2	Y^2
25	22	550	625	484
26	19	494	676	361
32	23	736	1024	529
28	25	700	784	625
34	24	816	1,156	576
40	26	1,040	1,600	676
36	28	1,008	1,296	784
41	32	1,312	1,681	1,024
43	35	1,505	1,849	1,225
45	38	1,710	2,025	1,444
$\sum_{i=1}^{10} X_i = \mathbf{350}$	$\sum_{i=1}^{10} Y_i = 272$	$\sum_{i=1}^{10} X_i Y_i = 9{,}871$	$\sum X_i^2 = 12{,}716$	$\sum Y_i^2 = 7{,}728$

Step 3: Calculate $\bar{X}$ and $\bar{Y}$ values from the above table

$$\bar{X} = \frac{1}{n}\sum_{i=1}^{n} X_i = \frac{351}{10} = 35.1$$

$$\bar{Y} = \frac{1}{n}\sum_{i=1}^{n} Y_i = \frac{272}{10} = 27.20$$

Step 4: Calculate $SS_{XX}SS_{YY}\&SS_{XY}$

$$SS_{XX} = X_i^2 - \frac{1}{n}\left(\sum_{i=1}^{n} X_i\right)^2 = 12{,}716 - \frac{351^2}{10} = 464.90$$

$$SS_{YY} = Y_i^2 - \frac{1}{n}\left(\sum_{i=1}^{n} Y_i\right)^2 = 7{,}728 - \frac{272^2}{10} = 329.60$$

$$SS_{XY} = X_iY_i - \frac{1}{n}\left(\sum_{i=1}^{n} X_i\right)\left(\sum_{i=1}^{n} Y_i\right) = 9{,}871 - 351 * \frac{272}{10} = 347.80$$

Therefore, based on the above calculations, the correlation coefficient is obtained as follows.

Step 5: Determine correlation coefficient 'r'

$$r = \frac{ss_{XY}}{\sqrt{SS_{XX}SS_{YY}}} = r = \frac{347.80}{\sqrt{464.90 * 329.60}} = 0.88$$

Step 6: Apply the test statistics 't' $t = \frac{r\sqrt{n-2}}{1-r^2}$, where r=correlation coefficient, r=correlation coefficient (0.80), n=the number of observations

$$t = \frac{0.80\sqrt{10-2}}{1-0.80^2} = 17.78$$

Extract table t-value at $n-2$ $df=10-2=8$ @ 5% level of significant

Step 6: Inference

Compare calculated critical value t v/s table t-value. If the calculated t-value is more than table t-*value*, reject the null hypothesis (H_0)

Step 7: Interpretation

The age and BMI are found to be significant correlation, the value of one variable linearly correlated to other variable of interest (Table 3.3).

The same hypothesis is tested by 'Z' test statistics

$$Z = \frac{1}{2}Log_e 10\left(\frac{1+r}{1-r}\right) = 1.153\, Log_e 10\left(\frac{1+0.80}{1-0.80}\right)$$

$$Z = 1.11, \text{Standard error} = \frac{1-r^2}{\sqrt{n}} = \frac{1-0.80^2}{\sqrt{10}} = 0.11$$

TABLE 3.3

Size of the Correlation and Interpretation Rule

Size of Correlation	Interpretation
0.90 – 1.00(–0.90 to –1.00)	Very high positive (negative) correlation
0.70 to 0.90 (–0.70 to –0.90)	High positive (negative) correlation
0.50 to 0.70 (–0.50 to –0.70)	Moderate positive (negative) correlation
0.30 to 0.50 (–0.30 to –0.50	Low positive (negative) correlation
0.00 to 0.30 (0.00 to –0.30)	Negligible correlation

3.3.1 Illustrations

Calculating the correlation coefficient must be followed by testing its significance. Even if the two variables are not associated in the population, we may, and often do, get non-zero values of r in samples. If the sample size is large, the distribution of r is approximately normal; therefore, if we can derive the standard error of r, we should be able to test the significance of departures from zero. In practice, however, the number of paired observations is seldom large enough (Erdfelder et al., 2009). A different test (using the t-distribution) is therefore used in such situations. The procedure is to enter the table with the degrees of freedom being 2 less than the number of paired observations, and read the values listed against the 5%, 1%, and 0.1% levels. If the sample r is larger than the r tabulated against the 0.05 probability level, we say that the correlation coefficient is significantly different from zero at the 5% level; similarly for the 1% and 0.1% levels. Suppose we considered, n=7 and r=0.98. From the table, the critical value of r for 5 degrees of freedom is 0.951 for P=0.001. Since the observed correlation coefficient (0.98) is larger, we conclude that P <0.001. In the absence of the table, or if the probability value is required more precisely, the quantity may be calculated and referred to the t-table with (n – 2) degrees of freedom. In the above case, r=0.98 and n=7, so the P-value for t with 5 degrees of freedom can be read off from the t-table as 0.001. The significance test is valid only if the two variables are approximately normally distributed (Hu et al., 2024) (Tables 3.4 and 3.5).

3.4 Calculation of Correlation Coefficient with Grouped Data

The calculation of the correlation coefficient from grouped data can be undertaken along the same lines as the calculation of means and standard deviations of frequency distributions (Table 3.6 and 3.7).

The expressions $\sum f=n$, $\sum fx$, $\sum fy$, $\sum fx^2$, $\sum fy^2$, and $\sum fxy$ are computed, and r is given by the expression

$$r = \frac{n\sum fxy - \sum fx \sum fy}{n\sum fx^2 - \left(\sum fx\right)^2 * n\sum fy^2 - \left(\sum fy\right)^2} \quad (3.1)$$

Since the correlation coefficient is unaffected by multiplication or division by, or addition or subtraction of, a constant factor, we will replace 'X' with 'x' given by the expression

TABLE 3.4

Sample Data to Find the Correlation Matrix

Age	CD_4 Counts	ESR	HB (%)	RBS
32	355	35	11.5	145
34	154	36	12.36	98
38	485	32	9.89	97
45	563	28	10.25	147
44	455	20	11.85	165
54	547	22	11.36	136
65	324	23	11.69	144
63	362	26	11.85	169
47	447	27	13.25	147
48	398	36	12.58	163
56	369	40	11.47	84
53	366	45	9.88	195
57	387	43	9.87	158
54	387	40	10.74	169
61	369	39	11.45	170

TABLE 3.5

Correlation Matrices of Different Parameters of SARS-CoV-2 Subjects

Variables	Age	CD_4 Counts	ESR	HB (%)	RBS
Age	1.0000	0.045	0	−0.073	0.377
CD_4 Counts		1.0000	−0.397	−0.296	0.136
ESR			1.0000	−0.4	0.07
HB (%)				1.0000	−0.098
RBS					1.0000

TABLE 3.6

Distribution of Data Series with *X* and *Y*

Value of '*X*'	Value of '*Y*'	Frequency
X_1	Y_1	f_1
X_2	Y_2	f_2
–	–	–
–	–	–
X_k	Y_k	f_k

TABLE 3.7

Age-Wise (*X*) Distribution of Systolic BP(*Y*)

Age (*X*)	Systolic BP (*Y*)	Frequency (*f*)
20	80	15
30	100	22
40	110	30
50	130	23
60	140	10

$\left(\dfrac{X-20}{10}\right)$ and '*Y*' with '*y*' given by the expression $\left(\dfrac{Y-80}{10}\right)$ (Table 3.8).

The above table then becomes

TABLE 3.8

X and *Y* Linearly Correlated Continuous Data Series

x	*y*	*f*
0	0	15
1	2	22
2	3	30
3	5	23
4	6	10

$n=\sum f=100,\ \sum fx=191,\ \sum fy=309,\ \sum fxy=809\ \sum fx^2=509$ $\sum fy^2=1{,}293$

$$r=\frac{(100*809)-(191*309)}{\left[(100*509-(191)^2\right]\left[100*1{,}293-(309)^2\right]}$$

$$r=\frac{21{,}881}{14{,}419*33{,}919}$$

$$r=0.99$$

Note

A note of caution must be sounded in interpreting the significance of a correlation coefficient. It is purely a measure of association and does not imply a cause-and-effect mechanism. Indeed, there are many examples of highly significant correlations that are nonsensical. For instance, it has been pointed out that the annual issue of broadcasting licenses is highly correlated with the annual rate of admission to mental hospitals, and that the import of tobacco into the UK is highly correlated with the divorce rate. The only logical conclusion in such situations is that, quite by chance, both measures are increasing at the same rate. Even when a significant correlation seems reasonable, we should not assume cause and effect, as a third variable may be influencing both the first and the second and thus explain the significant correlation.

3.5 Partial Correlation

One of the variables is kept constant to know the effect of the other variable. For example, Age, BMI, and Exercise (*x*, *y*, *z*) will be correlated as r_{12}, r_{23} (Hu et al., 2024). The partial correlation coefficient assesses the degree of association between two variables X_1 and X_2, when controlling (keeping constant) a third variable X_3 (Tables 3.9 and 3.10). Mathematically, the partial correlation coefficient *r* between X_1 and X_2, when controlling for X_3, is written as $r_{12.3}$, and it is computed using the

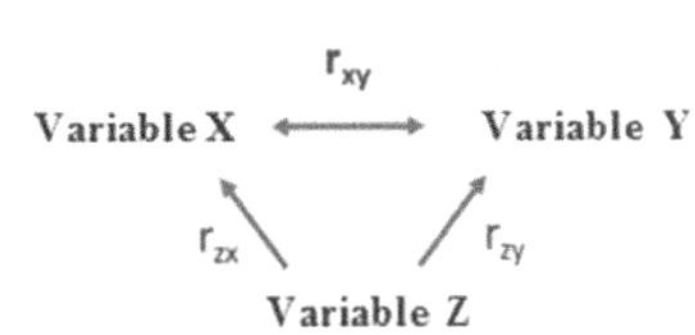

$$r_{12.3}=\frac{r_{XY}-r_{XZ}r_{YZ}}{\left(\sqrt{1-r_{XZ}{}^2}\right)\left(\sqrt{1-r_{YZ}{}^2}\right)} \quad (3.2)$$

TABLE 3.9

Correlative Variable for Partial Correlation

Correlative Variable	r	r^2
XY	0.25	0.063
XZ.Y	0.17	0.029
YZ.X	0.65	0.423

TABLE 3.10

First-Order Correlation

Correlative Variable	r	r^2	t	P
XY	0.186	0.035	0.50	0.632
XZ.Y	0.01	0.00	0.03	0.976
YZ.X	0.637	0.405	2.18	0.065

3.6 Second-Order Partial Correlation

A second-order partial correlation occurs when two variables are subtracted from the correlation of two variables, rather than just one variable. The equation to calculate the second-order partial correlation is:

$$r_{XY}.Z_1Z_2 = \frac{r_{XY\ Z_1} - r_x Z_2 Z_1 * r_y\ Z_2 Z_1}{\sqrt{\left(1 - r_x{}^2 Z_2 Z_1\right)} * \sqrt{\left(1 - r_y{}^2 Z_2 Z_1\right)}} \quad (3.3)$$

3.7 Correlation Matrices

A Correlation Matrix is a table in which the pairwise correlations between several variables are conveniently organized in a matrix form (Kaushal et al., 2024). The value in the ith row and jth column corresponds to the correlation between the variables X_i and X_j (Tables 3.11 and 3.12).

Steps

Step 1: List the variables you have, say X_1, X_2,..., etc. Each of these variables has a sample associated with it

Step 2: Take the ith and jth variables from your list, X_i and X_j, and compute the correlation coefficient for them. Call it r_{ij}

Step 3: Take the value r_{ij} and that will be the value of e row i, column j of the correlation matrix

Mean age of rural site (X_1): 24, 26, 24, 28, 26, 23, 27, 26, 28, 29

Mean age of urban site (X_2): 26, 28, 30, 29, 28, 19, 22, 24, 25, 32

Mean BMI of rural sites (X_3): 18, 18, 19, 21, 21, 22, 22, 23, 24, 26

Mean BMI of urban sites (X_4): 35, 32, 36, 34, 36, 33, 34, 30, 29, 20

TABLE 3.11

Heritability Values of ABO Blood Groups

Heritability (h^2) of ABO Blood Group				
X_1	X_2	*(AB)* X_3	*(O)* X_4	*(Discordant)* X_5
0.9	0.44	0.24	0.63	0.44
0.57	0.65	0.36	0.58	0.58
0.63	0.42	0.22	0.69	0.63
0.44	0.32	0.24	0.63	0.47
0.54	0.41	0.36	0.74	0.44
0.52	0.52	0.47	0.44	0.36
0.32	0.38	0.33	0.79	0.37
0.47	0.33	0.44	0.68	0.63
0.54	0.45	0.49	0.75	0.54
0.62	0.62	0.54	0.63	0.69

TABLE 3.12

Correlation of Heritability (h^2) on ABO Blood Group

	X_1	X_2	X_3	X_4	X_5
X_1	1.0000	0.316	−0.206	−0.256	0.195
X_2		1.0000	0.45	−0.474	0.296
X_3			1.0000	−0.165	0.25
X_4				1.0000	0.1
X_5					1.0000

	X_1	X_2	X_3	X_4
x_1	1.0000			
x_2	0.4422	1.0000		
x_3	0.6229	−0.0350	1.0000	
x_4	−0.6225	−0.2940	−0.7747	1.0000

Portfolio Variance Matrix Formula:

$$\text{Portfolio Variance} = w_{12}\sigma_{12} + w_{22}\sigma_{22} + 2w_1w_2\text{Cov}$$

w_1 = the first asset's portfolio weight

w_2 = the second asset's portfolio weight

V_1 = the first asset's standard deviation

V_2 = the second asset's standard deviation

Cov(1,2) = The covariance of two assets, denoted by p(1,2)12, where p(1,2) is the correlation coefficient between the two assets.

Correlation of Two Variables = Covariance of (Variable X_1, Variable X_2)/(Standard Deviation of X_1 × Standard Deviation of X_2)

3.8 Univariate Regression

When two variables are associated, it is frequently useful to study the way in which one depends on the other (Khuan et al., 2024). For example, we may be interested in variations in diastolic blood pressure (Y) that are associated with variations in age (X). Alternatively, given the age of an individual, we may wish to predict the expected diastolic blood pressure. In such a situation, we undertake a regression of Y on X. To take a simple relationship, let $Y=a+Bx$, i.e. 'Y' is a linear function of 'X'. When $X=0$, Y takes the value 'a', and 'b' gives the change in 'Y' for a unit change in 'X'. The above equation is called a regression line of 'Y' on 'X'. The coefficients a and b of this equation are determined in such a way that the sum of the squares of the deviations of the observed values from the corresponding expected values on the regression line is as small as possible. These deviations are represented as the vertical distance of each point from the line (Lin et al., 2023). The regression line passes through the point $(\overline{xy})$. The slope of the line 'b' is computed using the formula.

$$b=\frac{\sum(X-\bar{X})(-\bar{Y})}{\sum(X-\bar{X})2}=\frac{n\sum XY-\left(\sum X\right)\left(\sum Y\right)}{n\sum X2-\left(\sum x\right)2} \tag{3.4}$$

And the intercept 'a' is obtained by the equation (3.4)

$$a=\bar{Y}-b\bar{X} \tag{3.5}$$

The regression line, after the estimation of 'a' and 'b', is of the form $\hat{Y}=a+bX$, where $\hat{Y}$ represents the estimated value of 'Y' corresponding to 'X'. The variance of the observations about the regression line is given by

$$S^2y.x=\frac{\sum(Y-\bar{Y})2}{n-2}=(1-r^2)\frac{\sum Y^2-\frac{\left(\sum Y\right)^2}{n}}{(n-2)} \tag{3.6}$$

$$=\frac{(n-1)}{(n-2)}(1-r^2)S^2y$$

Equation (3.6) is the measure of the residual variation among the observations after the effect of the linear relationship has been accounted for. The variance of the slope (b) is calculated by the equation (3.7)

$$\frac{S2\ y.x}{\sum(X-\bar{Y})^2}=\frac{S^2\ y.x}{\sum X^2-\frac{\left(\sum Y\right)^2}{n}} \tag{3.7}$$

The difference between b and any postulated value B may be tested using a version of the t-test.

$$t=\frac{b-B}{\sqrt{\text{variance of } b}} \tag{3.8}$$

with $(n-2)$ degrees of freedom.

Usually, the postulated value ($B=0$), so we examine whether the slope is zero in the population. However, this is algebraically equivalent to test ($r=0$). A simpler method is applied to calculate 'r', as obtained in equation (3.8). The test statistic was employed for testing the significance of the correlation coefficient 'r' at the desired level of significance

$$t=\frac{r\sqrt{n-2}}{\sqrt{1-r^2}} \text{ and examine it as a } t \text{ with } (n-2)df \tag{3.9}$$

Least Square Regression

It is a method of finding the best-fitting line or curve to a set of data points. The goal is to minimize the sum of the squared errors between the observed values and the predicted values

$$y=\beta_0+\beta_1x+\varepsilon \tag{3.10}$$

Calculate the residuals

$$\varepsilon=y-(\beta_0+\beta_1x) \tag{3.11}$$

Sum of squared errors $SSE=\sum\varepsilon^2$

Minimize the SSE by finding the partial derivatives of SSE with respect to the β_0 & β_1

Set the partial derivatives equal to zero and solve for β_0 & β_1

$$\beta_1=\sum[x_i-\bar{x}]*(y_i-\bar{y})/\sum(x_i-\bar{x})^2 \tag{3.12}$$

Intercept $\beta_0=\bar{y}-\beta_1\bar{x}$

$$\text{Co efficient of determination } R^2=1-\left(\sum y_i-\hat{y}_i\right)^2\Big/\sum(y_i-\bar{y})^2 \tag{3.13}$$

Predicted formula

$$\hat{y}=\beta_0+\beta_1x'+\varepsilon \tag{3.14}$$

$$\text{Residuals } \varepsilon=y_i-\hat{y}_i \tag{3.15}$$

$$\text{MSE}=\text{SSE}/n-k \tag{3.16}$$

Total variability=Explained variability+Unexplained variability

$$\sum_{i=1}^{n}(y_i-\bar{y})^2=\sum_{i=1}^{n}(\hat{y}_i-\bar{y})^2+\sum_{i=1}^{n}(\varepsilon_i)^2\sqrt{a^2+b^2},\text{SST} \tag{3.17}$$

$$=\text{SSR}+\text{SSE}$$

$$\text{Standard error of regression(SER)}=\sqrt{\frac{\text{SSE}}{n-k-1}} \tag{3.18}$$

The F-statistic serves as a particular type of test designed to assess the overall significance of a regression model (Sipio et al., 2024). It is employed to evaluate the null hypothesis, which posits that all regression coefficients within the model are equal to zero, thereby suggesting that the independent variable does not have a significant impact on the dependent variable. The null hypothesis asserts that beta 1 is equal to 0, while the alternative hypothesis contends that beta 1 is not equal to 0.

$$H_0=\beta_1=0$$

$$H_0=\beta_1\neq0$$

A beta value of 0 indicates that the independent variable does not influence the dependent variable within the regression model.

Consequently, the model is deemed statistically insignificant. The *F*-statistic can be computed using the formula provided below:

$$F = \frac{\text{MSR}}{\text{MSE}} \tag{3.19}$$

The mean sum of squares (MSR) is equal to the regression sum of squares (SSR) divided by the regression degrees of freedom:

$$\text{MSR} = \frac{\text{SSR}}{k} \tag{3.20}$$

The latter equals 1 because the number of coefficients (*k*) is 1.

MSE is the mean sum of squares due to error. It's equal to the sum of squares error (SSE) divided by the residual degrees of freedom:

$$\text{MSE} = \frac{\text{SSE}}{n-k-1} \tag{3.21}$$

The latter equals the total number of observations (*n*) minus *k* (the number of independent variables) minus 1.

ANOVA, or Analysis of Variance, is a statistical technique employed to assess whether the differences in means among two or more groups are statistically significant (Shi, 2024). This method is frequently utilized in experimental and research contexts to evaluate hypotheses and derive conclusions regarding population means. In the context of linear regression, ANOVA can be applied when the predictor variable is categorical, offering insights into the variability present within a regression model. Within a linear regression framework, the ANOVA table aggregates the sum of squares into distinct components, thereby elucidating the levels of variability inherent in the regression model. The table encompasses various statistical metrics

Model	*df*	SS	MSS
Regression	k	$\text{SSR} = \sum_{i=1}^{n}(\hat{y}_i - \bar{y})^2$	MSR= SSR/DFR
Residual	$n-k-1$	$\text{SSE} = \sum_{i=1}^{n}(\varepsilon_i)^2$	MSE=SSE/DFE
Total	$n-1$	$\text{SST} = \sum_{i=1}^{n}(y_i - \bar{y})^2$	*P* value=1−*cdf* (*Fd*1, *d*2) (*F* score)

3.8.1 Illustration

$$\bar{X} = \frac{28}{7} = 4 \quad \bar{Y} = \frac{63}{7} = 9$$

Suppose we are interested in obtaining the regression of *Y* on X_1. We have simulated '*b*' and '*a*' the numerical values.

$$b = \frac{n\sum XY - \left(\sum X\right)\left(\sum Y\right)}{n\sum X2 - \left(\sum X\right)2} = \frac{(7*344)-(28*63)}{(7*140)-)28)2}$$

$$= \frac{644}{196} = 3.29 \text{ and } a = (9 - 3.29 \times 4) = -4.16.$$

Hence, the regression equation is $\bar{Y} = -4.16 + 3.29\, x^1$

This equation can be used for prediction purposes. For example, if *X*=6.5.

$$Y = -\ 416 + 3.29 * 6.5 = 17.22$$

$$\Sigma(Y - \bar{Y})^2 = \Sigma Y^2 - \frac{\left(\sum Y\right)2}{n} = 887 - \frac{(63)2}{7}$$
$$= 887 - 567 = 320$$

$$S^2 y = \Sigma\frac{\left(Y - \bar{Y}\right)^2}{n-1} = \frac{320}{6} = 53.33$$

$$r = \frac{n\sum XY - \left(\sum X\right)\left(\sum Y\right)}{\sqrt{\left(n\sum X2 - \left(\sum X\right)2\right)(n\left(n\sum y2 - \left(\sum y\right)2\right)}}$$

$$= \frac{(7\times 344)-(28\times 63)}{\sqrt{(7\times 140-(28)2)\times(7\times 887-(63)2}} = 0.97$$

$$S^2 y.x = \frac{(n-1)}{(n-2)}\left(1-r^2\right) \qquad S^2 y = \frac{6}{5}\times\left(1-0.97^2\right)\times 53.33 = 3.78$$

$$\text{Variance of } b = \frac{S2\ \ y.x}{\sum X2 - \frac{\left(\sum X\right)2}{n}} = \frac{3.78}{140 - \frac{28}{7}} = 0.135$$

$$\text{SE}(b) = \sqrt{0.135} = 0.367,\ t = \frac{3.29}{0.367} = 8.96 \text{ with 5 } df.$$

Considering the testing of *r*=0, $t = \frac{r\sqrt{n-2}}{\sqrt{1-r2}} = 8.92$

The *t*-table with (*df*=5) gives the probability *P*<0.001. We reject the null hypothesis H_0 that there is no linear relationship between '*Y*' and '*X*'. The test described the above assumptions, and marginal deviations are seen in both variables (Table 3.13).

3.9 Existence of Two Regressions

If we consider an equation of the form (*X*=*c*+*dY*) and estimate '*c*' and '*d*' by minimizing $(X\text{-}c\text{-}d\ Y)^2$ with regression on r_{XY}, this line also passes through the point of $(\bar{X}, \bar{Y})$ but its slope is different. Unlike in mathematics, we have

TABLE 3.13

Calculate Correlation Coefficient for the Following Observations

X	*Y*	X^2	Y^2	*XY*
1	1	1	1	1
2	2	4	4	4
3	6	9	36	18
4	7	16	49	28
5	10	25	100	50
6	16	36	256	96
7	21	49	441	147
28	63	140	887	344

$n = 7 \sum X = 28 \sum Y = 63 \sum X^2 = 140 \sum Y^2 = 887 \sum XY = 344.$

used a pragmatic equation for representing the relationship between two variables because we are not dealing with an exact relationship between *X* and *Y* but we are approximating the equation. For the prediction of '*X*' from '*Y*', the regression equation will be ($X = c + d\, Y$), where $c = \bar{X} - \mathrm{d}\bar{Y}$ and

$$d' = \frac{n\sum XY - \left(\sum X\right)\left(\sum Y\right)}{n\sum Y^2 - \left(\sum Y\right)^2}$$

The numerators for the expressions for '*b*' and '*d*' are the same. In the first instance (i.e., for getting the regression line for '*Y*' on '*X*'), we minimize the sum of squares of deviations of '*Y*' from its expected values, whereas in the second instance (*i.e.*, for getting the regression line for '*X*' on '*Y*'), we minimize the sum of squares of deviations of '*X*' from its expected values. An interesting relationship exists between the two regression coefficients and the correlation coefficient; the latter is the geometric mean of the two regression coefficients. The choice of the regression line is automatically decided if there is a cause-and-effect relationship. For example, optical density and drug concentration (Figures 3.3 and 3.4). In situations where it is not easy to say which is dependent and which is independent, the best method is to calculate the correlation coefficient (Tables 3.14–3.16).

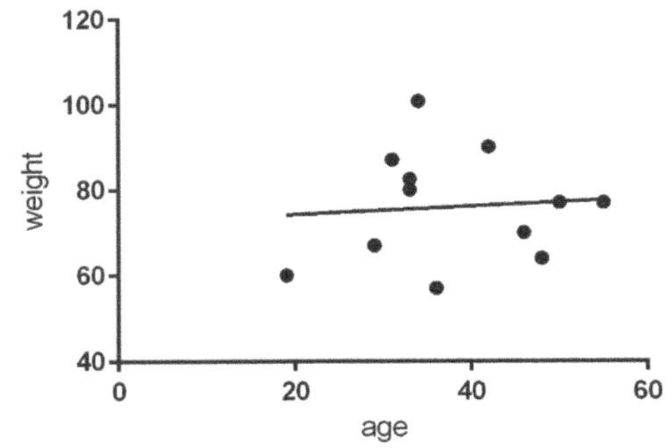

FIGURE 3.3 Scatter diagram of the regression plot for age and body weight.

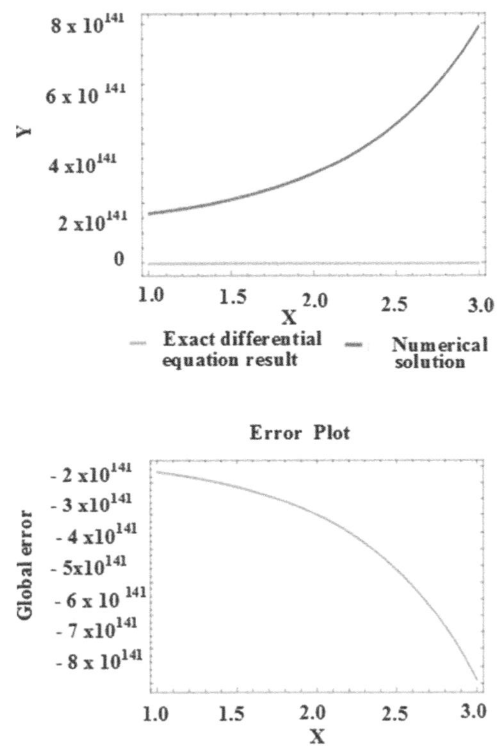

FIGURE 3.4 Estimation of global error of the regression model using Thompson iteration techniques.

Illustration

The simple regression model was optimized using the Range and Kutta method, with the incremental change of '*y*' on the *x* variable ($dy/dx = 0.39\, xy$, $y\,(72) = 2$, from 1 to 3, $h = .25$). From Table 5.8, the exact differential equations were simulated. A global error plot was created by iterating the error for the 'x' variable at the *n*th time.

↷ **Intercept**

$$y'(x) = f(x, y) = 0.39\ xy(x), y(72) = 2$$

↻ Error component

$$y_{n+1} = y_n + h\left(\frac{k_1}{6} + \frac{k_2}{6} + \frac{k_3}{6} + \frac{k_4}{6}\right) \quad (3.10)$$

TABLE 3.14

Distribution of Age and Weight to Determine Scattered Plot

Age (years)	33	33	34	42	29	19	50	55	31	46	36	48
Weight (kg)	80	82.5	101	90	67	60	77	77	87	70	57	64

$$y_{n+1} = y_n + h \quad (3.11)$$

$$k_1 = f(x_n, y_n) \quad (3.12)$$

$$k_2 = f\left(x_n + \frac{h}{2}, y_n + \frac{hk_1}{2}\right) \quad (3.13)$$

$$k_3 = f\left(x_n + \frac{h}{2}, y_n + \frac{hk_2}{2}\right) \quad (3.14)$$

TABLE 3.15

Regression Model Output

Best Fit Values	Line
Slope	0.09480 ±0 .3980
'Y' intercept	72.42±15.63
'X' Intercept	−763.90
1/Slope	10.55
CI-95%	
Slope $\left(\frac{\Delta y}{\Delta x}\right)$	−0.792 to 0.981
'Y' intercept	37.60 to 107.20
'X'-Intercept	$-\infty$ to -38.88
Goodness of Fit	
R Square (%)	0.0056
Sy. x	13.64, (*df*=1,10)
P-value	0.8162
R. equation	$Y = 0.0948 + 72.42\ X^1$

TABLE 3.16

Regression Local and Global Error Model Output

Step	*X*	*Y*	Local Error	Global Error
0	3.00	7.91×10^{141}	0.00	-7.91×10^{141}
1	2.75	5.97×10^{141}	0.00	-5.97×10^{141}
2	2.5	4.62×10^{141}	0.00	-4.62×10^{141}
3	2.25	3.67×10^{141}	0.00	-3.67×10^{141}
4	2.00	2.98×10^{141}	0.00	-2.98×10^{141}
5	1.75	2.48×10^{141}	0.00	-2.48×10^{141}
6	1.5	2.12×10^{141}	0.00	-2.12×10^{141}
7	1.25	1.85×10^{141}	0.00	-1.85×10^{141}
8	1.00	1.66×10^{141}	0.00	-1.66×10^{141}

$$k_3 = f\left(x_n + h, y_n + hk_3\right) \tag{3.15}$$

where $y_0 = 2,\ x_0 = 72,\ h = 0.25\ n = 0,1,2\ldots8$

The exact solution of estimation is $y(x) = 1.92\times10^{-439}$ $\text{expon}^{0.195x^2}$

3.10 Multiple Linear Regression

Regression is used to study the relationship between measurable variables. Linear regression is used for a special class of relationships, namely, one that can be described by a straight line. Regression techniques are widely used in every field of study. We have studied in the above section (Univariate regression) the case where a variable (*Y*) is expressed as linearly related to another dependent variable '*X*'.

$$Y_i = a + bX_j + \epsilon_{ij} \tag{3.16}$$

Usually, Y_i is called the dependent (outcome/response) variable, and 'X_j' is called the independent (explanatory/predictor) variable.

Life is rarely simple, and any outcome variable in the field of medicine or public health is usually affected by several variables. Univariate linear regression can be extended to take care of multiple independent variables. In this case, the corresponding model is

$$Y_{ijk} = b_0X_0 + b_1X_1 + b_2X_2 + \ldots.b_3X_3 + \epsilon_{ijk} \tag{3.17}$$

Here, '*Y*' is the response variable, $b_0, b_1, b_2, \ldots.. b_k$ are unknown coefficients, $X_1, X_2, \ldots.. X_k$ are the '*k*' explanatory variables, and ϵ_{ijk}'s are statistical errors to represent all other uncontrolled variables. From Table 5.10, we consider that the hemoglobin level (Hb) was determined in terms of correlation with age and packed cell volume (PCV). These measurements were made on 20 women. We have estimated multiple regression as the following equation (3.18):

$$Y_{ijk}(\text{HB\%} = b_0X_0 + b_1(\text{age})_1 + b_2(\text{PCV})_2 + \ldots.b_3X_3 + \epsilon_{ijk} \tag{3.18}$$

3.11 Multicollinearity of Regression

It is the statistical phenomenon that occurs when two or more independent variables in a regression model are highly correlated with each other (Table 3.17). In other words, multicollinearity indicates a strong relationship among the predictor variables. For example, paddy yield depends upon soil type, fertilizer, rainfall, water use efficiency, etc.

$$E(y) = \beta_0 + \beta_1X_1 + \beta_2X_2\ldots.\beta_kX_k \tag{3.19}$$

The following are indicators of multicollinearity:

1. Significant correlations between pairs of independent variables in the model.
2. A non-significant *F*-test for overall model adequacy $H_0 = \beta_1 = \beta_2 = \beta_3\ldots = \beta_k = 0$ is significant.
3. Opposite signs in the estimated parameters.
4. A variance inflation factor for a β parameter > 10, where

TABLE 3.17

A VIF1 Indicates Two Variables Not Correlated, a VIF between 1 and 5 Indicates Moderate Correlation and >5 Indicates High Correlation

Variables	BP	Age	Weight	Body Surface Area	Duration of Hypertension	Pulse
Age	2.93					
Weight	20.00*	1.69				
Body surface area	7.46*	1.61	8.00*			
Duration of Hypertension	1.41	1.52	1.25	1.15		
Pulse	3.58*	2.62*	2.93*	1.87	1.67	
Stress	1.20	1.58	1.04	1.02	1.45	2.02*

* Significant at 5% level ($p < 0.05$).

$$\text{VIF} = \left(\frac{1}{1-R_i^2}\right) \tag{3.20}$$

And R_i^2 is the multiple coefficient of determination for the model.

$$F(X_i) = \alpha_0 + \alpha_1 X_1 + \alpha_2 X_2 + \ldots.\alpha_{i-1}X_{i-1} + \alpha_{i+1}X_{i+1} + \ldots.\alpha_k X_k \tag{3.21}$$

Blood Pressure Variance Inflation Factor Matrix

3.12 Ridge Regression

Ridge regression is a specialized form of regression technique used to handle multiple regression data that exhibits multicollinearity. It is a fundamental regularization technique, though it is not used very widely due to the complex science behind it. The main idea of ridge regression is that it focuses on fitting a new line that does not overfit (Figures 3.5 and 3.6).

Ridge regression penalize the size of the regression coeffcient based on their l^2 forms

$$\text{argmin}\beta \sum_i (y_i - \beta' x_i)^2 + \lambda \sum_{k=1}^{k} \beta_k^2 \tag{3.22}$$

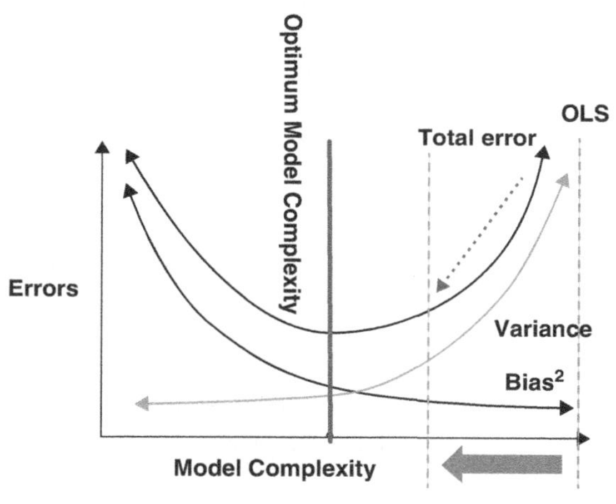

FIGURE 3.5 Ridge regression fitting line.

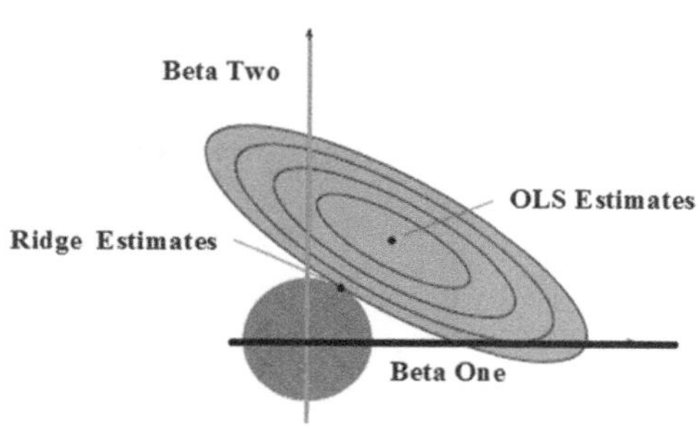

FIGURE 3.6 Geometric representation of ridge regression.

3.12.1 Working Rule and Application of Ridge Regression

It is carried out using L_2 regularization. In this, the penalty is equivalently added to the square of the magnitude of its coefficient. With response vector $y \epsilon R_n$ and a predictor matrix $y X \in R_{\text{nxp}}$ we can define the coefficients in the following ways.

λ is the tuning factor that controls the strength of the penalty term.

When $\lambda=0$, the objective is similar to simple linear regression. You will get the same coefficients as simple linear regression.

When $\lambda=\infty$, the coefficients you get would be zero due to infinite weightage on the square of coefficients, as anything less than zero makes the objective infinite.

When $0<\lambda<\infty$, the magnitude of λ decides the weightage allotted to the various parts of the objective.

The minimization objective=LS Obj+λ (the sum of the quare of coefficients)

Here, LS Obj is the Least Squares Objective. This is the linear regression objective without regularization.

When ridge regression in '*r*' shrinks the coefficients down towards zero, it tends to introduce some level of bias. However, it can also reduce the variance to a great extent, which gives you a better mean-squared error. λ multiplies the ridge penalty and controls the amount of shrinkage. A large λ denotes a greater level of shrinkage, and varying coefficient estimates can be obtained for varying values of λ (Figures 3.7 and 3.8).

3.12.2 Merits

- It protects the model from overfitting.
- It does not require unbiased estimators.
- There is only enough bias to make the estimates reasonably reliable approximations of the true population values.
- It performs well when there is a large multivariate dataset with the number of predictors (p) larger than the number of observations (n).

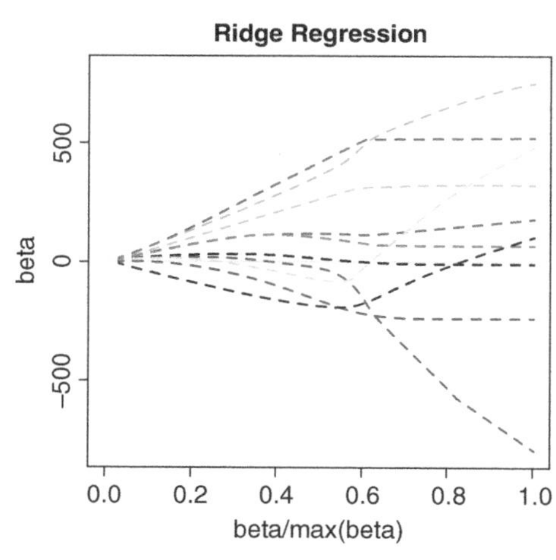

FIGURE 3.7 Ridge regression shrinking factors for the estimation of regression coefficients toward zero.

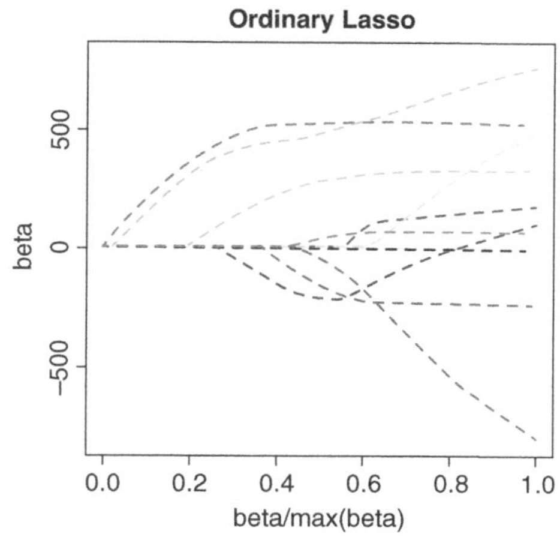

FIGURE 3.8 Lasso regression used for the coefficients (transition from positive to negative values).

- The ridge estimator is very effective in improving the least-squares estimate in situations where multicollinearity exists.
- Model complexity is reduced.

3.12.3 Demerits

- It includes all the predictors in the final model.
- It is not capable of performing feature selection.
- It shrinks coefficients toward zero.
- It trades variance for bias.

3.13 Adjusted *R*-Square

The Adjusted *R*-Square is the modified form of *R*-Square that has been adjusted for the number of predictors in the model. It incorporates the model's degrees of freedom. The Adjusted *R*-Square only increases if the new term improves the model's accuracy.

$$R^2_{\text{adj}} = 1 - \frac{(1-R^2)(N-1)}{N-p-1} \tag{3.23}$$

where, R^2 = sample R square

p = Number of predictor

N = Total sample size

3.14 Lasso Regression

This is a subtle but important change. Some of the coefficients may be shrunk exactly to zero. The **l**east **a**bsolute **s**hrinkage and **s**election **o**perator, or *lasso*, as described in Tibshirani (1996), is a technique that has received a great deal of interest (Figures 3.7 and 3.8).

$$\text{Ridge subject to} \sum_{j=1}^{p} (\beta_j)^2 < c, \text{ Lasso subject to} \sum_{j=1}^{p} |\beta_j| < c \tag{3.24}$$

As with ridge regression, we assume the covariates are standardized. Lasso estimates of the coefficients (Tibshirani, 1996) achieve

$$\min_\beta (y - X\beta)' + \lambda \sum_{j=1}^{p} |\beta_j| < c \text{ so that the L2 penalty of ridge regression} \tag{3.25}$$

$$\sum_{j=1}^{p} (\beta_j)^2 < c \text{ replaced by an L1 penalty} \sum_{j=1}^{p} |\beta_j| < c. \tag{3.26}$$

Let $c_0 = \sum_{j=1}^{p} |\beta_{LS.j}|$ denotes the absolute size of the least square estimates value of $0 < c < \infty$. If, for example, $c = c_0/2$ the average shrinkage of the LS coefficient is 50%. If λ is sufficiently large, some of the coefficients are driven to zero, leading to a sparse model.

The lines are the paths of regression coefficients shrinking toward zero. If we draw a vertical line in the figure, it will give a set of regression coefficients corresponding to a fixed λ. (The x-axis actually shows the proportion of shrinkage instead of λ).

3.14.1 Stepwise Regression

Stepwise regression is the step-by-step iterative construction of a regression model that involves the selection of independent variables to be used in a final model. It involves adding or removing potential explanatory variables in succession and testing for statistical significance after each iteration using *F* and *t*-tests.

- The forward selection approach starts with nothing and adds each new variable incrementally, testing for statistical significance.
- The backward elimination method begins with a full model loaded with several variables and then removes one variable to test its importance relative to overall results.
- Stepwise regression has its downsides, however, as it is an approach that fits data into a model to achieve the desired result (Figure 3.9).

3.14.2 Criteria for Adding Variable

1. Adding the variables based on the smallest *P*-value for the *F*-test of the model as a whole
2. The highest increase in *R*-squared
3. The highest drop-in model residual sum of squares compared to other variables under consideration

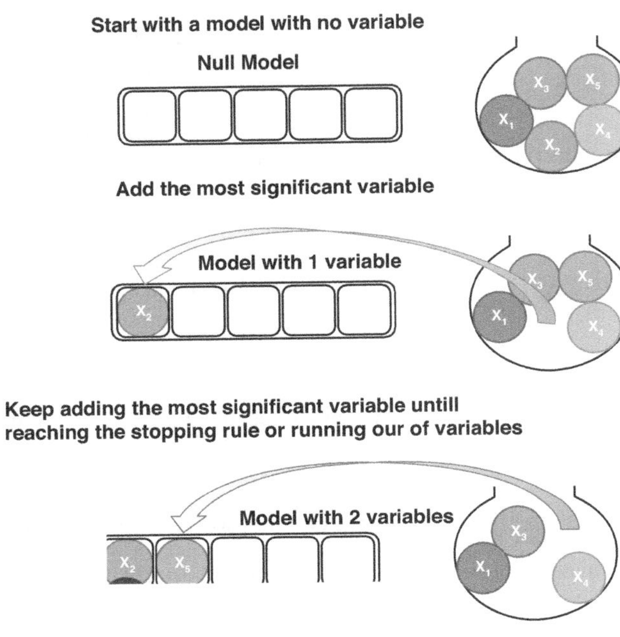

FIGURE 3.9 Flowchart of the stepwise regression model.

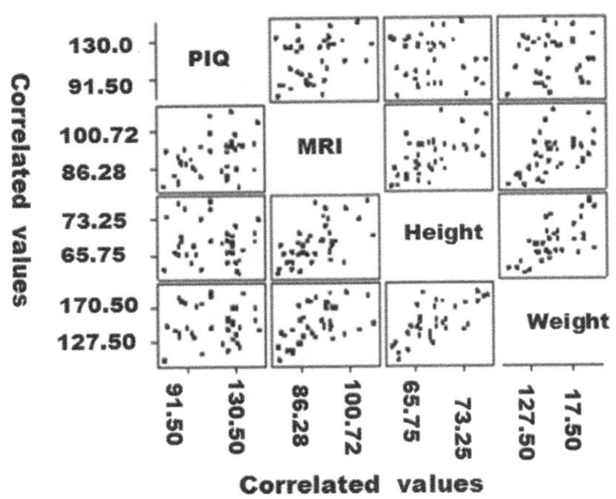

FIGURE 3.10 Stepwise regression modeling of PIQ.

Energy consumption of equipment at XYZ company is based on equipment run time, equipment age, size, staff, temperature outside, and time of the year, etc. It is the ability to manage a large number of predictor variables, fine-tuning the model to choose the better predictor variables from the available options (Figures 3.10 and 3.11). Watching the order in which variables are removed or added can provide valuable information about the quality of the predictor variables.

3.14.3 Steps

Step 1: Start with no predictors in the stepwise model. Regress y on $x_{1,}$ or y on $x_{2,}$ .., y on X_p

Step 2: At each step, enter or remove a predictor based on the t-tests.

Step 3: Stop when no more predictors can be justifiably entered or removed from the stepwise model. Use the smallest t and F value.

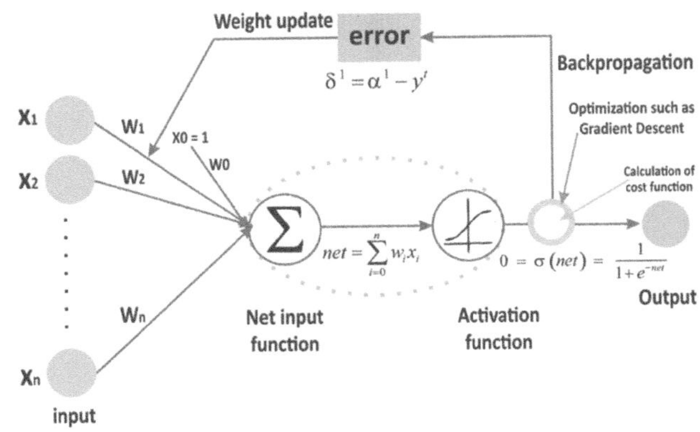

FIGURE 3.11 Gradient descent and backpropagation process.

TABLE 3.18

Stepwise Regression PIQ vs. MRI, Height, Weight Alpha to Enter: 0.50, Alpha to Remove :0.15, Response is PIQ on 3 Predictors with N=38

Steps	1	2
Constant	4.65	111.27
MRI	1.18	2.06
t-Value	2.45	3.77
P-value	0.019	0.001
Height		−2.73
t-Value		−2.75
P-value		0.0096
S	21.2	19.5
R-Sq	14.27	29.49
R-Sq(ADJ)	11.89	24.46
c-p	7.3	2.0

Step 4: Specify an alpha-to-enter (αE=0.15) significance level.

Step 5: Specify an alpha-to-remove (αR=0.15) significance level.

Step 6: Fit each of the one-predictor models, i.e., regress y on x_1, regress y on x_2, …, regress y on $xp-1$. The first predictor put in the stepwise model is the one that has the smallest P-value (below αE=0.15). If no P-value<0.15, stop.

Step 7: At each step, enter, remove, or add a predictor based on partial F-tests (i.e., the t-test)

Step 8: Suppose x_1 was the 'best' predictor. Fit each of the two-predictor models with $x1$ in the model, i.e. regress y on (x_1, X_2), regress y on (x_1, x_3), …, and regress y on (x_1, x_{p-1}). The second predictor added to the stepwise model is the one that has the smallest P-value (below α E=0.15). If no P-value<0.15, stop.

Step 9: Suppose x_2 was the 'best' second predictor. Step back and check the P-value for β_1=0. If the P-value for β_1=0 becomes not significant (above α R=0.15), remove x_1 from the stepwise model.

Step 10: Suppose both x_1 and x_2 made it into the two-predictor stepwise model. Fit each of the three-predictor models with x_1 and x_2 in the model, i.e. regress y on (x_1, x_2, x_3), regress y on (x_1, x_2, x_4), …, and regress y on (x_1, x_2, x_{p-1}) (Tables 3.18–3.20).

TABLE 3.19

Stepwise Regression Y vs. X_1, X_2, X_3 and X_4 to Enter: 0.50, Alpha to Remove: 0.15, Response is PIQ on 4 Predictors with $N=13$

Steps	1	2	3	4
Constant	117.57	103.10	71.65	52.28
X_4	90.738	−0.614	−0.237	
t-Value	−4.77	−12.62	−1.37	
P-value	0.001	0.000	0.205	
X_1		1.44	1.45	1.47
t-Value		10.40	12.41	12.10
P-value		0.000	0.000	0.000
X_2			0.416	0.662
t-Value			2.24	14.44
P-value			0.052	0.000
S	8.96	2.73	2.31	2.41
R-Sq	67.45	97.25	98.23	97.87
R-Sq(adj)	64.50	96.70	97.64	97.44
c-p	130.70	5.50	3.0	2.70

TABLE 3.20

Coefficient of Stepwise Regression Y vs. X_1, X_2, X_3, and X_4 to Enter: 0.50, Alpha to Remove: 0.15, Response is PIQ on 4 Predictors with $n=13$

Predictor	Coeff.	SE Coeff.	t-value	P-value
Constant	52.57	2.28	23.00	0.00
X_1	1.468	0.12	12.10	0.00
X_2	0.66	0.045	14.44	0.00
Predictor	Coeff.	SE Coeff.	t	P
Constant	71.65	14.14	5.07	0.001
X_1	1.45	0.11	12.41	0.000
X_2	0.41	0.18	2.24	0.052
X_4	-0.23	0.17	−1.37	0.205
Predictor	Coeff	SE Coeff.	t	P
Constant	48.19	3.91	12.32	0.000
X_1	1.69	0.20	8.29	0.000
X_2	0.65	0.04	14.8	0.000
X_3	0.250	0.18	1.35	0.209

3.15 Dummy Variable

A dummy variable is numerically expressed in regression analysis and represents subgroups within the sample study. In research design, a dummy variable is often used to distinguish different treatment groups. In the simplest case, we use values like 0 and 1 (e.g. control group coded as 0 and study group coded as 1). Dummy variables are useful because they enable us to use a simple regression equation to represent multiple groups. These variables act like switches that turn various parameters on and off in an equation

$$y_i = \beta_0 + \beta_1 Z_i + e_i \tag{3.27}$$

where y_i is the outcome of the ith unit

β_0 = coefficient for the intercept
β_1 =coefficient for the slope

$Z_i = 1$ if the ith unit in the treatment group
0 if the ith unit in the control group
e_i = residual for the ith unit

From equation (3.27), we determine effect for each group.
For control group, $(Z_i = 0)$

$$y_c = \beta_0 + \beta_1(0) + 0 \tag{3.28}$$

$y_c = \beta_0$, e_i is average to 0 across the group

For control group, $(Z_i = 1)$

$$y_T = \beta_0 + \beta_1(1) + 0 \tag{3.29}$$

$$y_T = \beta_0 + \beta_1$$

Then, find the difference between the two groups.

Treatment	Control
$y_T = \beta_0 + \beta_1$	$y_c = \beta_0$
$y_T - y_c = (\beta_0 + \beta_1) - \beta_0$	(3.30)
$y_T - y_c = \beta_0 + \beta_1 - \beta_0 = (\beta_0 \text{ cancelled})$	(3.31)
$y_T - y_c = \beta_1$	(3.32)

Whenever we have a regression model with dummy variables, we can always see how the variables are being used to represent multiple subgroup equations by following these steps

Create separate equations for each subgroup by substituting the dummy values

Find the difference between groups by finding the difference between their equations

3.16 The Individual Point Prediction (IPP) Method (Regression Point Displacement Design)

The central idea of this design is that, in the null case, one would not expect the treatment group to differ from the regression line of the population at greater than chance levels. Evidence for a treatment effect exists when there is a post-test (vertical) displacement of the treated group's point from the control group's regression line (thus leading to the name 'regression point displacement'). However, this evidence does not imply that the treatment of interest directly led to or 'caused' this vertical regression shift. As always, one must assess the plausibility of other rival potential causes for such a shift.

The RPDD is characterized by four major features: (i) the use of a single treated unit instead of many; (ii) the use of aggregate-level data instead of individual-level data; (iii) the absence of any need to ensure (or attempt to achieve by statistical adjustment) pre-treatment equality between treated and control groups; and (iv) the avoidance of 'regression artifacts' or under-adjustment due to 'errors in variables' by employing the observed regression line, rather than using it as a means of adjustment. Feature 4, and to a lesser extent Feature 3, are shared with the Regression Discontinuity (RD) design.

Traditionally, statisticians have treated data like that of the RPDD as a problem of testing whether an individual point deviates significantly from its regression line prediction, that is, whether its vertical distance from the regression line is statistically significant. Here, this approach is labeled the 'individual point prediction' method. While this approach is technically correct, it is a cumbersome procedure that requires applying a special t-test formula after computing the control group regression line to test the deviation hypothesis. A simpler computational procedure, which yields exactly the same result with greater potential generality, is recommended instead. This is discussed here as the 'analysis of covariance' method. The statistical proof of the equivalence of these two analytic methods is provided below.

The Individual Point Prediction (IPP) Method: To analyze the data from an RPDD using the IPP approach, following Fleiss and Tanur (1972), one first obtains the regression of post-test means onto pre-test means for the control units only. Given a set of pre-test values (e.g. means), X_i, and corresponding post-test values, Y_i, one typically (although not necessarily) fits the linear model.

From the simple regression equation,

$$y_i = \beta_0 + \beta_1 X_i + e_i \tag{3.33}$$

This equation can then be used to obtain the predicted post-test value for the treated unit. That is, if the treated unit has a pre-test value X_0 and a post-test value Y_0, then the predicted post-test value is

$$\hat{y}_0 = \hat{\beta}_0 + \hat{\beta}_1 X_i + e_i \tag{3.34}$$

To test the hypothesis that the observed treatment unit post-test value differs from the predicted one, a simple t-test can be constructed.

$$t = \frac{Y_0 - \hat{Y}_0}{\sqrt{\sigma^2_{Y_i - \hat{Y}_i}\left[1 + \frac{1}{N} + \frac{\left(X_0 - \overline{X}\right)^2}{\sum X^2_i}\right]}} \tag{3.35}$$

where Y_0 = The observed treated unit post-test value

$\hat{Y}_0$ = The predicted treated unit post-test value
$\sigma^2_{Y_i - \hat{Y}_i}$ = The residual mean square
N = The number of the control unit
X_0 = The observed treated unit pretest
$\overline{X}$ = The mean of the control unit pre-test values

$$\sum X^2_i = \sum \left(X_i - \overline{X_i}\right)^2 \tag{3.36}$$

This t-value can be tested in the usual way with $df = (N - 2)$. The numerator represents the treated unit's vertical displacement from the control unit regression line. Note that the treated unit's pre-test value enters into the formula through the term in the denominator, $X_0 - \overline{X}$, which is the horizontal distance from the treated unit's pre-test value to the pre-test mean for all control units respectively (Figures 3.12 and 3.13) (Table 3.21).

$$y_i = \beta_0 + \beta_1 X_i + \beta_2 Z_i + e_i \tag{3.37}$$

where y_i is the outcome of the ith unit

β_0 = Coefficient for the intercept
β_1 = Coefficient for the slope
Z_i = Dummy variable for treatment (0 control and 1 treatment ($n = 1$) groups
e_i = Residual for the ith unit

$$S = 0.0056 \; R\text{ Sq} = 72.60\%, \; R\text{ Sq.adj} = 70.60\%$$

$$Y = 0.0120 + 0.784X - 0.019\ Z$$

The Steps of Statistical Model Building: The four primary steps of statistical model building are

- Confirming data assumptions
- Building a model on training data
- Assessing the model's fit
- Analyzing model results

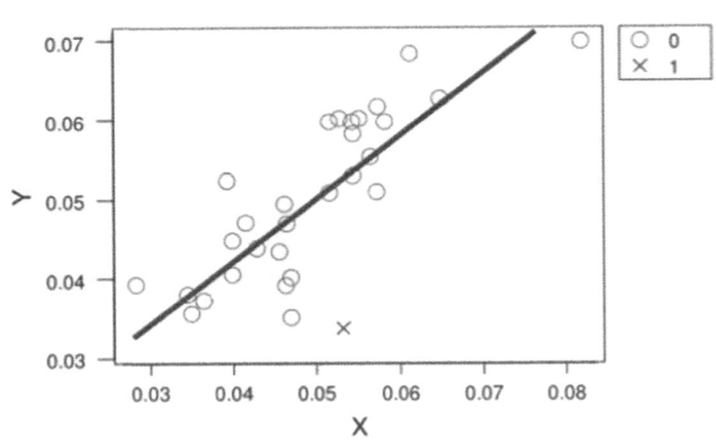

FIGURE 3.12 RPD scatter plot.

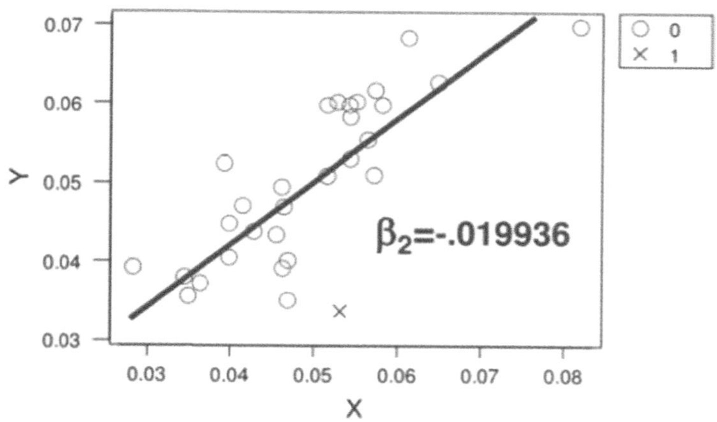

FIGURE 3.13 RPD scatter plot with regression analysis.

TABLE 3.21

Model Output

Predictor	Coeff.	St Dev	*t* Ratio	*P*
Constant	0.011	0.004	2.41	0.023
X	0.783	0.098	7.94	0.000
Z	–0.019	0.005	–3.44	0.002

R code for employee salary

```
# Create the data frame
data <- data.frame(
  Years_Exp = c(1.1, 1.3, 1.5, 2.0, 2.2, 2.9,
3.0, 3.2, 3.2, 3.7),
  Salary = c(39343.00, 46205.00, 37731.00,
43525.00,
             39891.00, 56642.00, 60150.00,
54445.00, 64445.00, 57189.00)
# Create the scatter plot
plot(data$Years_Exp, data$Salary,
     xlab = "Years Experienced",
     ylab = "Salary",
     main = "Scatter Plot of Years Experienced
vs Salary")
# Create a data frame with new input values
new_data <- data.frame(Years_Exp = c(4.0, 4.5,
5.0))

# Predict using the linear regression model
predicted_salaries <- predict(lm.r, newdata =
new_data)
 # Display the predicted salaries
print(predicted_salaries)
```

3.17 Weighted Regression Analysis

Let there be a set of n observation points $(x_{11}, x_{12}, x_{13} \ldots\ldots y_1)(x_{n1}, x_{n2}, x_{n3} \ldots\ldots x_{nk}, y_n)$. The objective is to find the coefficients, $\beta_0, \beta_1 \ldots \beta_k$

$$\hat{y}_i = \beta_0 + \beta_1 X_1 + \beta_2 X_2 \ldots \beta_k X_k + e_i \tag{3.38}$$

We have to minimize the equation (3.38)

$$\sum_{i=1}^{n} e_i^{\,2} = \sum_{i=1}^{n} (y_i - \hat{y}_i)^2 \tag{3.39}$$

$$e_i = y_i - \hat{y}_i \tag{3.40}$$

In weighted least squares, for a given set of weights $w_1, \ldots, w_n$, we seek coefficients $b_0, \ldots, b_k$ so as to minimize

$$\sum_{i=1}^{n} W_i (y_i - \hat{y}_i)^2 \tag{3.41}$$

Using the same approach as is employed in ordinary least square (OLS) analysis, we find that the $k+1 \times 1$ coefficient matrix can be expressed as

$$B = (X^T\ WX)^{-1} X^T\ WY \tag{3.42}$$

where W is the $n \times n$ diagonal matrix whose diagonal consists of the weights $w_1, \ldots, w_n$ (Table 3.22–3.24).

The $n \times 1$ matrix of predicted y values Y-hat$=[\hat{y}_i]$ and the residuals matrix $E=[e_i]$ can be expressed as

$$\hat{Y} = B^T X, E\ Y - \hat{Y} \tag{3.43}$$

An estimate of the variance of the residuals is given by

$$MS_{Res} = SS_{Res} / df_{Res} \tag{3.44}$$

where $df_{Res} = n - k - 1$

$$SS_{Res} = \sum w_i (y_i - \hat{y}_i)^2 = E^T WE \tag{3.45}$$

$$SST = Y^T\ WY - \frac{(1^T\ WY)^2}{1^T\ W1} \tag{3.46}$$

$$SS_{Reg} = SST - SS_{Res} \tag{3.47}$$

Illustration

TABLE 3.22

Demonstrated Simple Regression and Weighted Regression

X	*Y*	*W*
21	17.26	20
20	17.07	26
19	16.37	27
18	16.40	24
17	16.13	36
16	16.17	39
15	15.98	32

TABLE 3.23

Regression Analysis Model Output

Overall Fit		
Statistics	Simple regression	Weighted regression (value)
Multiple *R*	0.924	0.918
R Square	0.855	0.842
Adj *R* square	0.826	0.816
SE	0.204	1.07

TABLE 3.24

Regression of ANOVA

ANOVA	*df*	SS	MS	*F*	*P*
	Simple Regression				
Regression	1	1.23	1.23	29.57	0.0028**
Residual	5	0.208	0.041		
Total	6	1.44			
	Weighted Regression				
Regression	1	31.02	31.027	26.84	0.003**
Residual	5	5.778	1.1556		
Total	6	36.80			

** Significant at 5% level p < 0.05.

3.18 Lognormal Regression

When the observations are significantly affected by mean values, the log-normal distribution is a unique regression model that provides realistic statistical information for testing hypotheses

$$Y_i = \sigma\ \beta^{x_i} \varepsilon_i i = 12..n$$

where $\varepsilon_{i\sim \text{Lognormal}\,(0,1)}$ i.e. $\ln(\epsilon_i) \sim$ Normal (0, 1), σ is the known parameter and β is the unknown parameter we aim to estimate. The following steps are employed:

Step 1: Take the natural log transformation on both sides of the regression model. Solve the eqation ln (ϵ_i)

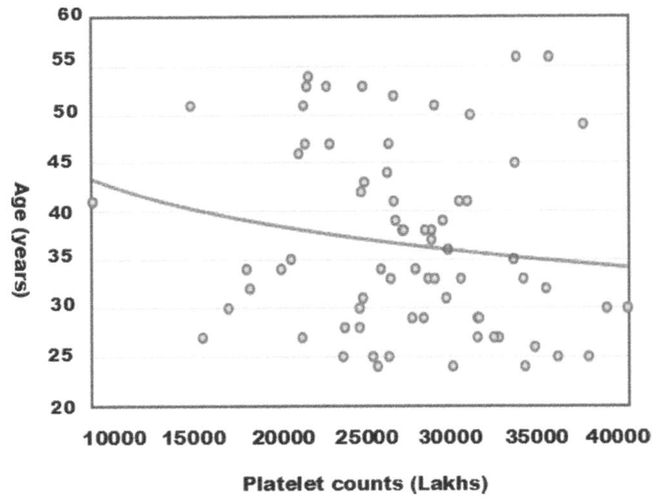

FIGURE 3.14 Log-normal distribution of CD_4 counts.

Step 2: Explore ln (ϵ_i)=ln Y_i − ln σ − $Xi \ln\beta$. By assumption, this means that ln Y_i − ln $\sigma - X_i \ln\beta \sim N(0,1)$ Use this fact to construct the likelihood function $L(\beta)$ for estimating β

3.19 Logarithmic Regression

Logarithmic regression represents scenarios in which growth or decline initially occurs at a rapid pace before gradually decelerating. To fit a logarithmic function to a collection of data points, the command 'LnReg' on a graphing calculator is utilized (Figure 3.14).

$$y = a + b \ln X$$

$$\text{Age} = 111.59 - 5.98.\ln(\text{Platlet counts}),\ \text{correlation}(r) = -0.162$$

3.20 Exponential Regression

Exponential regression is considered one of the most straightforward nonlinear regression models. A common application of exponential regression occurs when examining the relationship between the concentration of a substance (the dependent variable) and the passage of time (the independent variable) (Figure 3.15).

$$y = a.b^X \tag{3.48}$$

3.21 Logistic Regression

Logistic regression is used to determine the effect of independent variables on a dependent variable. The dependent variable should be categorical, yielding binary out comes (success vs. failure, censored vs. non-censored, survival vs. non-survival, positive vs. negative growth of microorganisms, etc.). If any events can be reduced to two outcomes, they are coded as 0 and 1, where 1 indicates that the event happened and 0 indicates it

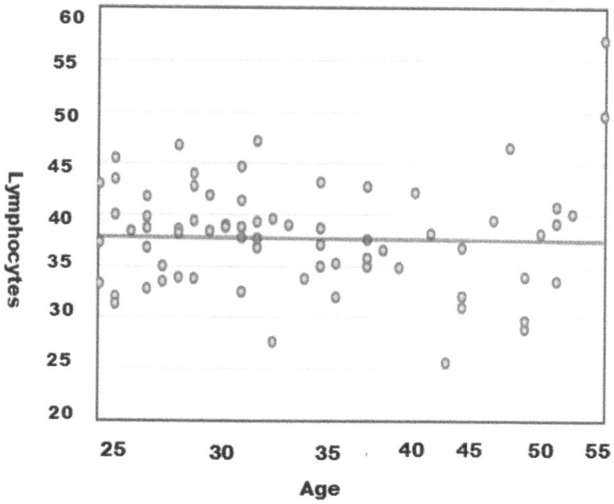

FIGURE 3.15 Exponential regression of lymphocyte counts.

TABLE 3.25

Output of Logistic Regression ($n=25$ subjects)

Variables	Coefficient	SE	*P*-value	CI-95%
Platelets	0.427	0.366	0.2433	[0.314,0.133]
Neutrophil	0.193	0.081	0.0175*	[0.702,0.966]
Lymphocytes	0.169	0.094	0.0712	[0.702,1.014]
TLC	0.000	0.000	0.009*	[0.998,1.003]
NLR	0.024	0.715	0.9726	[0.2399,3.968]
Survivability	16.85	0.680	0.0132	

* Significant at 5% level $p < 0.05$.

did not. The survival of patients can be assessed based on variables such as Platelets, TC, Neutrophils, Lymphocytes, NLR, etc. The logistic regression is formulated based on the following equation (Table 3.25).

$$P = \frac{1}{1+e^{-(\beta_0+\beta_1 X_1+\beta_2 X_2 \ldots \beta_K X_K)}} \quad (3.49)$$

$$P = \frac{1}{1+e^{-(16.85-0.427\ X_1-0.193\ X_2-0169\ X_3+0.004X_4-0.024\ X_5)}} \quad (3.50)$$

3.22 Polynomial Regression

It is a unique type of nonlinear regression method where the variables exhibit a curvilinear and nonlinear association between the independent and dependent variables at the *n*th degree. Typically, this corresponds to the least-squares method. The least-squares approach minimizes the coefficient variance (Table 3.26).

3.22.1 Assumptions

A dependent variable's behavior can be described by a linear or curved additive link between the dependent variable and a set of k independent factors.

TABLE 3.26

Polynomial Regression Equation

Polynomial	Form	Degree	Examples
Linear	$y = ax + \beta, a \neq 0$	1	$x+8$
Quadratic	$y = ax^2 + \beta x + c, a \neq 0$	2	$3x^2-4x+7$
Cubic	$y = ax^3 + \beta x^2 + c, a \neq 0$	3	$2x^2+3x^2+4x+6$
*n*th degree	$y = ax^n + \beta x^n + c, a \neq 0$	n	

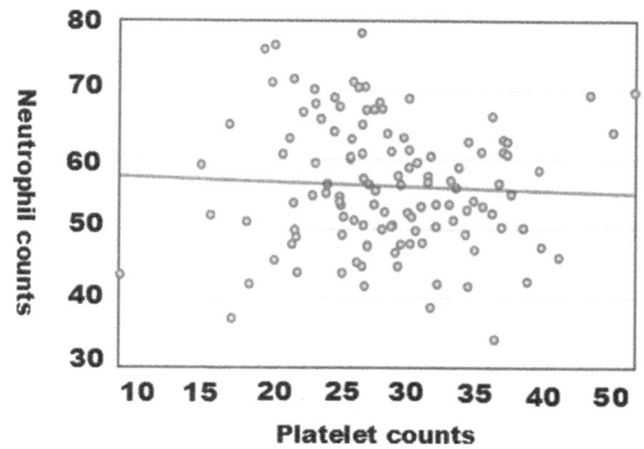

FIGURE 3.16 Methods of polynomial regression curve fitting.

TABLE 3.27

Second-Degree Polynomial Regression

Coefficient	Estimates	SE	*t*-statistics	*P*-value
β_0	58.23	11.21	5.19	–
β_1	–0.742	7.67	–0.09	0.923^{ns}
β_2	0.018	1.28	0.014	0.988^{ns}
		ANOVA		
	df	SS	MS	*F*
Regression	2	20.592	10.29	0.121^{ns}, $P=0.885$
Residual error	117	9,927.147	84.84	
Total	119	9,947.737	83.59	

The independent variables lack any interrelationship (Figure 3.16).

We employ datasets featuring independently distributed errors with a normal distribution, having a mean of zero and a constant variance (Table 3.27).

3.22.2 Python Code

```
import numpy as np
import matplotlib.pyplot as plt
from sklearn.linear_model import Linear
Regression
#Let randomly create some data's for two
variables,
x = 2 - 3 * np.random.normal(0, 1, 20)
y = x - 2 * (x ** 2) + 0.5 * (x ** 3) +
np.random.normal (-3, 3, 20)
```

```
#Visualize the variables spreads for better
understanding
plt.scatter(x,y, s=10)
plt.show()
Model out put
Neutrophil counts = 0.018.Platlet² - 0.741.
Platlet + 58.238, R square 21.0%
```

3.23 Sinusoidal Regression

Sinusoidal regression identifies a trigonometric model that provides the optimal curve fit. The concept parallels that of determining the line of best fit in linear regression. Nevertheless, rather than deriving a straight line, the objective here is to ascertain a sinusoidal curve (Figure 3.17).

$$y = a\sin(bx + c) + d \quad (3.51)$$

$$\text{Plantlets} = 9.23\sin(3.10\text{ age} - 3.1) + 2.9$$

Illustration

1. Campbell et al. (1985) invited women, in a pre-defined geographical area to have their hemoglobin (Hb) level and packed cell volume (PCV) measured. They were also asked about their age and whether or not they had experienced menopause. The response rate to the invitation was about 90%. Results from a random sample of (n=20) women from the group are given in Table 5.10.

We determined the multiple linear regression (MLR); the response variable should always be continuous, and the explanatory variables should also be continuous. Let ϵ_{ijk} be the value of the statistical error for ith case or patient. The basic assumptions made to determine MLR are as follows:

i. The mean of ϵ_{ijk} should be zero, and the variance of e's is a fixed unknown constant.
ii. The e_{ij}'s are independent with zero covariance.
iii. Errors are normally distributed (random variable) $\epsilon_{ij} \sim (\mu, \sigma^2)$.

$$Y = \beta_0 + \beta_1 X_1 + \beta_2 X_2 \ldots\ldots \beta_n X_n \quad (3.52)$$

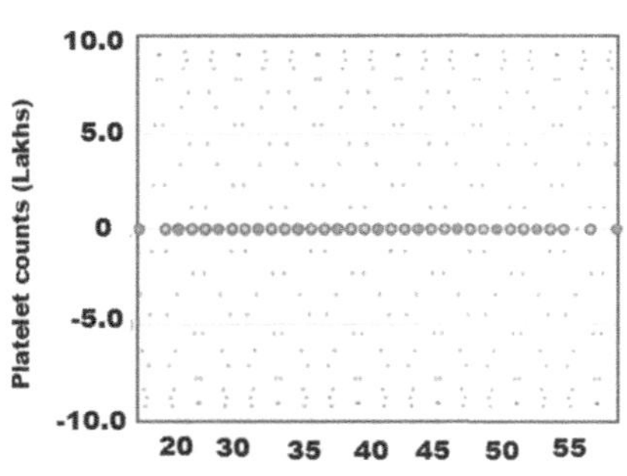

FIGURE 3.17 Sinusoidal regression on the platelet counts.

TABLE 3.28

Hemoglobin Level (Hb), Packed Cell Volume (PCV) Age, and Menopausal Status in a Group of 20 Women

Subject	Hb (%)	PCV (%)	Age (years)	Menopause (Yes=1, No=0)
1	11.1	35	20	0
2	10.7	45	22	0
3	12.4	47	25	0
4	14	50	28	0
5	13.1	31	28	0
6	10.5	30	31	0
7	9.6	25	32	0
8	12.5	33	35	0
9	13.5	35	38	0
10	13.9	40	40	0
11	15.1	45	45	1
12	13.9	47	49	0
13	16.2	49	54	1
14	16.3	42	55	1
15	16.8	40	57	1
16	17.1	50	60	1
17	16.6	46	62	1
18	16.9	55	63	1
19	15.7	42	65	1
20	16.5	46	67	1

In this context, the equation image indicator represents the anticipated or expected value of the dependent variable. The variables X_1–X_p denote p distinct independent or predictor variables, while β_0 signifies the value of Y when all independent variables (X_1–X_p) are set to zero. The coefficients β_1 through bp are the estimated regression coefficients, with each coefficient indicating the change in Y associated with a one-unit change in the corresponding independent variable. Specifically, in a multiple regression framework, β_1 illustrates the change in Y resulting from a one-unit increase in X_1, assuming that all other independent variables remain constant (i.e., held at a fixed value). Furthermore, statistical tests can be conducted to determine whether each regression coefficient significantly deviates from zero (Table 3.28).

As indicated above, multiple regression analysis serves as a tool to evaluate the presence of confounding factors. This method enables the estimation of the relationship between a specific independent variable and the outcome while keeping all other variables constant. Consequently, multiple linear regression offers a means to adjust for or account for potentially confounding variables that are incorporated into the model. For instance, consider a risk factor or exposure variable represented as X_1 (such as X_1=obesity or X_1=treatment) and an outcome or dependent variable denoted as Y. We can formulate a simple linear regression equation that connects the risk factor (the independent variable) to the dependent variable as follows:

$$\hat{Y} = \beta_0 + \beta_1 X_1 \quad (3.53)$$

where β_1 is the estimated regression coefficient that quantifies the association between the risk factor and the outcome.

In order to evaluate the relationship between body mass index (BMI) and systolic blood pressure, we will utilize data obtained from the seventh examination of the Framingham Offspring Study. The examination included a total of 3,539 participants, who exhibited a mean systolic blood pressure of 127.3 mmHg, accompanied by a standard deviation of 19.0 mmHg. Additionally, the average BMI recorded in this sample was 28.2, with a standard deviation of 5.3 (Table 3.29).

The simple linear regression model is:

$$Y = 102.28 + 0.67(\text{BMI})$$

where $\hat{Y}$ is the predicted or expected systolic blood pressure. The regression coefficient associated with BMI is 0.67; each one-unit increase in BMI is associated with a 0.67-unit increase in systolic blood pressure. The association between BMI and systolic blood pressure is also statistically significant ($p = 0.0001$).

We aim to evaluate whether age (a continuous variable expressed in years), male gender (binary: yes/no), and treatment for hypertension (binary: yes/no) may serve as potential confounders. If they are identified as such, we will appropriately adjust for these factors using multiple linear regression analysis. For the purposes of this analysis, treatment for hypertension is represented as 1 for yes and 0 for no, while gender is represented as 1 for male and 0 for female (Table 3.30). The results of the multiple regression analysis indicate the following:

$$\hat{y} = 68.15 + 0.58(\text{BMI}) + 0.65(\text{age}) + 0.95(\text{Male gender})$$
$$+ 6.44(\text{Treatment for Hypertension})$$

Evaluating solely the p-values indicates that these three independent variables possess equal statistical significance. The t-statistics' magnitude serves as a criterion for assessing the relative importance of the independent variables. In this instance, age emerges as the most significant independent variable, followed by BMI, treatment for hypertension, and then male gender. Notably, male gender does not achieve statistical significance ($p = 0.1133$) within the multiple regression model. Some researchers contend that even if a significant variable like gender does not reach statistical significance, it should still be included in the model to mitigate potential confounding effects. Conversely, other researchers advocate for the retention of only statistically significant variables. This exemplifies the intricacies associated with multivariable modeling. The multiple regression model yields an estimate of the relationship between BMI and systolic blood pressure, accounting for variations in systolic blood pressure attributable to age, gender, and treatment for hypertension. A one-unit increase in BMI correlates with a 0.58-unit increase in systolic blood pressure, while controlling for age, gender, and treatment for hypertension. Each additional year of age corresponds to a 0.65-unit increase in systolic blood pressure, with BMI, gender, and treatment for hypertension held constant. Males exhibit higher systolic blood pressures, approximately 0.94 units more, when controlling for BMI, age, and treatment for hypertension, and individuals undergoing treatment for hypertension show an increase of about 6.44 units in systolic blood pressure, with BMI, age, and gender held constant. The multiple regression equation can be utilized to estimate systolic blood pressures based on a participant's BMI, age, gender, and treatment status for hypertension. For instance, one can estimate the blood pressure of a 50-year-old male with a BMI of 25 who is not receiving treatment for hypertension.

$$\hat{Y} = 68.15 + 0.58\ (25) + 0.65(50) + 0.95(1) + 6.44(0) = 116.1$$

We can estimate the blood pressure of a 50-year-old female with a BMI of 25 who is on treatment for hypertension as follows:

$$\hat{Y} = 68.15 + 0.58\ (25) + 0.65(50) + 0.95(0) + 6.44(1) = 121.59$$

TABLE 3.29

Regression Model Output

Independent Variable	Regression Coefficient	T	P-value
Intercept	108.28	62.61	0.0001
BMI	0.67	11.06	0.0001

TABLE 3.30

Multiple Regression Model Results Output

Independent Variable	Regression Coefficient	t-Critical Value	P-Value
Intercept	68.15	26.33	0.0001
BMI	–0.58	10.30	0.0001
Age	0.65	20.22	0.0001
Male gender	0.94	1.58	0.1133
Treatment for hypertension	6.44	9.74	0.0001

3.23.1 Uses of MLR

i. It helps us identify a few important explanatory variables, out of several variables that are believed to influence the response variable.
ii. To determine the effect of a single explanatory variable on the response variable while several other explanatory variables are simultaneously considered (controlled or matched).
iii. To predict the response variable from initial observations made on new cases.

3.24 Determination of the Coefficients by *R*-Software

The coefficients in the model, namely b_0, b_1, b_2,.....b_k is determined on similar lines as done for the univariate regression method. That is, the sum of the squares of the deviations of the observed values from the corresponding expected values on the regression line is made to be small.

3.24.1 R-Code

```
input <- mtcars[,c("mpg","disp","hp","wt")]
print(head(input))
mpg    disp    hp      wt
Mazda RX4              21.0   160      110    2.620
Mazda RX4  Wag  21.0 160      110     2.875
Datsun 710              22.8   108       93    2.320
Hornet 4 Drive        21.4   258      110    3.215
Hornet Sport about  18.7   360      175   3.440
Valiant               18.1   225      105    3.460
input <- mtcars[,c("mpg","disp","hp","wt")]
# Create the relationship model.
Model<- lm(mpg~disp+hp+wt, data = input)
# Show the model.
Print(model)
# Get the Intercept and coefficients as vector
elements.
Cat("# # # # The Coefficient Values # # #
","\n")
a <- coef(model)[1]
print(a)
Xdisp <- coef(model)[2]
Xhp <- coef(model)[3]
Xwt<- coef(model)[4]
Print(Xdisp)
Print(Xhp)
Print(Xwt)
Call:
Lm (formula = mpg ~ disp + hp + wt, data =
input)
Coefficients:
(Intercept)      disp          hp            wt
 37.105505   -0.000937   -0.031157    -3.800891
# # # # The Coefficient Values # # #
(Intercept)      37.10551
         disp -0.0009370091
         hp -0.03115655
       wt -3.800891
```

3.25 Interpretation of Regression Output

Regression outputs require careful reading and interpretation. We have numerically illustrated this with the data given in Table 3.11.

$$\text{Hb}(\%) \;=\; b_0 + b_1(\text{Age}) + b_2(\text{PCV})_2 \;+\; e_{ijk} \quad (3.54)$$

If we had used one of the statistical packages, a typical output would be as follows:

$$R - \text{square}(\%) = 0.8512$$

The estimated value of the constant (b_0=5.24). The hemoglobin level estimated for someone with an age and PCV of zero has no clinical meaning. However, it is usually provided in multiple regression program outputs and is needed in the prediction equation (3.54). Age and PCV are the two independent variables, and the regression coefficients are the estimates of *b*'s in the regression following equation (3.55) (Table 3.31).

TABLE 3.31

MLR Output: Hb on Age and PCV (Response Variable: Hemoglobin Level)

Variable	Regression Coefficient	SE	*t*-value	*P*-value
Constant	b_0=5.24	1.21	4.34	0.0004
Age	b_1=0.110	0.016	6.74	0.0001
PCV	b_2=0.097	0.033	2.98	0.0085

$$\text{Predicted hemoglobin} = 5.24 + 0.110 * (\text{Age}) + 0.097 * (\text{PCV}) \quad (3.55)$$

3.26 Interpretation

The regression coefficient associated with age means that for a given PCV, hemoglobin increases by 0.110 g/dL for every unit increase in the individual's age. Similarly, the regression coefficient associated with PCV means that for a given age, hemoglobin increases by 0.097 g/dL for every unit increase in PCV. Every parameter estimate has an associated standard error (SE). The corresponding value is the regression coefficient estimate divided by its SE, and the number of observations minus the number of estimated parameters gives the degrees of freedom. In this case, n=20, the estimated parameters are 3, and hence (df=(20−3)=17). From these, we can derive the p-value by referring to the t-table. The p-values correspond to the probability of observing that particular regression coefficient or a larger one, under the null hypothesis that the true regression coefficient is zero. Since the coefficients associated with age and PCV are highly significant, we conclude that hemoglobin and PCV are related even after the effect of age is taken into account, and hemoglobin and age are related even after the effect of PCV is taken into account. The estimated $R^2(\%)$=0.8512. It is defined as the square of the correlation coefficient between the observed and estimated response variables. The value of R-square can range between 0 and 1 and is interpreted as the proportion of the variation in the response variable that is explained by the explanatory variables included in the model. In our case, 85% of the variation in hemoglobin is explained by the age and PCV parameters.

3.27 Adding Categorical Variable in the Regression

Suppose we wished to test whether women in Table 3.31 who had attained menopause have a different hemoglobin level than the control group. The mean and standard deviation of hemoglobin for pre-menopausal women are 12.29 and 1.57 g/dL, as compared to 16.36 and 0.63 g/dL for post-menopausal women. The difference in Hb for pre- and post-menopausal women is 4.07 g/dL. The t-test yields t=7.3, DF=18, and p<0.001 (highly significant). However, women who have attained menopause would be older than women who have not. It could be argued that if there were a steady rise in

TABLE 3.32

MLR Output for HB on the Age and Menopausal Status (Response Variable Hemoglobin Level)

Variable	Regression Coefficient	SE	*t*-value	*P*-value
Constant	$b_0=11.62$	1.99	5.81	≤0.000
Age	$b_1=0.081$	0.033	2.41	≤0.000
PCV	$b_2=1.88$	1.03	1.82	≤0.000

R-square (%)=0.8107.

TABLE 3.33

R-Square (%) Value on Menopausal Status-Response Variable: Hemoglobin Level

Model with Variables (*S*)	*R*-Square (%)
Age**	0.77
Age**, PCV*	0.85
Age*, PCV*, Menopause	0.87

*, ** Significant at 5% 1% level.

hemoglobin with age, this might account for the apparent importance of menopausal status, which can be added to the multiple regression model, taking two categorical values: '0' for menopause not attained and '1' for attained. We add age to the regression in order to adjust for differences in age between pre- and post-menopausal women (Table 3.32). The output from the multiple regression programs is given in Table 3.33 remains the same as in Table 3.32, i.e., the estimated hemoglobin increases by 0.81 g/dL for each unit increase in age when the effect of menopausal status is allowed. The interpretation of the coefficient for menopause is different. Since menopausal status takes values 0 and 1, the regression coefficient observed, 1.88, is interpreted as follows: after allowing for the effect of age, women who have attained menopause have 1.88 g/dL higher hemoglobin levels than women who have not. The *p*-value associated with menopausal status is 0.08, slightly exceeding the 5% significance level. The confidence interval for b_2 with 17 *df* is

$$1.88 - t_{0.05} \times 1.03 \text{ to } 1.88 + t_{0.05} \times 1.03 \qquad (3.56)$$

or −0.28 to 4.04. This interval includes zero, so the inference drawn previously that there was a difference between pre- and post-menopausal women is discounted. It is the relative ages of women in the two groups that account for most of the difference in hemoglobin levels. This analysis assumes that the relationship between Hb and age remains the same in women before and after menopause.

3.28 Alternative Predictive Model

When there are several explanatory variables, we use multiple regression analysis to choose a model that predicts the response variable satisfactorily with the least number of explanatory variables. For instance, those explanatory variables that have biological or medical importance or those that are cheaper to collect in terms of money or time are meaningful inclusions in the model. The ideal final model is expected to have several significant regression coefficients, along with an *R*-square that is significantly different from zero and numerically estimated to be 0.5 or more. The usefulness of adding a new explanatory variable to the regression equation is judged by the observed change in *R*-square. For illustrative purposes, we use the data given in Table 3.1 to judge alternative models. We have three explanatory variables, i.e. age, PCV, and menopausal status. Among these three, what constitutes the least number of explanatory variables? The following table provides the estimated R-square and the significance of the regression coefficients (Table 3.33).

The model with age alone has an *R*-square of 0.77. The model with age and PCV has an *R*-square of 0.85, an increase of 8% over the model with age alone, and both regression coefficients are significantly different from zero. When we add menopausal status to the model, which already includes age and PCV, there is only a 2% increase in *R*-square, which is not statistically significant, i.e., it is not significantly different from zero. Therefore, we do not gain much from the model having age, PCV, and menopausal status, and hence we are left with two models: the model with age alone or the model with age and PCV. The choice between these two depends on further considerations of practicability, the cost of data collection, etc.

3.29 Concluding Remarks

Sometimes, an explanatory variable is transformed (e.g., log, square root) into a new variable, which may make the model more appropriate. Multiple linear regression analysis is a vast subject and is widely used in the analysis of data. One requirement of data for use in multiple regression is that the response variable should be continuous. If it is not, alternative methods, such as logistic regression, need to be considered.

3.30 Logistic Regression

Frequently, we encounter observations made on individual experimental units that take one of two possible forms. For example, a patient in a clinical trial may or may not experience relief of symptoms; an insect in an insecticidal trial may survive or die when exposed to a particular dose of insecticide; a seed may germinate or fail to germinate under experimental conditions. Such data are said to be binary. In statistics, we are interested not in just one individual but in a group of individuals that have all been treated in the same manner. For instance, the individual responses from each of a number of patients in a clinical trial receiving the same treatment may be combined to give the proportion of patients obtaining relief from symptoms. The resulting data are then referred to as grouped binary data and represent the proportion of 'successes' out of the total number of units exposed to a particular set of experimental conditions. The estimated proportion of successes is considered a dependent variable.

3.31 Some Examples Where the Dependent Variable is a Proportion

- The probability of survival from cancer, associated with the duration or stage of the disease at the time of detection.
- The probability of adopting a family planning method, associated with parity.
- The Chance of infant mortality, associated with the age of the mother, parity, and educational level.
- The occurrence of coronary heart disease (CHD), associated with age, sex, and serum cholesterol level.
- The probability of developing byssinosis, associated with race, sex, smoking habits, length of employment, and dust at the workplace.

In the above examples, the estimated probability is considered the dependent variable (response variable), and the other variables are called independent variables (explanatory variables). Our aim is to find the relationship between the dependent variable and the explanatory variables.

3.32 Comparison with Multiple Regression Analysis

In a previous section, we considered multiple regression, in which the response variable is always a continuous variable, and the explanatory variables are generally continuous, through a few variables may be categorical. Therefore, multiple regression analysis with the estimated probability as the response variable is not an appropriate analysis. If we still undertake a multiple regression analysis on such data, we may sometimes get nonsensical results, such as negative probabilities. Logistic regression analysis is the method of choice in such situations.

3.33 Logistic Regression Analysis

This analysis is similar to multiple regression analysis, except for the fact that the response variable is not an estimated probability but a function of it (called the logit), defined as $\log\left(\frac{P}{1-P}\right)$. In logistic regression analysis, the explanatory variables can be continuous variables or categorical variables. We shall first illustrate the case of a continuous explanatory variable and then the case of a categorical explanatory variable, interpreting the regression coefficients in each case.

3.34 Continuous Independent Variable

The binary response (Y) is patients with retinopathy; the denominator (n) is the total number of patients in each duration category. Consider the relationship of the form

TABLE 3.34

Prevalence of Retinopathy in Diabetic Male Patients According to the Duration of Diabetes

Duration (years)	Mid-Point (d)	Retinopathy With	Retinopathy Without	Total	P
0–1	0.5	8	42	50	0.16
1–3	2	9	27	36	0.25
4–6	5	26	39	65	0.41
7–9	8	15	13	28	0.54
10–12	13	6	19	25	0.68

TABLE 3.35

Distribution of Cholesterol vs. CHD

Cholesterol	X	CHD	Total
<250	0	71	1,058
>250	1	57	395

$$\text{Log}\left(\frac{P}{1-P}\right) = b_0 + b_1 d \quad (3.57)$$

$$\text{Log}\left(\frac{P}{1-P}\right) = -1.623 + 0.22\ d \quad (3.58)$$

'b_0' is the outcome of a purely mathematical process of calculation and is necessary for prediction purposes; however, it does not necessarily have a clinical meaning. Log-odds increase by (0.22 if odds <1) for every year that the patient has diabetes (Table 3.34).

3.35 Categorical Independent Variable

The number of patients who experience coronary heart disease (CHD) over a period of time with cholesterol level as a risk factor is given below:

Let us code the variable 'X' = 0 as the unexposed group and 'X' = 1 as the exposed group. We fitted the logistic regression model to this data and the equation with the estimated coefficients is shown in the equation (Table 3.35).

$$\text{Logit}(P) = b_0 + b_1 X \quad (3.59)$$

$$\text{Logit}(P) = -2.739 + 1.061\ {}_1X$$

The interpretation of b_1: The log-odds of developing CHD for a person with a cholesterol level of ≥250 kg/100 mL, relative to a person with a cholesterol level <250 mg, is 1.061. We interpret b_1 only after taking its antilogarithm, which is 2.89. We conclude that the odds of developing CHD for a person with a cholesterol level of ≥250 kg/100 mL, relative to someone with a cholesterol level <250 mg, are 2.89.

3.36 Interpretation of Coefficients

The log-odds of developing CHD at the baseline value of cholesterol, i.e., for someone <250 mg/100 mL, is b_0 (= 2.739) when X=0. We may take the antilog of b_0 and interpret this as the odds of developing CHD in those with serum cholesterol of <250 mg/100 mL. The numbers in brackets in the above equation are the standard errors (se) of the coefficients, i.e., SE (b_0)=0.122, and SE (b_1) is 0.189. The SEutilize is used for assessing whether the observed regression coefficient is significantly different from 0. For instance, is the observed b_1 (1.061) significantly different from zero? This is done by finding the ratio of b_1 to se (b_1) (which is normally distributed); if the absolute value of the ratio is greater than or equal to 2, we conclude that the estimated b_1 is different from zero at the 5% level of significance.

3.37 The Framingham Study

A cohort study began in 1948 in an industrial town, Framingham, 20 miles west of Boston (USA). The aim was to determine which of a number of potential risk factors were related to the occurrence of Coronary Heart Disease (CHD). The factors studied were age, serum cholesterol level, systolic blood pressure, smoking history, and the results of ECG. Individuals free of CHD were followed for 12 years.

The deviance (D) is obtained by D

$$= 2\Sigma\left\{ y_i \log\frac{y}{y_i} + (n_i - y_i) \log \right\} \quad (3.60)$$

where $Y_i = n_i P_i$ and P_i is the estimated proportion.

The terms age and sex in the table below represent the interaction effect of age with sex, and similarly, cholesterol. Age represents the interaction effect of cholesterol with age (Table 3.36).

To determine whether both age and sex are associated with the probability of CHD, a logistic model containing a constant term is first fitted. The deviance for this model is then compared with that of other models that employ one, two, or three factors. Before proceeding further, let us see how the deviance and *degrees of freedom (df)* are used to construct a given model. In general, suppose that two logistic models, model (1) and model (2), are to be compared, These models are as follows:

$$\text{Model}(1):\ \text{Logit}(P) = b_0 + b_1x_1 + \ldots\ldots\ b_kx_k \quad (3.61)$$

TABLE 3.36

Regression Model Deviance and Degrees of Freedom

Terms Fitted in Model	Deviance	*df*
Constant	301.21	15
Age	171.05	14
Sex	221.50	14
age+sex	88.38	13
age+sex+age.sex	77.44	12
age+sex+age.sex+chol	22.39	9
age+sex+age.Sex+chol+chol Age	7.58	6
age+sex+age.sex+chol+chol sex	18.28	6
age+sex+age.sex+chol+chol age+chol.sex	3.47	3

$$\text{Model}(2):\ \text{Logit}(P) = b_0 + b_1x_1 + \ldots\ldots\ldots b_kx_k + \ldots\ldots\ b_px_p \quad (3.62)$$

Model (2) contains $(p-k)$ more variables than Model (1). Denote the deviance under each model by D_1 and D_2, so that D_1 and D_2 have $(n-k-1)$ and $(n-p-1)$ degrees of freedom, respectively, where n is the number of binomial observations. Model (1) is nested within Model (2). Since Model (2) contains more terms than Model (1), it will inevitably fit the data better, and D_2 will be smaller than D_1. The difference in deviance D_1-D_2 reflects the combined effect of the variables $X_{k+1} \ldots X_p$ after $X_1, X_2, \ldots X_K$ have already been included in the model. This difference in deviance is described as the deviance of fitting $X_{k+1} \ldots X_p$ adjusted for, or eliminating, $X_1, X_2 \ldots X_k$. Further, (D_1-D_2) is approximately distributed as a chi-square distribution with degrees of freedom equal to the difference of the two degrees of freedom, i.e., $(p-k)$. Returning to the example, the addition of age is associated with a deviance of 130.16 (301.21 − 171.05) with 1 (15 − 14) degree of freedom. This observed deviance is approximately distributed as a chi-square with 1 *df*. It is highly significant implying that age is associated with the probability of CHD. Similarly, sex is also associated with the probability of CHD, as the deviance associated with sex is 79.71 (301.21 − 221.50) with 1 *df*. Therefore, both age and sex are potential confounding variables and need to be included in the model. A term corresponding to the interaction between age and sex is also needed, as the change in deviance on adding age×sex to the model that contains age and sex is 10.9 with 1 *df*, which is significant at the 0.1% level. When cholesterol is added to the model containing age, sex, and age×sex, the reduction in deviance is 55.0 on 3 *df*. This highly significant change in deviance means that cholesterol is a risk factor for CHD, after allowing for the effects of age, sex, and their interaction. At this stage, the estimated coefficients are given below. Now, 0.2351 is interpreted as the log-odds for level 2 of the factor cholesterol (190–219 mg), relative to level 1 (less than 190 mg), adjusted for age, sex, and age×sex. The antilog of 0.2351 is 1.26, which is denoted by ψ_{21}.

$$\psi_{21} = 1.26 \quad \psi_{31} = 1.96 \quad \psi_{41} = 3.04\left(\begin{array}{l}\text{with adjustment for age,}\\ \text{sex and age*sex}\end{array}\right)$$

$$= 1.41 = 2.36 = 3.81(\text{without adjustment})$$

Further, to determine whether there is an interaction between cholesterol and age, and between cholesterol and sex, the terms cholesterol×age and cholesterol×sex are each added to the model containing cholesterol, age, sex, and age '×' sex. The reduction in deviance on including cholesterol×age is 14.8 on 3 *df* (p=0.002), while that on adding cholesterol×sex is 4.1 on 3 *df* (p=0.251). These results suggest that there is an interaction between cholesterol and age but not between cholesterol and sex. The deviance for the most appropriate model is 7.6 on 6 *df*. This contains age, sex, age×sex, cholesterol, and cholesterol '×' age (Table 3.37).

For persons aged 30–49, the effect due to the interaction of age with cholesterol is defined as zero, and hence the respective coefficients of cholesterol with antilog

TABLE 3.37

Estimated Coefficients

Terms Fitted in Model	Estimates	SE
Constant	−3.195	0.239
Age	0.056	0.315
Sex	0.924	0.276
Age+sex	1.540	0.332
age+sex+age * sex	1.420	0.188
age+sex+age * sex+chol	−1.363	0.243
age+sex+age * sex+chol+chol age	0.572	0.422
age+sex+age*sex+chol+chol sex	0.227	0.388
age+sex+age * sex+chol+chol*age+chol*sex	−0.550	0.371

3.38 Concluding Remarks

The study of observed proportions as a function of several explanatory variables or exposure variables can be handled through logistic regression. Such an application is logistic regression need not necessarily conform to the actual observed data set but could include interactions as well. The data may contain some observations, termed outliers that are not well fitted by the model, or observations, termed influential values, that have an undue impact on the conclusions to be drawn from the analysis. It could be that the logistic regression analysis for the data set may not be appropriate. Some procedures like the Poisson regression model or proportional hazards model may be more suitable. These are intricate problems in data analysis, and statistical consultation must be sought in such circumstances. Due to the availability of statistical packages like SPSS, SAS, BMDP, GLIM, and EGRET, the computational aspects of logistic regression need not worry us. We must learn the required commands of the package and understand the output of such programs.

3.39 Diagnostic Tool for Accuracy Testing

The accuracy of any test is evaluated by two indices: sensitivity and specificity. These indices are calculated by performing the test on a group of individuals who have already been evaluated by a 'Gold standard' and then comparing the results as described below:

Sensitivity is defined as the percentage of test positives among those who have the disease (Table 3.38).

$$\text{That is, Sensitivity}(\%) = \left[\frac{A}{A+C}\right] \times 100$$

$$= \left(\frac{\text{True negatives}}{\text{Disease}}\right) \times 100$$

Specificity is defined as the percentage of test negatives among those who do not have the disease.

$$\text{That is, Specificity } (\%) = \left[\frac{D}{B+D}\right] \times 100$$

$$= \left(\frac{\text{True negatives}}{\text{Disease}}\right) \times 100$$

TABLE 3.38

Matrices of Sensitivity and Specificity

Test Result	Gold	Standard	Total
+	Disease+A	Disease−B	(A+B)
-	(True positive) C	(False positives) D	(C+D)
Total	(False negatives) (A+B)	(True negatives) (B+D)	

TABLE 3.39

Results with EEC Screening

	Brain	Tumor
EEG result	Present	Absent
+	36	54,000
–	4	306,000
Total	40	360,000

Sensitivity=36/40−90% & Specificity=306,000/360,000=85%.

TABLE 3.40

Results with CAT

	Brain	Tumor
EEG result	Present	Absent
+	36	18,000
–	01	342,000
Total	40	360,000

Sensitivity=36/40−97.5% & Specificity=342,000/360,000=95%

As an illustration, let us compute and compare the sensitivity and specificity of two methods of screening (viz. EEG and CAT) for diagnosing brain tumors (Tables 3.39 and 3.40).

Hence, it can be concluded that both sensitivity and specificity of CAT are better than EEG in the diagnosis of brain tumor.

3.40 Relationship between Sensitivity and Specificity

When we are dealing with laboratory tests such as blood sugar, the situation is different. There will be overlapping of the distributions of a variable for diseased and non-diseased populations. False positives and false negatives comprise the area of the overlap. The following table gives the sensitivity and specificity of a 2-hour postprandial blood test for glucose for 70 true diabetics and 510 true non-diabetics at different levels of blood glucose (Table 3.41).

It may be noted from the above table that sensitivity and specificity are inversely related. As one increases, the other decreases. An ideal test should have both sensitivity and specificity at about 100%, but this seldom occurs.

3.41 Predictive Values

The sensitivity and specificity provide information about the accuracy of a test, but they do not provide information about the meaning of a positive or negative test result. The positive predictive value (+PV) of a test is the percentage of disease present given a positive test result.

$$+PV = \left[\frac{A}{A+C}\right] \times 100 = \left(\frac{\text{True positives}}{\text{Test} +}\right) \times 100 \quad (3.63)$$

The negative predictive value (–PV) of a test is the percentage of disease negative given a negative test result.

$$-PV = \left[\frac{D}{C+D}\right] \times 100 = \left(\frac{\text{True negatives}}{\text{Test} -}\right) \times 100 \quad (3.64)$$

Unlike sensitivity and specificity, the positive and negative predictive values of a test depend on the prevalence of disease in the population, as illustrated in the following table (Table 3.42).

TABLE 3.41

Distribution of Glucose Level

Blood Glucose (mg%)	Sensitivity (%)	Specificity (%)
80	100	1
90	99	7
100	97	25
110	93	48
120	89	68
130	81	82
140	74	91
150	64	96
160	56	99
170	53	99.6
180	50	99.8
190	44	99.8
200	37	100

Source: Lilienfeld, A. M. & Lilienfeld, D. *Foundations of Epidemiology*, Second Edition (Oxford Publications, New York, 1979).

3.42 Case–Control Studies

1. A group of individuals affected by a disease (cases) is compared with a group of individuals who do not have the disease (controls).
2. Information is obtained, usually retrospectively, from each group about the frequency of various personal and/or environmental factors that might be associated with the disease.
3. Large and statistically significant differences are studied carefully to determine whether they could have been caused by some bias or are artifacts (Krivicich et al., 2023). If this is not the case, then the association is considered genuine, and a causal hypothesis may be suggested.

3.42.1 Advantages of Case–Control Studies

1. Well-suited for the study of rare diseases or diseases with long latency. A cohort study would require too many subjects or take far too long to complete.
2. Relatively quick to initiate and conduct.
3. Relatively inexpensive.
4. May require fewer personnel to conduct.
5. Existing records (if reliable) can be used.
6. No risk to subjects (e.g. drugs or chemicals suspected of increasing the risk of disease may be temporarily discontinued).
7. Particularly appropriate when classification by disease is simple (+ or –) but many possible etiological factors need to be studied.

3.42.2 Disadvantages of Case–Control Studies

1. Relies on recall or records for information on past exposure.
2. Validation of data is difficult or impossible.
3. Control of extraneous variables may be incomplete.
4. Selection of an appropriate control group may be very difficult.

TABLE 3.42

Relation between the Prevalence and Predictive Values (Sensitivity 50%, Specificity 90%)

	Prevalence 5%			Prevalence 15%			Prevalence 25%		
Test	Disease		Total	Disease		Total	Disease		Total
	+	–		+	–		+	–	
+	25	95	120	75	85	160	125	75	200
–	25	855	880	75	765	840	125	675	800
Total	50	950	1,000	150	850	1,000	250	750	1,000
Sensitivity	25/50=50%			75/150=50%			125/250=50%		
Specificity	855/950=90%			765/850=90%			675/750=90%		
+PV	25/120=21%			75/160=47%			125/200=63%		
–PV	855/880=97%			765/840=91%			675/800=84%		

TABLE 3.43

Different Sources of Case and Controls

Cases	Controls
All cases diagnosed in the population	Sample of general population
All cases in a sample of the population	Non-cases (all or sub-group) in the sample of the population
All cases from all hospitals	Sample of patients from all hospitals, who do not have the disease under study
All cases in a single without the specified disease	Sample of patients in the same hospital
All cases diagnosed in some but not all hospitals	Sample of individuals in the same neighborhood/relatives of cases
	Controls can also be spouses, siblings, or associates of cases

5. Disease incidence rates in exposed and unexposed individuals cannot be determined (only odds ratio, which approximates relative risk under certain circumstances, can be calculated).
6. The possibility of different types of bias affecting the conclusion exists and must be carefully eliminated or adjusted for.

3.43 Eligibility Criteria

3.43.1 Case

Objective criteria for the diagnosis of the study disease are needed (e.g. Rheumatoid Arthritis)

It is better to include new cases for the study (incident cases). Exclusion criteria need to be specified, such as associated illnesses that may affect the study conditions (Andrade, 2022)

3.43.2 Control

Set out ineligibility criteria (for example, pregnant women cannot be eligible as controls in a study of whether recent (e.g. last month) use of oral contraceptives increases the risk of myocardial infarction) (Lutsey, 2023) (Table 3.43).

3.44 Choice of Controls

1. The likelihood of obtaining relevant information must be the same as in cases (i.e., non-response must not differ between cases and controls).
2. Controls must be derived from a population similar to that which gives rise to cases, with respect to factors such as age, sex, ethnic group, and socio-economic status.
3. Make allowances for confounding variables. For example, in a study of whether lactation protects against breast cancer, breast cancer cases have low parity and hence less lactation. It is important, therefore, to match for parity.
4. Do not select patients with diseases that may be associated with the exposure factor, especially if the likelihood of association is high.
5. Spouses—provided male and female cases are equal; siblings, relatives, and associates (e.g., at school or office)—Ensure similar ethnic and socio-economic status.
6. Take practical and economic considerations into account.

3.45 Matching

a. Frequency matching – Choose controls such that their distribution with respect to important factors is similar to the distribution of cases. However, this can be done only when the final distributions are obtained for cases. As time differentials may arise in interviewing, this method is not always advisable (Chidambaram & Josephson, 2019).
b. Individual matching – Match based on criteria such as age (±5 years), sex, socio-economic status, ethnic status, and parity (e.g., lactation versus breast cancer).

3.46 Precautions in Matching

1. Do not match for variables that are intermediate in a causal pathway. Example: Smoking → Enhanced blood cholesterol → High risk of cardiovascular disease (Thiese, 2014). Matching for blood cholesterol may diminish or destroy the association between smoking and cardiovascular disease.
2. Do not match for variables that are related to the cause but not the disease. Example: Use of contraceptives associated with but not caused by religion

Do not match for religion because it will result in a loss of statistical efficiency. However, if there is uncertainty about the non-association between the variable (e.g., religion) and the disease, it may be better to match for the variable and accept the loss of statistical efficiency rather than risk spurious associations (Noordzij et al., 2009). Overmatching, i.e., matching for too many variables, is likely to be logistically infeasible and may lead to problems in obtaining suitable controls. If matching is done at the design stage, it should be accounted for in the analysis.

3.47 Number of Cases and Controls

The number of cases and controls to be studied is determined by the magnitude of the odds ratio that is of practical interest (e.g., 2 or more), the magnitude of Type 1 error (usually 5%), and the power of the study (usually 90%). If feasible, it would be valuable to have different types of controls (e.g. hospital controls, neighborhood controls, general population controls) for each case studied. If differences are consistently observed between the cases and each of the different types of controls, the inferences drawn would be more robust. There can be more than one control for each case: however, having more than four controls per case does not improve the statistical efficiency of the design.

3.48 Conduct of Study

The tool used for data collection could be an oral interview (using a proforma), combined with clinical and/or biochemical and other investigations. Data collection from existing medical records can be considered if the quality of such data is satisfactory.

3.49 Problems of Measurement

Interviewer Bias

Ideally, the interviewer should be unaware of or 'blind' to the hypothesis under investigation. In reality, this is very difficult to achieve; therefore, a few precautions can be taken, such as the use of a formal questionnaire and standardized interviewing techniques.

3.50 Responder Bias

Neither the cases nor the controls should be aware of the hypothesis, and both should be subjected to the same type of probe to recall past events.

3.51 Analysis

If the odds ratio is high (i.e. more than 3), the risk factor is considered important. Such analysis can be done for each of the factors under study. Multivariate analytical methods are available for considering several factors simultaneously. With such methods, the effect of other confounding factors (such as age, sex, race, and alcoholism) can be controlled. Finally, the effect of chance in such associations must be considered by testing for the significance of the odds ratio and estimating its 95% confidence interval (Table 3.44).

TABLE 3.44

Configuration of Risk Factors

Risk Factor	Cases	Controls	Total
Present	a	b	$(a+b)$
Absent	c	d	$(c+d)$
Total	$(a+c)$	$(b+d)$	N

Odds of having risk factor among cases $=\frac{a}{c}$.

Odds of having risk factor among controls $=\frac{b}{d}$.

Odds ratio $=\frac{a}{c}+\frac{b}{d}=\frac{ad}{bc}$.

3.52 Cross-Sectional Studies

Cross-sectional studies are studies of the prevalence of disease (and its correlates) measured at one point in time or over a short period in the population under study (Srikanth & Doddamani, 2013). They are quick, relatively inexpensive, and logistically simpler to undertake. Therefore, they are often used for public health planning and the formulation of etiologic hypotheses.

3.52.1 Objective

The study objectives should be stated clearly. The answers to the following questions would greatly assist the process:

What are the specific questions to be answered by the study?
Are they answerable questions?
Do they meet the purpose of the study?
Are they expressed clearly?
Are they expressed in measurable terms?

Example: To estimate the prevalence of acute diarrhea disease in children less than 5 years in Village A and to study its association with a contaminated water supply, controlling for the effect of age, sex, and nutritional status.

3.53 Choice of Study Population

The following aspects must be considered before initiating a cross-sectional survey:

Is the study population representative?
Is it suitable for the attainment of the study objectives?
Will it be possible to obtain the required information?
Is it possible or necessary to survey the whole population or will a sample be adequate? If a sample would suffice, what sampling design should be employed?

To achieve the aforementioned objectives, the cohort of children under 5 years of age chosen for the study must accurately reflect the demographics of Village *A*. It is essential that the children and their mothers from the village are willing to participate in the data collection process. In instances where the village has a population of fewer than 5,000 residents, it would be feasible to survey all children in this age group. Conversely, in larger populations, a representative sample of these children would be adequate. Data regarding acute diarrhea episodes can be gathered at a single time point utilizing a weekly recall approach.

3.54 Measurement of Variables

The exposure variables should be clearly listed before the start of the survey. Variables are to be hypothesized based on existing literature, the experience of the investigators, and insights from other scientists working in related areas.

3.54.1 Questions to Be Answered Relating to the Variables

1. Have the disease and exposure variables been clearly defined?
2. How many variables will be included?
3. Have possible confounding variables been considered?
4. (Demographic variables such as age, sex, marital status, and social class are generally collected, as they can sometimes serve as useful explanatory variables.)

Care is needed in selecting variables that may require respondents to recall past events, as this demands considerable ability to recall on their part.

3.55 Methods of Data Collection

1. Questionnaires for oral interviewing (self-administered/interviewer-administered/postal).
2. Documentary sources (e.g. corporations, municipalities, primary health centers, and other reliable sources of population-based health registers).
3. Laboratory tests (e.g. blood, sputum, urine samples).
4. Physical measurement (e.g. anthropometry).
5. Special procedures (e.g. special endoscopic surveys on large populations for cancer registers).

3.56 Quality of Data

Repeating questionnaires/interviews by a more experienced person in a random sample of subjects, setting up quality control for laboratory tests and other measurements, and employing similar methods can help check the validity and reliability of the data collected and allow for mid-course corrections if necessary.

3.57 Sample Size

Cross-sectional surveys are usually conducted on representative samples selected using statistical sampling techniques. The size of the sample depends on the degree of precision required (as expressed by the 95% confidence interval) for the prevalence estimate and should be appropriate to provide a Relative Risk (RR) of at least 2 for associated factors.

3.58 Non-Response

The aim of the survey should be to collect data from all selected respondents. However, in real life, this may not always be possible. If the non-response rate is very high, the survey findings cannot be regarded as representative of the population under study. Adequate measures must be taken to minimize non-response. These could include repeated attempts or offering some incentives to those who are uncooperative. Another approach could involve taking a random sample of the non-responders, collecting information (by all available means) on all of them, and using this to extrapolate to all non-responders.

3.59 Prevalence

Prevalence may be defined as the number of cases detected in the cross-sectional survey expressed as a proportion of the population surveyed. This is known as 'point prevalence'. In situations where data collection extends over a long period (e.g. 6 months), the term 'period prevalence' is used.

3.60 Relative Risk

Prevalence measures (e.g. diarrhea) can also be calculated separately for those with exposure to a risk factor (e.g., contaminated water supply) and those without exposure (Table 3.45).

The comparison of prevalence among the exposed and the unexposed will provide an estimate of the risk of disease associated with exposure. This measure is called relative risk and is computed as:

Relative risk = Prevalence of disease among those exposed to the risk factor/Prevalence of disease among those unexposed to the risk factor

$$\mathrm{RR} = \frac{a}{a+b} \div \frac{c}{c+d}$$

A relative risk of 2 or more is usually considered significant for the purpose of an etiological hypothesis. Several risk factors may be considered one at a time, and the appropriate relative risk computed (Aggarwal & Ranganathan, 2019). However, many risk factors may be interrelated. Consequently, a more realistic picture of the importance of each factor can be obtained by considering all relevant factors simultaneously using multivariate analytical techniques (e.g. logistic regression).

3.61 Limitations of Cross-Sectional Studies

1. A major limitation is determining whether the exposure occurred prior to the disease or subsequently. For example, in a cross-sectional study of peptic ulcers, it is possible that the recent dietary pattern could have changed to soft food consumption due

TABLE 3.45

Relative Risk of Diarrhea

Contaminated Water Supply	Diarrhea Present	Diarrhea Absent	Total
Yes	*a*	b	$(a+b)$
No	*c*	*d*	$(c+d)$
Total	$(a+c)$	$(b+d)$	*N*

Prevalence rate of disease among exposed is $= \frac{a}{a+b}$.

Prevalence rate of disease among unexposed is $= \frac{c}{c+d}$.

to symptoms of abdominal pain increasing with the consumption of hot foods.

2. It is difficult to determine cause and effect. For instance, if people in a low socio-economic status have a high incidence of mental illness, we cannot be certain whether the low socio-economic status (and factors associated with it) leads to mental illness or whether mental illness leads to a lowering of socio-economic status.
3. Cross-sectional studies are not suitable for the study of rare diseases.

To conclude, cross-sectional studies are surveys conducted to generate hypotheses about the association of one or more factors with the disease, which need to be confirmed by prospective or intervention studies subsequently.

3.62 Cohort Studies

A cohort is defined as a group of individuals born in the same year or attending the same clinic for the first time in a specified period. At intake to a cohort study, all potential study subjects must be free from the disease of interest (Yang et al., 2021). Eligible participants are then followed over a period of time to assess the occurrence of the disease in relation to whether or not a suspected risk factor was present initially.

3.62.1 Advantages of Cohort Studies

1. Temporal sequence between exposure and disease can be more clearly established.
2. Well-suited for assessing the effects of rare exposures, e.g. occupational health hazards.
3. Allows for examination of multiple effects of a single exposure. Thus, information on a full range of health effects of a single exposure can be obtained.
4. Minimizes the potential for selection bias.
5. Can be used to study multiple outcomes (diseases) attributable to a common exposure.

3.62.2 Disadvantages of Cohort Studies

1. Involves the follow-up of a large number of individuals over many years.
2. Expensive and difficult to organize.
3. Losses in follow-up could be a major problem.

3.63 Types of Cohort Studies

Cohort studies are of three types depending on whether the follow-up is forward, backward, or both:

1. Prospective → Forward (Active study)
2. Retrospective ← Backward (Based on records)
3. Ambidirectional ↔ Both ways (Active study+ Records)

Retrospective and ambidirectional studies have a greater chance of bias if the hypothesis under study is known to the investigators collecting the data. Additionally, if records are in complete or of poor quality, such studies may not be productive. Finally, information must be available for other factors that affect both the disease and the risk factor, e.g. age and socio-economic status.

3.64 Design of Cohort Studies

A cohort study generally begins with a group of subjects who are exposed to a risk factor (e.g. employees of a factory exposed to asbestos) and a corresponding group of individuals who are not exposed to that risk factor (e.g. families of the employees of this factory). Both groups are followed for a sufficiently long period (e.g. 5–10 years) to observe the development of the disease (Baxter & Jack, 2015). The follow-up needs to be intensive and rigorous, ensuring that very few individuals are lost to follow-up. The cohort size should be large enough to demonstrate a statistically significant difference in disease occurrence between the two groups. (Details of sample size determination in such studies will be described later in the course.)

3.65 Issues for Design Formulation

Selection of Exposed Population

1. The exposure factor must be fairly common.
2. The cohorts chosen must be amenable to follow-up.
3. Collection of accurate and reliable information on the duration and quantum of exposure must be feasible.

Examples of common exposures include cigarette smoking, alcohol consumption (duration and quantum), occupational hazards in asbestos or cement factories, and radiation exposure in X-ray plants. For common risk factors such as smoking, the study population should consist of a general cohort (mix of exposed and unexposed individuals) drawn from a geographically well-defined area belonging to a specific age and sex group. For example, resident adult males of a particular area were studies in the classical Framingham Heart Study. Advantages of Choosing Such Cohorts are:

1. They are a stable population.
2. Numerous occupations and industries are included.
3. Subjects have common health facilities.

3.66 Study of Comparison Group (Unexposed Cohort)

A single, well-defined reference population as the comparison cohort (unexposed cohort) is preferable (Superchi et al., 2022). A group of unexposed subjects should constitute the comparison group. The subjects in this group must be very similar to those in the study group with respect to all other factors related to the disease, except the exposure under investigation. For instance, in a study of the smoking habits of patients with lung cancer, the comparison group could be adult males working in the same area/office and of similar economic status but nonsmokers.

3.67 Sources of Data for Cohort Studies

Exposure data can be collected from:

1. Existing medical records
2. Employment records

Data on study subjects themselves may be collected by:

1. Direct personal interviews
2. Mailed questionnaires
3. Medical examination or lab tests
4. Direct measurement of the environment of the cohort

Items 3 and 4 can provide only the current level of exposure; hence, periodic re-examination or re-surveys may need to be conducted to estimate the relationship between the quantum of exposure and the disease.

3.68 Sources of Outcome (Disease) Data

A precise definition of what constitutes the 'disease' under study is essential. If mortality due to the disease is the outcome measure, death certificates can be used, provided their reliability is good (Levin, 2006). Data on other morbidity conditions may be collected from:

- Hospital discharge sheets
- Private physician's records (Nursing Homes)
- Population-based disease registers

(e.g. leprosy case registers)

Apart from these, data can also be collected through periodic surveys using questionnaires, reviews of routine medical/lab reports, employment insurance records or direct medical examinations of the study subjects. Overdiagnosis or underdiagnosis can cause bias; therefore, the definition of the disease needs to be highly specific.

3.69 Follow-Up

Follow up is the most difficult aspect of a cohort study. The ascertainment of changes in levels of exposure and the occurrence of disease requires that all individuals be meticulously followed up. Different outcomes on follow-up of the study subjects are illustrated below:

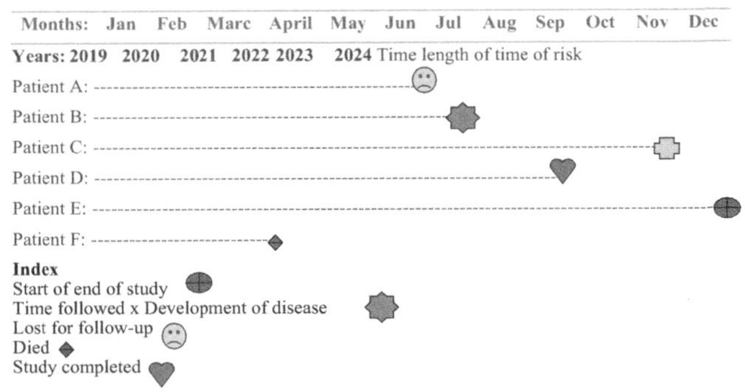

The exposure time varies from subject to subject, and this should be accounted for when computing the risk of disease development. The frequency of data collection on exposure and disease during follow-up depends on the following:

i. The type of disease under study.
ii. The likelihood of changes in exposure during the course of the study
iii. The magnitude of recall lapse for relevant data.

3.70 Data Analysis

Data analysis in cohort studies is mainly based on disease incidence rates. An incidence case is a new case of disease observed during the study period. Since such cases are the outcome of interest in cohort studies, it is usual to compare incidence rates of disease among the exposed and unexposed cohorts to obtain measures of risk for the disease.

$$\text{Incidence rate (IR)} = \left(\frac{\begin{array}{c}\text{No. of new cases identified} \\ \text{(usually in one year)}\end{array}}{\begin{array}{c}\text{No of Individuals} \\ \text{follow up attended}\end{array}} \right)$$

$$\text{Cumulative I (IR)} = \left(\frac{\begin{array}{c}\text{Proportion of people who develop} \\ \text{disease during a specified interval of time)}\end{array}}{\text{Total population at risk}} \right)$$

Total person-years is computed by summing up the individual periods of follow-up (expressed in years) for each study subject, separately for the exposed and the unexposed cohorts. Disease rates in the exposed and unexposed cohorts are compared using measures such as relative risk or rate ratios.

3.71 Measure of Effect in Cohort Studies

3.71.1 Ratio Measure

Incidence rate in the exposed/Incidence rate in the unexposed

Also known as relative risk, rate ratio, or risk ratio. It is a pure number

- Often used as a measure of etiological force/strength
- Not a good measure of public health impact

3.71.2 Difference Measure

Incidence rate in the exposed - Incidence rate in the unexposed. Also known as rate/risk difference or excess rate/risk.

3.71.3 Attributable Risk

- A measure of the absolute effect of exposure.
- Guides individual needs/benefits for each risk factor.
- Not used in etiological research but only for public health plans and policy.

3.71.4 Interpretation of Findings

The interpretation of findings depends on the roles played by chance, bias, and confounding in the study. This can be examined by performing appropriate statistical tests and obtaining a confidence interval for the relative risk. The effect of confounding (i.e. the influence of other factors affecting disease and exposure simultaneously) is controlled in the analysis. Sufficient care should be taken during the planning of the study to collect reliable data on such factors. Multivariate analytical techniques are very useful for controlling the effects of confounders.

3.72 Bias in Cohort Studies

3.72.1 Selection Bias

The differences in criteria employed for the selection of exposed and unexposed cohorts, e.g. the exposed may all be older individuals, and the unexposed relatively younger.

3.72.2 Misclassification

The accuracy with which subjects are classified as 'diseased' or 'not' may differ between the exposed and unexposed.

3.72.3 Lost to Follow-Up

This is a major problem in cohort studies due to the long-time duration involved in such studies. This will affect the validity of the results, especially if the proportion lost to follow-up is more than 20%.

3.72.4 Effects of Non-Participation (Refusals)

As the study progresses, it is likely that some subjects may not participate wholly or in certain items of the study (e.g. repeated investigations, X-rays). This will result in a loss of generalizability of the study findings. Bias can be controlled in cohort studies by

1. Minimizing losses to follow-up.
2. Comparing the incidence of confounding variables for exposed and unexposed groups and correcting the differences.

3.73 Implication of Human Intervention Study

Cohort studies are the best type of observational studies possible, which can lead to the estimation of causal associations. They fall short only of randomized controlled trials but can often provide valid results on associations for rare exposures. However, they are very expensive, time-consuming, and require a great deal of careful organization.

REFERENCES

Andrade, C. Research design: Case-control studies. *Indian Journal of Psychological Medicine* **44**, 307–309 (2022).

Aggarwal, R. & Ranganathan, P. Study designs: Part 2- Descriptive studies. *Perspectives in Clinical Research* **10**, 34–36 (2019).

Bhutto, A. H. et al. Correlation and regression analysis for yield traits in wheat (*Triticum aestivum* L.) genotypes. *Natural Science* **8**, 96–104 (2016).

Boller, B. Teaching correlation and regression in three European countries. *Teaching Mathematics and Computer Science* **21**, 161–183 (2024).

Baxter, P. & Jack, S. Qualitative case study methodology: Study design and implementation for novice researchers. *The Qualitative Report* (2015). https://doi.org/10.46743/2160-3715/2008.1573

Cui, L. et al. Correlation between ultrafiltration rate and hemoglobin level and erythropoietin response in hemodialysis patients. *Renal Failure* **46,** 2296609 (2024).

Chidambaram, A. G. & Josephson, M. Clinical research study designs: The essentials. *Pediatric Investigation* **3,** 245–252 (2019).

Durgadevi, M., Sai Lalitha, V., Supraja, P. & Lalithaparameswari, K. Correlation and regression analysis on savings and loans. *Pramana Research Journal* **8**, 85–100 (2018).

Erdfelder, E., FAul, F., Buchner, A. & Lang, A. G. Statistical power analyses using G*Power 3.1: Tests for correlation and regression analyses. *Behavior Research Methods* 41, 1149–1160 (2009).

Fleiss, J. L. & Tanur. On conditional and partial correlation. *The American Statistician* **30**(3), 146–149 (1972).

Hu, J., Chen, Y., Leng, C. & Tang, C. Y. Applied regression analysis of correlations for correlated data. *Annals of Applied Statistics* **18**, 184–198 (2024).

Kaushal, J., Mantri, A., Shalini & Goyal, H. Linear correlation and regression analysis of tap water quality parameters in rural area of Malwa Punjab. *Arabian Journal of Geosciences* **17**, 97 (2024).

Khuan, H., Judijanto, L., Rahmiyati, N., Yusuf, R. & Karyasa, T. B. Study of the effect of green investment policy, technological innovation, and sustainable development on industrial competitiveness. *West Science Interdisciplinary Studies* **2**, 146–154 (2024).

Krivicich, L. M., Jan, K. & Nho, S. J. Case-control study, in *Translational Sports Medicine*, 183–185 (Elsevier, 2023). https://doi.org/10.1016/B978-0-323-91259-4.00069-2

Lin, F., Huang, Y., He, X. & Liu, Z. Research on the socioeconomic factors that influence the development of voluntary, non-remunerated blood donation in China—A correlation and regression analysis based on data from 2012 to 2018. *Health Science Reports* **6**, e1341 (2023).

Lutsey, P. L. Case-control studies: Increasing scientific rigor in control selection. *Research and Practice in Thrombosis and Haemostasis* **7**, e1341 (2023).

Levin, K. A. Study design III: Cross-sectional studies. *Evidence-Based Dentistry* **7**, 24–25 (2006).

Noordzij, M., Dekker, F. W., Zoccali, C. & Jager, K. J. Study designs in clinical research. *Nephron - Clinical Practice* **113**, c218–c221 (2009).

Robert Tibshirani, Regression shrinkage and selection via the lasso. *Journal of the Royal Statistical Society. Series B (Methodological)* **58**(1): 267–288 (1996).

Sipio, W. D. et al. Correlation and regression analysis in different genotypes of sunflower (*Helianthus annuus* L.). *Pakistan Journal of Biotechnology* **21**, 172–177 (2024).

Shi, J. Research on the current situation and countermeasures of student employment management in higher education institutions based on multiple regression analysis. *Applied Mathematics and Nonlinear Sciences* **9**, 265–272 (2024).

Srikanth, D. & Doddamani, P. K. Overview of study designs. *International Journal of Pharmacy and Pharmaceutical Sciences* **5**, 1011–1015 (2013).

Superchi, C. et al. Study designs for clinical trials applied to personalised medicine: A scoping review. *BMJ Open* **12**, e052926 (2022).

Thiese, M. S. Observational and interventional study design types; an overview. *Biochemia Medica* **24**, 199–210 (2014).

Yang, B. et al. Study designs for comparative diagnostic test accuracy: A methodological review and classification scheme. *Journal of Clinical Epidemiology* **138**, 128–138 (2021).

4

Exploring the Applications of the Theory of Probability

4.1 Introduction

The answers to almost all questions asked about statistical data are, to some degree, couched in uncertainty (Bordalo et al., 2023). For example, if treatment *A* is, on average, better than treatment *B* in a clinical trial involving ($n = 100$) patients, can we be certain that *A* is better than *B*? Is it not possible that both treatments are actually equal, and that just by chance treatment *A* appears to be better than treatment B in this particular trial? Questions like these can be answered not with a simple 'Yes' or 'No', but only in terms of a statement of probability, such as 'There is less than a one in a thousand chance that treatments *A* and *B* are equally effective' (Batanero et al., 2024). Various definitions of probability are available. For simplicity, we shall define probability as a proportionate frequency (Ackerman et al., 2019). For instance, if a variable can take any of 'N' values and n of these constitute the event of interest to us, the probability of the event is (n/N). A few examples will be of value.

a. Tossing a coin: There are two possible outcomes: head or tail, and the probability of each is 1/2.
b. Throwing a die: There are 6 possible outcomes, namely a score of 1, 2, 3, 4, 5, or 6, and the probability of each is 1/6. If our event is obtaining a score of 3 or 4, the probability is, by definition, 2/6.
c. Drawing a card at random from a deck: There are ($n=52$) cards in a deck, and these include 4 Aces, 2 red and 2 black. The probability that a card randomly drawn is an Ace is therefore 4/52, a black Ace is 2/52, and a red Ace is 2/52.
d. Selecting a subject for inclusion in a morbidity survey: In a small community, there are 128 children aged less than 5 years ($n=128$), 192 children aged 5–14 years ($n=192$), and 480 adults aged 15 years or more ($n=480$). The probability that a subject randomly selected is (i) a child under 5 years is 128/(128+192+480), i.e. 128/800=0.16; (ii) a child under 15 years is (128+192)/(128+192+480), i.e., 320/800=0.4.

There are two kinds of probability: (i) Experimental probability and (ii) Theoretical probability (Table 4.1).

4.2 Rule of Multiplication

4.2.1 Independent Events

If $P(A)$ denotes the probability of an event A occurring and $P(B)$ denotes the probability of an event B occurring, and the two events are independent (i.e., the occurrence of one has no implications on the occurrence of the other), the probability of

TABLE 4.1

Difference between Experimental and Theoretical Probability

Experimental	Theoretical
It is found by repeating an experiment and observing outcomes	What expected to happen based on Mathematics
$P(\text{Event}) = \frac{\text{Number of times event occur}}{\text{Total number of trails}}$	$P(\text{Event}) = \frac{\text{No of favorable outcomes}}{\text{Total number of possible outcomes}}$
For example, a coin is tossed 10 times, a head is recorded 7 times and tail 3 times. $P(\text{Head}) = \frac{7}{10}, P(\text{Tail}) = \frac{3}{10}$	For example, a coin is tossed. $P(\text{Head}) = \frac{1}{2}, P(\text{Tail}) = \frac{1}{2}$
One way to find the probability of an event is to conduct an experiment. **Example:** A bag contains 10 red marbles, 8 blue marbles, and 2 yellow marbles. Find the experimental probability of getting a blue marble.	A bag contains 10 red marbles, 8 blue marbles, and 2 yellow marbles. Find the theoretical probability of getting a blue marble. There are 8 blue marbles. Therefore, the number of favorable outcomes=8. There are a total of 20 marbles. Therefore, the number of total outcomes=20. $P(\text{Event}) = \frac{8}{20}$

DOI: 10.1201/9781032713816-4

both A and B occurring at the same time is simply the product of the two separate probabilities (Bstianello et al., 2020).

$$\text{In other words}, P(AB) = P(A) \times P(B) \tag{4.1}$$

For instance, if event *A* obtains a Head on tossing a coin and event *B* obtains a score of 4 on throwing a die, we have $P(A)=½$ and $P(B)=1/6$. Since the outcome of the toss has no implications on the result of the throw of the die, the probability of getting Head and a score of 4, if a coin is tossed and a die is thrown at the same time, is obtained using the above formula as $½\times 1/6=1/12$. This can be verified from first principles. There are 12 outcomes of the experiment, namely Head associated with the scores 1, 2, 3, 4, 5, or 6 (six possibilities) and Tail associated with the scores 1, 2, 3, 4, 5, or 6 (six possibilities): of these only one (Head and score of 4) is of interest to us, so the required probability is 1/12.

4.2.2 Non-Independent Events

If two events *A* and *B* are not independent,

$$\begin{aligned} P(AB) &= P(A) \times P(B, \text{ given } A) \\ &= P(B) \times P(A \text{ given } B) \end{aligned} \tag{4.2}$$

For instance, consider a group of 5 persons consisting of 3 males (M) and 2 females (F). Suppose one person is selected at random, and then a second is selected, again at random, from the remaining 4 persons. Let us consider the probability of selecting a male on both occasions. Of the 5 persons, 3 are males, so the probability of selecting a male on the first occasion is 3/5. If the first selected person is male, there remain 4 persons, of which only 2 are males, so the probability of selecting a male on the second occasion is 2/4. Applying the above formula, the probability of selecting a male on both occasions is $3/5\times 2/4=6/20=3/10$.

This result can also be checked from first principles. The first person can be selected in 5 ways (M or M or M or F or F), and associated with each of these, the second can be selected in 4 ways (e.g. M or M or F or F, following M in the first selection), so the total number of ways in which 2 persons can be selected is $5\times 4=20$. There are 3 ways of selecting a male on the first occasion, and associated with each of these, there are 2 ways of selecting a male on the second occasion. The total number of ways in which 2 males can be selected is therefore $3\times 2=6$. Thus, the required probability is $6/20=3/10$.

4.3 Rule of Addition

4.3.1 Mutually Exclusive Events

If two events '*A*' and '*B*' are mutually exclusive, that is they cannot occur together, the probability that one or the other occurs is simply the sum of their two separate probabilities, that is

$$P(A \text{ or } B) = P(A) = P(B) \tag{4.3}$$

Consider a single throw of a die. The probability of getting a score of 3 is 1/6; the probability of getting a score of 4 is also 1/6. Since 3 and 4 cannot occur on the same throw, the probability of getting 3 or $4=1/6+1/6=2/6$, which agrees with the value obtained from the basic definition.

4.3.2 Events Not Mutually Exclusive

In the preceding paragraph, we considered mutually exclusive events. When this is not the case, if '*A*' and '*B*' can occur together on some occasions, there is a need to modify the previous formula (Table 4.2).

$$\begin{aligned} &\text{The modified version reads as } P(A \text{ or } B \text{ or both}) \\ &\quad = P(A) + P(B) - P(AB) \end{aligned} \tag{4.4}$$

Suppose, in a clinical trial, the proportion of males, *P*(*A*) is 0.6, and the proportion of patients aged less than 45 years (young), *P*(*B*) is 0.8. We wish to determine the probability that a patient is either male or young (or both), as such patients are believed to be more tolerant to the drugs proposed for administration. If we apply the formula from the preceding section, we would get $(0.6+0.8=1.4)$, which is absurd because probability values must always lie between 0 and 1. The absurdity is because patients who are male and young (*AB*) are counted twice, once in the proportion of 0.6 and again in the proportion of 0.8. If we know the proportion (male and young), we can subtract it from the value of 1.4 above (0.8 +0.6) and get the required probability of male or young (or both). If we do not know the proportion but can assume that the sex and age are independent, the probability of male and young can be derived as $0.8\times 0.6=0.48$ (from the rule of multiplication). Thus, the probability of either male or young (or both) becomes $1.4-0.48=0.92$. We shall next consider a laboratory example. The following table gives the number of contaminated cultures from 00 patients, each of whom had three sputum specimens examined by culture for TB (Table 4.3).

Of the $(n=500)$ patients, the number who had no contaminated culture is $(n=364)$; the proportionate frequency for 0 contaminated cultures is $364/500=0.728$, so the probability of

TABLE 4.2

Age-Wise Distribution of SARS–CoV-2 Cases

Age Sex	Young	Old	Total
Male	0.48	0.12	0.60
Female	0.32	0.08	0.40
Total	0.80	0.20	1.00

TABLE 4.3

Distribution of SARS–Cov-2 Cases

No. of Contained Culture	No. of Patients	Proportionate Frequency (*f*)
0	364	0.728
1	122	0.244
2	13	0.026
3	1	0.002

having '0' contaminated culture is 0.728. The proportionate frequency for 2 or 3 contaminated cultures is (13 + 1)/500 = 0/028, i.e., the probability of obtaining at least 2 contaminated cultures out of 3 examined is 0.028.

Illustrations

1. A man is given 10 keys of which only one fits the lock on his door. He tries them successively; no key is tried twice. What is the probability that he will be successful on the third attempt? If the same key is tried more than once, how would this probability change?

Solution

Success on the third attempt implies failure on the first and second attempts.

Attempt No. 1: Probability (Failure) = 9/10, as 9 keys do not fit the lock.
Attempt No. 2: Probability (Failure) = 8/9, as there are 9 keys left, and 8 of these are unsuitable.
Attempt No. 3: Probability (Success) = 1/8, as there are 8 keys left, and only one is suitable.

Probability (Failure, Failure, Success) = $9/10 \times 8/9 \times 1/8 = 1/10$; that is, success on the third attempt has a probability of 1/10 = 0.1.

If the same key is tried more than once (for instance, during an electricity breakdown), the probability corresponding to (a) will remain 9/10, that corresponding to (b) will be 9/10, and that corresponding to (c) will be 1/10. Thus, the probability of success at the third attempt = $9/10 \times 9/10 \times 1/10 = 0.081$.

2. Among a group of 10 persons (6 males and 4 females), one is randomly selected, and then a second person is randomly selected from the remaining 9. What is the probability that the person selected the second time is male?

Solution

The event of interest is selecting a male on the second occasion. This can materialize in two ways; it could be preceded by the selection of a male on the first occasion, or by a female. We must compute the probabilities of these two ways, and add them up.

The probability of selecting a male on both occasions = $6/10 \times 1/9 = 30/90$ (a)
The probability of selecting a female on the first occasion and a male on the second occasion = $4/10 \times 6/9 = 24/90$ (b)
The probability of selecting a male on the second occasion (a) + (b) = 30/90 + 24/90 + 54/90 = 6/10 = 0.6

3. Of 500 tuberculosis patients who received a regimen of streptomycin and isoniazid, 100 complained of side effects to one or both drugs. It is known that, in all, 75 had side effects from streptomycin and that 25 of these had side effects from isoniazid also. Can you estimate the total number of patients who had side effects from isoniazid?

Solution

Let us denote side effects of streptomycin as event A and side effects of isoniazid as event B.

$P(A \text{ or } B \text{ or both}) = 100/500$; $P(A) = 75/500$; $P(AB) = 25/500$.
We know that
$P(A \text{ or } B \text{ or both}) = P(A) + P(B) - P(AB)$
i.e. $100/500 = 75/500 + P(B) - 25/500$
$P(B) = 100/500 - 75/500 + 25/500$
$= (100 - 75 + 25)/500$
$= 50/500$
Therefore, the total number of patients with side effects of isoniazid = 50 (Table 4.4).
Steps or working rule
$P(A') = 1 - P(A)$
$= 1 - 0.5$
$= 0.50$
$P(B') = 1 - P(B)$
$= 1 - 0.4$
$= 0.60$
$P(A \cap B) = P(A).P(B) = 0.50 * 0.40$
$= 0.70$
$P(A \Delta B) = P(A) + P(B) - P(A \cap B)$
$= 0.50 + 0.40 - 2*0.20$
$= 0.50$
$P((A \cup B)') = 1 - P(A \cup B)$
$= 1 - 0.7$
$= 0.30$
$P(A \text{ occur but Not } B) = P(A)*1 - P(B) = (1 - 0.50)* 0.40 = 0.20$
$P(B \text{ occur but Not } A) = 1 - P(A)) \times P(B) = (1 - 0.50)* 0.40 = 0.20$

TABLE 4.4

Probability Laws and Venn Diagrams

Index No	Law of Probability	*P*-Value	Diagram
1	Probability of A not occurring: $P(A')$	0.5	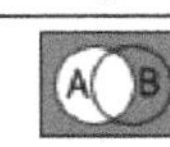
2	Probability of B not occurring: $P(B')$	0.6	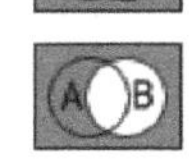
3	Probability of A and B both occurring: $P(A \cap B)$	0.2	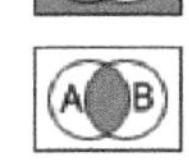
4	Probability that A or B or both occur: $P(A \cup B)$	0.7	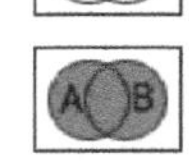
5	Probability that A or B occurs but not both: $P(A \Delta B)$	0.5	
6	Probability of neither A nor B occurring: $P((A \cup B)')$	0.3	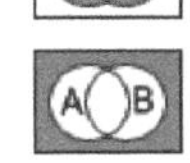
7	Probability of A occurring but not B:	0.3	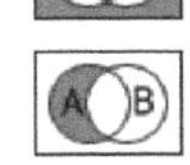
8	Probability of B occurring but not A:	0.2	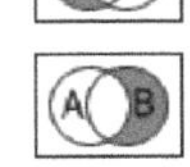

4.4 Probability of Two Events

Probability is the measure of the likelihood of an event occurring. It is quantified as a number between 0 and 1, with 1 signifying certainty and 0 signifying that the event cannot occur. It follows that the higher the probability of an event, the more certain it is that the event will occur (Dar et al., 2020). In its most general case, probability can be defined numerically as the number of desired outcomes divided by the total number of outcomes. This is further affected by whether the events being studied are independent, mutually exclusive, or conditional, among other factors (Garcia et al., 2022). The calculator provided computes the probability that an event *A* or *B* does not occur, the probability *A* and/or *B* occur when they are not mutually exclusive, the probability that both events *A* and *B* occur, and the probability that either event *A* or event *B* occurs, but not both (Horsten et al., 2023.

4.4.1 Complement of *A* and *B*

Given a probability **A**, denoted by *P*(*A*), it is simple to calculate the complement, or the probability that the event described by *P*(*A*) does not occur, *P*(*A*′). If, for example, *P*(*A*)=0.65 represents the probability that Bob does not do his homework, his teacher Sally can predict the probability that Bob does his homework as follows:

$$P(A')=1-P(A)=1-0.65=0.35$$

Given this scenario, there is, therefore, a 35% chance that Bob does his homework. Any *P*(*B*′) would be calculated in the same manner, and it is worth noting that in the calculator above, it can be independent; i.e. if *P*(*A*)=0.65, *P*(*B*) does not necessarily have to equal 0.35, and it can equal 0.30 or some other number (Novaes et al., 2017).

4.4.2 Intersection of *A* and *B*

The intersection of events *A* and *B*, written as $P(A \cap B)$ or *P*(*A* and *B*), is the joint probability of at least two events, shown below in a Venn diagram. In the case where *A* and *B* are mutually exclusive events, $P(A \cap B)=0$. Consider the probability of rolling a 4 and a 6 on a single roll of a die; it is not possible (Lopez et al., 2022). These events would therefore be considered mutually exclusive. Computing $P(A \cap B)$ is simple if the events are independent. In this case, the probabilities of events A and B are multiplied. To find the probability that two separate rolls of a die result in 6 each time (Figure 4.1)

$$P(A\cap B)=P(A)*P(B) \tag{4.5}$$

$$P(A\cap B)=\frac{1}{6}*\frac{1}{6}=0.0278$$

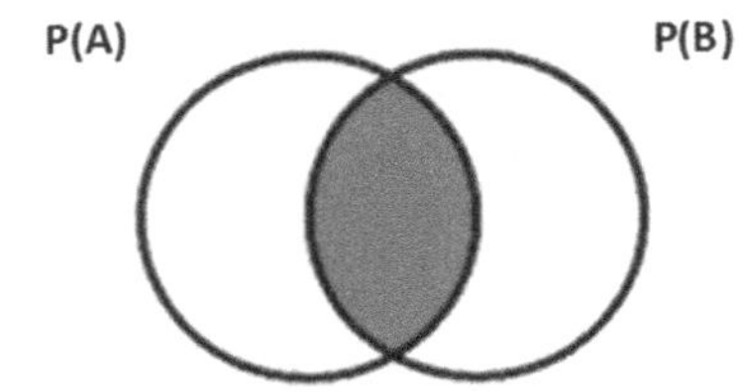

FIGURE 4.1 Venn diagram of intersection of events *A* and *B*.

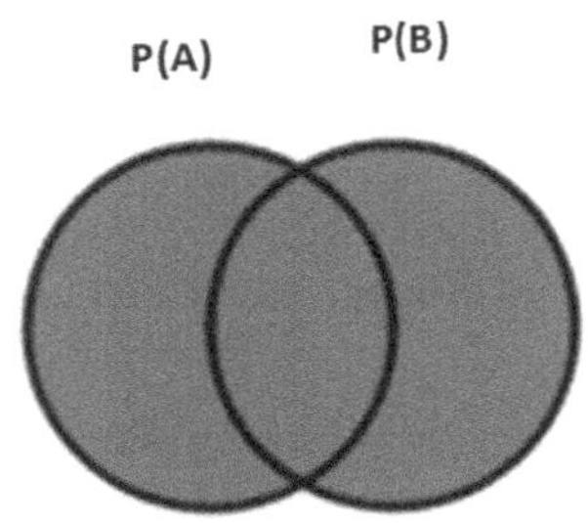

FIGURE 4.2 Venn diagram of the union of events *A* and *B*.

The calculator provided considers the case where the probabilities are independent. Calculating the probability is slightly more involved when the events are dependent and involves an understanding of conditional probability, or the probability of event A given that event B has occurred, P(A|B) (Lista et al., 2023). Take the example of a bag of 10 marbles, 7 of which are black, and 3 of which are black. Calculate the probability of drawing a gray marble if a black marble has been withdrawn without replacement (the black marble is removed from the bag, reducing the total number of marbles in the bag):

Probability of drawing a blue marble: *P* (*A*)=3/10
Probability of drawing a black marble: *P* (*B*)=7/10
Probability of drawing a black marble given that a blue marble was drawn: *P* (*B*|*A*)=7/9

As can be seen, the probability of drawing a black marble is affected by any previous event which is a black or blue marble was drawn without replacement. Thus, if a person wanted to determine the probability of drawing a blue and then a black marble from the bag,

the probability of drawing a blue and then a black marble using the probabilities calculated above:

$$P(A\cap B)=P(A)\times P(B\mid A)=(3/10)\times(7/9)=0.233$$

4.4.3 Union of *A* and *B*

In probability, the union of events, $P(A \cup B)$, essentially involves the condition where any or all of the events being considered occur, as shown in the Venn diagram below (Nevzoro et al., 2021). Note that $P(A \cup B)$ can also be written as *P*(*A* or *B*) (Figure 4.2). In this case, the "inclusive OR" is being used. This means that while at least one of the conditions within the union must hold true, all conditions can simultaneously be true. There are two cases for the union of events: the events are either mutually exclusive or not mutually exclusive. In the case where the events are mutually exclusive, the calculation of the probability is simpler (Figure 4.2)

$$P(A\cup B)=P(A)+P(B) \tag{4.6}$$

A basic example of mutually exclusive events would be the rolling of a die, where event A is the probability that an even number is rolled, and event B is the probability that an odd number is rolled. It is clear in this case that the events are mutually exclusive since a number cannot be both even and odd. Thus, $P(A \cup B)$ would be $3/6+3/6=1$, since a standard die only has odd and even numbers (Page et al., 2022).

The calculator above computes the other case, where events A and B are not mutually exclusive. In this case:

$$P(A\ U\ B) = P(A) + P(B) - P(A \cap B) \tag{4.7}$$

Using the example of rolling a die again, find the probability that an even number or a number that is a multiple of 3 is rolled. Here, the set is represented by the 6 values of the die, as shown in Table 4.5:

4.4.4 Exclusive OR of A and B

Another possible scenario that the calculator above computes is $P(A\ X \text{ or } B)$, as shown in the Venn diagram below (Rotondi et al., 2022). The "Exclusive OR" operation is defined as the event that A or B occurs, but not both simultaneously (Figure 4.3). The equation is as follows:

$$P(A \text{ or } B) = P(A) + P(B) - 2 * P(A \cap B) \tag{4.8}$$

As an example, imagine it is Halloween, and two buckets of candy are set outside a house, one containing Snickers and the other containing Reese's. Multiple flashing neon signs are placed around the buckets of candy, insisting that each trick-or-treater takes only one: Snickers OR Reese's, but not both! It is unlikely, however, that every child adheres to the flashing neon signs. Given a probability of Reese's being chosen as $P(A)=0.65$, or a probability of Snickers being chosen as $P(B)=0.349$, and a probability of **P**(unlikely)$=0.001$ that a child exercises restraint while considering the detriments of a potential future cavity, calculate the probability that Snickers or Reese's is chosen, but not both:

$$0.65 + 0.349 - 2 \times 0.65 \times 0.349 = 0.999 - 0.4537 = 0.5453$$

Therefore, there is a 54.53% chance that Snickers or Reese's is chosen, but not both.

TABLE 4.5
Distribution State of Sample Space

SAMPLE SPACE	$S=\{1,2,3,4,5,6\}$
Probability of an even number:	$P(A)=\{2,4,6\}=3/6$
Probability of a multiple of 3:	$P(B)=\{3,6\}=2/6$
Intersection of A and B:	$P(A \cap B)=\{6\}=1/6$
	$P(A \cup B)=3/6+2/6-1/6=2/3$

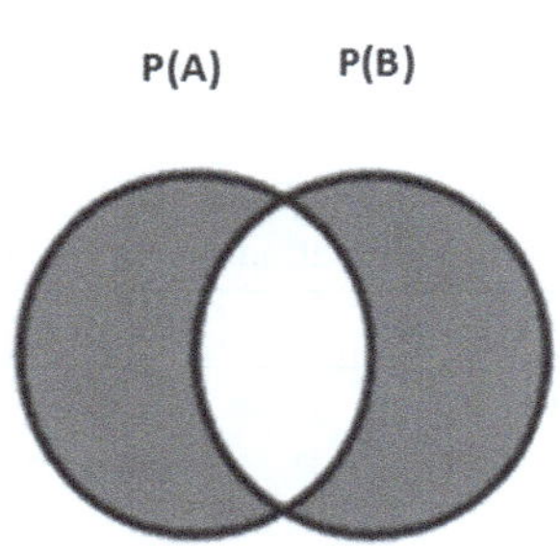

FIGURE 4.3 Venn diagram of exclusive OR ratio between events A and B.

4.5 Bayesian

The following is a general setup for a statistical inference problem: There is an unknown quantity that we would like to estimate (Rudas et al., 2009). We obtain some data and from the data, we estimate the desired quantity. In the previous chapter, we discussed the frequentist approach to this problem. In that approach, the unknown quantity θ is assumed to be a fixed (non-random) quantity that is to be estimated using the observed data. Consider a communication channel as shown in (Figure 4.4). We can model the communication over this channel as follows: at time n, a random variable X_n is generated and transmitted over the channel. However, the channel is noisy. Thus, at the receiver, a noisy version of X_n is received. More specifically, the received signal is

$$Y_n = X_n + W_n \tag{4.9}$$

$W_n \sim N\left(0,\sigma_i^{\,2}\right)$ is the noise added to X_n assumed that the receiver knows the distribution of X_n. The goal is to recover the value of X_n based on the observed value of Y_n.

The aim is to draw inferences about an unknown variable X by observing a related random variable Y. The unknown variable is modeled as a random variable X, with a **prior** distribution.

$$f_x(X) = \text{If, } X \text{ is continious}$$

$$P_x(X) = \text{If, } X \text{ is discrete}$$

After observing the value of the random variable 'Y', we determine the posterior distribution of 'x'. This is conditional PDF of X given $Y=y$.

$$f_{x/y}\left(\frac{x}{y}\right) = P_{x/y}\left(\frac{x}{y}\right) \tag{4.10}$$

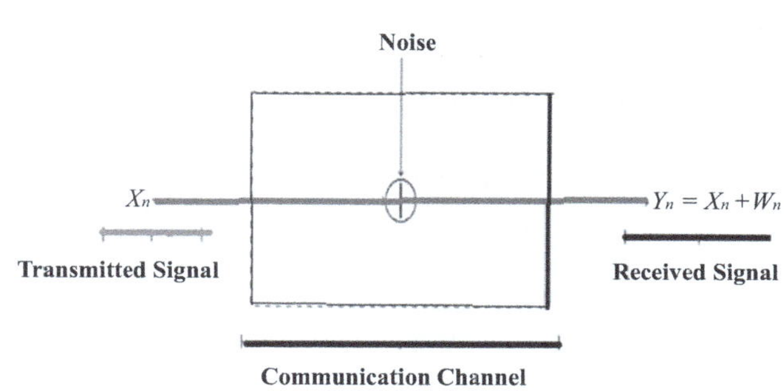

FIGURE 4.4 Noisy communication channel for random walk model.

The posterior distribution is usually found using Bayes' formula. Using the posterior distribution, we can then find point or interval estimates of 'X'.

Table 4.6 presents the Bayesian probability approach for a 2×2 cross table—numbers of patients with and without the disease who test positive and negative (Taori et al., 2019).

From Table 4.6, we estimated disease prevalence, test sensitivity, and test specificity (and, optionally, sample size): Prevalence = 0.432, Sensitivity (%) = 0.676, Specificity (%) = 0.208, LR+ = 0.85, and LR− = 1.56. Disease prevalence, positive likelihood ratio, and negative likelihood ratio (and, optionally, sample size) were simulated in Table 4.7 and Figure 4.5.

The probability of without replacement or dependent probability will depends upon mutually exclusive event (Figure 4.6).

Step 1: Draw the probability tree diagram and write the probability of each branch. (Remember that the objects are not replaced.)

Step 2: Identify all the available paths (or branches) for a particular outcome.

Step 3: Multiply along the branches and add vertically to find the probability of the outcome.

TABLE 4.6

Diagnostic Test for Disease Status 2×2 Table

Test status	Present	Absent	Total
Test +ve	25	38	63
Test −ve	12	10	22
Total	37	48	85

TABLE 4.7

Output of the Probability Model

Statistics	Estimated Values
Prior probability (odds)	44% (0.8)
Positive Test (+ve)	
Positive Likelihood ratio	0.85
95% confidence interval	[0.65, 1.11]
Posterior probability (odds)	40% (0.7)
95% confidence interval	[33%, 46%]
Negative Test (−ve)	
Negative Likelihood ratio	1.56
95% confidence interval	[0.76, 3.20]
Posterior probability (odds)	55% (1.2)
95% confidence interval	[37%, 71%]
(~1 in 2.2 with negative test	are well)

Odds = Probability/(1 − Probability), +LR = Sensitivity/(1 − Specificity), −LR = (1 − Sensitivity)/Specificity, Posterior Odds = Prior Odds × LR

Note: ~ 1 in 2.5 with positive test are sick.

Illustration

A jar consists of 21 sweets: 12 are green and 9 are blue. William picked two sweets at random (Table 4.8).

a. Draw a tree diagram to represent the experiment.
b. Find the probability that
   i. both sweets are blue.
   ii. one sweet is blue and one sweet is green.

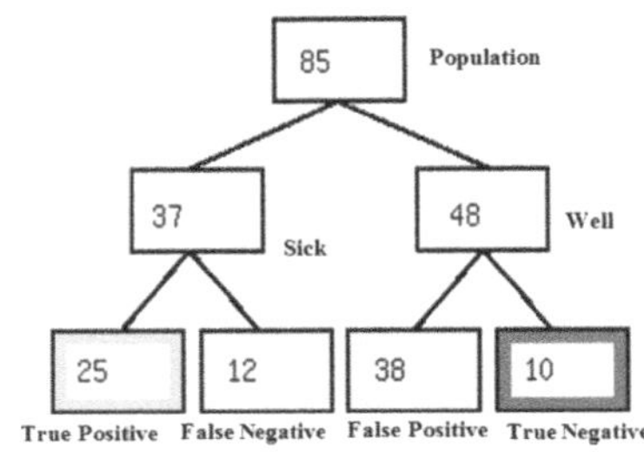

FIGURE 4.5 Flow chart of Bayes' probability working rule.

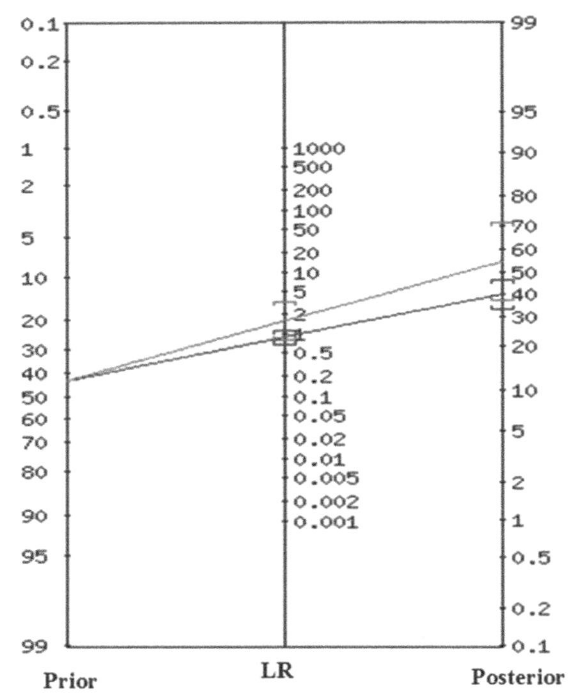

FIGURE 4.6 Likelihood ratios (LR) along with prior and posterior probabilities.

TABLE 4.8

Probability of Binding of AA for the Formation Alpha Helix

A type (*N*)	*B* type (*N*)
10	08
odds 5/4	odds 4/5

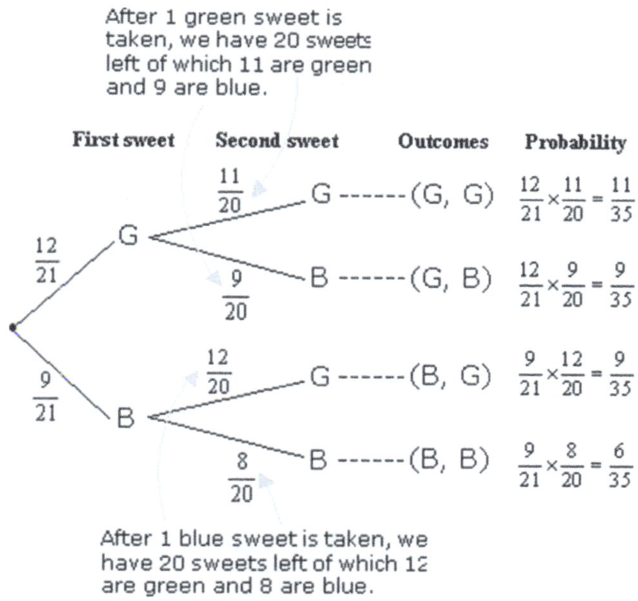

FIGURE 4.7 Flow tree chart that outlines the probable chances associated with the selection of an edible item (case 1).

c. William randomly took a third sweet. Find the probability that:
   i. all three sweets are green.
   ii. at least one of the sweets is blue.

Solution

a. Although both sweets were taken together, it is similar to picking one sweet and then the second sweet without replacing the first (Figure 4.7).
b. P (both sweets are blue) $= P(B,B)$

$$P(B,B) = \frac{9}{21} * \frac{8}{20} = \frac{6}{35}$$

ii. P(one sweet is blue and one is green) $= P(G, B)$ or $P(B, G)$

$$P(G, B) \text{or } P(B, G) = \frac{9}{35} + \frac{9}{35} = \frac{18}{35}$$

c. i. P(all three sweets are green) $= P(G, G, G)$

$$P(G, G, G) = \frac{12}{21} * \frac{11}{20} * \frac{10}{19} = \frac{22}{133}$$

ii. P(at least 1 sweet is blue) $= 1 - P$(all three sweets are green)

$$1 - \left(\frac{22}{133}\right) = \left(\frac{111}{133}\right)$$

Probability of binding
$P_b = A/(A + B)$
$= 10/(10 + 8)$
$= 10/18$
$= 0.556$
Chance = 55.6%
Probability of non-binding
$P_n = B/(A + B)$
$= 8/(10 + 8)$
$= 8/18$
$= 0.444$
Chance = 44.4%

Illustration

Bag A contains 10 marbles, of which 2 are red and 8 are black. Bag B contains 12 marbles, of which 4 are red and 8 are black. A marble is drawn at random from each bag.

a. Draw a probability tree diagram to show all the outcomes of the experiment.
b. Find the probability that:
   i. both marbles are red.
   ii. both marbles are black.
   iii. one marble is black and one is red.
   iv. at least one marble is red.

Solution

a. A probability tree diagram that shows all the outcomes of the experiment (Figures 4.8 and 4.9).
   i. Both are red.
      $P(R, R) = 1/15$
   ii. Both are black.
      $P(B, B) = 8/15$
   iii. One is black and one red.
      $P(R, B)$ or $P(B, R) = 2/15 + 4/15 = 2/5$
   iv. At least one is red.
      $1 - P(B, B) = 1 - 8/15 = 7/15$

Illustration: A box contains 4 red and 2 blue chips. A chip is drawn at random and then replaced. A second chip is then randomly drawn (Figures 4.10 and 4.11).

a. Show all the possible outcomes using a probability tree diagram.
b. Calculate the probability of getting:
   i. At Sleast one blue.
   ii. One red and one blue.
   iii. Two chips of the same color.
      An above figure denotes the success of outcome of an experiment depends on exhaustive condition (case 2)
b. The probability of getting:
   i. at least one blue
      $P(R, B)$ or $P(B, R)$ or $P(B, B) = 2/9 + 2/9 + 1/9 = 5/9$
   ii. one red and one blue
      $P(R, B)$ or $P(B, R) = 2/9 + 2/9 = 4/9$
   iii. two of the same color
      $P(R, R)$ or $P(B, B) = 4/9 + 1/9 = 5/9$

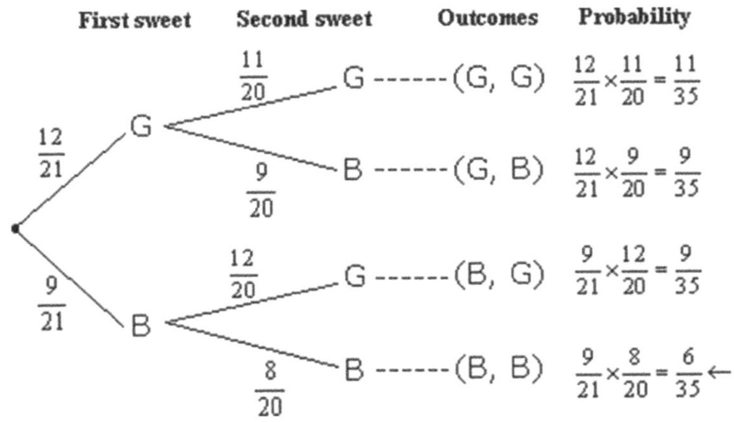

FIGURE 4.8 Probable chances for selecting an edible sweet (case 2).

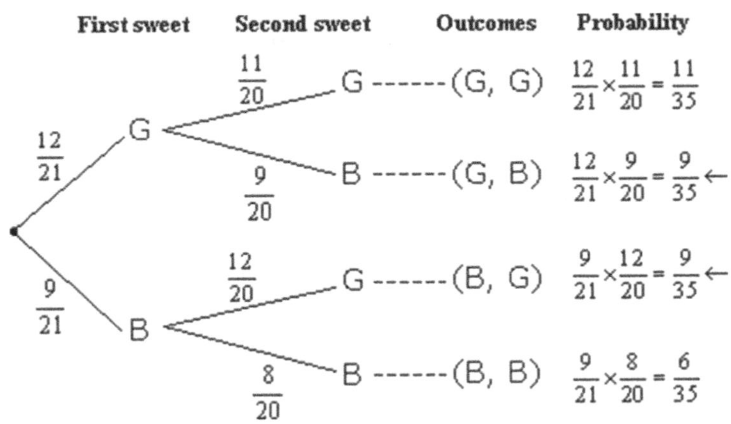

FIGURE 4.9 Highlights the probable chances for selecting an edible sweet (case 3).

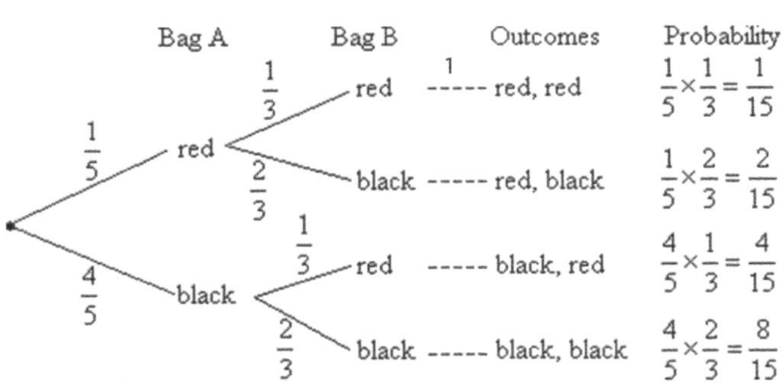

FIGURE 4.10 Anticipated success of the outcome of Experiment (Case 1).

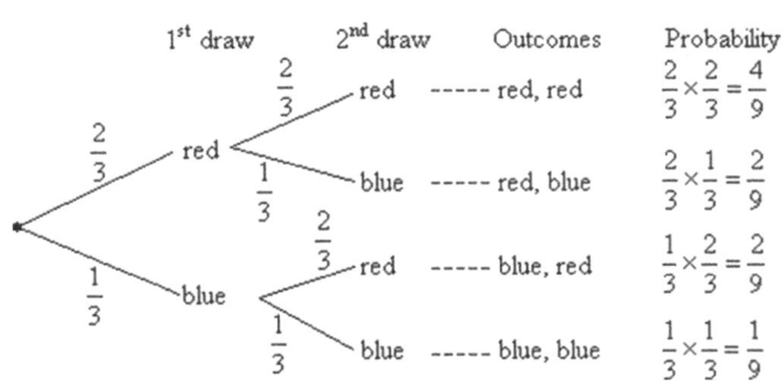

FIGURE 4.11 Expected success of the outcome of Experiment (Case 2).

The list gives the formulas for the probability of independent and dependent events.

Independent event: The outcome of one event does not affect the outcome of another. *A* and *B* are independent and the probability of occurrence of P(AB) is $P(AB) = P(A) * P(B)$

Dependent event: The outcome of one event affects the outcome of another. *A* and *B* are dependent and occurring is $P(A \text{ and } B) = P(A) * P(B / A)$

Illustration

A purse contains four \$5 bills, five \$10 bills, and three \$20 bills. Two bills are selected without the first selection being replaced. Find *P*(\$5, then \$5).

Solution

There are four \$5 bills.

There are a total of 12 bills.

P(\$5)=4/12

The result of the first draw affected the probability of the second draw.

There are three $5 bills left.

There are a total of 11 bills remaining.

P($5 after $5) =3/11, P($5, then $5)=$P$($5) · P($5 after $5)=

$$\left(\frac{4}{12} * \frac{3}{11} = \frac{1}{11}\right)$$

4.6 Practical Application of Theory of Probability

The theory of probability is one of the important statistical measures for estimating the likelihood and degree of uncertainty. It is a way of quantifying how likely something is to happen in the future. A probability of 1 for an event means that it is guaranteed to happen, while a probability of '0' for an event means that it will not happen. In some genetics problems, we may need to calculate the probability that any of several events will occur. In this case, we need to apply another rule of probability, the sum rule. According to the sum rule, the probability that any of several mutually exclusive events will occur is equal to the sum of the individual probabilities of the events. As an example, let's use the sum rule to predict the fraction of offspring from an $Aa \times Aa$ cross that will have the dominant phenotype (AA or Aa). In this cross, three events can lead to a dominant phenotype. Every day, we experience the practical application of probability without noticing them ourselves. Examples of the practical applications of probability include predicting the outcome when we flip a coin, choosing a card from a deck, throwing a die, pulling a green candy from a bag of red candies, and winning a lottery among many millions. Another important milestone we achieved in the 21st century is the forecasting and assessment of how likely an event can occur in terms of percentage and recording the risk associated with weather using artificial intelligence tools (AIs), seismic waves of earthquakes, sports strategies, agricultural risk assessment, survival of patients after therapy, evidence-based medicine, quantum theory, collision estimation in space research, psychological intervention research, aqtuarial science, public health, animal and veterinary science, computational biology, bioinformatics, medical and engineering science, the development of large language models in machine learning, and, irrespective of the interdisciplinary field, probability may be applied (van Schuppen et al., 2021; Weirich et al., 2021 & Zorzos et al., 2023).

Practical Exercise

Illustration 1

An insurance company insured 2,000 scooter drivers, 4,000 car drivers, and 6,000 truck drivers. The probabilities of an accident involving a scooter driver, a car driver, and a truck driver are 0.01, 0.03, and 0.015, respectively. One of the insured persons meets with an accident. What is the probability that they are a scooter driver? **Answer 5/7**

Illustration 2

In a study, physicians were asked what the odds of breast cancer would be in a woman who was initially thought to have a 1% risk of cancer but who ended up with a positive mammogram result (a mammogram accurately classifies about 80% of cancerous tumors and 90% of benign tumors). Ninty-five out of a hundred physicians estimated the probability of cancer to be about 75%. Do you agree?

Illustration 3

Suppose we have 3 cards identical in form, except that both sides of the first card are colored red, both sides of the second card are colored black, and one side of the third card is colored red and the other is colored black. The 3 cards are mixed up in a hat, and 1 card is randomly selected and put down on the ground. If the upper side of the chosen card is colored red, what is the probability that the other side is colored black? **Answer -1/3**

Illustration 4

It is estimated that 50% of emails are spam emails. Some software has been applied to filter these spam emails before they reach your inbox. A certain brand of software claims that it can detect 99% of spam emails, and the probability of a false positive (a non-spam email detected as spam) is 5%. **Answer 5/104**

Illustration 5

A coin is thrown 3 times. What is the probability that at least one head is obtained? **Answer 7/8**

Illustration 6

There are 5 green and 7 red balls. Two balls are selected one by one without replacement. Find the probability that the first is green and the second is red? **Answer 35/132**

Illustration 7

Three dice are rolled together. What is the probability of getting at least one 4? **Answer 91/216**

Illustration 8

A problem of breeding selection involves three genotypes, P, Q, R, whose respective chances of solving it are 2/7, 4/7, and 4/9, respectively.What is the probability that the problem is solved by selecting the best genotype? **Answer 122/147**

Illustration 9

Two dice are thrown together. What is the probability that the number obtained on one of the dice is a multiple of the number obtained on the other die? **Answer 11/18**

Illustration 10

Find the probability that a leap year has 52 Sundays. **Answer 5/7**

Hint: A leap year can have 52 Sundays or 53 Sundays. In a leap year, there are 366 days, out of which there are 52 complete weeks and 2 remaining days. These two days can be (Sat, Sun), (Sun, Mon), (Mon, Tue), (Tue, Wed), (Wed, Thur), (Thur, Friday), and (Friday, Sat). So there are 7 cases, out of which (Sat, Sun) and (Sun, Mon) are the two favorable cases. Therefore, P(53 Sundays)=2/7

$$P(52) + P(53) = 1$$

$$P(52S) = 1 - 2/7 = 5/7$$

REFERENCES

Ackerman, N. L., Freer, C. E. & Roy, D. M. On the computability of conditional probability. *Journal of the ACM* **66**, 1–40 (2019).

Batanero, C. & Álvarez-Arroyo, R. Teaching and learning of probability. *ZDM - Mathematics Education* **56**, 5–17 (2024).

Bordalo, P., Conlon, J. J., Gennaioli, N., Kwon, S. Y. & Shleifer, A. Memory and probability. *Quarterly Journal of Economics* **138**, 265–311 (2023).

Bastianello, L. & Vergopoulos, V. About subjective probability. *Revue Economique* **74**, 601–612 (2023).

Dar, P. A. & Afroz. in *Machine Learning and Big Data: Concepts, Algorithms, Tools and Applications*, 31–51 (Wiley, 2020). https://doi.org/10.1002/9781119654834.ch2

García-Bertrand, R., Baringo, L. & García-Cerezo, Á. in *Encyclopedia of Electrical and Electronic Power Engineering: Volumes* 1–3 **1**, V1-734–V1-746 (Elsevier, 2022).

Horsten, L. & Brickhill, H. Sets and probability. *Erkenntnis* (2023). https://doi.org/10.1007/s10670-023-00670-x

Novaes, M. Probability distributions. *Revista Brasileira de Ensino de Fisica* **44**, e20210424 (2022).

Mwangi, S. W. Probability and non probability sampling. *Asian Research Journal of Business Management* **4**, 547–553 (2017).

López-Astorga, M., Ragni, M. & Johnson-Laird, P. N. The probability of conditionals: A review. *Psychonomic Bulletin and Review* **29**, 1–20 (2022).

Lista, L. in Roberta Citro (ed.), *Lecture Notes in Physics* **1010**, 17–35 (Springer Science and Business Media Deutschland GmbH, 2023).

Nevzorov, V. B., Ahsanullah, M. & Ananjevskiy, S. *Probability Theory: A Logic of Science. Probability Theory: A Logic of Science*, 1–235 (Nova Science Publishers, Inc., 2021). https://doi.org/10.1063/1.1825273

Page, D. N. Possibilities for probabilities. *Journal of Cosmology and Astroparticle Physics* **2022**, (2022).

Rotondi, A., Pedroni, P. & Pievatolo, A. in *UNITEXT - La Matematica per il 3 piu 2* **139**, 1–37 (Springer-Verlag Italia s.r.l., 2022).

Rudas, T. in *International Encyclopedia of Education*, Third Edition, 378–382 (Elsevier, 2009). https://doi.org/10.1016/B978-0-08-044894-7.01359-2

Taori, P., Mamidipudi, S. & Agrawal, D. in *International Series in Operations Research and Management Science* **264**, 945–963 (Springer New York LLC, 2019).

van Schuppen, J. H. in *Communications and Control Engineering*, 13–56 (Springer Science and Business Media Deutschland GmbH, 2021). https://doi.org/10.1007/978-3-030-66952-2_2

Weirich, P. Subjective probability. *Rivista di Filosofia* 2023, 53–64 (2023).

Wijaya, A., Elmaini & Doorman, M. A learning trajectory for probability: A case of game-based learning. *Journal on Mathematics Education* **12**, 1–16 (2021).

Zorzos, M. & Avgerinos, E. Research on visualization in probability problem solving. *Eurasia Journal of Mathematics, Science and Technology Education* **19**, em2247 (2023).

5

Recent Probability Distribution and Statistical Methods in Life Science Research

5.1 Introduction

Because life science can study living things and their biochemical structure, environmental functions, growth, origins, evolution, and biological distribution status, it is regarded as one of the leading scientific disciplines. It is defined as 'the life process of a group or category of living organisms', to put it another way. According to the practical uses of science, life science will be expanding its frontiers for cutting-edge discoveries using modern analytical techniques (equations derived from various sciences and supplementary statistical tools and methods) to address practical issues in a variety of scientific fields, including genetics cell biology, developmental biology, computational biology, biomedical engineering, and translational health sciences. Advanced statistical techniques are crucial for researchers to introduce new concepts in statistical research globally, especially given the extraordinary growth of life science. Life science research has no boundaries and is conducted in a vast structure encompassing the following areas or subjects (Figure 5.1). Too many postulates and theorems will be demonstrated and derived by a new scientific formation every day. This life science area is broad in scope and focuses mostly on theory and real-world applications. Every section mentioned above has a strong connection to life science (Figure 5.1); science is always changing and developing to explore and identify its mother root. We will start the important subsection of this book by thinking about the topics mentioned above. Currently, the book's structure and design are based on five main subjects: genetics, bioinformatics, molecular biology, biochemistry, and biotechnology. New statistical techniques were used to demonstrate the actual data, and their derivations are presented methodically with no overlap in the analytical formulas.

5.1.1 Application of Statistical Science

It is the field of study that is a master of none and a jack of all disciplines when it comes to statistical science. Our experimental results will not be valid if the hypothesis is not tested and statistical analyses are not performed. We will try to explore new interventions for the alteration of data in biological sciences. These days, single nucleotide polymorphisms (SNPs) are used to identify rare diseases in both human and animal populations, and genome-wide analyses (GWA) and genomics research are among the micro-level research areas that heavily utilize modern statistical methods. According to next-generation sequencing (NGS) for HIV/AIDS and cancer detection, among other diseases, the assessment of epigenetic factors on various disease traits (malignancy, genetic disorders) is a scourge. Additionally, by integrating cell divisions,

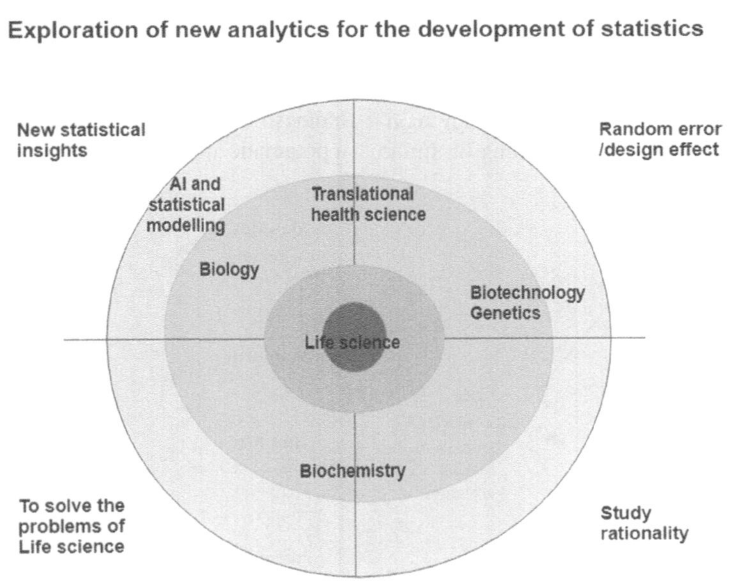

FIGURE 5.1 Tree chart of Life Science.

DOI: 10.1201/9781032713816-5

the model-based study will quantify proteomics for genetic disorders and spatial phenotyping for the identification of mutant genes causing cancer. For example, statistical methods or tools will be critical to describe the following: mapping of sequence DNA, sequence RNA, and mi-RNA; traits assessment in population genetics, spatial phenotyping for geotagging of breeds in plant and animal breeding programs; evolutionary biology for translational health sciences; and image processing techniques for decision support. This practical strategy will support sophisticated statistical techniques so that scientists can discover new information and act appropriately when making decisions. We try to formulate the most appropriate and dependable statistical methods for data compilation and interpretation in relation to the five major subjects listed above to improve science at the reader level and target better compilation.

5.2 Why Is Advanced Statistics Necessary for Life Science?

We have asked all life scientists to think about this. There is a really important reason for this. Statistics are used in practically all fields (medicine, molecular biology, computational biology, bioinformatics, etc.) in real life and can be highly sophisticated. Understanding the various experiments and studies related to biology is sometimes essential to performing precise and high-quality statistical inferences about those experiments and studies. Each of the aforementioned fields has unique methodologies and experiments that address various research issues. Advanced statistical techniques will be crucial for analytical purposes to create a distinct field of computational biology and statistics that can be included in life science. To be proficient in statistical procedures (biostatistics), one must be aware of both life science and statistical methods. In biostatistics, biological knowledge is frequently underestimated, despite its potential importance in academic or research settings. Proficiency in statistical applications entails proficiency in both life science and statistics. Here's an illustration of how statistics are used in life sciences, such as genetics. While performing analysis in this field, any analyst or statistician would be completely unaware of the practical applications of genetic codes, homology models of codons for amino acid identification, binding of amino acids with retention time, grouping of seq-DNA, seq-RNA omics data, NGS and gene identification by spatial phenotyping, cross-validation of seq-DNA, and Chip-seq by modeling techniques, which are all understood by many life scientists. For the most part, the life sciences share this scenario. This is particularly significant for human biology and biomedicine, which are crucial for today's pharmaceutical sector and medical research. Numerous variables can confuse biochemical experiments, such as the thousands of relevant genes, phenotypes, metabolites, and physical traits. Any practicing scientist must comprehend them. Any of these could be a significant latent variable, confounding factor, or variable that needs to be adjusted for or disregarded when using statistical techniques. High throughput multifaceted wide genomic sequencing and molecular sequential big datasets have revolutionized biomedical research and are promising for evidence-based medicine in many life-threatening diseases like HIV, cancer, and other genetic disorders, according to the technological transformation and advancement of molecular studies. This breakthrough in biotechnological research has made it possible for us to generate large-scale, structured, and unstructured datasets on cloud platforms, which poses a significant quantitative challenge for biomedical and bioinformatics research. To address these issues, the field of computational biology and bioinformatics (CBB) was founded in 1990. It consists of multiple biological and quantitative scientists working together on a single platform to manage massive amounts of both qualitative and quantitative data series. The vast amount of information they hold has been extracted, and they have also developed future methodological work-based environments based on the statistical principles clarified to address biological research problems more comprehensively by using all available data to unearth novel biological insights. In addition to varying probability and various statistical measures, which are crucial in these endeavors, data scientists have developed a strong commitment to understanding variability and quantification. In light of this, the current chapter has made an effort to investigate novel analytical techniques to address pressing issues related to transcriptomics, NGS, proteomics, WGAs, SNPs for specific diseases, modeling of potential genes for fatal illnesses, etc. Our predictive model-based approach will not only help researchers comprehend different scientific challenges in computational biology but will also initiate several research programs based on a pragmatic approach (Figure 5.2).

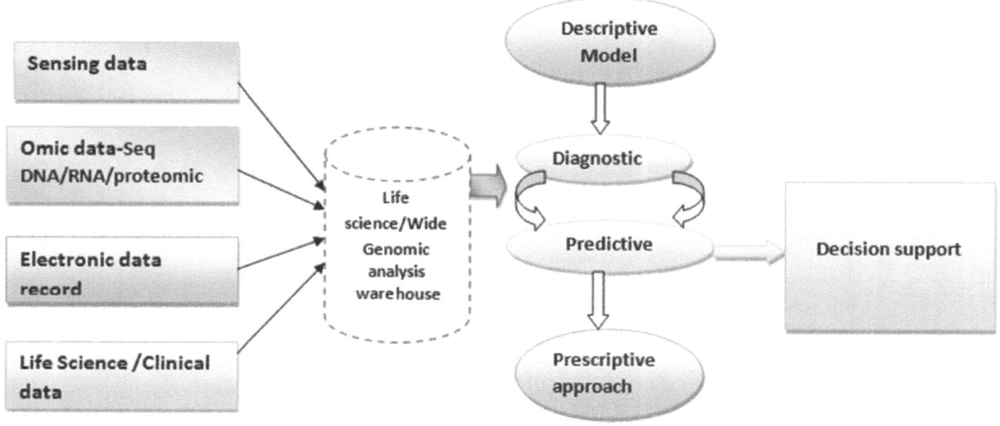

FIGURE 5.2 Flowchart detailing the application of statistical methods and their role in decision support.

5.1.1 Objectives of This Chapter

i. The aim of life science is to establish fundamental statistical principles that eagerly govern biological systems. Such instruments have been necessary for life science to determine biological systems. In the fields of genetics, bioinformatics, molecular biology, biochemistry, and biotechnology, statistics can be used to describe the correlation between many biological and life-defining variables. These variables should be tested using the appropriate techniques necessary for testing the hypothesis. In this regard, the currently proposed chapter aims to develop several statistical models. These developed techniques can be of great use to life scientists in planning different types of research projects and performing more accurate testing of hypotheses.

ii. The aforementioned fields—seq-DNA, seq-RNA, and seq-peptides—are widely employed, and the advancement of science greatly depends on current analytical research methods like Bayesian naive, machine learning, and simulation techniques. The formulated models need to be developed by life scientists to estimate gene expression variation in homology systems. This type of omics data expression may require advanced novel statistical techniques. The secondary goal of this pragmatic approach is to advance current science and open up new avenues of inquiry.

iii. Life science can make use of the gathering of qualitative and quantitative traits. During the investigation, some haphazard deviations were observed. These deviations have been attributed to significant changes in the experimental results. Early detection of variability in omics data for proteomics (expressive, functional, and structural), NGS for rare diseases, and pyrosequencing of high-throughput data (Illumina) are challenging tasks for specialists. The third objective is to develop suitable stochastic convergence modeling in this context by means of multiple mathematical iterations. With these methods, researchers will be able to ascertain the variability and derive valuable interpretations in a timely manner.

iv. By utilizing the Bootstrap model to extract WGA and multiple sequencing alignments (MSA) from omics data using the programming languages C++ and Java, the fourth goal aims to present new research and insights at the conclusion.

v. The fifth goal is to outline the key elements and various statistical measures, such as nested models, fixed, random, mixed, and multivariate analysis with repeated measures, as well as generalized linear and nonlinear models illustrated for the reader. This objective helps us build confidence among life scientists and other professionals. This model has been treated as a remarkable tool around the world.

There are two parts to this chapter's section. Advanced statistical measures and their application to the life sciences are the sole focus of the first section. The fundamentals of experimental design, hypothesis testing, estimation theory, and descriptive methodologies are covered in the forthcoming section in separate, connected chapters. Using weighted parameters, this section will be useful for testing hypotheses in small or large observations. Through the use of numerous illustrations, eye-catching diagrams, and figures, each section in all the chapters accurately depicts the entire crust and analytical techniques. Readers are also provided with an overview of the literature that has been published in the last 10 years. Each illustration makes use of very basic mathematical derivations. The concept of statistics should be simple for the reader.

5.2.2 Previous Work Conducted (Review of Literature)

This section makes an effort to correlate the various research articles that have been published globally; both stacked and unstacked published articles are carefully reviewed in light of the aforementioned key goals. The following clear presentation of relevant and significant literature describes the general structure of the book. A deep learning model (DLM) was created by Zhang et al. (2021a) to predict the depth of NGS using DNA sequences. The researchers developed statistical models that comprised a bidirectional recurrent neural network that receives as input the identities of DNA nucleotides as well as the estimated probabilities of nucleotides; it has been discovered that the Seqexpression is unpaired. The DLM was used simultaneously in three different NGS panels: a 2000-plex panel for human long non-coding RNA (lncRNA), a 7373-plex panel for non-human sequences targeted for storing DNA information, and a 39145-plex panel for human SNPs. According to the cross-validation, the developed DLM model predicted sequencing depth by a factor of three and had 99% accuracy for the non-human and 93% accuracy for the SNPs. In the independent testing method, the DLM predicted the lncRNA panel with (89.0%) accuracy after being trained on the SNPs. It was demonstrated that they could reasonably forecast the single-plex kinetic rate constants of DNA hybridization and strand displacement using the same model. This study is model-based and will help the reader classify high-throughput DNA sequencing using the primary genomics approach. Morris & Baladandayuthapani (2017) investigated high-throughput multi-platform genomics technologies providing whole genome molecular (WGM) biological samples in another study along these lines. This newer technology has produced large datasets at high throughput. They suggested that the research study should focus on the following four main areas: (i) experimental design and reproducibility; (ii) pre-processing and feature extraction; (iii) unified modeling; and (iv) structure of integrations. They highlighted some key contributions in each of these interventions and described the various statistical ideas that form the basis of these strategies. In addition to highlighting the important contributions statisticians have made to bioinformatics, the research goal has motivated statisticians to get involved early in creating new technologies that will resemble methodological works based on statistical principles in the future. Baladandayuthapani et al. (2010) carried out a high-throughput comparative genomic hybridization analysis to investigate the genetic causes of cancer. The resulting datasets give a cytogenetic representation of relative DNA copy number and are composed of log fluorescence ratios as a function of genomic DNA location. When analyzing such data, several state variables were usually expressed at each location, and different DNA segments with a similar copy number of states were analyzed.

One sample's worth of state variables at a time was mathematically modeled, and common regions on the copy of several aberrations were investigated. To model aCGH data, the current study suggested a hierarchical Bayesian random segmentation approach that employed cross arrays from a shared population to produce various segments of a shared copy of several changes. The underlying population characterizes these changes, which made it possible for the researcher to compare the various population aCGH profiles and determine which genomic regions have differential alterations. The hypothesis was proved based on a unified Bayesian hierarchical model approach to obtain the transient probabilities of alteration of states and corresponded to the local false discovery rates (LFD) of both single and multiple groups. This Bayesian detection method identified shared aberrations in a-CGH (BDS AScgh) for long-read datasets. Dias & Torkamani (2019) studied artificial intelligence (AI) systems that can perform tasks that normally require human intelligence. Advances in AI software and hardware, especially deep learning algorithms and the graphics processing units (GPUs) that power their training have led to a recent and rapidly increasing interest in medical AI applications. In clinical diagnostics, AI-based computer vision approaches are poised to revolutionize image-based diagnostics, while other AI subtypes have begun to show similar promise in various diagnostic modalities. In some areas, such as clinical genomics, a specific type of AI algorithm known as deep learning is used to process large and complex genomic datasets. In this review, they summarized the main classes of problems that AI systems are well suited to solve and described the clinical diagnostic tasks that benefited from these solutions. Next, they focused on emerging methods for specific tasks in clinical genomics, including variant calling, genome annotations, variant classifications, phenotype-to-genotype correspondence, etc. Finally, the research intervention discussed the future potential of AI in individualized medicine applications, especially for risk prediction in common complex diseases, and the challenges, limitations, and biases that must be carefully addressed for the successful deployment of AI in medical applications, particularly those utilizing human genetics and genomics datasets.

Rajkomar et al. (2018) demonstrated predictive modeling using electronic health record (EHR) data, which is anticipated to drive personalized medicine and improve healthcare quality. The predictive statistical model's construction typically requires the extraction of curetted predictor variables from normalized EHR data. This labor-intensive process discards the vast majority of information in each patient's record. The study proposed a representation of patients' entire raw EHR records based on the Fast Healthcare Interoperability Resources (FHIR) format, and EHR demonstrated deep learning methods using this representation, which are capable of accurately predicting multiple medical events from multiple centers without site-specific data harmonization. Their formulated model showed a validated approach using de-identified EHR data from two selected medical centers with (n=216, 221) adult patients hospitalized for at least 24 hours. In the sequential format proposed, this voluminous EHR dataset was unrolled into a total of (N=46, 864, 534, 945) data points, including clinical notes. The deep learning models achieved high accuracy for tasks such as predicting in-hospital mortality (area under the receiver operator curve [AUROC] across sites 0.93–0.94), 30-day unplanned readmission (AUROC 0.75–0.76), prolonged length of stay (AUROC 0.85–0.86), and all of a patient's final discharge diagnoses (frequency-weighted AUROC 0.90), etc. These models outperformed traditional, clinically used predictive models in all cases. The author believed that this approach could be used to create accurate and scalable predictions for a variety of clinical scenarios.

Dou et al. (2015) developed a Bayesian naive model for many RNA-Seq data analyses and correction algorithms. Although these newer methods and algorithms have steadily improved transcriptomic analysis, greater prediction accuracy was needed to guide better experimental designs for computational results. This study identified a new tool was identified named G-Exposer to explore differentially expressed genes with RNA-seq data. This tool introduced a local normalization algorithm to reduce the bias of non-randomly positioned read depth. The naive Bayes classifiers were employed to integrate fold change, transcript length, and GC content to identify differentially expressed genes. The results of several independent tests showed that G-Exposer outperformed other methods. The combination of the local normalization algorithm and naive Bayes classifiers with three attributed variables achieved better results, as both false-positive rates and false-negative rates showed a negative trend. However, only a small portion of genes was affected by the local normalization and GC content correction.

Zhang et al. (2021b) stated that N7-methylguanosine (m7G), a well-studied posttranscriptional modification of RNA, is crucial for controlling the expression of certain genes. Precise determination of m7G sites within the transcriptome is essential for enhancing the understanding of their putative functional mechanisms. High-throughput experimental techniques are costly and time-consuming, despite being able to locate m7G sites precisely. Therefore, it is essential to develop a computationally effective method that can precisely identify m7G sites. In this study, he proposed a novel method by incorporating a BERT-based multilingual model in bioinformatics to represent the information of RNA sequences. Firstly, he treats RNA sequences as natural sentences and then employs bidirectional encoder representations from the Transformers (BERT) model to transform them into fixed-length numerical matrices. Secondly, a feature selection scheme based on the elastic net method is constructed to eliminate redundant features and retain important features. Finally, the selected feature subset is input into a stacking ensemble classifier to predict m7G sites, and the hyperparameters of the classifier are tuned with a tree-structured Parzen Estimator (TPE) approach. By 10-fold cross-validation, the performance of BERT-m7G is measured with an ACC of 95.48% and an MCC of 0.9100. The experimental results indicate that the proposed method significantly outperforms state-of-the-art prediction methods in the identification of m7G modifications.

Vestal et al. (2020) studied the potential understanding of transcriptional changes within an individual over time, which differs depending on the treatment or environmental conditions. As per the past literature, the authors reviewed several methods for dealing with repeated measures within RNA-Seq analyses. Earlier methods were restricted to handling only paired measurements. Also, they can be tested for differences between two groups, and/or had issues with maintaining nominal false positive and false discovery rates. To address these pitfalls, they proposed a Bayesian hierarchical negative binomial generalized linear mixed model framework that can flexibly model RNA-Seq counts from studies with arbitrarily many repeated observations, including covariates, while maintaining nominal false positive (FPR) and false discovery rates (FDR) in its posterior inference. The simulation results showed that the proposed method (MCM-Seq) best combines high statistical power (i.e. sensitivity or recall) with managing nominal false positive and false discovery rates compared to other available strategies, especially at smaller sample sizes investigated. This behavior was replicated in an application to real RNA-Seq data where MCM-Seq was able to find previously reported genes associated with tuberculosis infection in a cohort with longitudinal measurements. The study concluded that failing to account for repeated measurements when analyzing RNA-Seq experiments significantly inflates false positive and false discovery rates. Of the methods they investigated, whether modeling RNA-Seq counts directly or working on transformed values, the Bayesian hierarchical model implemented in the MCM-Seq R package best combined sensitivity and nominal error rate control.

Caudai et al. (2021) reviewed the current applications of artificial intelligence (AI) in functional genomics. They opined that the recent explosion of AI has gained remarkable achievements which can be solved by deep learning methods along with a burst of 'big data analytics'. Bioinformatics disciplines that contribute to this bulk of data are collectively known as functional genomics, proteomics, transcriptomics, etc. It will be necessary to study the following components using more advanced bioinformatics tools: (i) genomics, the study of DNA information; (ii) epigenomics, the study of reversible changes to DNA; (iii) transcriptomics, the study of RNA transcripts originating from a genome; (iv) the ensemble of chemical modifications decorating different types of RNA transcripts (epi transcriptomics); (v) the small molecules made from cell metabolism (metabolomics); and (vi) the results of protein-coding transcripts (proteomics) that are present in an organism or system at a specific moment under physiological or pathological circumstances. Following an overview of the primary uses of AI in functional genomics, they also discussed significant ancillary topics, such as the importance of explainability and ethical, legal, and economic concerns. Patel et al. (2020) explored how DNA methylation, histone modification, post-transcriptional modifications, and epigenetic changes in genes and pathways at various levels are linked to cancer, a complex disease that results in the development of malignant phenotypes. Nevertheless, an accurate statistical insight is obtained with a multitude of datasets for epigenetic status, which may offer some semblance of evidence for phenotypic alteration and specific DNA behavior that remain absent in breast cancer. The authors took into account four distinct datasets: RNA-Seq, methylation, histone, and human genome data, which have all been utilized to distinguish between cancerous and non-cancerous cells in breast cancer cases. An internal r-Script was created to preprocess the data, and the Weka tool was used to select and classify the features. Four types of feature selection methods namely PCA, Gain Ratio, Relief, and CFS, and eight different types of classifiers with 10-fold cross-validations, were used to accurately classify datasets. Different combinations/ratios of data sizes for training and testing were used to achieve optimized classification accuracy.

The entire data processing was conducted for predicting and testing breast cancer. With the help of the machine learning method (ML), the epigenetic data shows the prediction of breast cancer in the given set of cells. The results were supported by various data tables and graphs, which demonstrate that CFS is the best feature selection method and that linear support vector machine (SVM) is the best classifier out of all the selected ones since it gives 100% accuracy the greatest number of times with the combination of different feature selection methods. However, for the CFS feature selection, the random forest classifier opted for the best model for different characteristics exploring by giving 100% accuracy the greatest number of times. The total number of features was ($n = 1{,}564$) spanning the four categories of CpG methylation, histone H3 data, nucleotide composition, and RNA-Seq data. The CFS feature selection method selected the best 245 features. CFS and linear SVM classification techniques were chosen as the best overall model with various characteristics, with an accuracy of 100% using 10-fold cross-validation. With the help of multiple tests and methods, it was observed that different classifiers exhibit different results when tested among various training-testing data. Also, the outcome varies with the variations in the dataset and data processing techniques. However, the feature selection method plays a vital role in driving the results. With the help of various tests and methods, it was observed that different classifiers exhibit different results. As per the past literature at the global level, the MSA methods refer to a series of algorithmic solutions for the alignment of evolutionarily related sequences while taking into account evolutionary events such as mutations, insertions, deletions, and rearrangements under certain conditions. These methods can be applied to DNA, RNA, or protein sequences. A study in Nature by Van Noorden et al. (2014) revealed that the MSA is one of the most widely used modeling methods in biology, with the publication by Thompson et al. (1994) ranking at #10 among the most cited scientific papers of all time. Indeed, a large number of in silico analyses depend on MSA methods.

These include domain analysis, phylogenetic reconstruction, motif finding, and a whole range of other applications, extensively described by Kemena & Notredame (2009). MSA is indeed an important modeling tool whose development has required addressing a complex combination of computational and biological problems. The computation of an accurate MSA has long been known to be an NP-complete problem (Chatzou et al., 2016), a situation that explains why over 100 alternative methods have been developed over the past three decades (Kemena & Notredame, 2009). Original MSA methods (MSAMs) and their applications have been extensively covered by several reviews (Kemena & Notredame, 2009; Edgar & Batzoglou, 2006). To avoid redundancy, we will focus here on the main developments that have taken place over these past 10 years and put them in a broader historical context when needed (Wallace & Higgins, 2005). The first three sections will detail the general algorithmic framework of MSAMs and show how it relates to the newest methods and their application to all sorts of biological sequences (proteins, RNA, DNA) (Kemena & Notredame, 2009). The fourth part will cover method validation and available benchmarks, with a special emphasis on the newest generation designed to cater to evolutionary and structural modeling (Edgar & Batzoglou, 2006). The last part of this review will deal with the quantification of local reliability within MSAs. This task had long been identified as instrumental, and possibly more important than the computation of the models that are necessarily approximate. It is, however, only recently that systematic approaches have been developed with the explicit aim of quantifying local reliability, thus allowing systematic filtering and weighting for downstream modeling. We will review these methods in light of the latest study by Hong et al. (2022).

5.3 Plan of the Study

A purposive study design was adopted to formulate model building, and high-throughput Seq DNA and RNA library datasets with different annotations were downloaded from the NCBI online platform. The annotation IDs were systematically cross-checked in the individual cases, and exomes, peptides, and gene-sequencing data were also gathered from the omics data library. Further, the rarest of the rare genetic disorder diseases with different clinical attributes were collected from selected private and referral hospitals in Bangalore city. The following standards for inclusion and exclusion were used when creating the statistical instruments.

5.3.1 Criteria for Inclusion

The sequencing of particular diseases, such as cancer (adenocarcinoma), SARS-CoV-2 gene sequencing, and different types of protein sequencing datasets, were included (2017–2022). Only the adult age group (>18 years) was included.

5.3.2 Criteria of Exclusion

Overlapping annotated data series were excluded.

Gene sequencing of children (aged 1–17 years) was not included.

5.4 Model Formulation and Research Methods

Because bioinformatics uses sophisticated statistical tools, clinicians use them to diagnose disease stages, and health policymakers use them to plan new policy interventions, life scientists will need more statistical modeling in healthcare to make accurate interpretations. Various frequency-based techniques, such as network analysis of disease risk assessment, artificial neural networks (ANN), random forests, regression estimators, etc., were employed in the current scenario to simulate the models. The conventional statistical models mentioned above have already been discussed in relation to explaining disease intervention and resolving biological issues. We have developed more recent models in this area (Figure 5.3). Wide genomic sequencing and electronic record (ER) life datasets were obtained from various tertiary care hospitals in selected sites in accordance with standard operating procedures (SOP) in terms of methodology. The pertinent health-related and quality-related issues were gathered using pre-tested questionnaires; adverse drug reactions (ADR) and severe adverse drug reactions (SDR) were methodically gathered for specific diseases. To gather qualitative data (COVID and cancer-related risk), regular focus groups, direct interviews, and in-depth interviews

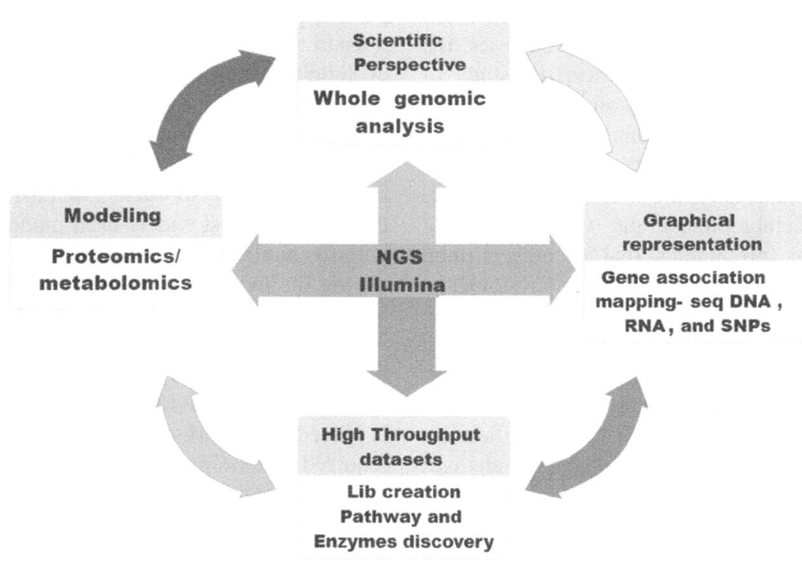

FIGURE 5.3 NGS Illumina of genetic arrays.

were held. R software, an advanced statistical package, was used to analyze the data gathered. To describe the pathways of various diseases, several mathematical and predictive models were developed, including large random walk entropy (LRWE), Bayesian network model (BNM), Gaussian process (GP) regression model, and Gaussian process risk assessment for gene expression, proteomics, exomes, and amino acid homology models. It was shown that the compartment state simulation models (comprising both ordinary and higher-order differential equation methods) could determine disease tolerance based on phylogenetic tree and gene association mapping. Additionally, benchmark analysis was conducted, and the risk of EHR sets was determined through the modeling process using Bayesian posterior and prior probability network models. The unit under test (UUT) demonstrated drug disease tolerance, and measurement reference standard limits were calibrated with a specified hospital stay time interval using information from the patient record. Using the guard band limit theorem and axioms, the current study simultaneously tested the limits inside and outside of risk tolerances (to increase true false positive accept risk and decrease false negative acceptance risk). Conversely, we minimized risk error variations by optimizing disease risk using the assumed false rejection risk (FRR) and false acceptance risk (FAR), which were modified based on containment limits. Bayesian approaches for phylogenetic analysis, sparsity priors for protein-protein interaction predictions, Markov modeling techniques for CGH array data, and Bayesian networks for gene expression data were also applied.

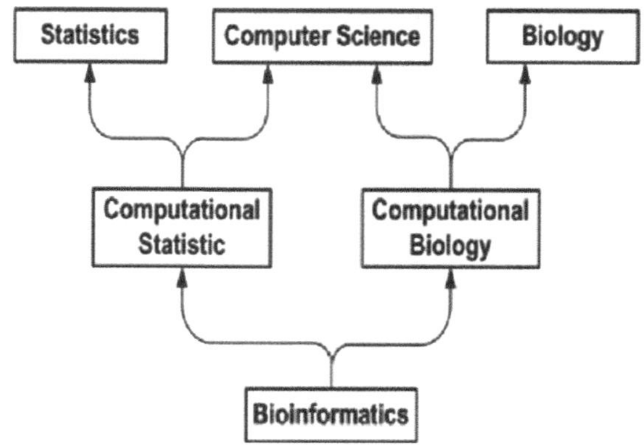

FIGURE 5.4 Flow diagram of bioinformatics.

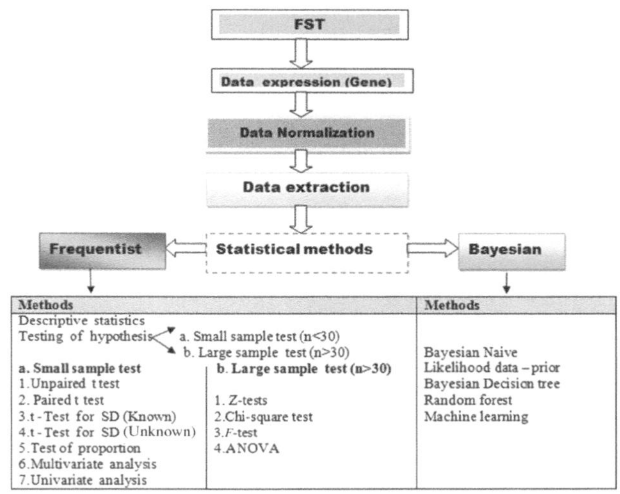

FIGURE 5.5 Flowchart of statistical science.

5.5 Statistical Applications

5.5.1 Applications of Bioinformatics

Paulien Hogeweg, a Dutch complex systems researcher and theoretical biologist, is credited with coining the term 'bioinformatics', which refers to the study of biological systems using appropriate statistical techniques and computer programming languages. To put it another way, bioinformatics is an interdisciplinary field that develops and enhances various techniques for organizing, retrieving, storing, and analyzing biological datasets to address pressing issues in the fields of medical microbiology, biotechnology, and life sciences. Creating software tools to produce data that scientists can use is a key function of bioinformatics. The bioinformatics flow chart is shown in Figures 5.3 and 5.4.

It shows the interconnection between different interdisciplinary sciences to form bioinformatics. The flow chart describes the relationship between biology, computer science, and statistics for applying bioinformatics tools (Figure 5.5).

TABLE 5.1

Distribution of Data on Spikes

Sample (X_1)	Spikes fmol (X_2)	Comparison LR
1	100	P_1 v/s $P_2 = 1$
2	50	P_1 v/s $P_3 = 2$
3	25	P_1 v/s $P_4 = 3.3$
4	10	P_4 v/s $P_6 = 3.3$
5	5	P_6 v/s $P_9 = 3.3$
6	1	P_1 v/s $P_9 = 10$
7	0.5	
8	0.21	
9	0.1	

5.6 Illustration

The MaxQuant (MQ) is one of the most popular quantified machines for discovering the proteome (PD) (a recently introduced minor tool). Here, we present a comparative evaluation of six MS1-based quantifications available in MQ and PD intensity. The area of the precursor ion peaks was estimated for the simulation of sequencing proteins (differential proteomics), which covers a wide range of protein abundance ratios. PD outperformed MQ in terms of quantification yield and the dynamic range of reproducible capacity. As per the output of MQ, we aim to test the following results hypothetically using a suitable statistical method (Table 5.1).

5.6.1 Procedure for Testing Hypothesis

The procedure of testing of hypothesis involved various steps

Step 1: State the null hypothesis $\mu_1 = \mu_0$ (there is no significant differences between spikes proteome discoverer and sampled data)

Step 2: To fix the level of significance ($\alpha = 0.01$)

Step 3: Apply t-test to determine critical value

$$t_{\text{cal value}} = \frac{\bar{X}_1 - \bar{X}_2}{\sqrt{\text{SP}^2\left(\frac{1}{n_1} - \frac{1}{n_2}\right)}} \tag{5.1}$$

Average of sample-I $\bar{X}_1 = \frac{\sum X_1}{n_1}$; Average of sample -II $\bar{X}_2 = \frac{\sum X_2}{n_2}$

$$\text{SP}^2(\text{Pooled variance}) = \frac{S_1^2(n_1 - 1) + S_2^2(n_2 - 1)}{n_1 + n_2 - 2}$$

$$S_1^2(\text{Variance}) = \sqrt{\frac{\sum_{i=1}^{n}(\bar{X}_1 - \bar{X}_1)^2}{n_1 - 1}};$$

$$S_2^2(\text{Variance}) = \sqrt{\frac{\sum_{i=1}^{n}(\bar{X}_2 - \bar{X}_2)^2}{n_2 - 1}}$$

Step 4: Out put

$t = 1.442^{ns}$, $df = 16$, SE difference = 11.31, CI-95%: −40.305−7.68, $P = 0.168$. The difference is not considered statistically significant ($P > 0.01$)

Step 5: Compare t_{cal} v/s t_{table} value

Step 6: Decision - If the $t_{cal} > t_{table}$ value, the differences will be found to be significant or else the mean difference is statistically insignificant (reject the null hypothesis H_0) (Table 5.2)

5.7 Approximation of *t*-Test by Binomial Distribution (BD)

The probability of success of the proteome was discovered by MaxQuant MQ ($p = 0.168$ obtained from the above tool), and the probability of unsuccessfula results $q = (1 - p)$, is 0.832 with

TABLE 5.2

Results Output by *t*-Test

Group	Sample ($n=9$)	Spikes fmol ($n=9$)
Mean	5.00	21.31
SD	2.73	33.84
SEM	0.912	11.28

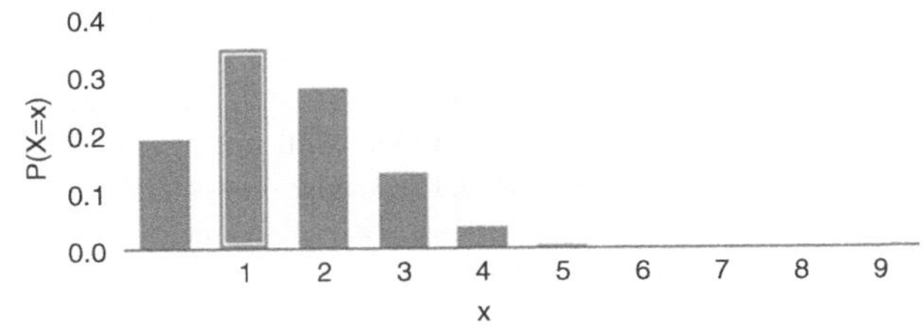

FIGURE 5.6 Density plot of the binomial distribution for the proteome—MaxQuant (MQ).

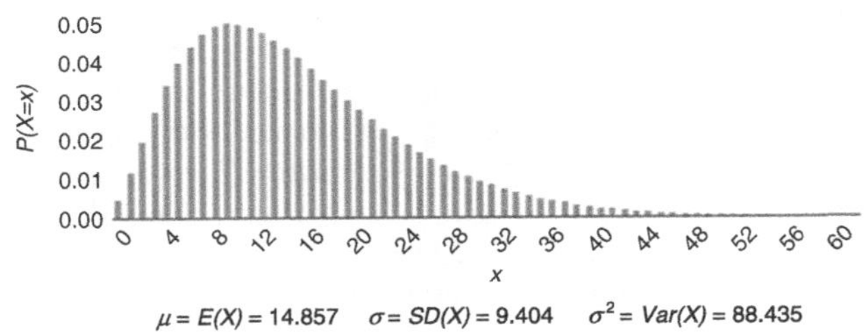

FIGURE 5.7 Density plot of the negative binomial distribution for the proteome.

TABLE 5.3

Model Output by Binomial Distribution Approximation

($X=x$)	$\mu = E(x)$	$\sigma = \text{SD}(x)$	Variance = σ^2	P-value
0 day	1.512	1.22	1.258	0.1912
1 day	1.512	1.12	1.258	0.3473
2 days	1.512	1.22	1.258	0.2804
3 days	1.512	1.12	1.258	0.1321
4 days	1.512	1.22	1.258	0.0402
5 days	1.512	1.12	1.258	0.0081

a fixed number of ($n = 9$) sample. If the number of successes is desired, the distribution is closely approximated to the binomial distribution (Figures 5.6 and 5.7). The density function of BD is given by

$$P(x) = \binom{n}{x} P^x q^{n-x} = \frac{n!}{(n-x)!\,x!} P^x (1-p)^{n-x} \tag{5.2}$$

where $x = 0, 1, 2, \ldots 5$ days, P=Probability of success for proteome, $q = (1-p)$ (Table 5.3)

$$P(x = 3\text{days}) = \binom{9}{3} 0.168^3 0.832^{9-3}$$

$$= \frac{9!}{(9-3)!\,3!} 0.16^3 (1 - 0.832)^{9-3}$$

The number of failure 'X' before a specified number of successes will be the maximum outcome of independent trials. its probability of success is treated as ($p = 0.168$) with 'r' success, and the probability distribution is given by (Figure 5.6) (Table 5.4)

TABLE 5.4

Negative Binomial Distribution Probability Density Function Estimated Values (PDF)

No. of Trial	1	2	3	4	5	6	7	8	9	10	11	12
−ve BD $P(X=x)$	0.012	0.02	0.027	0.0341	0.0397	0.044	0.047	0.049	0.0498	0.0497	0.049	0.047

TABLE 5.5

Distribution of Data Series for Both Groups (Non-Malonate/Malonate)

Seq No	1	2	3	4	5	6	7	8	9	10	11	12	13	14
Non-Malonate	0.026	0.052	0.045	0.09	0.012	0.084	0.006	0.078	0.016	0.005	0.12	0.055	0.055	0.02
Malonate	0.026	0.047	0.005	0.088	0.02	0.087	0.007	0.074	0.015	0.001	0.08	0.035	0.035	0.018

$$f(x)=f\left(\frac{x}{r},p\right)=\begin{pmatrix} r+x-1 \\ x \end{pmatrix}p^r q^x \quad (5.3)$$

Number of successes on each trial $p=0.168$, $(q=1-p)=0.832$

5.8 Paired *t*-Test

It is used to compare two groups of experimental data series in which a particular individual measurement has been expressed in one sample, and related individual measurements have been taken from another sample. Our goal is to minimize all sources of variation rather than experimental effects. This will be done using standard procedures and a uniform design of the experimental area in standard form (as similar as possible among the treatment groups). However, in many cases, it is not possible to eliminate random sources of variation because causes and effects are spurious in nature.

5.8.1 Steps for Hypothesis Testing

Step 1: State the null hypothesis $H_0: \mu_1 = \mu_0$ (there is no significant difference between the variables or groups)

Step 2: To fix the level of significance ($\alpha = 0.01$).

Step 3: Apply the statistical formula to determine critical value of t_{cal}

$$t_{Cal} = \frac{\sum d}{\sqrt{\frac{n\left(\sum d^2\right)-\left(\sum d\right)^2}{n-1}}} \quad (5.4)$$

where μd is the difference value on each paired observation, 'n' is the number of samples -Reaction rates (1/min) Figure 5.6 (Table 5.5)

Figure 5.6 depicts the effect of a competitive inhibitor, illustrating the mean reaction velocity (1/min) for trials conducted with and without malonate

Step 4: Compare t_{cal} v/s t_{table} value

TABLE 5.6

Descrpitve Statistcs of G_1 and G_2

Group	G_1 ($n=14$)	G_2 ($n=14$)
Mean	0.047	0.038
SD	0.035	0.031
SEM	0.009	0.008

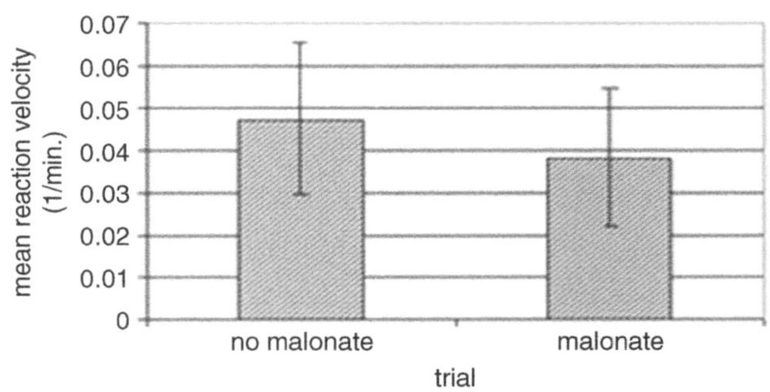

FIGURE 5.8 The effect of a competitive inhibitor, illustrating the mean reaction velocity (1/min) for trials conducted with and without malonate.

Step 5: Decision - if the $t_{cal} > t_{table}$ value, the differences will be found to be significant or else, results is statistically insignificant (reject the null hypothesis) (Table 5.6)

Step 6: Output (Figures 5.8 and 5.9)

$t=2.21$, $df=13$, $P=0.0453$. This group is statistically significant at 5%, CI 95%: 0.00022–0.01778, with a mean difference of 0.009.

$$Z_{cal} = \frac{\bar{x}-\mu_0}{\sigma / \sqrt{n}} \sim \text{BD}\left(\mu,\sigma^2\right) \quad (5.5)$$

5.9 Test for One Proportion (Single Variable)

The test of proportion has been applied to test the hypothesis that an observed proportion is equal to a pre-specified proportion. For example, if a leprosy case was diagnosed in a specific provenance, the proportion is expressed within the geographical

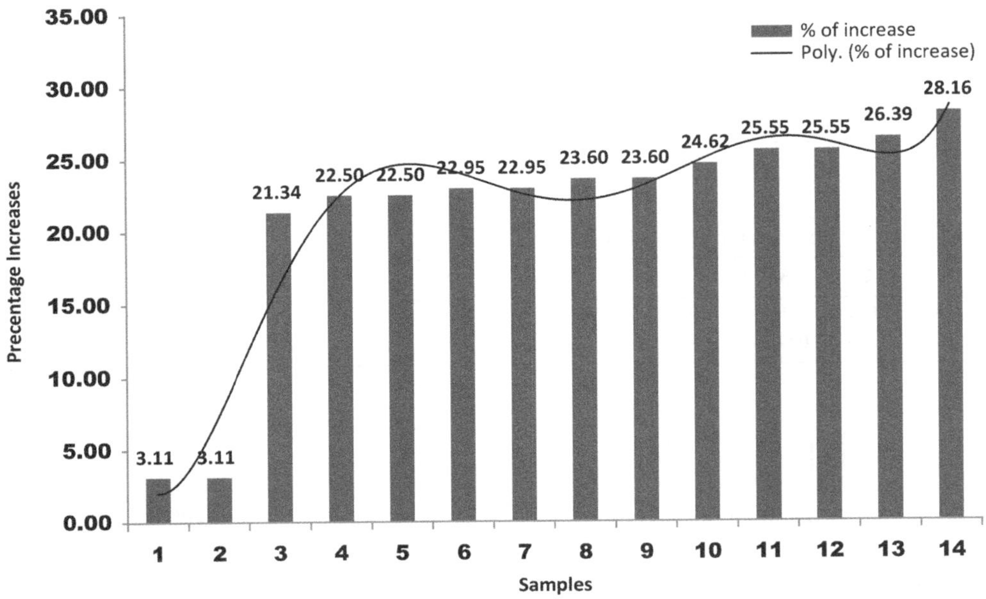

FIGURE 5.9 Probability density function (PDF) of a normal distribution with a mean (μ) of 0.04 and a standard deviation (SD) of 0.013.

TABLE 5.7

Observed Proportion of Leprosy

Observed Proportion of Leprosy	N	H_0 (%)
25%	113	38.73
Z-test value	2.996, $p=0.0027$, CI95%: 17.33% to 34.02%	

area (sampled observations can be very small). The pre-specified proportion (the value compared to the observed proportion) is expressed as a percentage. The proportion of leprosy disease progression was tested by equation (5.6).

$$Z_{\text{cal}} = \frac{p - p_{\text{exp}}}{\text{SE}(p)} \quad (5.6)$$

where 'p' is an observed proportion; p_{exp} is the null hypothesis (H_0) (or expected) proportion; and SE(p) is the standard error of the expected proportion.

$$\text{SE}(p) = \frac{p_{\text{exp}}(1 - p_{\text{exp}})}{n} \quad (5.7)$$

The P-value of the normal distribution falls outside $\pm z$. If the P-value is <0.05, the hypothesis that the observed proportion is equal to the pre-specified proportion value is rejected, and the alternative hypothesis (H_1) that there is a significant difference between the two proportions can be accepted (Figures 5.7 and 5.8) (Table 5.7).

Hypothesis was tested (two proportion) by

$$z = \frac{\hat{P}_1 - \hat{P}_2}{\sqrt{P_0(1-P_0)\left(\frac{1}{n_1} + \frac{1}{n_2}\right)}}. \quad (5.8)$$

The proportion of sample 1 and 2 was found to be statistically significant $p<0.01$ (Table 5.8).

TABLE 5.8

Proportion of Sample 1 and 2

Proportion of Sample 1 ($n=153$)	Proportion of Sample 2 ($n=90$)
0.09	0.013
Z-test value	2.43, $p=0.0159$

TABLE 5.9

Distribution of Seqpeptide

Sequences			
S_1	S_2	S_3	S_k
S_1	S_1	S_1	S_1
S_2	S_2	S_2	S_2
S_3	S_3	S_3	S_3
.	.	.	.
.	.	.	.
S_{i1}	S_{i2}	S_{i3}	S_{ik}

5.10 Rolling Correlation Model for Seq Peptide

The rolling correlation (R) is a powerful full-scale statistical tool to test the association between seq peptides in proteomics. We know the correlation between two or more seq peptides and seq DNA simultaneously. This model is very useful to understand the association between various seq peptides on a sequential basis since it may be used for the construction of latent models for library datasets for predictive purposes. It can take more time to categorize the series of sequences for grouping (Table 5.9).

5.10.1 Assumptions of Rolling Correlation

The variables must be classified as either interval or ratio measurements.

The variables must be approximately distributed.

There is a linear relationship between two variables: $r\,xy = r$ and $yx = ryz$

Outliers are either kept to a minimum or removed entirely.

There is homogeneity in the data (homoscedasticity).

5.10.2 Steps for Rolling Correlation of Seq Peptide

Step 1: Scrutinize the sequenced data series (Figure 5.10)

Step 2: Convert ratio or interval scale by assigned ranks of repeated sequences

Step 3: Determine covariance of sequences $S_1, S_2 \ldots S_n$

Step 4: Calculate Var (S_1), Var (S_2), … Var (S_k),

Step 5: Apply the following rolling correlation formula

$$R = \frac{\operatorname{Cov}(S_1,S_2)}{\operatorname{Var}(S_1),\operatorname{Var}(S_2)}; \frac{\operatorname{Cov}(S_1,S_3)}{\operatorname{Var}(S_1),\operatorname{Var}(S_3)}; \frac{\operatorname{Cov}(S_2,S_3)}{\operatorname{Var}(S_2),\operatorname{Var}(S_3)} \cdots\cdots\cdots \frac{\operatorname{Cov}(S_1,S_k)}{\operatorname{Var}(S_1),\operatorname{Var}(S_k)} \quad (5.9)$$

$$R = \frac{n_{k1}\sum_{i=1}^{niseq} S_{1i}S_{2i} - n_{k1}\sum S_{1i}\sum S_{2i}}{\sqrt{\left(n\sum (S_{1i})^2\right) * \left(n\sum (S_{2i})^2\right)}} - \frac{n_k\sum_{i=1}^{niseq} S_iS_j - n_k\sum S_i\sum S_j}{\sqrt{\left(n\sum (S_i)^2\right) * \left(n\sum (S_j)^2\right)}} \quad (5.10)$$

Step 6: Model output

Step 7: Test the hypothesis by z-test

$$Z = \frac{1}{2}\log_e\frac{1-R_1}{1+R_1}; \text{at } Z_k = \frac{1}{2}\log_e\frac{1-R_{1k}}{1+R_{1k}} \quad (5.11)$$

Step 8: Draw your conclusion based on the Z cal and Z value table

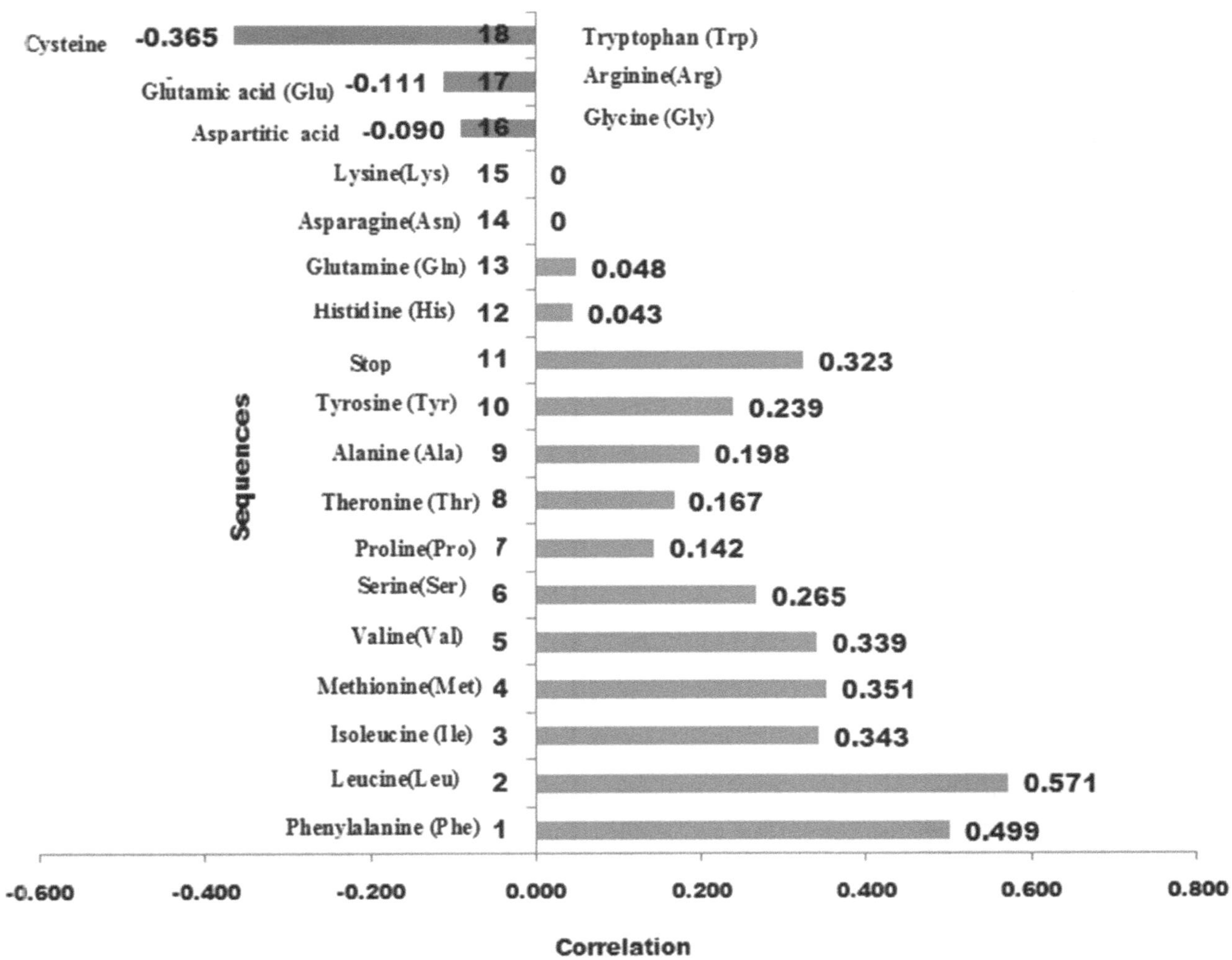

FIGURE 5.10 Rolling correlation analysis.

5.10.3 Statistical Method for Seq Proteins on Repeated Measures

Protein statistics return the number of occurrences of each residue in the sequence entered. The sequence percentage total is determined from the FASTA. This will allow us to quickly correlate the relative comparison of each peptide. As a rationalized comparison goal, we should formulate a homology model for effective interpretation.

$$Y_{ij} = \mu + \alpha_i + \varepsilon_{ij} \tag{5.12}$$

All sequences are omitted as randomly distributed with mean μ and common variance σ^2.

The sequences are identical and independent in nature.

Errors will be associated with the ith sequence and the jth repeated process.

Step 1: Correction factor $CF = \frac{(GT)^2}{nk} = \frac{\left(\sum X\right)^2}{N}$

Step 2: Sum of square of total (SST)

$$= \sum X^2 - \frac{\left(\sum X\right)^2}{N}$$

Step 3: Sum of square for the treatment (SST)

$$= \sum \left(\bar{X}_{\text{Treat}} - \bar{X}_{\text{Grand}}\right)^2$$

Step 4: Sum of square for the Subjects (SS_{Sub})

$$= \sum \left(\bar{X}_{\text{Subjects}} - \bar{X}_{\text{Subjects}}\right)^2$$

Step 5: Sum of square for the error (SSE) $= \text{SST} - SS_{\text{Treat}} - SS_{\text{Sub}}$

Step 6: $MS_{\text{treat}} = SS_{\text{Total}} / df_{\text{treat}}$

Step 7: $MS_{\text{Error}} = SS_{\text{Error}} / df_{\text{error}}$

Step 8: F_b=MSBG/MSE

Step 9: F_w=MSWG/MSE

Step 10: F_t=MST/MSE

Step 11: To take decision based on F_t calculated, compare with table F value at $(k-1)$, $(n-1)$ $(k-1)$ df

Step 12: If the calculated F value is≥Table F value, reject H_0 (Table 5.10)

5.11 ANOVA Illustration

The following seq-proteins were downloaded from the NCBI database. Different patterns were estimated from the repeated sequences of proteins and also correlated for each peptide based on the retention time for the formation of the primary and secondary structures of proteins. The following data expression was used to determine whether the experiment is feasible or not (H_0 tested by the Fisher F-value) at a 1% level of significance (Tables 5.11 and 5.12).

5.12 Friedman Test for Testing Repeated Sequences

The Friedman test will be used to test the Ho hypothesis for one-way repeated measures assigned by ranks. Its use of ranks is similar to the Kruskal–Wallis one-way ANOVA by ranks. Many statistical applications in life science and genetic studies widely support the Friedman test.

$$F_R = \left[\frac{12}{NK(K+1)} \sum_{j=1}^{k} R_j^2\right] - 3N(K+1) \tag{5.13}$$

where, R_j is the sum of the ranks in each treatment of j (j=1, 2…, k)

N=Number of groups

k=Number of repeated peptides

As per the analysis $F_R = 0.1 * 1400 - 120 = 20$, $p<0.001$. The sequences between and within groups were found to be statistically significant at the 1% level.

5.13 Random Walk Model for Repeated Measures

A stochastic process is described by the equation

$$Y_t = Y_{t-1} + \varepsilon_t \tag{5.14}$$

TABLE 5.10

ANOVA

Source of Variation	df	SS	MSS	F-Critical Value
Between the group (BG)	$(k-1)$	SSBG	MSBG=SSBG/$(k-1)$	F_b=MSBG/MSE
Within the group (WG)	$n(k-1)$	SSWG	MSWG=SSWG/$n(k-1)$	F_w=MSWG/MSE
Subjects (treatment)	$(k-1)$	SST	MST=SSt/$(k-1)$	F_t=MST/MSE
Error	$(n-1)(k-1)$	SSE	MSE=SSE/$(n-1)(k-1)$	
Total	$(nk-1)$			

TABLE 5.11

Protein Patterns Distributed with Respect to Groups

Group	Pattern	Times Found	Group	Pattern	Times Found	Group	Pattern	Times Found
1	B	0	2	S,C,M	55	3	Aromatic F,W,Y	152
1	X	0	2	D	60	3	L	83
1	Z	0	2	Q	65	3	Basic K,R,H	192
1	C	17	2	A	73	3	Aliphatic hydroxyl S,T	218
1	W	23	2	R	23	3	Acidic B,D,E,N,Q,Z	260
1	H	28	2	G	83	3	Aliphatic G,A,V,L,I	559
1	M	38	2	E	85	3	tRNA synthetase class II B,G,A,P,S,T,H,D,N,K,F	727
1	P	40	2	K	83	3	tRNA synthetase class I Z,E,Q,R,C,M,V,I,L,Y,W	749

TABLE 5.12

ANOVA Repeated Measures-Model Output

Source	SS	*df*	MS	F_{Cal}
Between-group	608,124	2	304,062	15.54^{Sig}
Within-group	458,386	21	21,827	
Error	273,842	14	19,560	

where Y_t is the dependent variable, ε_t is the errors or noisy data. Frequently, the protein sequence is repeated multiple times at '*t*'. The version of stochastic process will yield the following equation (3.14):

$$Y_t = Y_{t-1} + \delta + \varepsilon_t \tag{5.15}$$

The random walk with drift and trend is equated by the following equation (3.15):

$$Y_t = Y_{t-1} + \delta + \gamma t + \varepsilon_t \tag{5.16}$$

5.13.1 Modeling of Seq-DNA and RNA

Using various probabilistic models adaptable to explain the consensus of primary and secondary structures, the general concept of modeling can describe a variety of DNA and RNA sequencing analyses. These models provide structural information by constructing theoretical forms and insights. Readers will learn the following details, which are fundamental, about the composition and purpose of genetic material (DNA and RNA).

5.14 Structure and Function of DNA

Deoxyribonucleic acid (DNA) is the genetic material of living organisms. In 1869, Friedrich Miescher separated a cellular substance from the nuclei of pus cells and called it 'nuclein'. Subsequent studies discovered that this material had acidic properties and hence it was renamed 'nucleic acid'. The term nucleic acid is derived from two components or types: deoxyribonucleic acid (DNA) and ribonucleic acid (RNA), the two major types of nucleic acids found in cells. Nucleotides are the structural units of nucleic acids. Each nucleotide has three components. A pentose sugar called deoxyribose or ribose, which is a five-carbon compound with a pentagonal ring structure. A phosphate group derived from phosphoric acid, which has three active groups, two of which two are involved in strand formation. The formation of strands is called nitrogen bases; these are cyclic compounds made up of carbon, hydrogen, oxygen, and nitrogen atoms (CHO_2N). Bases were named adenine (A), thymine (T), cytosine (C), guanine (G), and uracil (U). Furthermore, they are classified into two groups: purines and pyrimidines. Adenine and guanine are doubling compounds called purines, whereas thymine, cytosine, and uracil are single-ring compounds referred to as pyrimidines. Uracil is absent in DNA and present in RNA (Figures 5.9 and 5.10).

According to the above figure, the model was proposed by Watson and Crick. The DNA molecule consists of two long strands coiled around a common base. It looks like a twisted ladder and shows alternating major and minor grooves. Each strand of DNA consists of several nucleotides and is made up of deoxyribose sugar. In double-helix DNA, the nitrogen bases present on one strand pair with the nitrogen bases of the opposite-strand purines. Pyrimidine Adenine=thymine (A=T or T=A). Guanine=cytosine (G=C or C=G). As the base sequence present on one strand of DNA determines the base sequence of the other strand, the two strands are regarded as complementary strands. The distance between two successive steps on rungs is 0.31 nm. Due to complementary base pairing, the total number of purine bases is always equal to the total number of pyrimidine bases (1:1). This is called Chargaff's rule and can be represented as (A+G=T+C). Each strand of DNA has a specific polarity. One end is considered the 5′ end, whereas another end is the 3′ end. This polarity is caused by the third and fifth carbon atoms of the deoxyribose sugar. At the 5′ end, there is a free phosphate group, while at the 3′ end, there is a free OH group. One of the strands runs in a 5–3′ direction, while the other is in a 3–5′ direction. This arrangement of the two strands of DNA is considered antiparallel because the 5' end of one strand lies close to the 3' end of the other strand, and vice versa. Figure 5.10 shows the schematic representation of double-stranded DNA. The double-helix structure is illustrated on the left side, while a more detailed description is shown on the right side. The chemical structure of the four different bases is presented at the bottom. The circled symbols H (hydrogen bond) and P (phosphate) are represented. The 3' and 5′ symbols indicate the asymmetry of DNA strands. Major and minor grooves are normally seen; the strands undergo right-handed coiling around a central imaginary axis. A double helix results in information about major or deep grooves, and the twisting of two strands around one another forms minor or shallow grooves. They found alternating patterns. The diameter of the molecule is 2.0 nm. The spiral, ladder-like arrangement of the molecule is due to the deflection of the pitched angle between two successive steps of 36°. It makes one complete 360° turn after covering a distance of 3.4 nm. All four types of nucleotides are found in the nucleoplasm in the form of their monophosphates. They are activated into triphosphates, such as deoxyadenosine triphosphate (dATP), deoxyguanosine triphosphate (dGTP), deoxythymidine triphosphate (dTTP), and deoxycytidine triphosphate (dCTTP), using adenosine triphosphate (ATP). Replication begins at certain specific sites situated on the molecule. Such sites are called origins or initiation points. In eukaryotes, there are several origin points. During replication, molecules are split by the activity of initiator proteins at the origin, resulting in the formation of an incision. This initiates the breaking of hydrogen bonds between the two strands. After the unwinding of DNA strands by the activity of initiator proteins, the strands of DNA are free from each other but not separated. This separation of the strands is achieved by unwinding proteins. Now the molecule appears as an inverted '*Y*'-shaped structure called a replication fork. The separated strands are prevented from coiling by SSBP (single-strand DNA binding protein) or helix-destabilizing protein synthesis of complementary new strands. It takes place with the help of a small RNA molecule called RNA primer. DNA polymerase catalyzes only in one direction that is 5–3; consequently, one-strand replication continues the structure and function of DNA. In this way, for each old strand, a new complementary strand is constructed. Simultaneously, both strands (new and old) undergo coiling, and two identical daughter molecules are formed at the end of the process. After replication, each molecule has one old and one new strand. This shows that 50% of the mother molecule is retained or conserved, while the remaining 50% is newly constructed. Hence, the process is referred to as semi-conservative replication. Meselson and Stahl (1958) experimentally proved the semi-conservative nature of replication using heavy isotopes of nitrogen and *E. coli*. In prokaryotes, circular DNA is present. There is only one origin, and the replication is called zero replication, as it looks like the Greek letter theta (Figures 5.11 and 5.12).

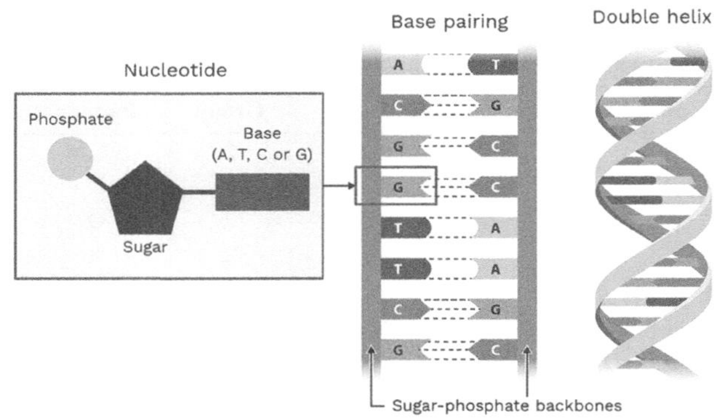

FIGURE 5.11 Primary Structure of deoxyribonucleic acid (DNA).

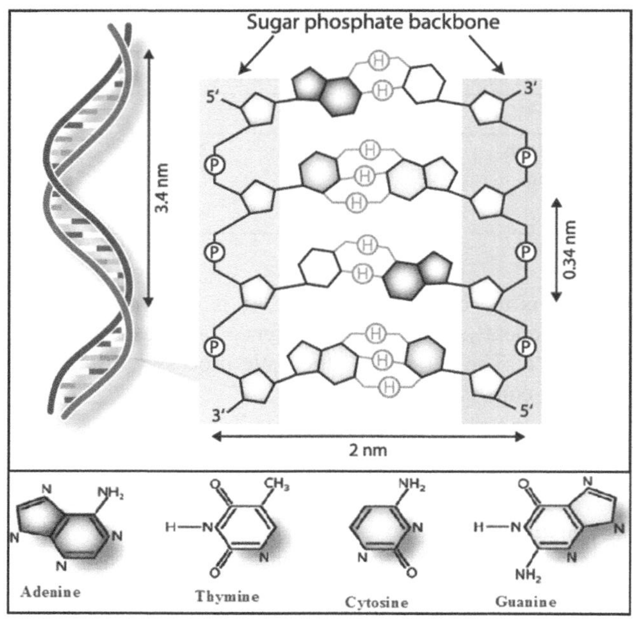

FIGURE 5.12 Structure of DNA.

5.14.1 Mathematical Formulation of DNA

We have integrated DNA polymerization and synthesis, which represent the formation of complementary strands constructed based on new and old under-closing and two identical DNA that are formed by newer and older strands. Mathematically, we have derived two processes considered for the formation of molecules, integrated by the following integral equations:

$$I = \int f(x)g(x)dx = f(x)\int g(x)dx - \int f'(x)\int g(x)dx \quad (5.17)$$

Equation (5.17) shows the definite formation of new and old strands, old new strand $g(x)$=formation of new strand.

An integrated product of any two functions=[I function *Integrated of the II function]–(Integral of differentaied of the first function) *Integral of the second function

In case of mutation under different stages of cell division, the formation will take place at different infinite time intervals.

$$I = \int_{-\infty}^{\infty} f(x)g(x)dx = f(x)\int_{-\infty}^{\infty} g(x)dx - \int_{-\infty}^{\infty} f'(x)\int_{-\infty}^{\infty} g(x)dx \quad (5.18)$$

By changing the independent variable '*x*' to '*t*' in the given form of the integral function, we consider time at time '*t*' taken for the formation of the DNA strand (complementary strand). The equation (5.18) becomes:

$$I = \int f(x) \quad (5.19)$$

Let us substituted the value of independent value $X=g(t)$ for the integral function $\int f(x)$; we will get:

$$\frac{dx}{dt} = g'(t);\ dx = g'(t)dt \quad (5.20)$$

$$I = \int f(x) = \int g(t),\ g'(t)dt \tag{5.21}$$

$$I = \int_b^a f(x)dx = fg(t)g'(t)dt \tag{5.22}$$

5.14.2 Silhouette Method

The Silhouette score measures how close each Seq-DNA is to its cluster and how far it is from the closest to the position. For each point i: a_i is the average distance of point 'I' from all his cluster's points. b_i - the average distance of point i from all points in the closest cluster.

$$S_i = \frac{b_i - a_i}{Max\ (a_i, b_i)} \tag{5.23}$$

High Silhouette value range was [–1, 1]. The point is close to its cluster and far from other clusters. Zero Silhouette value means moving the point to the closest cluster will not cause a big change in the SSE. Negative Silhouette value means the point is considered an outlier or was assigned to the wrong cluster (Table 5.13).

The data k=3 was divided into 3 clusters, which explained 68.26% of the variance. We will increase 'k' to get a higher percentage of variance explained (SS between groups/SS total). Clusters: 0, 0, 0, 0, 1, 2, 1, 1, 2, 1, 2 (see the cluster column above). The explained variance ratio was 0.68. The sum of squared distances from all the points to the centers (SSE) was 6.34, SSE within was 13.65, and the SSE group was 2.56, with a maximum of four iteration. The centers were [–1.0553, 0.207], [0.4523, 0.6948], and [0.804, –1.2023].

5.15 Ribose Nucleic Acids RNA

Ribose Nucleic acids are also polynucleotides, but the pentose ribose has a free hydroxyl group at position 2′. It is a long-chain polynucleotide that exists in a regular conformation, like a double-chain DNA, although some viruses, e.g. reovirus and wound tumor virus have double-stranded RNA. The single RNA strand folds upon itself, entirely or in certain regions. Approximately 50% RNA is present in the ribosome and rough endoplasmic reticulum, 25% in cytoplasm, 15% in mitochondrion, and 10% in nucleus. In the folded region, most bases are complementary and are joined by hydrogen bonds (Figure 5.5). This helps in the stability of the molecule. In the unfolded regions, the bases have no complements. Due to this, RNA does not have purine–pyrimidine equality found in DNA. It is understandable that on a single chain of RNA, the molar proportion of purines and pyrimidine can vary considerably. RNA does not contain the pyrimidine base thymine; instead, it has the pyrimidine uracil. In a region where purine–pyrimidine pairing takes place, adenine pairs with uracil and guanine pairs with cytosine. This base pairing is of more importance when RNA is synthesized from DNA and when RNA is involved in protein synthesis. In addition to that, the four bases mentioned above RNA also contains some unusual bases. There are more unusual bases in RNA than in DNA. All normal RNA chains either start with adenine or guanine.

TABLE 5.13

Nucleotide Position and Base Pair Frequency Distribution

Position (bp)	680	690	700	710	720	730	740	750	760	770	780
Number of base pairs	8	11	10	09	11	8	10	10	08	10	06

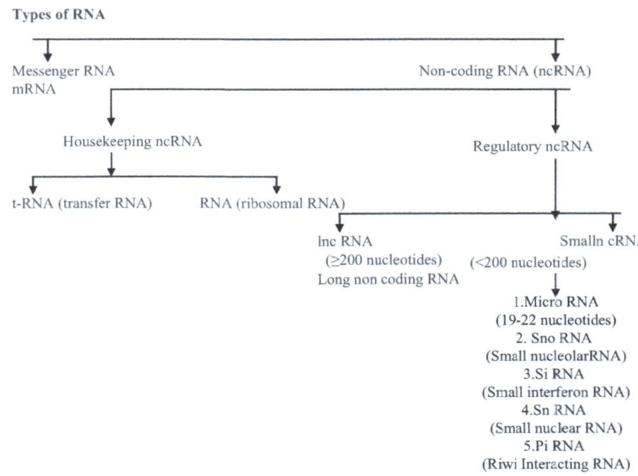

[Figure] Flow Chart of RNA classifications

5.16 Messenger RNA (mRNA)

In eukaryotic cells, mRNA accounts for about 5% of the total RNA. As per the literature survey, the length of ribonucleotides in mRNA molecules varies widely depending on the type of protein they account to code. Therefore, mRNA molecules are considered the most heterogeneous of the three types of RNA present in the cell. It is synthesized from a DNA molecule by the action of the enzyme RNA polymerase, and this process is termed transcription. At the end of transcription, the mRNA has been synthesized. This newly synthesized mRNA is further edited or modified to remove non-coding regions and to make the molecule stable. This process is called post-transcriptional modification and occurs only in the eukaryotic cell. During the post-transcriptional modification, along with cleaving out of non-coding regions from mRNA molecules, their 5′′ end is capped with a guanosine triphosphate nucleotide to make it recognizable to ribosomes during the protein synthesis. Additionally, during this process, the 3′ end of the mRNA molecule is stabilized to prevent enzymatic degradation. It is more sensitive to heat and can degrade on prolonged exposure to high temperatures. However, for some manipulations, it can be heated for a specific period of time at distinct high temperatures.

The mRNA is a single-stranded RNA molecule running from the 5′ to 3′ direction. In a prokaryotic cell, there is a lack of a distinct nucleus; thus, the mRNA synthesized contains a copy of DNA sequences with a terminal 5′ triphosphate and 3′ hydroxyl group. Prokaryotic mRNA at its 5′ end has a Shine-Dalgarno sequence, rich in purine nucleotides and essential for facilitating the mRNA binding to the ribosome. In the case of prokaryotic mRNA, it starts at 5′ with the triphosphate group, followed by the Shine-Dalgarno sequences (untranslated region). As we move along to the 3′ direction, the coding regions begin. The coding regions begin with a start codon and end when they reach a stop codon. After encountering a stop codon, protein-coding regions end, followed by another untranslated region that terminates at the end of 3′. However, in a eukaryotic cell, the nucleus is distinct and has numerous enzymes that enable synthesized mRNA molecules to undergo modifications at their terminals 5′ and 3′ to maintain mRNA stability (post-transcriptional modification). The 5′ triphosphate group of eukaryotic mRNA is esterified, forming a cap structure. This process is referred to as 5′ capping, which occurs when a 7-methylguanosine cap is passed to a 5′ free triphosphate group via a 5′-5′ phosphate linkage with the help of an enzyme called guanylyltransferase. The 5′ capping is important for the recognition of mRNA by the ribosomes during protein synthesis. Likewise, adenine monophosphates are added to make the RNA molecule more stable and prevent it from degradation. This tail of numerous adenine nucleotides is referred to as the Poly-A tail. The Poly-A tail added to the 3′ end of eukaryotic mRNA is usually 100–200 bases long; the addition is catalyzed by process of addition it was facilitated by an enzyme known as poly (a polymerase), which identifies the sequence 'AAUAAA' as a signal for the addition. The Poly-A tail is vital during transporting mRNA from the nucleus to the cytoplasm for protein synthesis. Eukaryotic mRNA starts at 5′ with a cap followed by untranslated regions (5′ UTR), and as we move along in the 3′ direction, the coding regions begin with the start codon and continue till they reach the stop codon, which marks the termination of coding regions. After there is another untranslated region (3′UTR) that terminates when the Poly-A tail begins, which further terminates at the end of 3′. There are three types of non-genetic RNA. They play an important role in protein synthesis in both pro and eukaryotic cells. These are m-RNA, r-RNA, and t-RNA. All three types of RNA are produced on DNA templates by the process called transcription. In the case of prokaryotic cell protein formation, the looping of RNA starts and ends coded as 1 and 0, and we can formulate the Bayes probability distribution model (Table 5.14).

$$\text{Prob}(A/B) = P(B/A) * P(B)/P(A) \tag{5.24}$$

$P(A/B)$=Likelihood of protein synthesis; $\text{Prob}(B/A)$= Posterior probability

$P(B)$=Prior probability of protein synthesis and $P(A)$= Prior probability of protein formation is done

TABLE 5.14

Amino Acid Codons for Binding

Event	Amino Acids		
	UAG	UAA	UGA
Binding/formation of new	Start (A) Stop (D)	Start (B) Stop (E)	Start (C) Stop (F)

$$\text{Prob}(A/D) = P(D/A) * P(A)/P(D) \quad \text{Bindings of UAG} \tag{5.25}$$

$$\text{Prob}(B/E) = P(E/B) * P(B)/P(E) \quad \text{Bindings of UAA} \tag{5.26}$$

$$\text{Prob}(C/F) = P(F/C) * P(C)/P(F) \quad \text{Bindings of UGA} \tag{5.27}$$

An overall binding of anticodons is presented as follows:

$$\begin{aligned}\text{Prob}(A/D)/B/E/C/F) &= P(D/A) * P(A)/P(D) \\ &+ P(E/B) * P(B)/P(E) + P(F/C) * P(C)/P(F) = 1\end{aligned} \tag{5.28}$$

5.16.1 Application of Binomial Distribution with Respect to mRNA

The 'x' is random variable (amino acids), 'n' is the fixed number of amino acids (aa) binding at a locus, and the probabilities of success and failure are equally ($p=q=1/2$) contributed to the completion of binding in the different sets of codons (Figures 5.13 and 5.14). The density function of the Binomial distribution for amino acid binding is modeled by the following equation (5.16):

$$P(x=r) = \binom{n}{r} p^r q^{n-r}, \; r = 0,1,2, \ldots n \tag{5.29}$$

where n=the number of amino acids, P=probability of Success, and q=the probability of failure. The mean of amino acids $\mu = np\text{SD} = \sqrt{npq}$.

$$\text{for } x = 0,1,2,..n; \; \mu = E(X) = np;$$

$$\sigma^2 = \text{Var}(X) = np\,(1-p) \tag{5.30}$$

$$\sigma = \text{SD}(X) = \sqrt{np(1-p)}; X \sim \text{Bin}(n,p) \tag{5.31}$$

where n is the number of codons and np is the mean of codons.

5.17 Transfer RNA (tRNA)

The tRNA molecule is also single-stranded. About 10%–20% of the total cellular RNA is of this type. It is the smallest of all the RNA types, containing only 73–93 nucleotides, with a molecular weight ranging between 23,000 and 30,000. It shows two patterns of floating upon itself. The structure of tRNA can be explained by the hairpin model (proposed by Hoagland) and the anticodon model. In the hairpin model of tRNA, folding results in the formation of one loop having a triplet of unpaired bases called the anticodon. The 5′ end has a G-nucleotide, while 3′ is a sequence of CCA nucleotides. tRNA is also called sRNA (soluble) because it cannot be easily separated even by the ultracentrifugation technique. It carries a specific type of amino acid at the CCA end to the ribosomes during protein synthesis. It places the required amino acid properly in the sequence and translates the coded message of mRNA into amino acids (Figure 5.15).

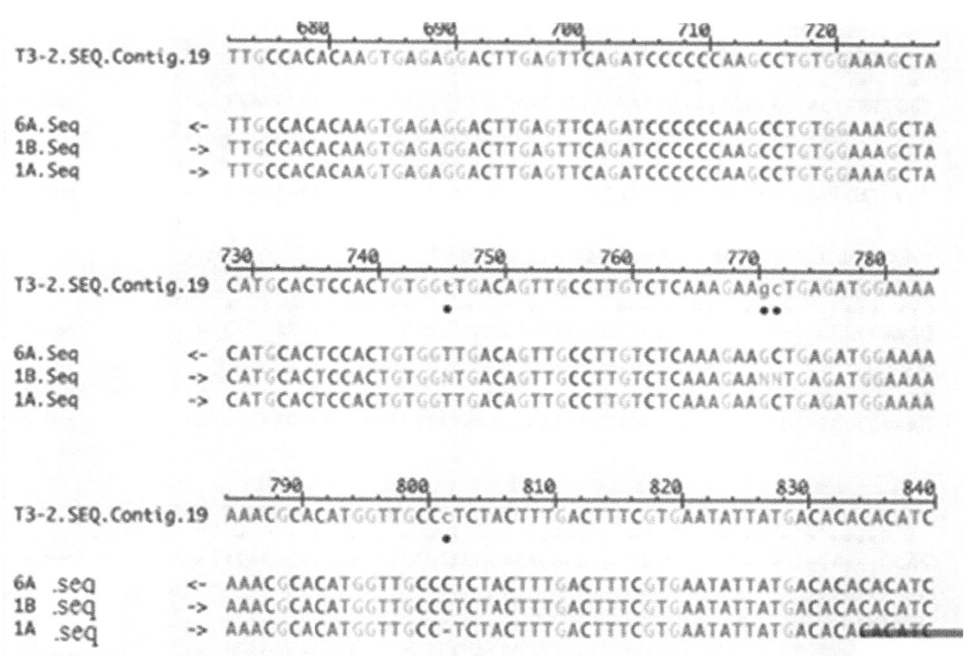

FIGURE 5.13 The distribution of DNA sequences indicates the DNA synthesis with a base pair count of 1,500.

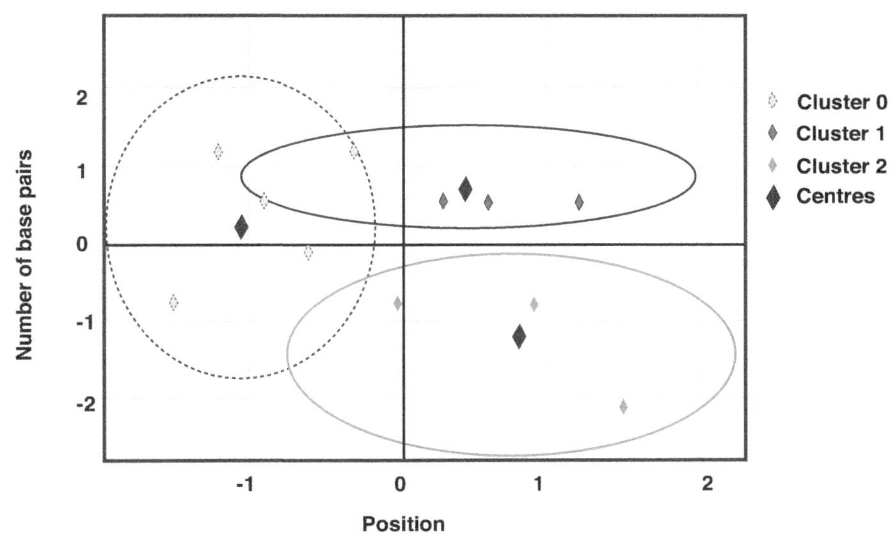

FIGURE 5.14 Cluster analysis of DNA sequences.

5.18 Function of mRNA

5.18.1 Messenger RNA (mRNA)

The mRNA molecule is always simple and straight, without any folds. It carries a message for protein synthesis from DNA to ribosomes. About 3%–5% of the total cellular RNA is of this type. It has a molecular weight of up to 500,000. It has two ends: a 5′ end and a 3′ end. A sequence of three nucleotides on mRNA is called a codon. Each 'triplet' codon on mRNA codes for specific amino acids. This is called mRNA language, genetic code, or cryptogram. The codon present at the 5′ end of mRNA is called the start codon or initiation codon. It is usually AUG; in some cases, it may be GUG. The codon at the 3′ end is a stop or nonsense codon. It is UAA, UAG, or UGA (Figures 5.16 and 5.17).

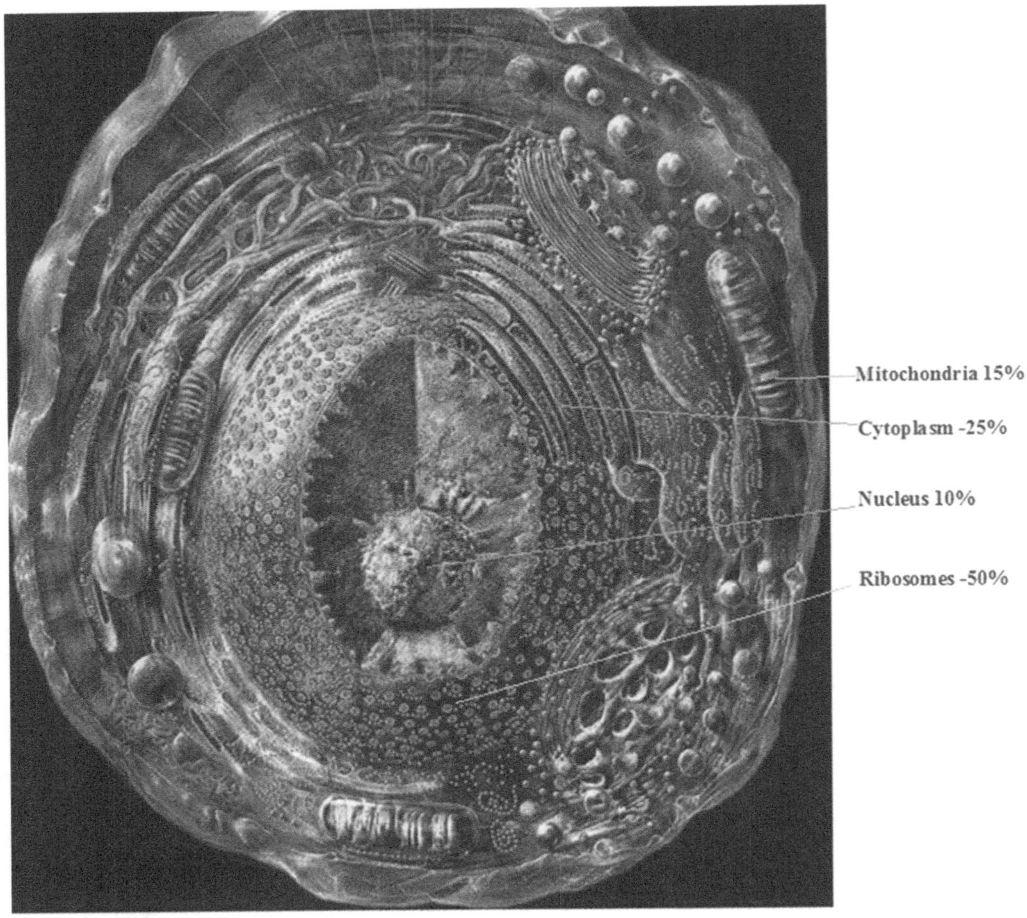

FIGURE 5.15 Structure of a human cell, showcasing a cross-sectional view.

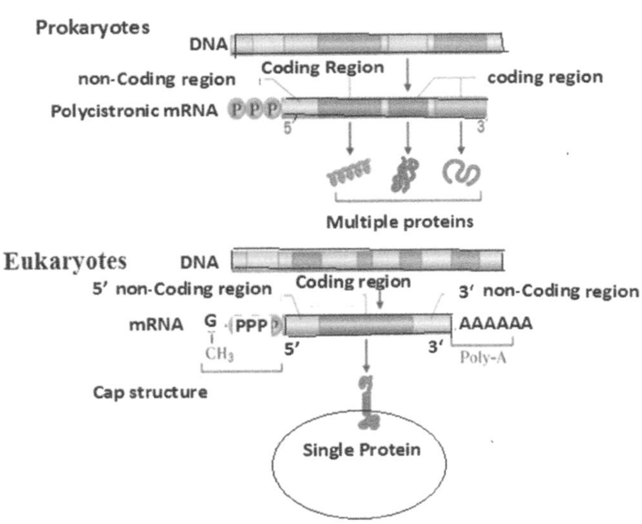

FIGURE 5.16 Messenger RNA in prokaryotic and eukaryotic cells distinctions between prokaryotic and eukaryotic mRNA.

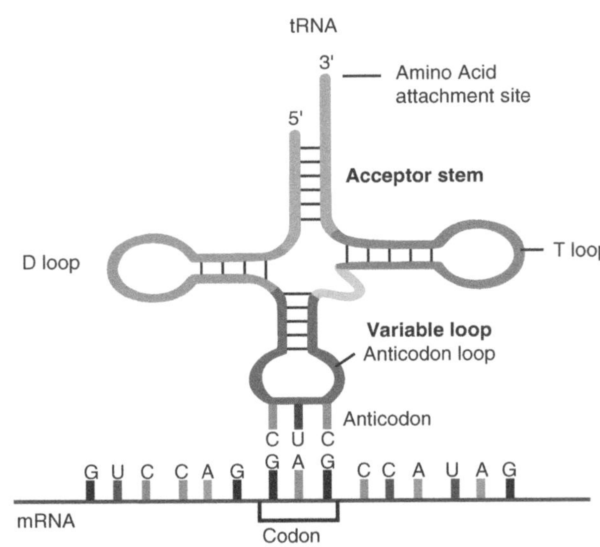

FIGURE 5.17 Structure of tRNA.

5.19 Functions of RNA

Protein synthesis is the most important and significant metabolic activity in the living world, taking place continuously in every living cell. For the actual phenotypic expression of any character in a living cell or the body of an organism, biochemical reactions are essential. Each biochemical reaction is essential and needs a specific enzyme for its initiation and completion. Almost all enzymes are proteins. The cell also needs many other structural proteins. Thousands of different types of structural and catalytic proteins are continuously required within the cell at any moment.

5.20 Protein Synthesis

The information required to build proteins resides in the DNA in the form of sequences of bases and has triplets of bases called codons, following the genetic code. Once the genetic information describing how to build a protein is transcribed to RNA, it has to be 'translated' into the language of proteins. The sequence of bases on the RNA forms a code for the amino acid building blocks of the protein. The word 'translation' seems particularly appropriate for this process since the information contained in the DNA is translated from the four-character alphabet of the bases to the 20-character alphabet of amino acids. The construction of the protein is carried out on the ribosomes of the host cell. There are free-floating ribosomes, where proteins are built for functions in the cytoplasm. Proteins that are to be used as part of the cell membrane or transported outside the cell are manufactured on ribosomes attached to the endoplasmic reticulum (Figure 5.18) (Table 5.15).

$$\mu = E(X) = 9.45;\ \sigma = 1.87, \text{variance Var}(x) = 3.46$$

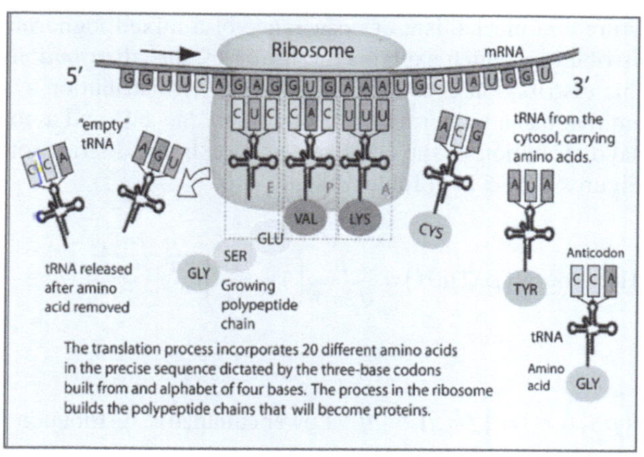

FIGURE 5.18 The process of protein synthesis in prokaryotic cell.

5.21 Application of Negative (−ve) Binomial Distribution with Respect to seqRNA

The number of successes and failures (seqRNA) requires independent trials before we can obtain the specific number of successes that occur. The following characteristics of the negative binomial distribution are: (i) the experimentation should be homogeneous and repeated with a number of random trials 'x'; (ii) each trial has two distinct possible outcomes (one is success and the other is failure); (iii) the outcome of one trial is independent of the outcome of another trial; (iv) the trial should be carried out until 'r' successes are obtained. An infinite number of amino acids will be responsible for the binding process; however, a very small number of amino acids take more retention time for the completion of binding. Since the success of bindings will depend on the repeated number of trials, each individual binding is naturally independent and continuously occurs in the events of success and failure until we get the success of a specific amino acid binding. For example, Robert is a football player, and his success rate in goal-hitting is 70%. The probability of success on the fifth attempt is 0.185. In this process of amino acid binding, we will formulate the following probability density function (Figure 5.10).

$$P(x = r) = n + r - 1\ C_{r-1}\ P^r q^n \tag{5.32}$$

$$P(x = r) = \sum_{i=1}^{r} \left(\begin{matrix} n + r - 1 \\ r - 1 \end{matrix} \right) P^r q^n \tag{5.33}$$

Mean $= \dfrac{pr}{1-p}$ Variance $= \dfrac{pr}{(1-p)^2}$ where $p(x) =$ Negative BD; $n =$ number of finite trials;

$P =$ probability trial for each trials $q = (1-p)$ negative probability (Figure 5.19)

5.22 Poisson Distribution

The 'n' number of amino acids can be responsible for increasing the capacity of binding over an indefinite time 't', but the

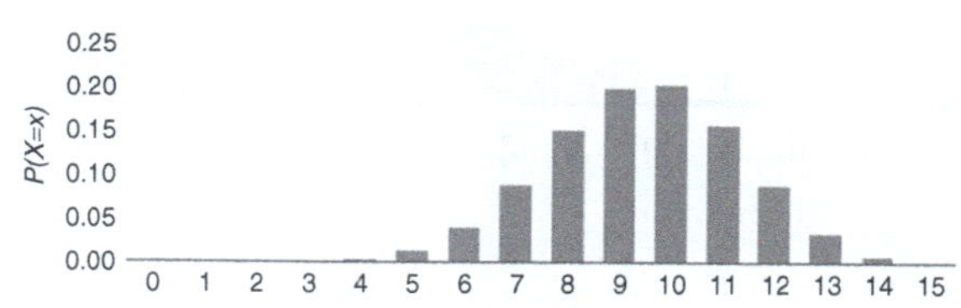

FIGURE 5.19 Binomial Distribution across various codons of amino acids.

TABLE 5.15

Codons and Its Density Function

Codons	1	2	3	4	5	6	7	8	9	10	11	12	13	14
$P(X = x)$	0.001	0.001	0.0075	0.03	0.001	0.04	0.08	0.15	0.235	0.212	0.158	0.09	0.03	0.009

probability of success upon the completion of the binding process is very low (p is very small). The density function is given by

$$\text{PDF} = f(x,\lambda) = \frac{e^{-\lambda}\lambda^{x}}{\Gamma(x+1)} \quad (5.34)$$

$$P(x,\lambda) = \sum_{t=0}^{x} f(t,\lambda) \text{ Lower cumulative distribution} \quad (5.35)$$

$$Q(x,\lambda) = \sum_{t=x}^{\infty} f(t,\lambda) \text{ Upper cumulative distribution} \quad (5.36)$$

An event that occurs 0, 1, 2....n times at intervals. The average number of events in an interval is designated as λ (parameter). The probability of 'x' events of an interval is given by the equations (5.39) and (5.40).

5.23 Application of Logarithmic Normal Distribution with Respect to Massive Data of Seq DNA

A log-normal distribution is a continuous probability distribution that is obtained by taking the natural logarithm of random variables. Usually, LND is used in the stock markets since stock prices tend to follow a log-normal distribution, as well as in the binding of amino acids, proteomics, etc. A logarithmic normal distribution is a type of normal distribution that is characterized by a logarithmic scale (x-axis) and a linear scale (y-axis). In the distribution often used for modeling biological research datasets, the data series shows skewness to the right or left (Figure 5.21) (Table 5.17).

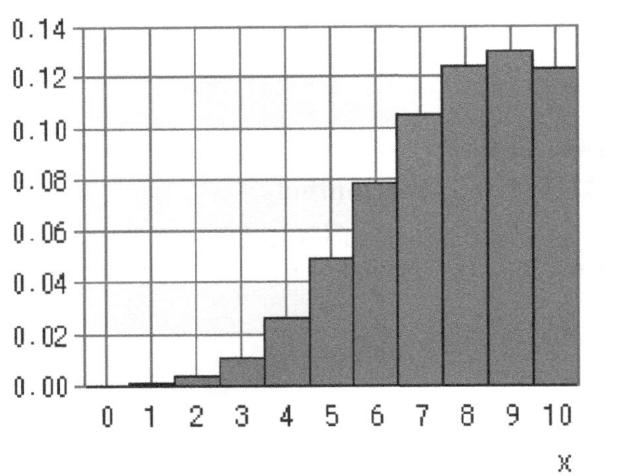

FIGURE 5.20 Probability Density Function (PDF) of the Poisson distribution.

$$\text{Log N}(X,\mu,\sigma) = \frac{1}{\sqrt{2\pi\sigma x}} e^{-\frac{1}{2}\left(\frac{\ln(x)-\mu}{\sigma}\right)^2} \quad (5.37)$$

$$p(x,\mu,\sigma) = \int_0^x f(t,\mu,\sigma)dt \text{ Lower cumulative distribution} \quad (5.38)$$

$$q(x,\mu,\sigma) = \int_x^\infty f(t,\mu,\sigma)dt \text{ Upper cumulative distribution} \quad (5.39)$$

$$\text{Mean} = e^{\mu+\frac{\sigma^2}{2}} \quad \text{Median} = e^{\mu} \quad \text{Mode} = e^{\mu-\sigma^2} \quad (5.40)$$

5.24 Application of Hybrid Lognormal Distribution

A hybrid lognormal distribution is the distribution function that changes the percentile 'x' to ln $(x)+x$ in the normal distribution function. It is determined by the random variable hyb $(\rho x)=\rho x+\ln(\rho x)$, including the parameter 'x'. The distribution in which the variable, after the transformation ln $\rho X+p$ 'X', includes the parameter 'ρ', follows the normal distribution N (μ, σ^2) for the positive random variable X. The dose distribution of workers engaged in the work has been affected by exposure reduction management and has approached the lognormal distribution side when the management is weak and the normal distribution side when the management is strong. In this paradigm, we describe the nature and mechanism of occurrence of a mixed lognormal distribution, which explains the fitting of dose distributions. This distribution constitutes a new type of distribution system that has a lognormal distribution on one side and a normal distribution on the other side as its extreme distributions (Figures 5.21–5.24) (Table 5.18).

$$\text{Hyb Log } N(\rho X,\mu,\sigma) = \frac{p}{\sqrt{2\pi\sigma}}\left(1+\frac{1}{\rho x}\right)e^{-\frac{1}{2}\left(\frac{\rho x+\ln(\rho x)-\mu}{\sigma}\right)^2} \quad (5.41)$$

$$P(\rho X,\mu,\sigma) = \int_0^x f(t,\mu,\sigma)dt \quad \text{Lower cumulative distribution} \quad (5.42)$$

TABLE 5.16
Probability Density Function of Poisson Distribution

X	0	1	2	3	4	5	6	7	8	9	10
PMF Poisson distribution	0.001	0.002	0.004	0.011	0.026	0.049	0.078	0.105	0.124	0.130	0.123

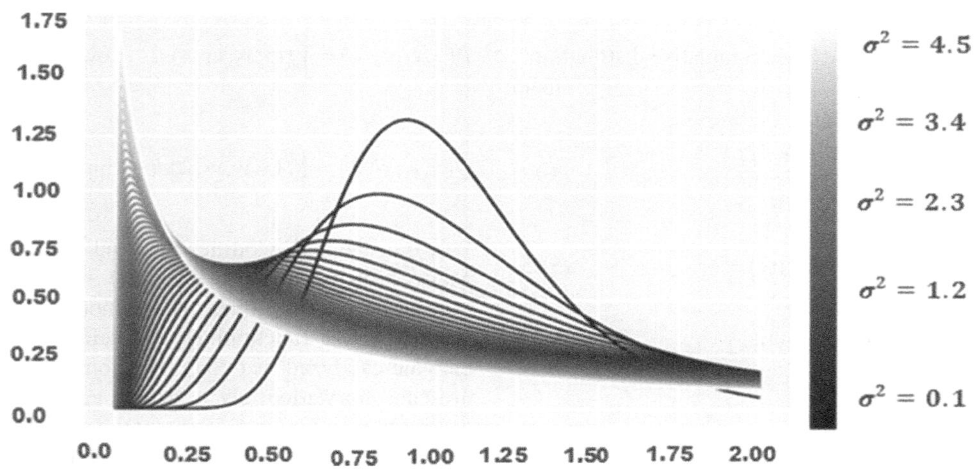

FIGURE 5.21 The lognormal distribution (1.0).

TABLE 5.17

Probability Density Function of Logarithmic Distribution

PDF	Lower Cumulative P	Upper Cumulative Q	Mean	Median	Mode
6.07E-7	2.16 E-7	1	73,017.2	12,708.2	384.945

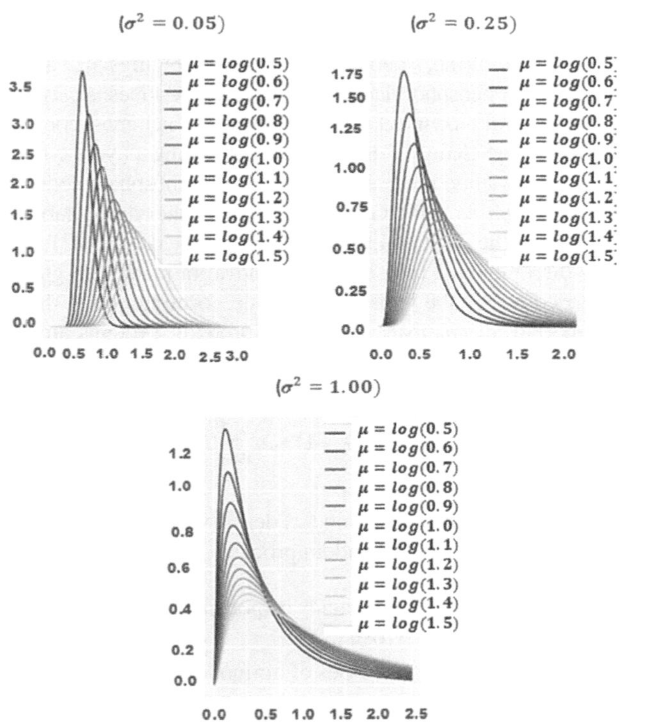

FIGURE 5.22 Lognormal distribution for variances of (σ^2=0.05), (σ^2=0.25).

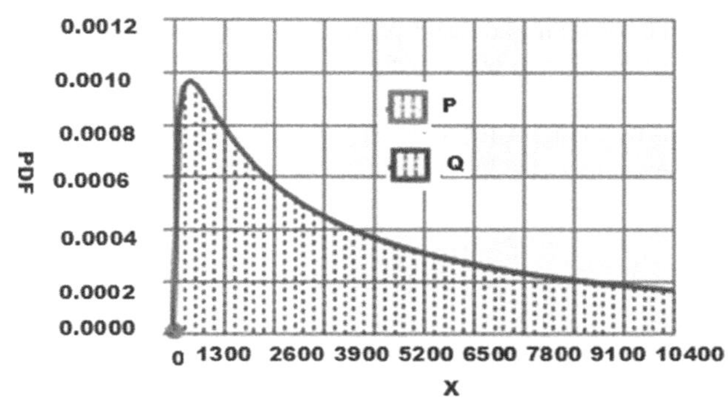

FIGURE 5.23 PDF of the logarithmic normal distribution.

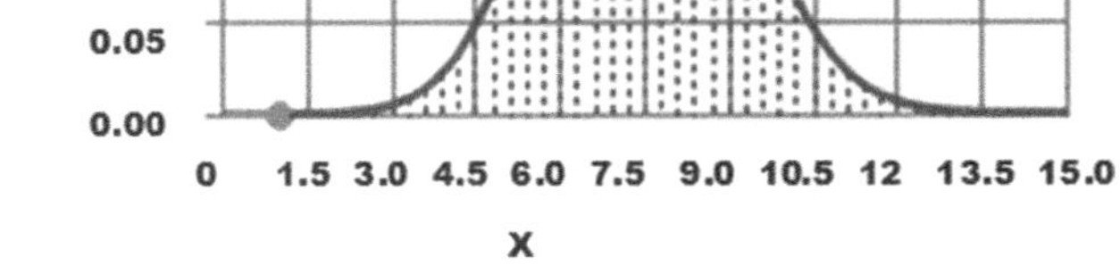

FIGURE 5.24 PDF of the hybrid lognormal distribution.

TABLE 5.18

Probability Density Function of Hybrid Logarithmic Distribution

PDF	P	Q	Median
1.577 E-5	3.118 e-6	0.999	7.442

$$Q(\rho X,\mu,\sigma)=\int_X^\infty f(t,\mu,\sigma)dt \quad \text{Upper cumulative distribution} \tag{5.43}$$

$$\text{Median}: x_c=\frac{\text{hyb}^{-1}(\mu)}{\rho} \tag{5.44}$$

$$\text{hyb}(\rho x)=\rho x+\ln(\rho x)=y \tag{5.45}$$

$$\rho x=\text{hyb}^{-1}(y)=\text{cyb}(y) \tag{5.46}$$

5.25 Application of Noncentral *F* Distribution

This distribution will need to be computed using the power of tests with multiple outliers. Non-linear models were applied to fix the location and scale parameters (no central generalized *F* distributions) (Figure 5.25). The F-distribution was used to explore the new genomics of terminal illnesses or malignant stages by changing the location and scale parameters at selected WGAs, using the following mathematical equations: A with random variables is formulated by equation (1.28) (Table 5.19).

$$f(x,v_1,v_2,\lambda)=\sum_{j=0}^{\infty}\frac{e^{-\lambda/2}\left(\frac{\lambda}{2}\right)^j v_1^{\frac{v_1}{2}} v_2^{\frac{v_2}{2}} x^{\frac{v_1}{2}-1}}{j!\,B\left(\frac{v_1}{2},\frac{v_2}{2}\right)\left(v_2+v_j x\right)^{\frac{v_1+v_2}{2}}} \text{ where } v_j \tag{5.47}$$

$$=v_1+2j$$

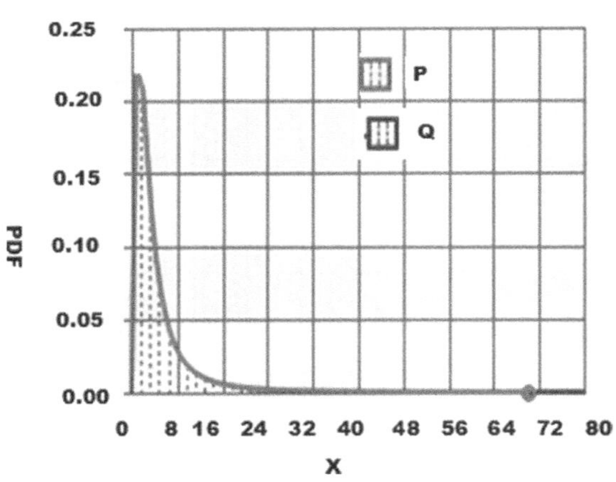

FIGURE 5.25 Non-central *F* distribution, denoted as $F(k_1, k_2, \lambda)$.

TABLE 5.19

Probability Density Function of Non-Central *F* Distribution

PDF	*P*	*Q*
1.02	0.99	0.003

$$P(x,v_1,v_2,\lambda)=\int_0^x f(x,v_1,v_2,\lambda)dt \quad \text{Lower cumulative} \tag{5.48}$$

$$Q(x,v_1,v_2,\lambda)=\int_X^\infty f(x,v_1,v_2,\lambda)dt \quad \text{Upper cumulative} \tag{5.49}$$

Post-translational modification and structure of proteins: Many proteins undergo chemical modifications with certain amino acid residues. These modifications are essential for the normal functioning of proteins through one or more enzyme-catalyzed reactions. The commonly observed PTMs are phosphorylation, glycosylation, acylation, alkylation, and hydroxylation. Figure 5.26 shows the gene sequence at the top, followed by an arrow pointing to the left and the blue structure. This must be visualized to show the inset below. A red cross must appear on this arrow. Next, an arrow pointing to the right must appear, followed by the pink structure, which must again be zoomed in to show the inset below. The final structure of functional proteins will more often not correlate directly with the corresponding gene sequences because PTMs occur at various amino acid residues in the protein, which cause changes in interactions between the amino acid side chains, thereby modifying the protein structures. This further increases the complexity of the proteome compared to the genome.

5.26 Phosphorylation

Phosphorylation is the process by which a phosphate group is attached to certain amino acid side chains in the protein, most commonly serine, threonine, and tyrosine. Figures 5.27 and 5.28 depict that phosphorylation of amino acid residues is carried out by a class of enzymes known as kinases, which most commonly modify side chains of amino acids containing a hydroxyl group. Phosphorylation requires the presence of phosphate donor molecules such as ATP, GTP, or other phosphorylated substrates. Serine is the most commonly phosphorylated residue, followed by threonine and tyrosine. The removal of phosphate groups is carried out by the phosphate enzyme, forming one of the most important mechanisms for regular proteins. Proteins are mathematically expressed in equation (5.31).

$$f(x)=(n-k+1)+a_i\sum_{i=1}^{ni}\beta_{ij}+\epsilon_{ij} \tag{5.50}$$

where, $f(x)$ = The function $f(x)$ describes the process of phosphorylation; (x) is the random process.

- n = Number of amino acids responsible for phosphorylation (serine, tyrosine)
- k = Number of successes of amino acids are repeated
- a_i = Regression intercept (point of inflection)
- β_{ij} = Regression co effcient with respect to ATP and ADP molecules during the kinase (At ith enzyme with jth amino acid)
- ϵ_{ij} = Phosphorylated residual error with ith ATP/ADP molecule with jth amino acid phosphorylated

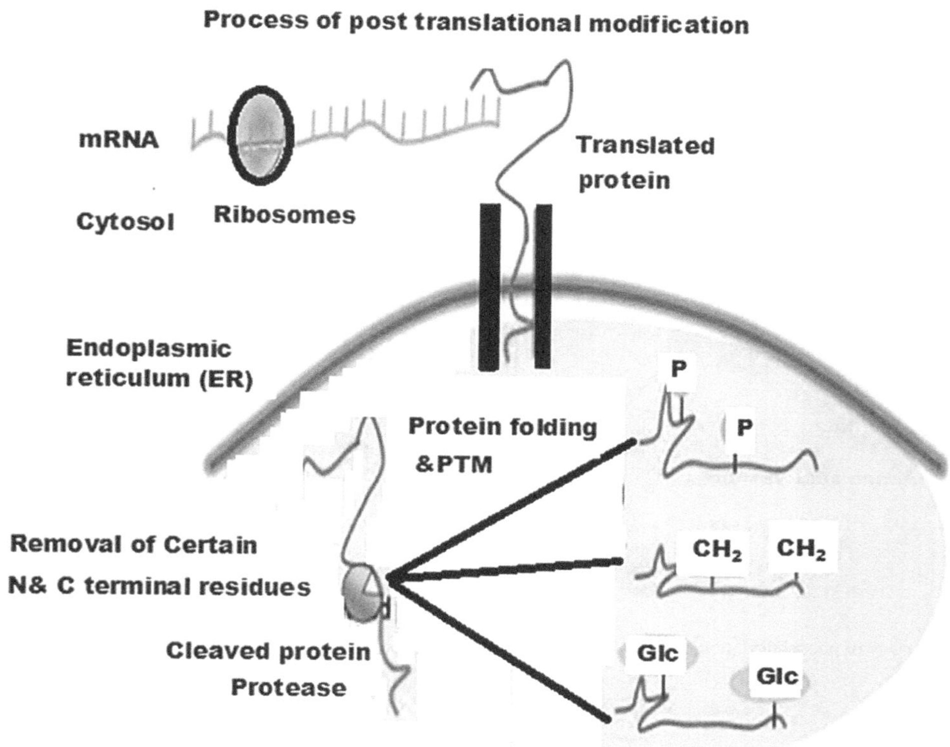

FIGURE 5.26 Pathway of post-translational modifications (PTMs).

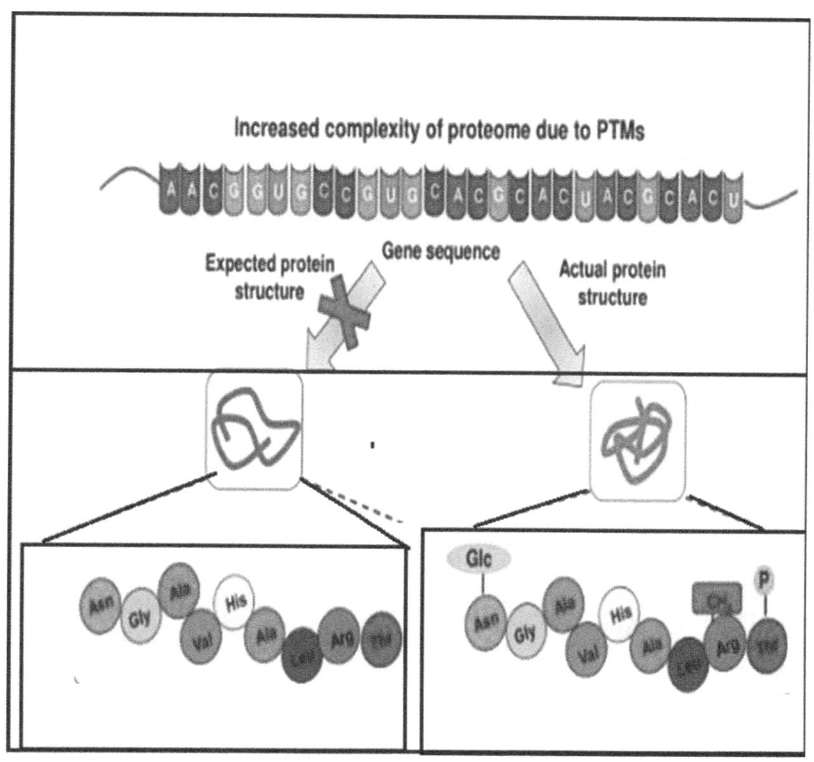

FIGURE 5.27 Complexity of the proteome resulting from PTMs.

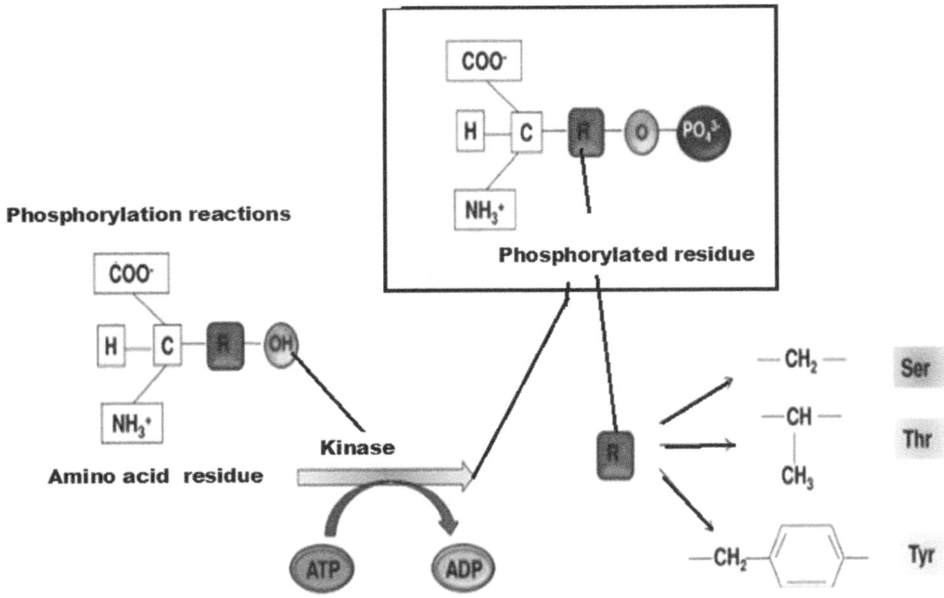

FIGURE 5.28 The pathway of phosphorylation reactions.

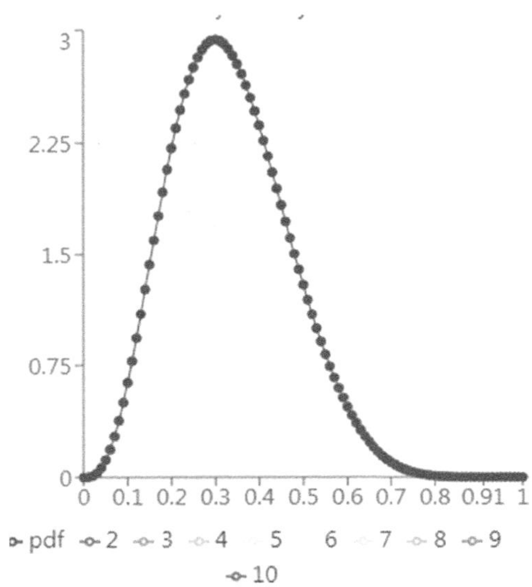

FIGURE 5.29 Density function of phosphorylation.

$$(\alpha = k+1); (\beta = n-k+1) \text{ Example } \alpha = k+1$$

$$= (7+1) = 8,\ \beta = 10-7+1 = 4$$

The random behavior of the proportion of phosphorylation rate or percentage was modeled by $\beta-$distribution. It provides a powerful tool to quantify the tasks that have kinase enzymatic reactions with ATP and ADP molecules (Figures 5.29 and 5.30). The probability density function is

$$f(x,\alpha,\beta) = \frac{x^{\alpha-1}(1-x)^{\beta-1}}{B(\alpha,\beta)} \alpha > 0; \beta > 0; 0 \le x \le 1 \tag{5.51}$$

Beta distribution is the conjugate of Binomial distribution.

$$f(x) = \binom{n}{x} \theta^x (1-\theta)^{n-x} \text{ BD} \tag{5.52}$$

$$g(\theta) = \frac{1}{B(\alpha,\beta)} \theta^{x-1} (1-\theta)^{n-x} \tag{5.53}$$

5.26.1 Pathway of Glycosylation

The attachment of sugar moieties to N_2 or O_2 atoms in the side chains of amino acids like asparginine, serine, or threonine.

5.26.2 Glycosylation

It involves the enzymatic addition of saccharide molecules to amino acid side chains. There are two primary forms of glycosylation: N-linked glycosylation, which involves the attachment of sugar residues to the amide group of asparagine, and O-linked glycosylation, which connects sugar moieties to the hydroxyl groups of serine or threonine. These reactions are facilitated by specific glycosyltransferase enzymes.. Sugar residues that are most commonly attached include galactose, mannose, glucose, N-acetylglucosamine, and N-acetylgalactosamine as well.

5.26.3 Acylation

The process by which an acyl group is linked to the side chain of amino acids like asparagine, glutamine, or lysine.

5.26.4 Alkylation

The addition of alkyl groups, most commonly a methyl group, to amino acids such as lysine or arginine. Other longer-chain alkyl groups may also be attached in some cases.

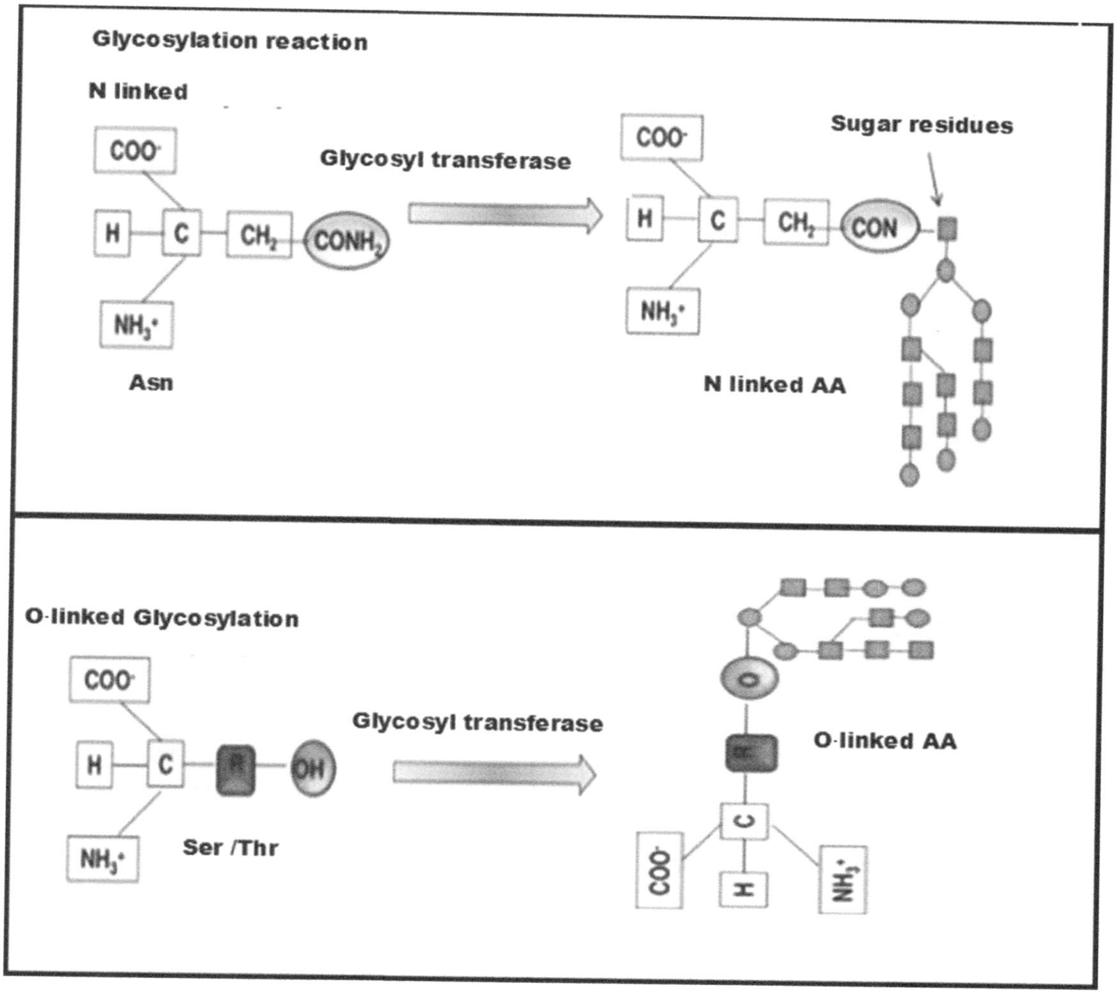

FIGURE 5.30 The process of glycosylation.

5.26.5 Hydroxylation

This PTM is commonly found on proline and lysine residues, which make up the collagen tissue. It enables cross-linking and, therefore, strengthens the muscle fibers (Table 5.20).

Compared to the two mean populations of mannose and glucose with respect to different Golgi (cis, medial, and trans), we use the following formula:

Step 1: Take the mean differences of mannose and glucose in the cis, medial, and trans Golgi

$\left(X_{1i}-\overline{X_{2i}}\right)^2$=Squared value of the mean differences of mannose and glucose with respect to the cis Golgi

$\left(X_{3i}-\overline{X_{4i}}\right)^2$=Squared value of the mean differences of mannose and glucose with respect to the medial Golgi

$\left(X_{5i}-\overline{X_{6i}}\right)^2$=Squared value of the mean differences of mannose and glucose with respect to the trans Golgi

Step 2: Determine the calculated t values of the cis Golgi

$$t_{\text{calvalue}}=\frac{\left(X_{1i}-\overline{X_{2i}}\right)^2}{\text{SP}_1^2\left(1/n_2+1/n_2\right)} \tag{5.54}$$

where, SP_1^2 = Pooled varaince

$$\text{SP}_1^2=\frac{S_1^2(n_1-1)+S_2^2(n_2-1)}{n_1+n_2-2} \tag{5.55}$$

where, $S_1=\sqrt{\frac{\sum_{i=1}^{n}\left(X_{1i}-\overline{X_{2i}}\right)^2}{n_1-1}}=S_1^2$ Variance of sample X_1

$$S_2=\sqrt{\frac{\sum_{i=1}^{n}\left(X_{1i}-\overline{X_{2i}}\right)^2}{n_1-1}}={S_2}^2 \text{ Variance of sample } X_2$$

Substituting the variance components into equation (5.35), we will get the pooled variance. By applying the t-calculated formula, we can obtain the t-value and simultaneously compare it to the table

TABLE 5.20
Model Formulation or Statistical Method

	Cis Golgi		Medial Golgi		Trans Golgi	
Glycosylation	**Mannose**	**Glucose**	**Mannose**	**Glucose**	**Mannose**	**Glucose**
Alpha–Man I	X_{11}	X_{21}	X_{31}	X_{41}	X_{51}	X_{61}
Gn TI	X_{12}	X_{22}	X_{32}	X_{42}	X_{52}	X_{62}
Fuc T	X_{13}	X_{23}	X_{33}	X_{43}	X_{53}	X_{63}
Xylosyltransferase Gn T II Gal TI	X_{1i}	X_{2i}	X_{3i}	X_{4i}	X_{5i}	X_{6i}
Mean $\bar{X}$	$\sum_{i=1}^{n}\frac{X_{1i}}{n_1}$	$\sum_{i=1}^{n}\frac{X_{2i}}{n_2}$	$\sum_{i=1}^{n}\frac{X_{3i}}{n_3}$	$\sum_{i=1}^{n}\frac{X_{4i}}{n_4}$	$\sum_{i=1}^{n}\frac{X_{5i}}{n_5}$	$\sum_{i=1}^{n}\frac{X_{6i}}{n_6}$
SD (σ_i)			$\sqrt{\frac{\sum_{i=1}^{n}(X_i - X_i)^2}{n-1}}$			
SEM $\left(\frac{\sigma}{\sqrt{N}}\right)$			$\frac{\sqrt{\frac{\sum_{i=1}^{n}(X_i - X_i)^2}{n-1}}}{\sqrt{n}}$			
CV (%)			$\frac{\sigma_i}{x_i}*100$			

t-value (at a 1% level of significance). Compare the calculated t-value and the table t-value. If we reject the null hypothesis (H_0), it means that the mean differences of mannose and glucose with respect to the cis Golgi are positively associated and statistically significant. If we accept the null hypothesis (H_0), that means that the mean differences of mannose and glucose with respect to the cis Golgi are negatively associated and statistically insignificant ($p > 0.05$).

Step 3: Determine the calculated t-value of the medial Golgi

$$t_{\text{calvalue}} = \frac{\left(X_{3i} - \overline{X_{4i}}\right)^2}{\text{SP}_2^2\left(\frac{1}{n_3} + \frac{1}{n_4}\right)} \tag{5.56}$$

where, SP_2^2 = Pooled varaince

$$\text{SP}_2^2 = \frac{S_3^2(n_3 - 1) + S_2^2(n_3 - 1)}{n_3 + n_4 - 2} \tag{5.57}$$

where, $S_3 = \sqrt{\frac{\sum_{i=1}^{n}\left(X_{3i} - \overline{X_{4i}}\right)^2}{n_3 - 1}} = S_1^2$ Variance of sample X_1

$$S_4 = \sqrt{\frac{\sum_{i=1}^{n}\left(X_{3i} - \overline{X_{4i}}\right)^2}{n_4 - 1}} = S_2^2 \text{ Variance of sample } X_2$$

Substitute the variance components into equation (1.37), and we will get the pooled variance. Follow the same procedure as above for testing the H_0

Step 4: Determine the calculated t-values of the trans Golgi

$$t_{calvalue} = \frac{\left(X_{5i} - \overline{X_{6i}}\right)^2}{SP_3^2\left(\frac{1}{n_5} + \frac{1}{n_6}\right)}$$

where, SP_3^2 = Pooled varaince

$$SP_3^2 = \frac{S^2{}_5(n_5 - 1) + S^2{}_6(n_6 - 1)}{n_5 + n_6 - 2} \tag{5.58}$$

where, $S_5 = \sqrt{\frac{\sum_{i=1}^{n}\left(X_{5i} - \overline{X_{6i}}\right)^2}{n_5 - 1}} = S_5^2$ Variance of sample X_1

$$S_6 = \sqrt{\frac{\sum_{i=1}^{n}\left(X_{5i} - \overline{X_{6i}}\right)^2}{n_5 - 1}} = S_6^2 \text{ Variance of sample } X_2$$

Substitute the variance components into equation (1.39), and we will get the pooled variance. Follow the same procedure as above for testing the H_0.

CI (99%): $\overline{X_i} - z\frac{SD}{\sqrt{n}}$ where table Z value = 1.96 Lower boundary

CI (99%): $\overline{X_i} + z\frac{SD}{\sqrt{n}}$ where table Z value = 1.96 upper boundary

Step 4: Hypothesis testing at population level (N)

When mannose and glucose are produced in different Golgi, we can apply the formula.

$$Z = \frac{(X_i - \mu_k)^2}{\sigma^2} \sim \chi^2(n) \tag{5.59}$$

where μ_k = Population mean at kth mannose and glucose, $\left(\frac{\sum_{i=1}^{N} X_i}{N}\right)$ $\sigma(\text{SD}) = \sqrt{\frac{(X_i - \mu_i)^2}{N}} = \sigma^2$ (Square of the SD)

Step 5: Population homogeneity of the cis, medial, and trans by chi-square (Table 5.21)

TABLE 5.21

Glycosylation Pathway Data Series

	Cis Golgi		Medial Golgi		Trans Golgi		
Glycosylation	**Mannose**	**Glucose**	**Mannose**	**Glucose**	**Mannose**	**Glucose**	**Row total**
Alpha–Man I	X_{11}	X_{21}	X_{31}	X_{41}	X_{51}	X_{61}	$\sum R_1$
Gn TI	X_{12}	X_{22}	X_{32}	X_{42}	X_{52}	X_{62}	$\sum R_2$
Fuc T	X_{13}	X_{23}	X_{33}	X_{43}	X_{53}	X_{63}	$\sum R_3$
Xylosyltransferase Gn T II Gal TI	X_{1i}	X_{2i}	X_{3i}	X_{4i}	X_{5i}	X_{6i}	$\sum R_4$
Column total	$\sum C_1$	$\sum C_2$	$\sum C_3$	$\sum C_4$	$\sum C_5$	$\sum C_6$	**Grand total (*N*)**

TABLE 5.22

One-Way ANOVA (Only One Factor)

Source of Variation	***df***	**Sum of Square (SS)**	**Mean Sum of Square (MSS)**	**F_{cal} (Fisher Value)**
Between the variation	$(K-1)$	SB^2(SSB)	MSB=SSB/$(K-1)$	F = MSB/MSE
Within the variation	$(N-K)$	SW^2 (SSW) or SSE	MSE=SSW/$(N-K)$	
Total	$(N-1)$	ST^2 (TSS)		

We can apply the chi-square test for testing the population homogeneity.

$$\chi^2 = \sum_{i=1}^{N} \frac{\left(O_{ij} - E_{ij}\right)^2}{E_{ij}} \tag{5.60}$$

where, O_{ij} = Observed value of ith row and jth column

E_{ij} = Expected value of ith row and jth column

$$E_{ij} = \frac{\text{Row total} * \text{Column total}}{\text{Grand total } (N)}$$

$\chi^2_{\text{Caluclated}} \geq \chi^2_{\text{table}df}$; df=(Column-1) (Row-1) reject the H_0 (production of mannose and glucose from Golgi found to be significant. $\chi^2_{\text{Caluclated}} \leq \chi^2_{\text{table}df}$ = df=(Column-1) (Row-1) reject the H_0 (production of mannose and glucose from Golgi found to be insignificant).

Step 5: Population homogeneity of cis, medial, and trans by ANOVA (analysis of variance)

Hypothesis setting (null and alternative hypotheses) (Table 5.22)

$$H_0 = \mu_1 = \mu_2 = \mu_3 = \mu_4 = \mu_5 = \mu_6$$

$$H_1 = \mu_1 \neq \mu_2 \neq \mu_3 \neq \mu_4 \neq \mu_5 \neq \mu_6$$

$$H_0 = \sigma^2{}_1 = \sigma^2{}_2 = \sigma^2{}_3 = \sigma^2{}_4 = \sigma^2{}_5 = \sigma^2{}_6$$

$$H_1 = \sigma^2{}_1 \neq \sigma^2{}_2 \neq \sigma^2{}_3 \neq \sigma^2{}_4 \neq \sigma^2{}_5 \neq \sigma^2{}_6$$

Sub-step 1: Determine degrees of freedom (df)= where k=4

Between the variation $(K-1)=(4-1)=3$

Within the variation $(N-K)=(24-4)=20$

Total $df=(N-1)$ $(24-1)=23$

Step 2: Determine the grand total (GT)=(Row total and column total)

Step 3: Correction factor (CF)=$\frac{(\text{GT})^2}{N}$

Step 4: Determine total sum of square (TSS)= $\sum X_{ij}^2 - CF$

Step 5: Calculate between the variations

$$SB^2(\text{SSB}) = \frac{\sum_{i=1}^{4} R^2{}_i}{t} - \text{CF}$$

Step 6: Calculate within the variation

SW^2 (SSW) or SSE=Total Sum of Square (TSS)−Sum of the square between the variation (SSB)

Step 7: Calculate the mean sum of squares between and within the variations

$$\text{MSB} = \text{SSB} / (K-1)$$

$$\text{MSE} = \text{SSW} / (N-K)$$

Step 8: Determine the Fisher-Snedecor critical calculated value (F-cal value)

$$F = \text{MSB/MSE}$$

Step 9: To test the hypothesis (H_0) by using a table and calculated F-value

TABLE 5.23

Two-Way ANOVA (Only Two Factor)

Source of Variation	*df*	Sum of Square (SS)	Mean Sum of Square (MSS)	F_{cal} (Fisher Value)	F_{table}
Treatment	$(t-1)$	St^2 (SSt)	$MSt = St^2/(t-1)$	F_t=MSt/MSe	$df\,(t-1),\,(t-1)(r-1)$
Replication	$(r-1)$	Sr^2 (SSR)	$MSr = Sr^2/(r-1)$	F_r=MSr/MSe	$df\,(r-1),\,(t-1)(r-1)$
Error	$(t-1)(r-1)$	Se^2 (SSE)	$MSe = Se^2/(t-1)(r-1)$		
Total	$(r*t-1)$	ST^2 (TSS)			

If the $F_{cal} \geq F_{table}$ Reject the null hypothesis extract table value $(K-1, N-K)$ *df* from the critical *F*-table value. If the $F_{cal} \leq F_{table}$ accept the null hypothesis (Table 5.23).

Sub-step 1: Determine degrees of freedom (*df*), where (k=4)

Treatment $df\,(t-1)=(4-1)=3$

Replication $df\,(r-1)=(24-4)=20$

Total $df=(rt-1)\;(24-1)=23$

Step 2: Determine the grand total (GT)=(Row total and column total)

Step 3: Correction factor (CF)=$\frac{(GT)^2}{N}$

Step 4: Determine the total sum of square (TSS) $=\sum X_{ij}^2 - \text{CF}$

Step 5: Calculate the treatment sum of square

$$St^2(\text{SST}) = \frac{\sum_{i=1}^{4} T_i^2}{r} - \text{CF}$$

Step 6: Calculate replication sum of square

$$Sr^2(\text{SSR}) = \frac{\sum_{i=1}^{4} R_i^2}{t} - \text{CF}$$

Step 7: Calculate the error sum of the square. Take the differences of TSS−SSt−SSR

Step 8: Mean sum of the square of treatment, replication, and errors

$$\text{MSt} = \text{St}^2/(t-1);\ \text{MSr} = \text{Sr}^2/(r-1);\ \text{MSe} = \text{Se}^2/(t-1)(r-1)$$

Step 9: Determine the Fisher *F*-calculated value

F_t=MSt/MSe

F_r=MSr/MSe

Step 10: Extract the critical value from the table

$df\,(t-1),(t-1)\times(r-1)$ with respect to treatment

$df\ (r-1),\ (t-1)\times(r-1)$ with respect to replication

Step 10: To test the hypothesis (H_0) by using a table and calculated *F*-value

If the $F_{cal} \geq F_{table}$ reject the null hypothesis extract table value $(t-1)$, $(t-1)\times(r-1)$ *df* from the critical *F*-value table. If the $F_{cal} \leq F_{table}$ accept the null hypothesis.

5.27 DNA Sequencing

It refers to the general laboratory technique for determining the exact sequence of nucleotides, or bases, in a DNA molecule. The sequence of the bases (often referred to by the first letters of their chemical names: A, T, C, and G) encodes the biological information that cells use to develop and operate. Establishing the sequence of DNA is key to understanding the function of genes and other parts of the genome. There are now several different methods available for DNA sequencing, each with its own characteristics, and the development of additional methods represents an active area of genomics research. DNA consists of a linear string of nucleotides, or bases, referred to, for simplicity, by the first letters of their chemical names—A, T, C, and G. The process of deducing the order of nucleotides in DNA is called DNA sequencing. Since the DNA sequence confers information that the cell uses to make RNA molecules and proteins, establishing the sequence of DNA is key to understanding how genomes work (Li et al., 2020). The technology for DNA sequencing was made faster and less expensive as a part of the Human Genome Project. Recent developments have profoundly increased the efficiency of DNA sequencing even further (Table 5.24).

From n' amino acids, we need to choose the amino acid (n_1=type$_1$, n_2=type$_2$, and nk is type kth) based on binding.

$$\binom{n}{n_1 n_2 \ldots n_k} = \frac{n!}{n_1!\ n_2! \ldots n_k!} \tag{5.61}$$

Let N_i denote the number of results in the category 'i' in a sequence of independent amino acids with probability p_i for a result in the ith category on each time binding $1 \leq i \leq m$, where $p_1 + p_2 \ldots p_m = 1$. Then, for every m tuple of non-negative integers $n_1, n_2 \ldots n_m$ with sum 'n'

$$P(N_1 = n_1,\ N_2 = n_2 .. N_m = n_m = \frac{n!}{n_1!\ n_2! \ldots n_k!} P_1^{n_1} P^{n_2}{}_2 P_3^{n_3} .. P_m{}^{n_m} \tag{5.62}$$

$$P(N_1 = n_1,\ N_2 = n_2 .. N_m = n_m = \frac{n!}{n_1!\ n_2! \ldots n_k!} \sum_{i=1}^{m} P_m{}^{n_m} \tag{5.63}$$

$$P(N_1 = n_1,\ N_2 = n_2 .. N_m = n_m = \binom{n}{n_k} \sum_{i=1}^{m} P_m{}^{n_m} \tag{5.64}$$

TABLE 5.24

Genetic Code Correlates on Different Types of Amino Acid

First Base Codons		Second Base Codons							Third base codons
	Codons	**U**	**Codons**	**C**		**A**	**Codons**	**G**	
U	UUU	Phenylalanine	UCU	Serine	UAU	Thyrosine	UGU	Cysteine	U
	UUC	Phe	UCC	Se	UAC		UGC	Cys	C
	UUA	Leucine	UCA		UAA	Stop	UGA	Stop	G
	UUG	Leu							
			UCG		UAG		UGG	Tryptophan	A
								Trp	G
C	CUU	Leucine	CCU	Proline	CAU	Histidine	CGU	Arginine	U
	CUC	Leu	CCC	Pro	CAC	His	CGC	Arg	C
	CUA		CCA		CAA	Glutamine	CGA		A
	CUG		CCG		CCG	Gln	CGG		G
A	AUU	Isoleucine	ACU	Threonine	AAU	Asparagine	AGU	Serine	U
	AUC	Ile	ACC	Thr	AAC	Asn	AGC	Ser	C
	AUA		ACA		AAA	Lysine	AGA	Arginine	A
	AUG	Methionine	ACG		AAG	Lys	AGG	Arg	G
		Met							
G	GUU	Valine	GCU	Alanine	GAU	Aspartic acid	GGU	Glycine	U
	GUC	Val	GCC		GAC	Asp	GGC	Gly	C
	GUA		GCA		GAA	Glutamic acid	GGA		A
	GUG		GCG		GAG	Glu	GGG		G

$$L(p)=\prod_{i=1}^{N} P_1^{n_{1,1}} \ldots P_m^{n_{11,m}} = p_m \sum_{i=1}^{N} m_{i\ 1\ m} \tag{5.65}$$

The product of 'N' probability mass functions (PMFs), each M $(1, p)$, depends upon the sample data only through the vector of 'k' frequency counts of amino acids. By using the factorization theorem, sufficient statistics were provided for the parameter of interest. The frequencies of P_1, P_2, and later p-values with multiple subscripts will refer to frequencies in a multidimensional table, such as the frequencies in row i (the frequencies of the ith amino acid with the jth row and kth column). The likelihood function becomes:

$$L(p)= p_1^{n_1} \ldots p_i^{n_1} \ldots = p_1^{n_1} \ldots p_{n-1}^{pi-1}\left(1-\sum_{i=1}^{j-1} p_i\right)^{N-\sum_{i=1}^{p-1} ni} \tag{5.66}$$

Likelihood ratio test $X_1 \ldots . X_N \sim F_\theta, \theta \in \Theta$; $H_0 : \theta \epsilon \Theta_0 H_A : \theta \epsilon \Theta_0 \cap \Theta_0$

Then, $G^2 = -2\ln\left(\frac{\max \theta \epsilon \Theta_0 L(\theta)}{\max \theta \epsilon \Theta L(\theta)}\right)$ under H_0 : $H_0 : \theta \epsilon \Theta_0 H_A : \theta \epsilon \Theta_0 \cap \Theta_0$ has an approximate chi-square distribution for large 'N', rejected when G^2 is large (Tables 5.25 and 5.26)

Wu et al. (2021) stated that protein engineering seeks to identify protein sequences with optimized properties. When guided by machine learning, protein sequence generation methods can draw on prior knowledge and experimental efforts to improve this process. In this review, they highlight recent applications of machine learning to generate protein sequences, focusing on the emerging field of deep generative methods (Table 5.27).

Table 5.24 shows the 66 residues (86% of the sequence) modeled with 99.3% confidence using the single highest-scoring template. The PDB header has a membrane protein, with receptor chain sequence modeling of SARS-CoV-2 (Table 5.28).

$$R = -0.136,$$

Peptide sequencing homology modelling:

$$S_n = a_i + (n-1)d_j \tag{5.67}$$

$$\text{Unstable Random walk (URW)} = \left[\frac{S_n}{1+\exp^{-a_i+b_j(X_0-X_n)}}\right] * F_i \tag{5.68}$$

where S_n = Sequencing of specific peptides of whey looping homology at nth amino acids, a_i = Homology of peptide sequencing looping takes place at ith amino acids position, $d_j = \frac{\Delta a_1}{\Delta a_2}$ incremental difference between the peptide looping from first and subsequent peptide sequencing responsible for looping; b_j = Regression coefficient; $X_0 - X_n$ = differences of Largest sequencing – smalest sequencing position; F_i = The frequency of particular amino acids for looping.

5.28 Calculation Summary

Sum of X=369; Sum of Y=33,630; Mean X=10.8529; Mean Y=989.1176

Sum of squares (SS_X)=1,006.2647; Sum of products (SP)=−12,964.4118

Regression equation=$\hat{Y}$=b X+a; b=SP/SS$_X$=−12,964.41/1,006.26=−12.8837

TABLE 5.25

Proportion and P-Value for Binding of Amino Acids with Repeatation of Uracil

Amino Acids	Codons	Frequency of Uracil	Total Codons	Proportion (n/N)	Multinomial P-Value
Phenylalanine (Phe)	UUU,UUC	5	6	0.83	1
Leucine (Leu)	UUA,UUG,CUU,CUC CUA,CUG	9	18	0.50	–
Isoleucine (Ile)	AUU,AUC,AUA	4	9	0.44	0.14
Methionine (Met)	AUG	1	3	0.33	0.39
Valine (Val)	GUU,GUC,GUA,GUG	5	12	0.41	0.12
Serine (Ser)	UCU,UCC,UCA,UCG	5	12	0.41	0.12
Proline (Pro)	CCU,CCC,CCA,CCG	1	12	0.08	0.38
Theronine (Thr)	ACU,ACC,ACA,ACG	1	12	0.08	0.38
Alanine (Ala)	GCU,GCC,GCA,GCG	1	12	0.08	0.38
Tyrosine (Tyr)	UAU,UAC	3	6	0.50	–
Stop	UAA,UAG,UAA,UAG	4	12	0.33	0.17
Histidine (His)	CAU,CAC	1	6	0.16	0.39
Glutamine (Gln)	CAA,CAG	0	6	–	–
Asparagine (Asn)	AAU,AAC	1	6	0.16	0.39
Lysine (Lys)	AAA,AAG	0	6	–	–
Aspartitic acid	GAU,GAC	1	6	0.16	0.39
Glutamic acid (Glu)	GAA,GAG	0	6	–	–
Cysteine	UGU,UGC	3	6	0.50	–
Tryptophan (Trp)	UGG,	1	3	0.33	0.39
Arginine (Arg)	CGU,CGC,CGA,CGG	1	12	0.08	0.38
Glycine (Gly)	GGU,GGC,GGA,GGA GGG	1	15	0.06	0.37
t-test	t=−6.42; CI-95%−4.50−8.63				

TABLE 5.26

Weibull Distribution of Uracil and Total Codons

	Total Codons (N)	Uracil Codons (n)
Distribution	Estimated ± SD	Estimated ± SD
Multinomial Distribution		
Shape (α)/scale (β)	11.25 ± 0.51	8.69 ± 1.38
Weibull Distribution		
Shape (α)	2.40 ± 0.40	1.80 ± 1.44
Scale (β)	10.02 ± 0.961	7.44 ± 1.45
Logistic Distribution		
Shape (α)	8.63 ± 0.93	5.92 ± 1.30
Scale (β)	2.38 ± 0.41	2.22 ± 0.62

TABLE 5.27

Protein Sequence Alphabetic Upper Letters

Letters	Amino Acid	Letters	Amino Acid
A	Alanine	R	Arginine
N	Asparagine	D	Aspartic acid
C	Cysteine	Q	Glutamine
G	Glycine	E	Glutamate
H	Histidine	I	Isoleucine
L	Leucine	K	Lysine
M	Methionine	F	Phenylalanine
P	Proline	S	Serine
T	Threonine	W	Tryptophan
Y	Tyrosine	V	Valine
*	Stop		

TABLE 5.28

SARS-CoV-2 Peptide Sequencing Alpha Helix Looping

Peptide Sequencing Homology	F_i		Position a_n	No. of Amino Acids S_n
veevlsearqhl	Glutamate	3	40	12
vayrkvll	Leucine	2	120	8
trelmrel	Leucine	2	170	8
lecikdllar	Leucine	4	200	10
tlseql	Leucine	2	220	6
ndnlleilqkekv	Leucine	3	450	13
safvetvk	Valine	2	490	8
kafkqives	Lysine	2	500	9
lyafasea	Alanine	2	540	8
arvvrsifsrtleta	Arginine	3	550	15
sqyslrlidammfi	Methionine	2	580	14
ltnifgtvyeklkpv ldwleek fkegveflrdgweiv	Glutamate	5	650	37
ikesvqtffklvnkf lalca	Lysine	3	720	
lne	–	0	850	3
facvvadaviktl	Valine	3	880	12
gvagalnk	Alanine	2	1,070	8
namqvesddyia	Aspartic acid	2	1,090	12
ediqllksayenf	Leucine	2	1,130	12
pilhlrvcvdtv	Valine	3	1,160	13
lydklvssf	Serine	2	1,190	9
ekqvieqk	Lysine	2	1,200	8
temlakalrk	Lysine	2	1,310	10

(*Continued*)

TABLE 5.28 (*Continued*)

SARS-CoV-2 Peptide Sequencing Alpha Helix Looping

Peptide Sequencing Homology	F_i		Position a_n	No. of Amino Acids S_n
k t v l k k c k	Lysine	4	1,350	8
w n l r e m l a h a e e	Glutamate	3	1,380	12
a i v s t i q r k y	Isoleucine	2	1,400	10
v a s l i n i l n d	Leucine	2	1,440	10
l e e a a r y m	Alanine	2	1,460	8
h f i e t i s l	Isoleucine	2	1,500	8
l k t l l s l r e v	Leucine	4	1,560	10
p s f l g r y m s a l	Leucine	2	1,650	11
c y l a t a l t l	Leucine	3	1,680	9
p a l q d a y y r a r	Tyrosine	2	1,700	12
n f c a l i lay	Leucine	2	1,710	9
v r e t m s y l f	-	0	1,730	9

$a=M_Y - b\ M_X=989.12-(-12.88*10.85)=1{,}128.94368$; $\hat{Y}=-12.8837X+1{,}128.94368$

Cohen's $d=(989.11-10.85)/44.72=21.87$; Glass's $delta=(989.11-10.85)/1.22=801.85$

Hedges' $g=(989.11-10.85)/44.711=21.87$.

The sequencing has been distributed geometrically with particular amino acid binding. The number of polypeptides was obtained and multiplied by a constant. This constant term is called the common ratio, "*r*." The equation for the *n*th term of geometric sequencing for polypeptide binding with particular amino acids is:

$$a_n = a_i r^{n-1} \tag{5.69}$$

a_n = Homology of peptide sequencing looping takes place at *n*th amino acid position

a_i = Homology of peptide sequencing looping takes place at *i*th amino acid position

r = Common ratio; n = Number of peptide

The formula of the sum of the *n*th peptide sequencing geometrically distributed is

$$S_n = a_i\left(\frac{1-r_n}{1-r}\right) * F_i \tag{5.70}$$

$$\text{Common ratio}(r) = \frac{\text{Repeated aminoacid for looping}}{\text{Total no of aminoacids favorable for looping}} = \frac{f}{N}$$

$$\text{Common ratio}(r) = \frac{3}{12} = 0.25$$

The first term a_1=Position of looping is 40

Number of terms n= 12 amino acid involved for first looping

a_1^2 Peptide sequencing looping of geometric progression $a_1 = 40$th position of the first looping, the common ratio is $\frac{3}{12} = \frac{1}{4} = r$ then

$$a_{12} = 40\,(1/4)^{12-1};\ a_{12} = 40\left(\frac{1}{4{,}194{,}304}\right);\ a_{12} = \left(\frac{5}{524{,}288}\right)$$

The first few terms of the amino acid sequences in looping are 40, 10, 5/2, and 5/8, respectively.

The minimum number of amino acids involved in looping is solved using the Cramer-Rao system of equations (2×2) matrices. From the table, we consider the system of equations based on the position, frequency, and total number of amino acids for homology looping.

$$\left.\begin{aligned} 40x + 3y &= 12 \\ 12x + 2y &= 8 \end{aligned}\right|^{x=0 \text{ and } y=4}$$

Step I: Determine the coefficient matrix (D), X-matrix (D_x), and Y matrix (D_y). In the above equation, we have

$$D = \begin{bmatrix} 40 & 3 \\ 12 & 2 \end{bmatrix} D_x = \begin{bmatrix} 12 & 3 \\ 8 & 2 \end{bmatrix} D_y = \begin{bmatrix} 40 & 12 \\ 12 & 8 \end{bmatrix}$$

Step II: Determine the coefficient matrix (D), X matrix (D_x), and Y matrix (D_y). In the above equation, we have

$$|D| = \begin{bmatrix} 40 & 3 \\ 12 & 2 \end{bmatrix} = 40*2-3*12 = 44$$

$$|D_x| = \begin{bmatrix} 12 & 3 \\ 8 & 2 \end{bmatrix} = 12*2-8*3 = 0$$

$$|D_y| = \begin{bmatrix} 40 & 12 \\ 12 & 8 \end{bmatrix} = 40*8-12*12 = 176$$

Step III: Determined X and Y

$$\frac{\det|D_x|}{\det|D|} = \frac{0}{44} = 0$$

$$\frac{\det|D_y|}{\det|D|} = \frac{176}{44} = 4$$

The minimum number of four amino acids is responsible for the sequencing of looping. The distance between two groups, $A=(X_a, Y_a)$; (3, 12) and $B=(X_b, Y_b)$; (2, 8) sequences, is calculated.

$$d(A,B) = \sqrt{(X_b - X_a)^2 + (Y_b - Y_a)^2} \tag{5.71}$$

$$A = (3,12) \Rightarrow X_a = 3, Y_a = 12;\ B = (2,8) \Rightarrow X_b = 2, Y_b = 8$$

$$d(A,B) = \sqrt{(2-3)^2 + (8-12)^2};\ d(A,B) = \sqrt{(-1)^2 + (-4)^2}$$

$$d(A,B) = \sqrt{1+16};\ d(A,B) = \sqrt{17} = 4.12$$

The distance of the peptide group (sequencing Group 1, Group 2) for looping is 4.12.

In the case of looping of amino acids, the heavily concentrated peptide is extracted and formulated (Figures 5.31–5.35). The $p(x)=3X^2-40x+12$ represents glutamate repeated three times in the first looping at the 40th position, with a total of 12 amino acids. This determines the polynomial looping of the peptide sequences, which is $x_1=13.02$, $x_2=0.30$ (Tables 5.29 and 5.30). Plug in the value by quadratic equation $(x_1,x_2)=\dfrac{-b\pm\sqrt{b^2-4ac}}{2a}$

Number of Codons (n)	Mean (X)	Mean (Y)	Intercept (a)	Slope	Regression Equation
64	15.24	0.38	0.177	0.009	$Y_t=0.177+0.00978X_1$

The codons homology model is described in the following illustrative figures and diagrams (Figures 5.36–5.52).

5.29 ANOVA Fixed Effect Model of Peptides/Codons

The model in which the value of independent variables is assumed to be fixed (constant) to determine the effect on dependent variables is called the fixed effect model. This model allows for extrapolating the variation between and within peptide groups or codons (expressed in homology).

$$Y_{ij}=\mu+\alpha_i+\varepsilon_{ij} \tag{5.72}$$

where Y_{ij} represents the observed value of percentage expression and the fraction of codons with the ith codon and jth amino acid (Tables 5.31 and 5.32).

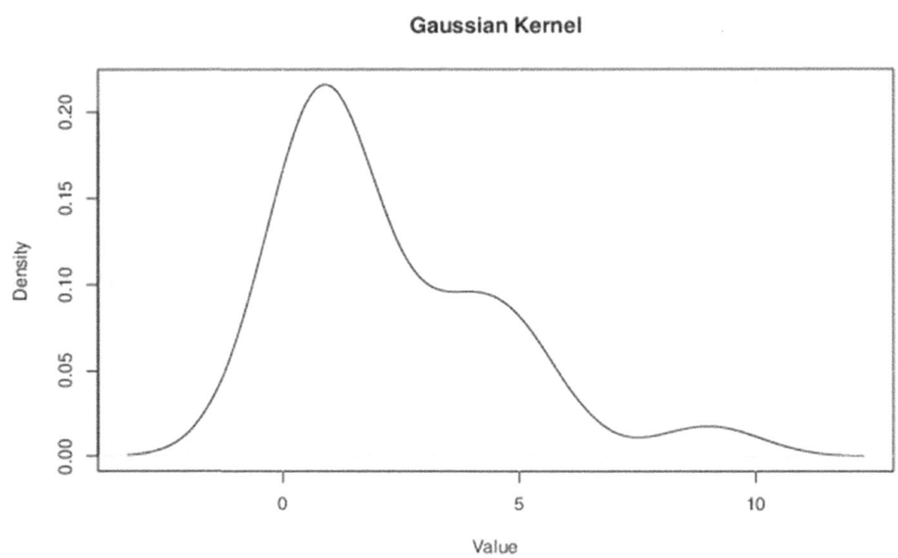

FIGURE 5.31 Kernel density of uracil (k=0.25).

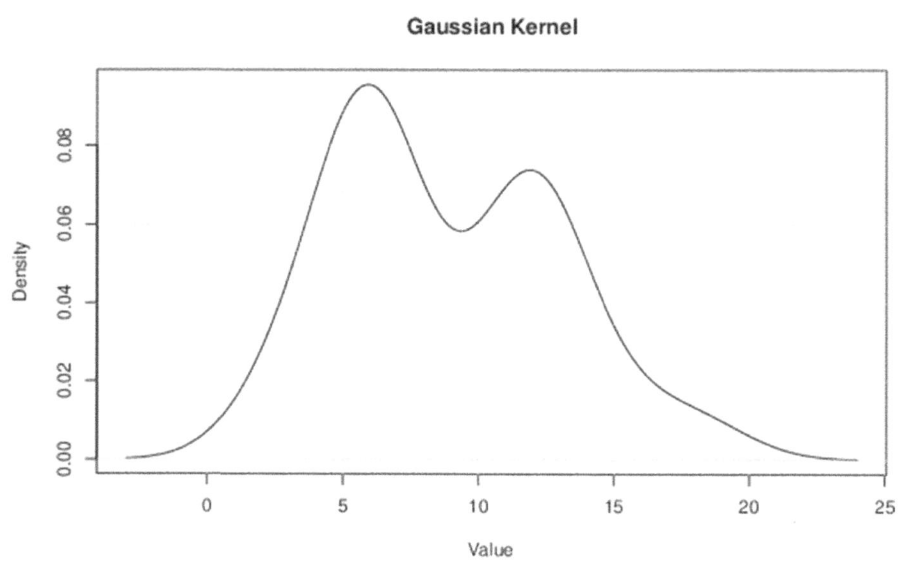

FIGURE 5.32 The total codons density.

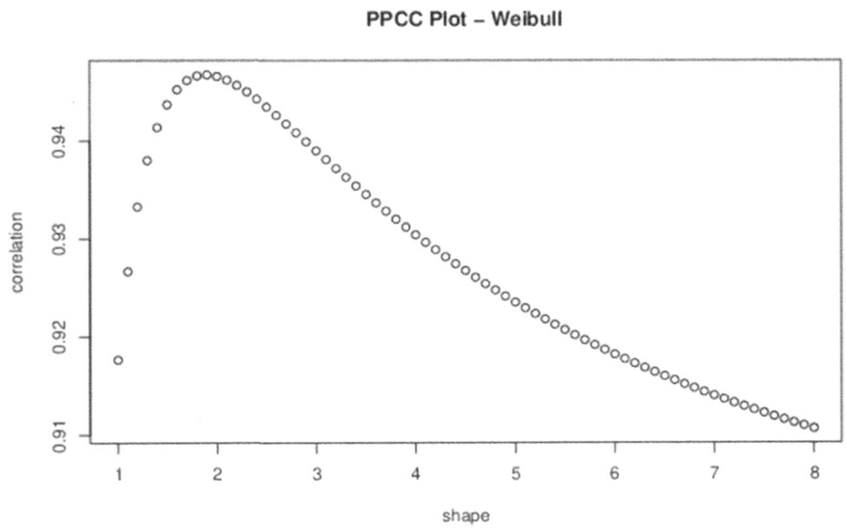

FIGURE 5.33 A PPC plot for total codons Weibull probability plot.

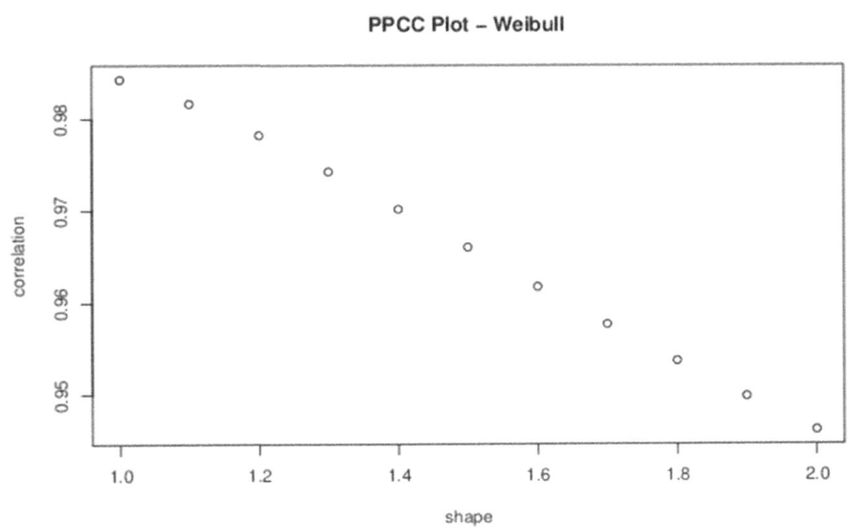

FIGURE 5.34 PPC plot for uracil.

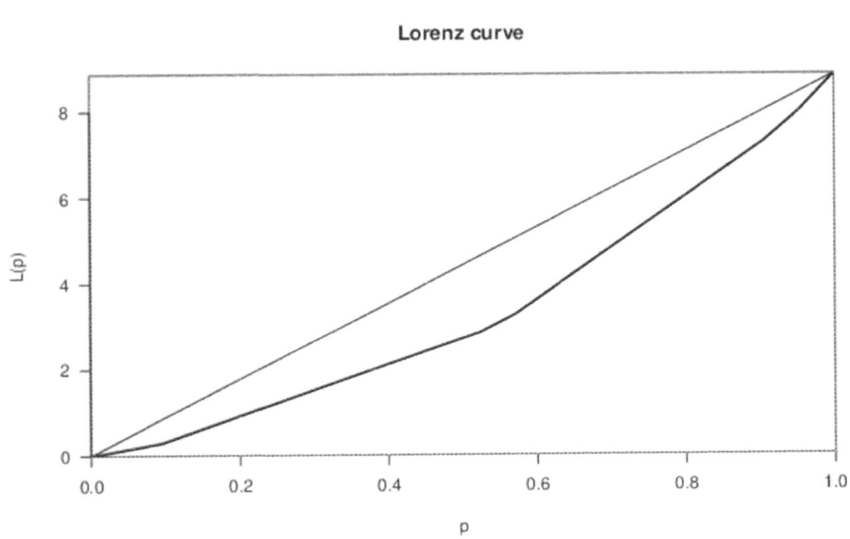

FIGURE 5.35 Lorenz curve for total codons.

TABLE 5.29

Homology Model of Peptides/Codons

AA	Codons	Per 1,000	Fraction
Gly	GGG	11.2	0.15
Gly	GGA	8.59	0.12
Gly	GGT	25.14	0.34
Gly	GGC	28.63	0.39
Glu	CAG	18.45	0.32
Glu	GAA	39.52	0.68
Asp	GAT	32.32	0.63
Asp	GAC	19.32	0.37
Val	GTG	25.47	0.36
Val	GTA	11.23	0.16
Val	GTT	19.06	0.27
Val	GTC	14.83	0.21
Ala	GCG	32.2	0.34
Ala	GCA	20.56	0.22
Ala	GCT	16.22	0.17
Ala	GCC	24.97	0.27
Arg	AGG	1.63	0.03
Arg	AGA	2.72	0.05
Ser	AGT	9.15	0.15
Ser	AGC	15.6	0.26
Lys	AAG	11.38	0.25
Lys	AAA	34.38	0.75
Asn	AAT	18.68	0.46
Asn	AAC	21.58	0.54
Met	ATG	2.721	1
Ile	ATA	5.34	0.09
Ile	ATT	29.92	0.5
Ile	ATC	24.63	0.41
Thr	ACG	14.02	0.26
Thr	ACA	7.93	0.15
Thr	ACT	9.52	0.18
Thr	ACC	22.7	0.42
Trp	TGG	14.45	1
End	TGA	0.97	0.3
Cys	TGT	5.17	0.45
Cys	TGC	6.33	0.55
End	TAG	0.26	0.08
End	TAA	1.98	0.62
Tyr	TAT	16.68	0.58
Tyr	TAC	12.3	0.42
Leu	TTG	13.28	0.13
Leu	TTA	13.83	0.13
Phe	TTT	22.05	0.57
Phe	TTC	16.49	0.43
Ser	TCG	8.71	0.15
Ser	TCA	7.89	0.13
Ser	TCT	9.33	0.16
Ser	TCC	8.89	0.15
Arg	CGG	5.67	0.1
Arg	VGA	3.66	0.07
Arg	CGT	20.7	0.37
Arg	CGC	21.15	0.37
Glu	CAG	28.82	0.66
Glu	CAA	14.78	0.34

(Continued)

TABLE 5.29 (*Continued*)

Homology Model of Peptides/Codons

AA	Codons	Per 1,000	Fraction
His	CAT	12.68	0.57
His	CAC	9.63	0.43
Leu	CTG	50.94	0.49
Leu	CTA	3.98	0.04
Leu	CTT	11.33	0.11
Leu	CTC	10.6	0.1
Pro	CCG	22.34	0.51
Pro	CCA	8.51	0.2
Pro	CCT	7.19	0.17
Pro	CCC	5.37	0.12

TABLE 5.30

Classification of the Codons Based on the Homology

Criteria	Codons Group	Coded
<Mean of $X - 1/2*$ (a) intercept	Low expressive fraction	1
= Mean of $X - 1/2*$ intercept	Medium expressive fraction	2
>Mean of $X - 1/2$ * (a) intercept	High expressive fraction	3

μ = Mean of the fraction of codons

α_i = Effect of ith codons

ε_{ij} = Errors associated with homology groups (AA) of codons ith row and jth column

Stage (i): The one-way biological process of information from DNA to protein is called the central dogma of molecular setup. The information stored in DNA copies replicated (transcription), which is used to assemble proteins (translation). Each DNA strand comprises subunits called nucleotides, or bases, for short. There are four types of base pairs: A-adenine, T-thymine, G-guanine, and C-cytosine. Adenine always pairs with T-thymine and G-guanine with C-cytosine. Thus, the two strands of DNA are complementary to each other. Nucleotides are arranged in a specific order on DNA strands; this is called the DNA sequence. This is further sectioned into genes, a short segment of DNA that is made into one polypeptide chain.

Stage (ii): Transcription of DNA to RNA occurs in the nucleus. These are processes in which RNA is assembled from a DNA template to transcribe a gene. The DNA is first unwound using the enzyme DNA helicase, which breaks the bonds between the nucleotide pairs. Then, an enzyme called DNA polymerase 'reads' one of the DNA strands and converts it into complementary strands called mRNA. The pre-mRNA is also made up of nucleotides but uses uracil (U) instead of thymine (T). Thus, an 'A' base on the DNA has been transcribed into a messenger RNA molecule during transcription; the mRNA must be translated to produce a protein. In translation, mRNA, transfer RNA (tRNA), and ribosomes work together produce transcriptomic chain.

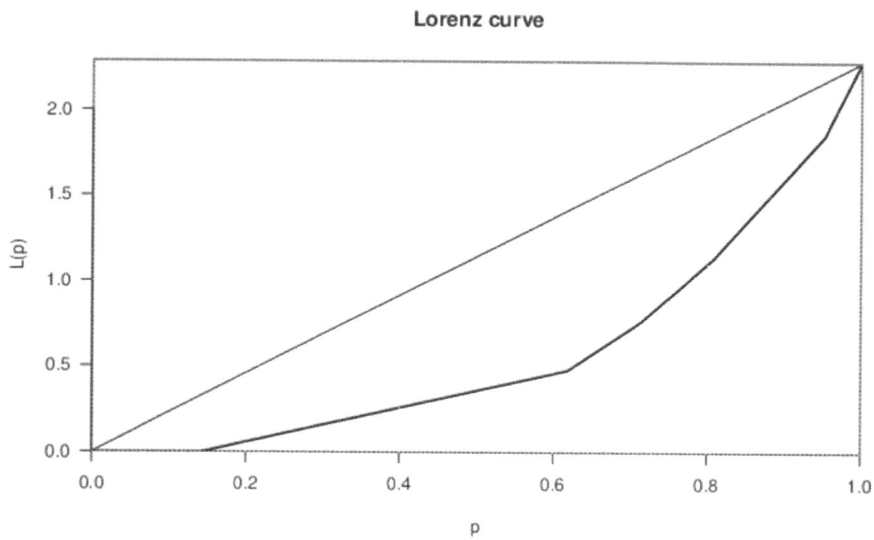

FIGURE 5.36 Lorenz curve for uracil codons.

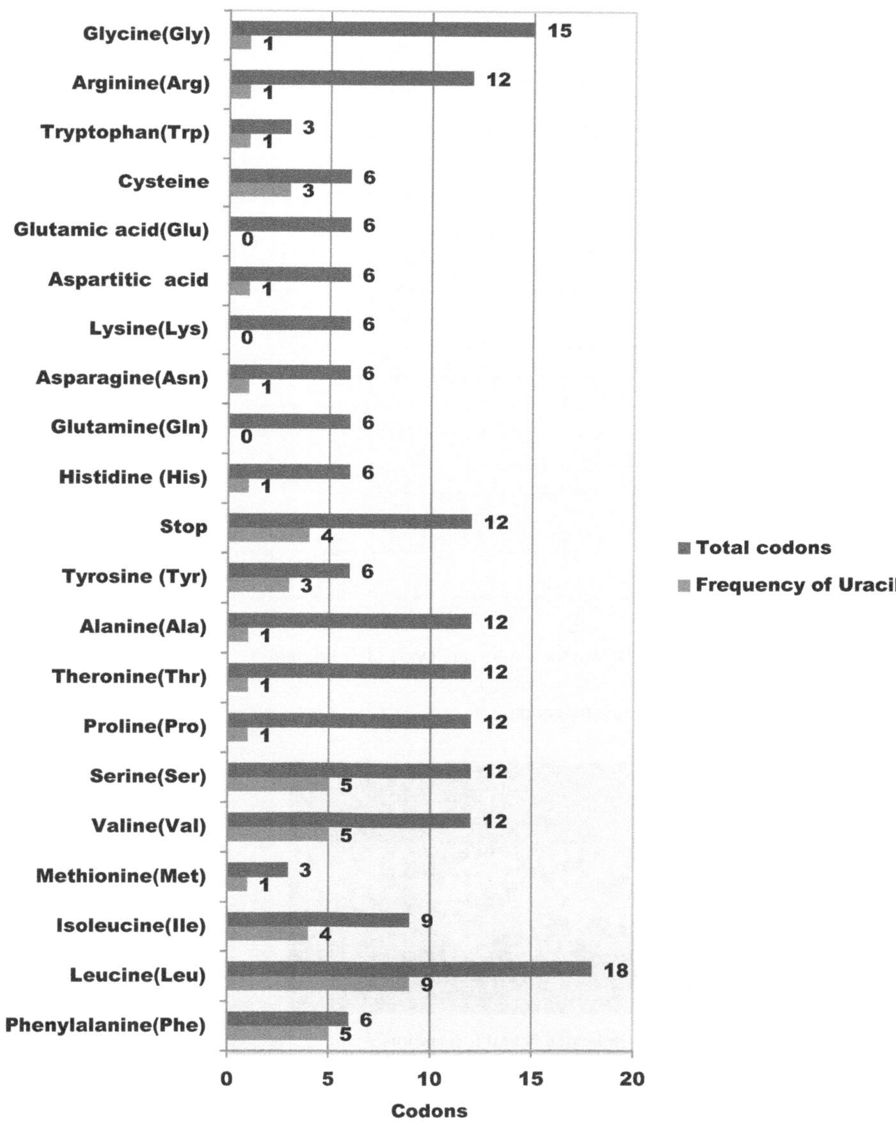

FIGURE 5.37 Bar diagrams of total codons and the frequency of uracil.

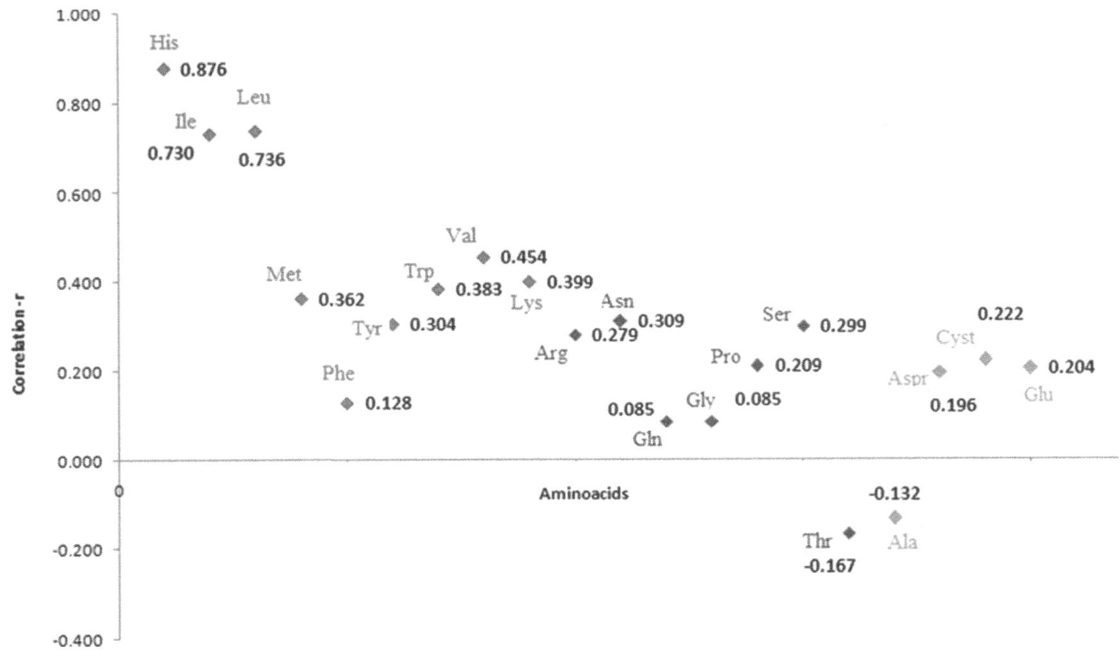

FIGURE 5.38 Simulation of rolling correlation concerning the repetition of uracil.

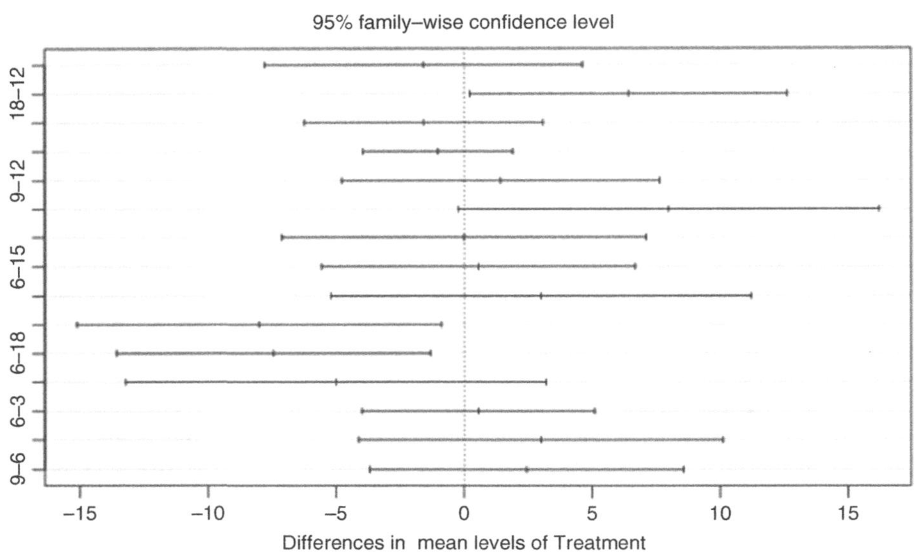

FIGURE 5.39 CI -95% family-wise confidence intervals for codons.

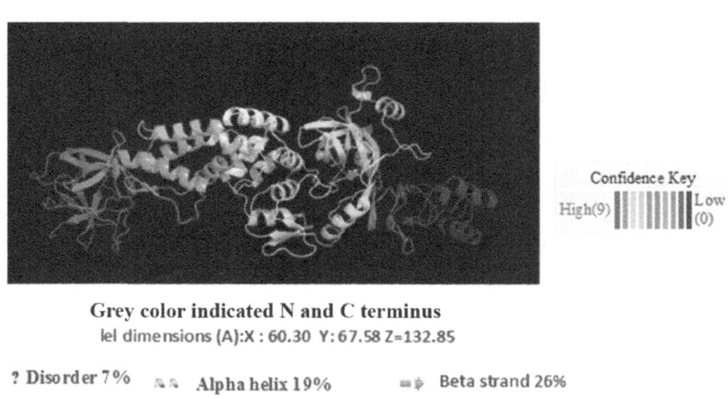

FIGURE 5.40 Protein secondary structure of the SARS-CoV-2 sequence.

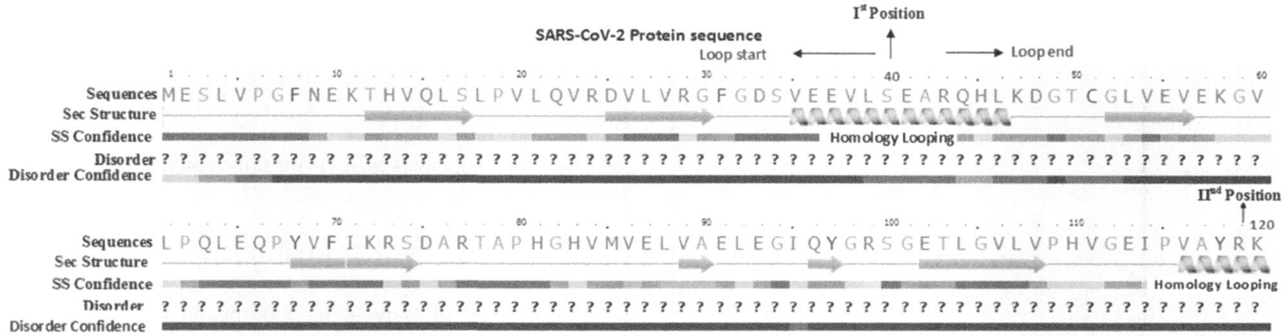

FIGURE 5.41 Homology of protein sequences for SARS-CoV-2.

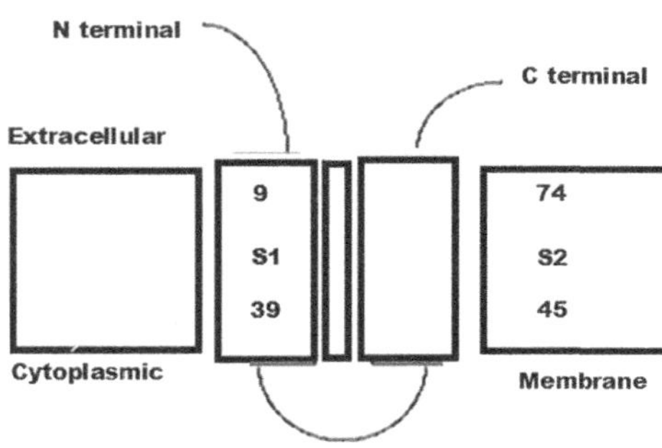

FIGURE 5.42 Predicted transmembrane helices based on the adopted topology of the sequences.

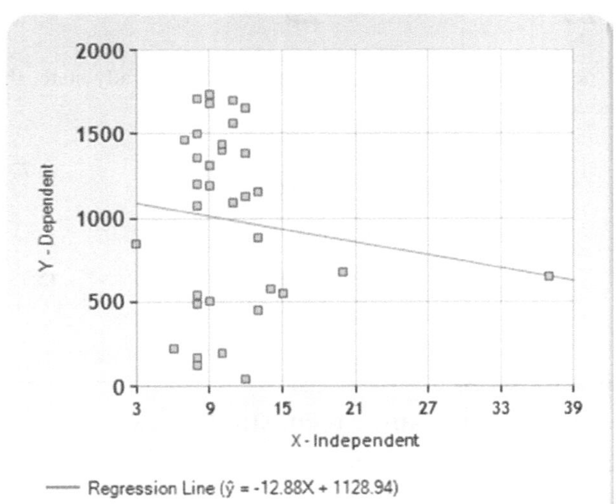

FIGURE 5.43 Regression line correlating amino acids (AAs) with retention time.

Stage (iii) Protein synthesis: Transfer RNA plays a crucial role in protein synthesis and translation. This process translates the message within the nucleotide sequence of mRNA into specific amino acid sequences. These sequences are joined together to form a protein, and transfer RNA is shaped like a cloverleaf with three loops. It contains an amino acid attachment site on one end and a special section in the middle loop called the anticodon site. This anticodon recognizes a specific area on mRNA called a codon. The last set of codons on mRNA molecules is called the 'stop' codons; they do not have a corresponding tRNA. When the ribosome reaches this sequence, it detaches from the mRNA. The polypeptide chain (100s to 1,000's amino acids long) is released and migrates to the Golgi apparatus, where it is modified and eventually formed into a 3D shape called a protein.

5.30 Markova Protein Sequencing Model

A Markov process is a stochastic model that describes a sequence of possible events or sequences of proteins in which the probability of each peptide depends only on the state attained in the previous peptide. A countable infinite number of peptide sequences, in which the peptide chain moves through states at discrete time steps, gives a discrete time Markov chain. The Markov chain is a random walk through study phase space, where we consider polypeptide sequences with Ni and Ck terminals with S_1s and $S_2 \ldots S_n$ cytoplasmic membrane to converge transition probability '*P*'. The model will converge to a semi-Markov decision process,$S_1 \rightarrow S_2 \rightarrow S_2 \rightarrow S_3 \ldots \rightarrow S_n$, where S_1 to S_n is the state of the peptide system. The transition probability, $P(S_n \rightarrow S_{n+1})$, is represented by a stochastic matrix.

$$P(S \rightarrow S') \geq 0 \sum P(S \rightarrow S') = 1 \tag{5.73}$$

In the above equation (5.73), the distribution of S_{n+1} depends only on S_n (*n*th peptide sequencing by real definition of a Markov chain).

We consider the frequency of a peptide repeated at the *n*th time; the equation will be $f_n(S)$. It evolves according to the different stages of insertion and deletion of peptides.

$$f_{n+1}(S') = \sum_S f_n(S) P\ (S \rightarrow S') \text{ or } f_{n+1} = Pf_n \tag{5.74}$$

An individual peptides sequences undergoes binding eigenfunction $P = p\varnothing = \varepsilon\varnothing$, where *p* represents the positive value,

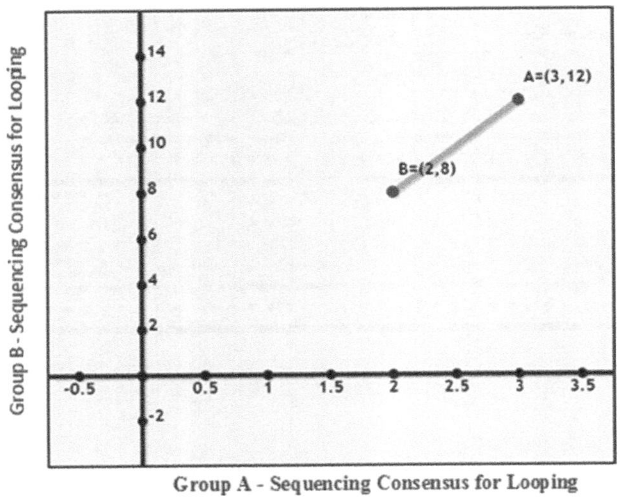

FIGURE 5.44 Polynomial looping model.

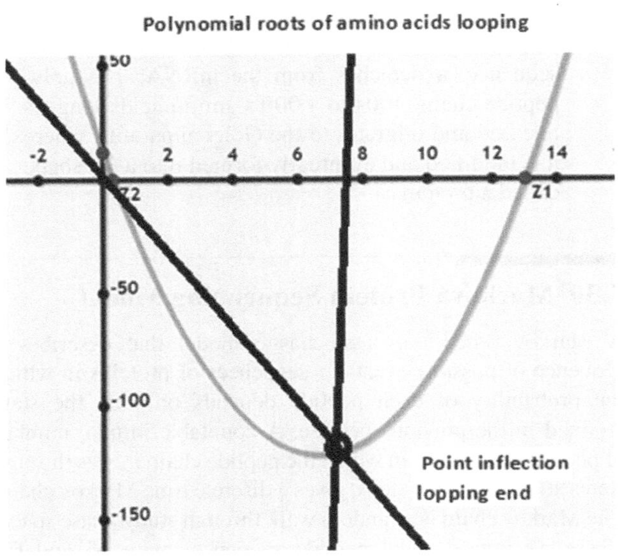

FIGURE 5.45 Polynomial roots associated with the inflection points of amino acid looping endpoints.

$\varepsilon \leq 1$; represents the negative value, and $\varepsilon = 1$ represents the equilibrium state.

The peptide sequences $S_n \to S_{n+1}$ converge to an ergodic distribution, implying that the convergence of multiple sequences of peptides is geometric and monotonic.

$$f_n(S) = \pi(s) + \sum \lambda^n C_\lambda \phi_\lambda \ (s) \text{ if } \ \varepsilon < 1 \tag{5.75}$$

after 'n' iteration then becomes zero

By using transition rules, we consider two state, S_m and S_n, linked by the transition probabilities $P(m \to n)$. Sppose we determined the reliability of each consensus of peptide sequences with or without loping on multiple arrays. Then, there is a 60% chance of being sequenced in the first step, followed by the next looping, and so on. The percentage variation changes will be observed in the subsequent sequencing process.

Looping Without Looping:

$$P(S \to S') = \begin{bmatrix} 0.60 & 0.40 \\ 0.30 & 0.40 \end{bmatrix} \text{Loop/without loop}$$

Suppose the sequences will be initially equally likely to loop or be without loop

$$\pi(1) = [0.50, \ 0.50 \ ]$$

The probability of sequencing looping on the next read of homology will be

$$\pi(2) = T \ (1 \to 2)\pi \ (1) = \begin{bmatrix} 0.60 & 0.40 \\ 0.30 & 0.70 \end{bmatrix} (0.50, 0.50)$$

$$= (0.45, \ 0.50)$$

$$\pi(2) = T \ (2 \to 3)\pi \ (1 \to 2) \ \pi \ (1) \ \pi \ (2) = (0.43, \ 0.56)$$

$$\pi = \log_{\text{Lim } x \to \infty} T^x \pi = \left(\frac{3}{7}, \frac{4}{7}\right) = (0.42, 0.57)$$

$$\pi = T\pi \text{ As a Eigenvalue problem}$$

The detailed sequences are described by $\pi(s)P(S \to S') = \pi \ (S')P(S' \to S)$

$$\sum_s \pi(s)P(S \to S') = \pi \ (S') \sum_s P(S' \to S) = \pi(S') \tag{5.76}$$

If peptides are possible to stay in the same steady state, the model becomes

$$P(S \to S) = 1 - \sum_{S' \neq S} P(S' \to S) \tag{5.77}$$

$$P(S' \to S) = T \ P(S' \to S) \left[\frac{\pi \ (S')}{\pi(S)}\right] S \neq S' \tag{5.78}$$

5.31 Discrete Time Stochastic Process

Now we have a discrete-time stochastic process, in which the state of the system is observed at discrete instant at time 't'. A group of peptides are sequenced at different time intervals and form a Markov chain model.

$$P(X_{t+1} = S \ / \ X_t = S_t, X_{t-1} = S_{t-1} \ldots\ldots X_1 = S_1, X_0 = S_0) \tag{5.79}$$

X_{t+1}= Sequencing of peptide at '$t+1$', where t=0, 1, 2...n, with different sequencing consonance. $S_0, S_1, S_2 \ldots S_t$ S is the discrete Markov chain containing random variables transitioning from one state to another, depending on certain assumptions and definite probabilistic rules that have the Markov property.

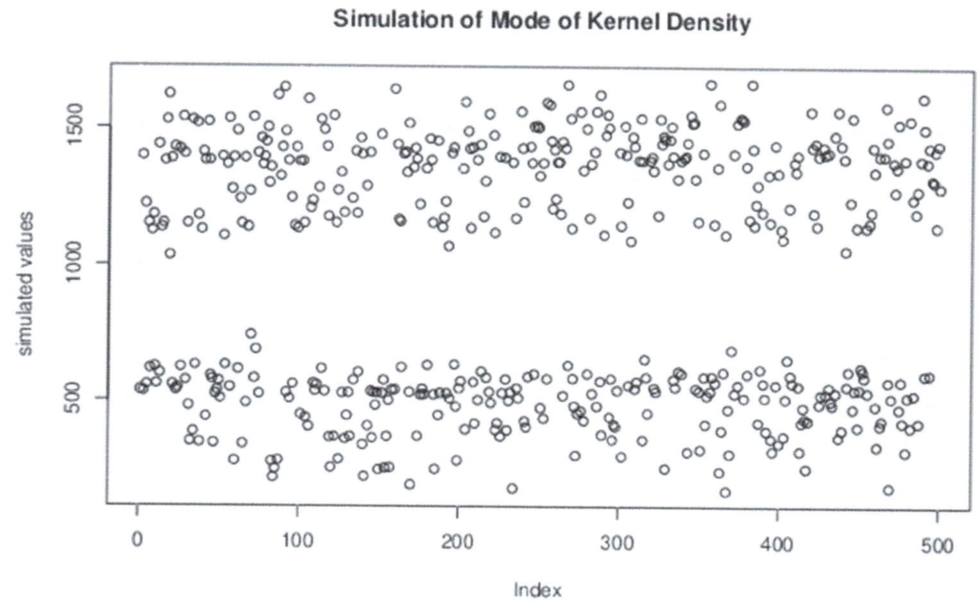

FIGURE 5.46 Simulation model of kernel density.

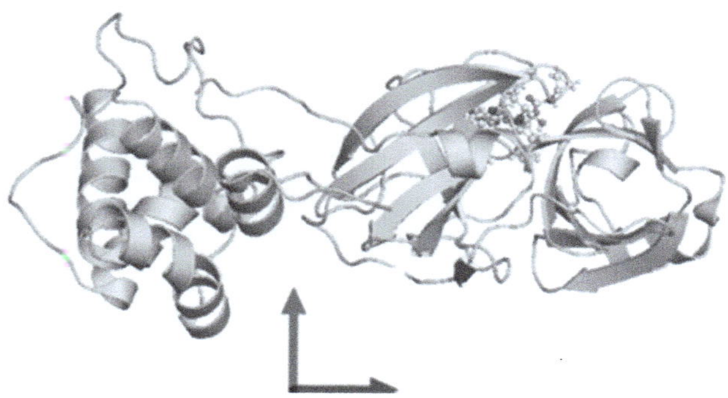

FIGURE 5.47 PDB of SARS-CoV-2, color-coded by chain and viewed from a lateral perspective.

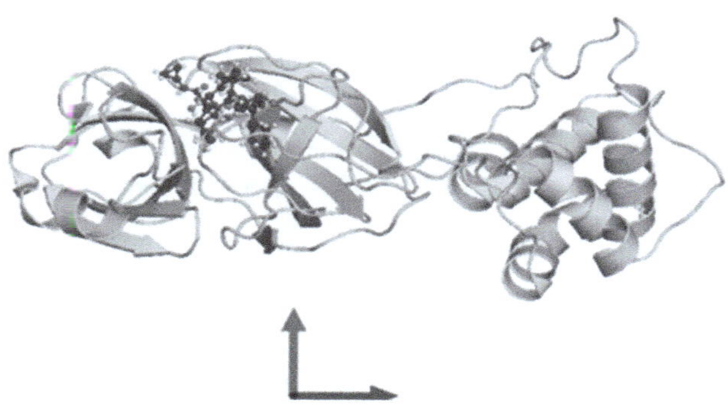

FIGURE 5.48 PDB of SARS-CoV-2, differentiated by chemically distinct molecules and observed from the side.

FIGURE 5.49 PDB of SARS-CoV-2 viewed from the frontal perspective.

5.31.1 Assumptions

The present peptide bonding is independent of the past; this property makes it a memoryless process,, as it is solely dependent on the current state and the randomness of transitioning to the next state of the binding. In model formulation there are three measures we consider: (i) States—All the states of peptide sequence occurrences fall within the state

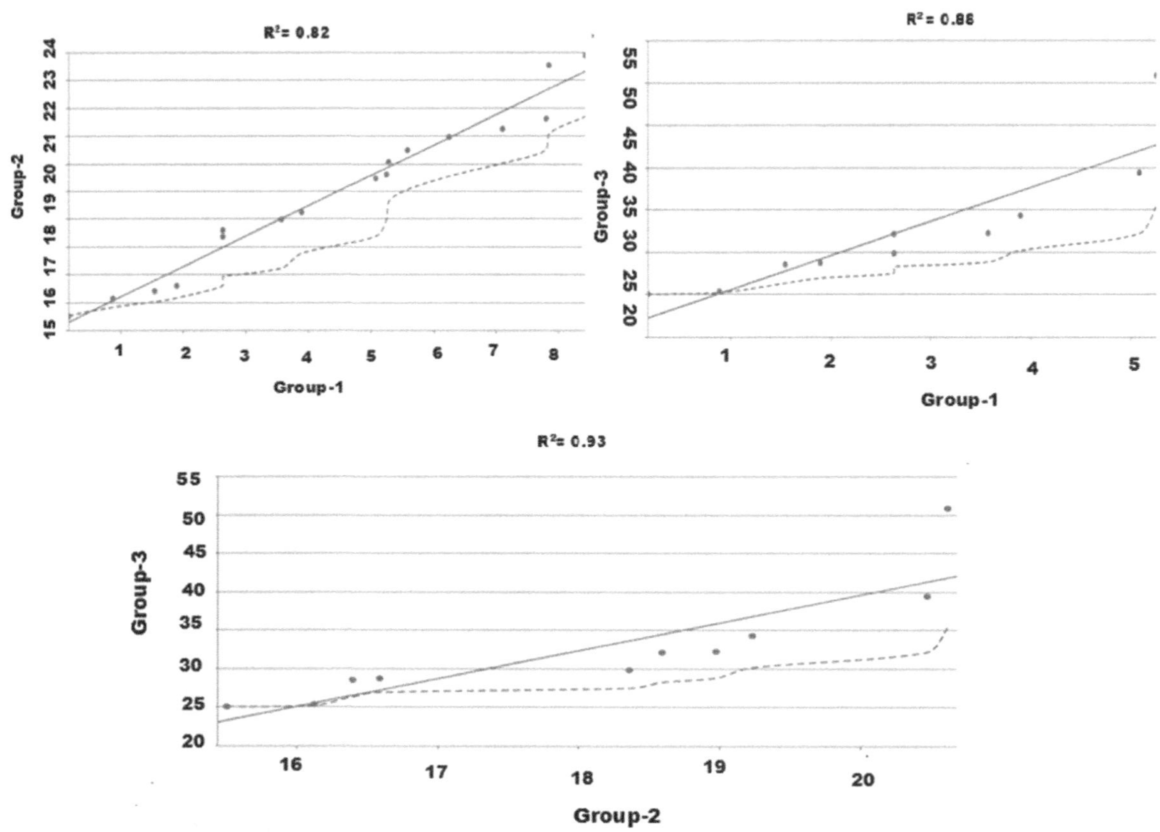

FIGURE 5.50 Homology model group correlation through scatter plots—homology genetic sequence model.

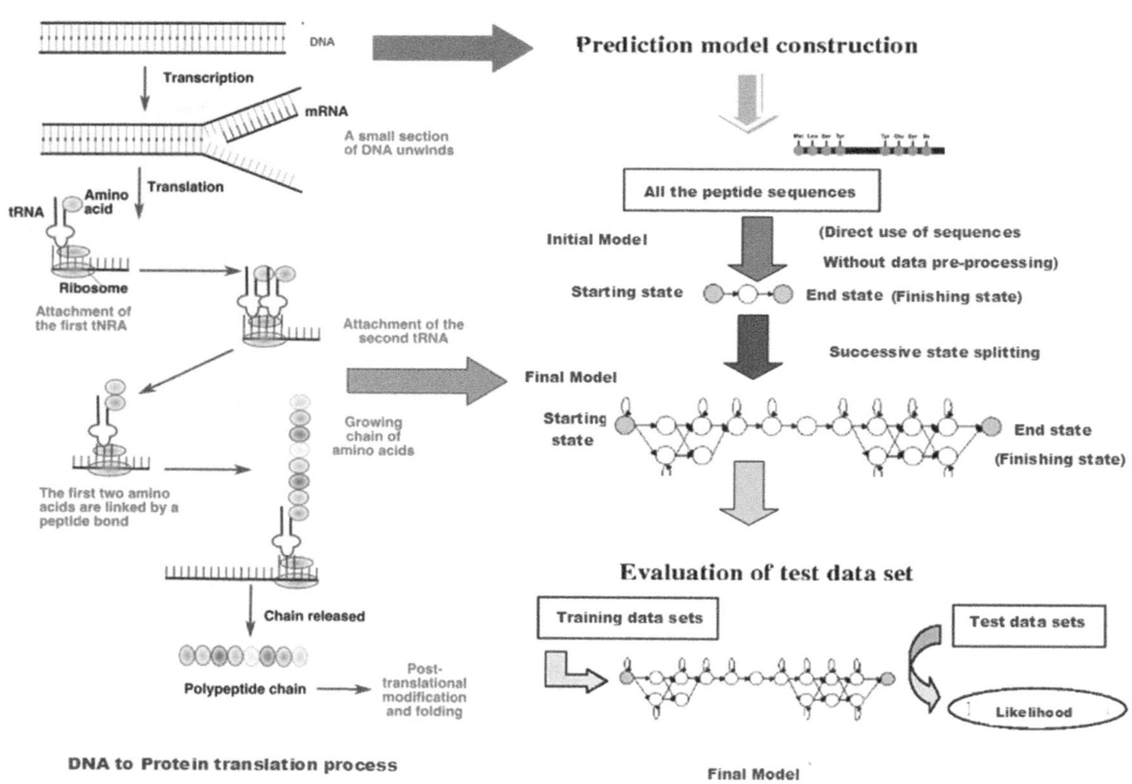

FIGURE 5.51 Simulation model of kernel density—PI.

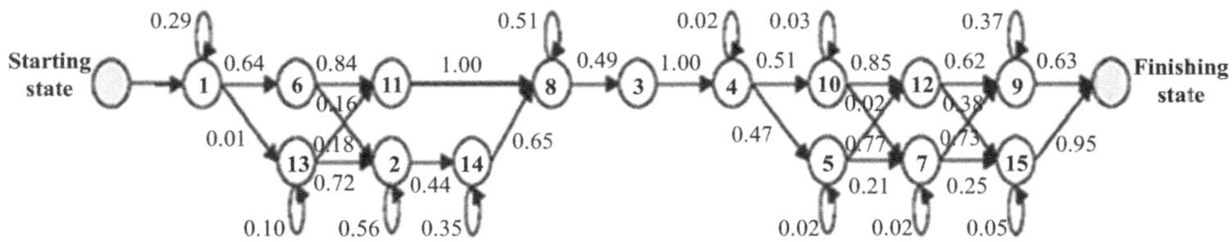

FIGURE 5.52 PDB structure of SARS-CoV-2.

TABLE 5.31

Homology Groups (Codons Expression)

	Homology Groups (Codons Expression)			
Results	**Low-G_1 ($n=37$)**	**Medium-G_2 ($n=17$)**	**High-G_3 ($n=10$)**	**Total ($n=64$)**
$\sum x$	303.90	339.73	326.4	970.13
$\bar{x}$	8.21	19.98	32.64	15.15
$\sum x^2$	3,133.29	6,919.53	11,190.76	21,243.58
Std.Dev	4.20	2.85	7.68	10.18
Skewness	−0.221	0.0483	1.645	Overall skewness-0.89

Table ANOVA

Source	**SS**	***df***	**MS**	**F_{cal}**
Between the groups	5,239.62	2	2,619.81	123.07 Sig, $p<0.001$
Within the groups	1,298.42	61	21.28	
Total	6,538.05	63		

TABLE 5.32

Tukey's Post Hoc Test for Pair-Wise Comparison

Pair Wise Comparisons		**HSD$_{.05}$=3.9012 HSD$_{.01}$=4.9143**	**$Q_{.05}$=3.3973 $Q_{.01}$=4.2795**
T_1:T_2	$M_1=8.21$ $M_2=19.98$	11.77	$Q=10.25$ ($p=0.00000$)
T_1:T_3	$M_1=8.21$ $M_3=32.65$	24.44	$Q=21.28$ ($p=0.00000$)
T_2:T_3	$M_2=19.98$ $M_3=32.65$	12.66	$Q=11.03$ ($p=0.00000$)

Model results

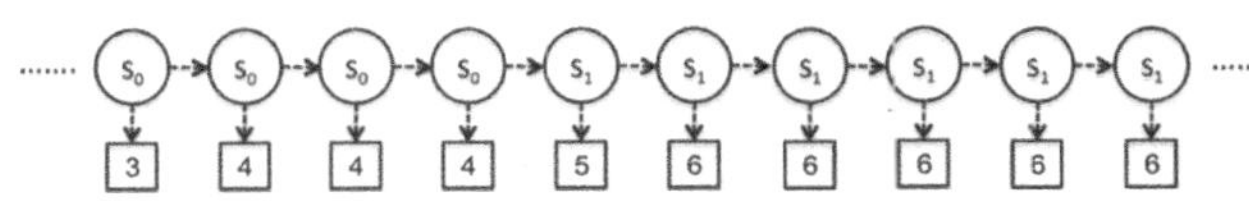

	0	1	2	3	4	5	6
S_0	0.000	0.000	0.004	0.1277	0.566	0.268	0.0367
S_1	0.000	0.000	0.001	0.0093	0.0726	0.2088	0.7092

S_0	S_1
0.989	0.0108

	S_0	S_1
S_0	0.998	0.002
S_1	0.023	0.976

spaces of the dynamic system; (ii) The initial state distribution—an initial state probability distribution of the starting state, encoded into a column vector denoted as 'q'; (iii) State transition probabilities—the transition probability of moving from one state to another, encoded in an 's' by 's' matrix denoted as 'P'. $S=\{S_0,S_1,S_2\ldots S_t\}$. The considered state space $q=\{q_0,q_1,q_2\ldots q_t\}$ is the initial probability distribution 'q'.

$$P=\begin{bmatrix} P_{11} & P_{12} & P_{1S} \\ P_{21} & P_{22} & P_{2S} \\ P_{31} & P_{32} & P_{3S} \end{bmatrix}_{S*S} \Rightarrow \text{Transition probability state matrix} \quad (5.80)$$

$$P=\begin{bmatrix} P_{11} & P_{12} & P_{1S} \\ P_{21} & P_{22} & P_{2S} \\ P_{31} & P_{32} & P_{3S} \end{bmatrix}_{S*S} = \sum_{j=1}^{S} P_{ij} = 1, \text{each row is equal to 1} \quad (5.81)$$

$$P_{ij}=P\left(X_{t+1}=j,\ X_i=i\right),\forall_{i,j}\in\ S\leq t=0,1,2\ldots n \quad (5.82)$$

5.32 The Modeling Formulation Steps

Assuming there are already assumptions suitable for HMM modeling, the basic ordered steps we suggest to designate the elements are as follows:

1. Define the sequence of observed symbols; the query or input of the HMM.
2. Define the alphabet of observed sequences. These are derived from the observed input defined earlier.

3. Define the alphabet of hidden symbols or hidden states. These define attributes whose most probable sequence we seek to determine in the query. These symbols represent the format of the output of the HMM.
4. Determine the vector of prior probabilities of the hidden states. This is the probability distribution of the first hidden state.
5. Establish the transition probabilities between hidden states.
6. Establish the emission probabilities from hidden states to an observed state.

5.33 Model Discussion

A Markov model is an advanced stochastic convergence model in machine learning that randomly changes systems possessing the Markov property. This means that, at any given time, the next state is only dependent on the current state and is independent of anything in the past. The Markov chain has found its place in various practical applications. It has been applied to chemical bonding in biochemistry and the Krebs cycle pathways. It is a process that evolves in steps, which could be in terms of time or sequence. Therefore, for example, at each step, the process may exist in various countable states. As the process evolves, the system may remain in the same state or change to a different state during the time epoch. These movements between states are normally described in terms of transition probabilities. These transition probabilities allow us to predict the future the possibility of the system being in a state many time epochs later. We will see a few examples of this in this model. The state X_n is the process of sequencing peptides that are binding to particular amino acids at the time instance "n." For example, consider the spread of SARS-CoV-2. X_n is the number of people who have COVID 19 at time 'n'. The number of those who have the virus at time $n+1$ is X_n+1. Following our Montecarlo simulation model, we can therefore write that the number of infectiond at time $n+1$ depends on those who were infected at time 'n'. $X_n+1 \Rightarrow X_n$, where X_n is the random process of the equation $\{X_{n-1}, X_{n-2} \ldots\ldots, X_0\}$. From the above formulated model, a Markov chain cannot be in two states at the same time. It can, however, change from one state to another. When this happens, it is called a transition from state S_n to S_n+1. This transition is represented with a probability value. A Markov chain transits and remains in the same state from 'n' to time $n+1$. Our fitted model is the state transition diagram for a two-state Markova chain. Initially, at state 1, the system was at the higher binding. Assume that the probability of binding is higher than the probability of the pathway of binding of amino acids. The probability that shows a lower risk was <0.5. Prediction with MC requires the knowledge of the probabilities of the initial state of the process and the state transition matrix. This model illustrated the binding of AA with ambient temperature and retention time. The always prediction=initial state * transition probability (state transition matrices). One of the best examples we have seen in day-to-day life is data analysis with Markov chain analysis of loan situations. Consider an "unfair bank". This bank has offered many housing loans, some to customers who pay back their loans regularly and others to loan defaulters. We will categorize unfair bank loan customers in four terms based on the types of loans the bank has offered to customers. The bank wants to understand which types of loans have a chance of being paid up and on time and which are bad loans. Bad loans eat into banks' revenue bases and therefore need to be avoided as much as possible.

5.34 Application of the Model

5.34.1 Personalized Medicine

Personalized medicine is a newer concept in medical science, experiencing changing periods with many attributes. In our realistic experience, this is driven by the health sector, where very advanced tools are precluded for disease-exposed populations and extensive molecular biological data sequences with individual patients' accurate diagnosis and treatment. This trend will change based on the development of newer, novel regimens and formulations. Evidence-based medicine has correlated in accordance with the stipulated regularity guidelines of the Food and Drug Administration (FDA). Bioinformatics magnifies the research for the manufacturing of drugs and their usage by using molecular techniques. This relationship has been estimated from real datasets and pertains to patients and their disease phenotypic expressions. Thus, the computer-based simulation approach will play a significant role in interpreting molecular datasets and also serve as a diagnostic tool for providing valid recommendations by medical practitioners.

5.34.2 Molecular Medicine

Nowadays, medical specialists and clinicians have used trial and error to find the best drugs to treat different types of patients, as those with the same clinical symptoms can show a wider range of responses to the treatment. With this advent of practical exposure, the physician is unable to analyze the patients based on their prior and posterior investigations. The physicians will prescribe the best available drugs or therapies based on dose-response analysis using advanced bioinformatics techniques.

5.34.3 Microbial Genomic Applications

Microorganisms are ubiquitous, and they are found everywhere. They have been found surviving and thriving in extremes of heat, cold, radiation effects, salt, acidity, pressure, etc. By studying the genetic material of these organisms, scientists can begin to understand the effects of microbes at a very fundamental level and isolate the genes that give them their unique abilities to survive under extreme conditions. With the advancement of science, we have been using bioinformatics tools to maintain the sustainability of organisms. The artificial neural network (ANN) and Markov chain model were used to fill the above research gap.

5.34.4 Antibiotic Resistance

Medical microbiologists have been examining the genome of *Enterococcus faecalis*, which is the leading cause of bacterial infections in hospital-admitted cases and outpatients. They have discovered a virulence region by using a number of antibiotic-resistant genes that may contribute to the bacterium's transformation from harmless gut bacteria to menacing invaders. The discovery of this region, known as pathogenicity, could provide useful biomarkers for detecting pathogenic strains and help us control their spread.

5.34.5 Preventive Medicine

With the specific details of the genetic mechanisms of diseases unraveled, the development of diagnostic tests to measure a person's susceptibility to different diseases will become a distinct reality. Preventive actions such as changes in lifestyle or having treatment at the earliest possible stages are more likely to be successful, resulting in huge advances in our struggle to conquer disease with bioinformatics tools.

5.34.6 Gene Therapy

Gene therapy has been used to treat many diseases, even cure or prevent them, by changing the expression of genes. The potential for using genes themselves to treat disease may become a reality. Genomics is a very complex problem due to the enormous amount of data generated for different reasons. The above-observed results are estimated using different statistical methods or programs. Each individual therapy should be integrated to produce one count of every chromosome of the genomic sequences due to the complex nature of the genomes and ideal climatic conditions. An artificial neural network (ANN) and advanced tools will solve this.

5.35 Applications of Central Limit Theorem to Life Science

If (n=9) mice out of (n=36) choose the right path when running a forked maze, would you conclude there is evidence of directional preference among the mice? We calculate the significance probability using the exact probability model and the normal approximation. Under one hypothesis concerning the inheritance of a certain characteristic, the proportion of offspring exhibiting the characteristic is 0.5, whereas under the alternative hypothesis, the proportion is 0.25. How many individuals must be examined such that the alternative hypothesis will be detected at least 60% of the time when it is true? Set α=0.05. Determine the sample size using the central limit theorem first and then using the binomial tables. A television program rating company would like to establish the proportion of Wednesday evening viewers who are watching Batman. The anticipated registered cases were simulated based on a proportional approach, with an initial approximation of 20% utilized to construct the central limit theorem model for establishment.

If 20 out of 60 tubes respond when inoculated with 1 cc of a suspension of bacteria, give a point and interval estimate, using the central limit theorem, of the proportion of tubes responding following this treatment. What are the corresponding point and interval estimates of the concentration of organisms in the original suspension?

In a test of the effect of a drug on the response rate, n tests are inoculated with 1 cc of the above suspension, and n tubes are given both 1 cc of the suspension and 0.01 cc of the drug. How large should n be if one wants to be 80% sure of detecting the difference when the proportion responding with the drug is only half that without the drug, using a=0.05?

A gambler wants to test the fairness of a coin. He proposes to flip the coin five times, and if he gets either 0 or 5 heads, he will reject the coin as unfair; otherwise, he will consider the coin fair. What is the probability of a Type I error (a)? What is the test against the alternative hypothesis that the coin is two-headed?

These data have been taken from the results of a large experiment designed to determine the effects of the prolonged use of heparin and fibrinolysin on the development of pulmonary metastases of carcinoma in the rabbit when non-irradiated cancer cells are injected via an ear vein. One group of 20 rabbits received 1,000 units of heparin intravenously 10 minutes before the inoculation of 50,000 non-irradiated cancer cells and thereafter daily injections of 1,000 units of heparin for 6 days. The number of rabbits with pulmonary metastases at the time of death or sacrifice at the end of 3 months in the two groups is given in the following table (Table 5.33).

Is there evidence that the proportion of rabbits with metastases is different at a 95% confidence interval difference? During a 2 years period, 761 premature infants 2 were admitted to the intensive care unit for newborns, of which (n=107) died. Of the (n=107) deaths, (n=88) were autopsied. Of the (n=88) autopsies (n=83) met the author's criteria of completeness and provided the basis for their study. The authors compare the incidence of hyaline membrane disease in those infants receiving air with that in those infants receiving oxygen continuously during their lives (n=32 infants received oxygen for only a portion of their lives before death and are omitted in this problem). These figures are given in the table below: oxygen treatment and hyaline membrane disease (Table 5.34).

Is there evidence that the incidence of hyaline-membrane disease is associated with oxygen therapy? In this problem, what issues must be kept in mind when interpreting the results of the significance test? A group of 30 mice consisting of 10 males and 20 females is divided by a random process

TABLE 5.33

Attributes for Mean Value Theorem

	Number of Rabbit		
Attributes	**With Metastases (n=18)**	**Without Metastases (n=18)**	**Total**
Single dose of heparin	13	07	20
Multiple doses of heparin	05	15	20
Total	**18**	**22**	**40**

TABLE 5.34

Attributes v/s Disease Status

	Number of Rabbit		
Attributes	**With Disease ($n=41$)**	**Without Disease ($n=10$)**	**Total**
Air	07	05	12
Oxygen	34	5	39
Total	**41**	**10**	**51**

into two groups of size. Is it likely that one group will receive two or fewer males? Let $a=0.05$. Use both the tables for the hypergeometric distribution and the central limit theorem in answering the question.

5.36 Application: Central Limit Applied to Binomial (BD)

We observe ($x=9$) successes in ($n=36$) trials. If the trials are independent and P (= probability of choosing the right path) is constant, then x will have a binomial distribution. We wish to test Ho: $P=0.5$ against H_a: $P=0.5$. To get an a of 0.05, we must find cut-off points 'x_1' and 'x_u' such that $p(x<x_1|p=.5)\leq 0.025$ and $p(x>x_u|p=0.5)\leq 0.025$; then reject Ho if the observed x is$\leq x_1$ or$\geq x_u$. A section of the cumulative binomial table with $n=36$ and $p=0.5$ is given below. (This table is in the lab.)

x	p (x>x)
10	0.998
11	0.994
12	0.986
13	0.967
.	–
.	–
24	0.033
25	0.014

From this, we see $p(x\geq 25|\ p=0.5)=0.014<0.025$ and that $p(x\leq 11\ |\ p=0.5)=1-p(x\geq 12\ |\ p=0.5)=1-0.986=0.014<0.0$ 25.thus we could reject H_0 if the observed x were less than or equal to 11 or greater than or equal to 25, yielding a test with an actual a of $0.014+0.014=0.028$. If one were satisfied with an a of 0.066, one could instead reject H_0 if the observed x were≤ 12 or≥ 24. Here the cases was 9 so one would reject H_0 in either case. The significance probability or p-value is the probability of a result as extreme as or more extreme than the result observed here, when H_0 is true. For a two-tailed test, we can get the p-value by taking twice the probability in the tail, including and beyond our observed result, under H_0. Here, we observed $x=9$ successes, which is in the lower tail of the binomial distribution with $n=36$ and $p=0.5$. Then, the p-value$=2p(x\leq 9|$ $p=0.5)=2[1-p(x\geq 10\,|\,p=0.5)]=2(1-0.998)=0.044$.

5.36.1 Illustration of Central Limit Theorem

Let x=number of successes in 'n' trials. If 'x' is binomial with p=probability of success, and if both 'np' and 'nq' are ≥ 5 a normal approximation may be used. $X=N\ (np,'npq)$ or $p=x\backslash n=N(p,\ p\ q\backslash n)$, then to do a two-sided.

Test of H_0: $p=0.5$ against H_a: $p=0.5$ one computes the test statistic. From the number of successes,

$$z=\frac{|x-np_o|-\frac{1}{2}}{\sqrt{np_0q_0}}-\frac{|x-36(0.50)|-\frac{1}{2}}{\sqrt{36\ (0.5)(0.5)}}$$

Or the test statistics from the fraction of success

$$Z=\frac{|\hat{p}-p_0|-\frac{1}{2n}}{\sqrt{\frac{p_0q_0}{n}}}=\frac{|\hat{p}-0.5|-\frac{1}{2(36)}}{\sqrt{\frac{0.5*0.50}{36}}}$$

To get $\alpha=0.50$, reject H_0 if the test statistics is $\geq Z_{0.975}=1.96$.

Here, the test statistics is $\dfrac{|9-18)|-\frac{1}{2}}{\sqrt{36\ (0.5)(0.5)}}=2.83$ and since $2.83>1.96$ reject H_0.

We can get the p-value using a procedure similar to that used in the exact case. For a two-sided test, the p-value is twice the probability of getting a test statistic>the observed value of the test statistic. When HO_o is true, $P=2p$ (test statistic≥ 0.83 $|H_o)=2p$ $[2\geq 2.83$, where $z=N$ $(0.1)]=2[1\ -f$ $(2.83)]=2(0.0023)=0.0046$ using normal tables.

We wish to test H_0: $p=0.5$, where p is the proportion of offspring with the characteristic. If the only alternative is that $p=0.25$ we will to debone – sided test rejecting only proportions smaller than 5. In finding the number of trials (n) required to get a power of $1-B\geq 0.6$ when $p=0.25$ $a=0.05$, we will assume that the binomial assumptions are valid.

5.37 Postulates

If we assume that that the sample size will be large enough for the normal approximation to apply, we can use the formula for n derived in the above section. For a one-sided test of H_0: $p=p_0$ against H_a: $p<p_0$ with given a, to get pPower '$(1-\beta)$' when $p=p_1$, where

$$p_1<p_0\text{usen}\geq\left[\frac{Z_{1-\alpha}\sqrt{p_0q_0}+z_{1-\beta}\sqrt{p_1q_1}}{p_1-p_0}\right]^2$$

$$n\geq\left[\frac{Z_{0.95}\sqrt{0.5*0.5}+z_{0.6}\sqrt{0.25\ (0.75)}}{0.25-0.5}\right]^2$$

$$n \geq \left[\frac{1.64\sqrt{0.25} + 0.25\sqrt{0.1875}}{-0.25}\right]^2 = 13.90 \text{ or } 14 \text{ individuals.}$$

Determining the sample size required for a given power using the exact binomial table is a tedious trial-and-error procedure. In general, one picks some n, finds cutoff points using that n, and computes the power at the given alternative using those cutoff points. If the power computed is less than desired, one picks a larger n and repeats the process; if the power computed is greater than desired, one picks a smaller n and repeats the process. Here, since we have found, using the CLT, that n shold be about 14, we will try only $n=15$ to see what power that gives using the exact test. X=number of successes in 15 trials will have a binomial distribution ($n=15$, $p=0.5$), if H_0: $p=0.5$ is true. To get an a of 0.05 for a one-sided test with H_a: $p<0.5$, we find a cutoff point, X_1, such that $p(x \leq x_1 \mid p=0.5) \leq 0.05$. Suppose we considered $n = 15$ and $p = 0.50$ one finds, for $n=15$ and $p=0.5$, that $p\ (x \geq 4) = 1-0.9408 = 0.0592$. This is as close to $a=0.5$ as we can get using $n=15$, as $p\ (x \leq 3) = 0.0176$. For the test procedure of rejecting Ho if the observed x is<4, with actual $a=0.0592$, the power of the test when $p=0.25$ is $1-B=p$ (reject H_0| p=0.25)$=p\ (x \leq 4 \mid p=0.\ 25) = 1-p\ (x \geq 5 \mid p=0.25)$. From the binomial table with $n=15$ and $p=0.25$, we find p $(x \geq 5) = 0.3135$ so $1-B -1-0.3135=0.6865$. This is greater than the required power of, so 15 individuals would be adequate. We want a 95% confidence interval for the percentage of viewers watching "Batman" to establish the value within=½%, thus to have a length no greater than 1%. This is equivalent to getting a 95% confidence interval for p, the proportion of viewers watching "Batman", which will have a length greater than 0.01. If we assume constant p and independent trials the number of "Batman" viewers among n viewers will be binomial. If n is large enough, the central limit theorem states that we can use a normal approximation. The 95% confidence interval for p is then $\hat{p}+1.96\sqrt{\frac{\hat{p}\hat{p}}{n}}$. We need to find the n required so that the length is 2. $1.96\sqrt{\frac{\hat{p}\hat{p}}{n}}$ is<0.01. We know that p, and hence p, will be about 2. Thus, we have $2(1.96)\sqrt{0.2(0.8)/n} \leq 0.01$ or $n \geq \left[\frac{2(1.96)}{0.01}\right]^2 (0.20)(0.80) = 24.58$. Nearly 25,000 persons would be needed in the sample to get the desired precision.

Alternatively, we can use the formula from lecture. The sample size such that the $1-\alpha$ confidence interval for 'p' has a half width of 'd' units is

$n = \dfrac{Z_{\left(1-\frac{\alpha}{2}\right)} pq}{d^2}$. Here, we want a 95% confidence interval with half width $d = \dfrac{0.01}{2} = 0.005$, and we know p is about 0.20.

$$\text{Then, } n = \frac{(1.96)^2(0.2)(0.8)}{0.05^2} = 24.58$$

Let x=the number of tubes responding among $n=60$ tubes. Assume x is binomial with p=proportion responding. '$n\ p$' seems to be large enough to justify the use of the central limit theorem in getting the confidence interval. The point estimate of the proportion responding is $\hat{p} = \frac{x}{n} = \frac{20}{60} = \frac{1}{3}$.

The 95% CI for p is $\hat{p}+1.96\sqrt{\frac{\hat{p}\hat{p}}{n}}$

$$\frac{1}{3} - 1.96\sqrt{\frac{\left(\frac{1}{3}\right)\left(\frac{2}{3}\right)}{60}} < p < \frac{1}{3} + 1.96\sqrt{\frac{\left(\frac{1}{3}\right)\left(\frac{2}{3}\right)}{60}}$$

$$0.21 < p < 0.45$$

If one assumes that a tube responds whenever there is at least one organism present in the 1 cc of suspension that it receives, then the proportion of tubes responding, p, is also the probability of at least one organism in 1 cc of the suspension. If the suspension is well-mixed, the number of organisms iIn 1 cc should have a Poisson distribution with y, the average number of organisms per cc, or concentration. Then, for a given y, p=probability of 1 or more organisms in 1 cc of a suspension with y organisms per cc or $p=p(x \geq 1)$ for any y. Using the cumulative Poisson table, we find that $P=p(x \geq 1)=1/3$ corresponds to m(= y here) of about 0.4 as $(x>1 \mid m=0.4)=0.3297$. thus $y=0.4$ organisms per cc. Similarly, we can find the values of y that correspond to the endpoints of the confidence interval for p to get the endpoints of the confidence interval for y. $p\ (x>1 \mid m=0.2)=0.1813$ and $p(x>1 \mid m=0.3)=0.2592$. So the y that corresponds to $p=0.21$, i.e. $p(x>1 \mid y)$, is somewhere between 0.2 and 0.3, say 0.2. Similarly, $p\ (x>1 \mid m=0.6)=0.4512$. Thus, we get $0.25<y<0.60$.

We can get more exact values for y using the table of natural logarithms. $P=p$ (at least one organism in 1 cc)$=1-e\ y$. p is then an estimate of $1-e-y$. We found $p=1/3=1-e-y$. e−y=2/3. y=in (2/3)=0.41 organisms per cc. To get a confidence interval for y, we have 0.21< p 0.45 cc=0.95 then $0.21<1-e-$, $y<0.45$ and $0.55<0.79$. −in $(0.79)<y<$−in (0.55) or $0.24<y<0.60$. If we are interested in any effect the drug might have on the response rate, we should do a two-sided test of H_0: $p_1=p_2$ using $a=0.50$. A total sample size is required to get power $(1-B)$ at the alternative values

p_1 and p_2 $n(\text{total}) \geq$

$$\left[\frac{Z_{1-\alpha/2}\sqrt{2\ p_0 q_0 + 2\ p_0 q_0} + Z_{1-\beta}\sqrt{2 p_1 q_1 + p_1 q_1}}{p_1 - p_2}\right]$$

If we let p_1=response rate without drug, we have established it to be 1/3. In the first part of this problem, if p_2 were only half as large, it would be about 1/6, and $p_0=1/2\ (p_1+p_2)$=½ (1/3+1+6)=¼. Then the total number (n) of subjects is as follows

N = [z.975 2(1) (3) + 2(1) (3)+ Z.8 2(1) (2)+2(1) (5)]2
4 4 3 3 6 6

$$= \frac{[1.96.75 + 84.7222]\ 2 = 52 \text{ or } 26 \text{ tubes in each group.}}{0.333}$$

Let x=number of heads in n=5 independent tosses of coin; x is binomial with p=probability of heads. We are testing H_0: p=0.5 against H_a: p=0.5 and are rejecting H_0 if x=0 or if x=5.

$a=p$ (type I error)$=p$ (Reject $H_0|H_0$ true)$=p$ (x=0 | p=0.5)+p(x=5 | p=0.5).

Using the binomial table with p=5, n=5, we find p (x=0)=p(x=5)=0.0312. Then a=0.0312+0.0312=0.0624. The power of the test against the alternative of a two-headed coin is the probability of rejecting H_0 when p=1=probability of heads.

$(1-B)=P$ (reject $H_0|$ p=1)=p(x=o| p =1)+p (x=5| p=1)=0+1=1

Since with a two-headed coin, one will always get x=5 heads in 5 tosses. One is 100% sure of detecting a two-headed coin. Let x_1=number of rabbits with pulmonary metastases among n_1=20 rabbits receiving a single dose of heparin, where the probability of metastases is p_1. Let x_2=number of rabbits with pulmonary metastases among n_2=20 rabbits receiving multiple doses of heparin, where the probability of metastases is p_2. Both x_1 and x_1 and x_2 should be binomial, and in both cases, '*np*' seems to be large enough to justify using the normal approximation.

5.38 Conclusion

The model is most robust when modelng an HMM. Hidden variables represent the elements of the answer we are looking for. The observed sequences are the query, and the variables provide the information about how the hidden states are related to each other and to the observed states. Finally, understanding the parameters of an HMM and the consequences of predictingf hidden attributes is a powerful tool for understanding the biological pathway of proteins. This modeling process of the model can be very useful for biologists and geneticists for the accurate study of hidden biological connections between peptides in unstacked datasets. The model can also extend to explore machine learning programs in econometrics for the calculation of inflation, the business of large venture banking sectors, weather forecasting, space science (collision effects), the spread of cancerous cells, the latent effect of growth of organisms, the estimation of flux in drug triathe l experimentation, cascading of marketing effects and linkage, etc.

REFERENCES

Baladandayuthapani, V., Ji, Y., Talluri, R., Nieto-Barajas, L. E. & Morris, J. S. Bayesian random segmentation models to identify shared copy number aberrations for array CGH data. *Journal of the American Statistical Association* 105(492), 1358–1375 (2010). https://doi.org/10.1198/jasa.2010.ap09250

Caudai, C. et al. AI applications in functional genomics. *Journal of Computational and Structural Biotechnology* **19**, 5762–5790 (2021). https://doi.org/10.1016/j.csbj.2021.10.009

Chatzou, M. et al. Multiple sequence alignment modeling: Methods and applications, *Briefings in Bioinformatics* **17**(6), 1009–1023 (2016).

Dias, R. & Torkamani, A. Artificial intelligence in clinical and genomic diagnostics. *Genome Medicine* **11**(3), 70–75 (2019). https://doi.org/10.1186/s13073-019-0689-8

Dou, Y., Guo, X., Yuan, L., Holding, D. R. & Zhang, C. Differential expression analysis in RNA-Seq by a naive Bayes classifier with local normalization. *BioMed Research International* **25**(3), 789–795 (2015). https://doi.org/10.1155/2015/789516

Edgar, R. C. & Batzoglou, S. Multiple sequence alignment. *Current Opinion in Structural Biology* **16**, 368–373 (2006).

Hong, Y. et al. A-Prot: Protein structure modeling using MSA transformer. *BMC Bioinformatics* **23**(1), 93–95 (2022). https://doi.org/10.1186/s12859-022-04628-8

Kemena, C. & Notredame, C. Upcoming challenges for multiple sequences alignment methods in the high-throughput era. *Bioinformatics* **25**, 2455–2465 (2009).

Li, L. F. et al. Genome sequencing and population genomics modeling provide insights into the local adaptation of weeping forsythia. *Horticulture Research* **7**, 130–138 (2020). https://doi.org/10.1038/s41438-020-00352-7

Meselson, Matthew, & Franklin W. Stahl. The replication of DNA in *Escherichia Coli. Proceedings of the National Academy of Sciences* **44**, 671–682 (1958).

Morris, J. S. & Baladandayuthapani, V. Statistical contributions to bioinformatics: Design, modeling, structure learning, and integration. *Statistical Modelling* **17**(4–5), 245–289 (2017). https://doi.org/10.1177/1471082X17698255

Patel, N., Passi, K. & Jain, C. K. Improved prediction of breast cancer on epigenomics data using feature selection and machine learning. *Advances in Proteomics and Bioinformatics* **3**, 115–118 (2020). https://doi.org/10.29011/2690-0092.100015

Rajkomar, A. et al. Scalable and accurate deep learning with electronic health records. *NPJ Digital Medicine* **1**, 18–22 (2018). https://doi.org/10.1038/s41746-018-0029-1

Thompson, J. D., Higgins, D. G., & Gibson, T. J. Clustal W: Improving the sensitivity of progressive multiple sequence alignment through sequence weighting, position-specific gap penalties and weight matrix choice. *Nucleic Acids Research* **22**, 4673–4690 (1994).

Van Noorden, R., Maher, B. & Nuzzo, R. The top 100 papers. *Nature* **514**, 550–553 (2014).

Vestal, B. E. et al. MCMSeq: Bayesian hierarchical modeling of clustered and repeated measures RNA sequencing experiments. *BMC Bioinformatics* **21**, 375–380 (2020). https://doi.org/10.1186/s12859-020-03715

Wallace, I. M. & Higgins, D. G. Evaluation of iterative alignment algorithms for multiple alignments. *Bioinformatics* **21**, 1408–1414 (2005).

Wu, Z., Johnston, K. E., Arnold, F. H., & Yang, K. K. Protein sequence design with deep generative models. *Current Opinion in Chemical Biology* **65**, 18–27 (2021). https://doi.org/10.1016/j.cbpa.2021.04.004

Zhang, J. X. et al. A deep learning model for predicting next-generation sequencing depth from DNA sequence. *Nature Communications* **12**, 4387 (2021a). https://doi.org/10.1038/s41467-021-24497-8

Zhang, L. et al. BERT-m7G: A transformer architecture based on BERT and stacking ensemble to identify RNA N7-methylguanosine sites from sequence Information. *Computational and Mathematical Methods in Medicine* **25**(1), 7764764 (2021b). https://doi.org/10.1155/2021/7764764

6

An Expert's Guide to Using Statistical Hypothesis for Life Science

6.1 Introduction

Tests of statistical significance are invariably applied nowadays by research scientists, and good medical journals refuse to accept papers for publication if the authors have not used the philosophy of significance testing in evaluating their results (Brownlee, 2018). It is not adequate to mechanically undertake significance tests (unfortunately, this has become far too easy with the advent of computers and programmed pocket calculators); the scientist/experimental worker must fully understand the basic concepts underlying a significance test, the assumptions involved, and the limitations, for making proper interpretations (Emmert-Streib & Dehmer, 2019). To regard a statistically significant difference as proof of the existence of an important difference, or to equate a non-significant difference with proof of no difference, is very naive, and such thinking deserves to be strongly discouraged (Chen et al., 2021). The statistical hypothesis consists of four important components: (i) Null hypothesis (H_0), (ii) alternative hypothesis (H_1), (iii) test statistics, and (iv) rejection of H_0.

1. Null hypothesis (H_0): The null hypothesis, which is denoted H_0, must specify that the parameter is equal to a single value.
2. Alternative hypothesis (H_a/H_1): The alternative hypothesis, denoted by H_a, answers the questions by specifying that the parameter meets one of the following criteria: (i) greater than the value shown in the null hypothesis, (ii) less than the value shown in H_0, or (iii) a different value from the one shown in the null hypothesis (Figure 6.1).

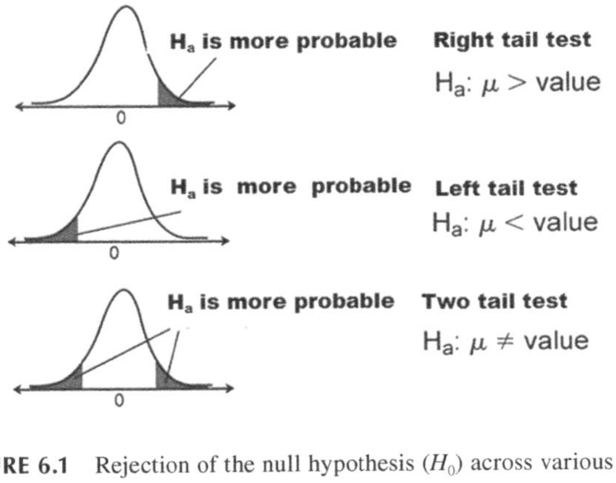

FIGURE 6.1 Rejection of the null hypothesis (H_0) across various critical regions.

3. Test statistics: The test statistic is the sample statistic upon which we base our decision to either reject or not reject the null hypothesis.
4. Rejection (CR): The rejection region is a range of values such that if the statistic falls into that range, we decide to reject the null hypothesis.

6.2 Sampling Fluctuation

To properly appreciate the role of significance testing, it is important to first understand the concept of sampling fluctuation. Consider a large population of, say, size 10,000, with a mean height of 65 and a standard deviation of 10. If, from this population, we draw a representative sample (by random sampling) of ($n = 100$) subjects, the mean height of these subjects will not necessarily be 65. For instance, in the first such sample, it may be 66. If random samples of ($n = 100$) are drawn repeatedly, the means could be 63, 65, 64, 67, etc. If a very large number of such samples are drawn, the mean of the means will be virtually the same as the population mean (Hung et al., 2020). The standard deviation of the mean, which is usually referred to as the sampling error of the mean, will, however, be much smaller – in this case, it will be a tenth of the original standard deviation. In general, if the size of the random sample is n, the sampling error of the mean will be SD and must obviously be taken into account when interpreting any observed difference (J & Lehmann, 1960). In general, we should take into account the possibility of sampling fluctuations when interpreting differences or associations.

6.2.1 Sampling Distribution

In our general discussion about the principal idea of a hypothesis test above, we mentioned that the connection between a test statistic and its sampling distribution is crucial for any hypothesis test. For this reason, we elaborate on this point in more detail in this section.

In this section, we aim to answer the following questions:

1. What is the sampling distribution?
2. How to obtain the sampling distribution?
3. How to use the sampling distribution?

First of all, the sampling distribution is a probability distribution. The meaning of this sampling distribution is that it is the distribution of the test statistic T, which is a random variable,

DOI: 10.1201/9781032713816-6

given some assumptions. We can make this statement more precise by defining the sampling distribution of the null hypothesis as follows:

Let $(n)=\{X_1,\cdots,X_n\}$ be a random sample from a population with $X_i \sim P(pop)$ and $T(X(n)$ be a test statistic. Then the probability distribution $(x|H_0\ true)$ of (n), assuming H_0 is true, is called the sampling distribution of the null hypothesis or the null distribution (Kawamoto et al., 2024).

Similarly, one defines the sampling distribution of the alternative hypothesis by $(x|H_1\text{true})$. Since there are only two different hypotheses, H_0 and H_1, there are only two different sampling distributions in this context. However, we would like to note that sampling distributions also play a role outside statistical hypothesis testing, for example, in estimation theory or data bootstrapping.

6.3 Concept of Test of Significance of the Mean

Suppose we have taken a random sample of n observations (e.g. 100 heights) from a large population ($n=10{,}000$), and we wish to use this sample to test some hypothesis about the population. In particular, we may believe that the population has a particular mean value μ (say 65″), and we may wish to use our sample to test the validity of this belief (Sharma, 2022). The procedure employed for this purpose is called a test of significance of the mean. Suppose our sample has a mean value of X (e.g. 68). We start with the assumption (null hypothesis) that the population mean is indeed μ (65) and then compute the probability of drawing a random sample of n (100) observations with mean (68″) from this population (see the next section). If the probability is small, we conclude that there is good reason to suppose that the population value is not μ (65). If the probability is not small, we conclude that the findings are consistent with a population mean of μ (65).

What constitutes a 'very small' probability is a subjective matter, but conventionally, a 5% significance level (i.e. probability of 0.05) is taken. A probability of 0.05 or less leads to a decision to reject the null hypothesis. In taking such a decision, however, there is a 1 in 20 chance that we may be making a mistake, i.e. we may be rejecting the null hypothesis when it is really true, that is, concluding that the population mean is not μ when it is indeed μ. This error is called a Type I error (Lehmann, 1997).

6.4 Computation of the Probability

For computing the probability that the sample mean of 68 (based on 100 subjects) could have originated from the population with a mean of 65, we shall assume that height follows a normal distribution. In this case, the mean of means from repeated random samples is expected to be 65″, and the standard deviation of the mean (sampling error) will be $10/\sqrt{100}=1$. For computing the probability that is of interest to us, we calculate a quantity (called the normal deviate) given by (Observed mean − Expected mean) = (68 − 65) = 3.1.

As this exceeds 1.96, which is the critical value for significance at the 5% level, we conclude that the observed difference is statistically significant and consequently reject the null hypothesis that the population mean is 65. (If this had been less than 1.96, we would have concluded that the observed difference is not statistically significant and therefore accepted the null hypothesis). From the table of areas under the normal curve, we can compute the exact probability of $0.00135\times2=0.0027$. In other words, the chance of obtaining a mean of 68 in a random sample of 100 observations if the true population mean is 65 is 0.27%, which is very small indeed. In the above, we have assumed that the measure of interest (height) follows a normal distribution. If the variable does not follow a Normal distribution, we may transform it to another scale (e.g. to logarithms in the case of drug concentrations or to square roots in the case of counts); the resultant distributions may then be 'Normal.' If this is not feasible, but the sample size is fairly large (e.g. more than 30), we can invoke an important statistical theorem that states the mean is approximately normally distributed and compute the probability as above. Only in this case, the probability would be an approximate one.

6.5 Applications for a Significance Test of the Mean

At first sight, it may appear that situations in which a sample mean is tested against a known population mean are extremely rare. However, if we consider any comparative study (e.g. a controlled clinical trial of two regimens, a laboratory experiment to evaluate a new culture medium against a standard medium, or a comparison between the mean systolic BP of laborers and sedentary workers), the null hypothesis is always that the two groups have the same mean. If so, the theoretical difference between the two means is zero. In practice, we test whether the observed difference between the two means is significantly different from the theoretical value of zero. Thus, this test finds frequent application in real-life situations (Lago et al., 2021).

6.6 Cautions about Interpretation of a Test of Significance

It should by now be clear that a difference being statistically significant does not necessarily mean that it is large or clinically important. It means that the observed difference is unlikely to be due to sampling fluctuation, and by unlikely, we usually mean a probability of less than 0.05 (5%). Equally, a finding that a difference is not statistically significant (non-significant) does not necessarily mean that there is no real difference (Liu et al., 2023). There could, in reality, exist some

TABLE 6.1
Postulates of Hypothesis Testing Assumptions

	Statement	
Hypothesis	**True**	**False**
Rejection of H_0	Type I error 'α' (False Positive)	Correct True Positive
Acceptance of H_1	Correct True negative	Type II error $\beta = (1 - \alpha)$ False negative

difference (perhaps even an appreciable difference), but due to inadequate sample size and/or vagaries of sampling fluctuation, we may have obtained a 'non-significant' result and consequently accepted the Null hypothesis of zero difference. This is called a Type II error (Yamamoto et al., 2023).

6.7 Occurrence of Type I and Type II Errors

Every decision-making process, including a significance test, carries with it both Type I and Type II errors. It is impossible to make both errors zero, but their magnitudes can be controlled by an appropriate increase in the sample size. The relative importance of the two types of errors varies from situation to situation, e.g., drug trials in tuberculosis, where many effective drugs are available (Type I error is more important), and drug trials in cancer, where very few effective drugs are known (Type II error is more important).The area given above is that lying between the particular value of the normal deviate and infinity, e.g. if we consider the normal deviate of 1.96, 0.025 represents the area from 1.96 to infinity. To obtain the area for two tails, we multiply the area by 2, i.e. $0.025 \times 2 = 0.05$ (Table 6.1).

Statistics is the science of analyzing huge amounts of data. In the real world, it is nearly impossible to deduce statistics about the entire population. This huge amount of data needs interpretation to draw meaningful conclusions. Hence, we take some random samples from the population, derive statistical measures (e.g. mean, standard deviation, variance), and draw conclusions about relationships from the data collected.

Data can be interpreted by assuming a specific outcome and using statistical methods to confirm or reject the assumption. This assumption is called a hypothesis, and the statistical test used for this purpose is called hypothesis testing (Murakami, 2014).

In statistics, a hypothesis is a statement about a population that we want to verify based on information contained in the sample data.

Hypothesis testing quantifies an observation or outcome of an experiment under a given assumption. The result of the test enables us to interpret whether the assumption holds true or false. In other words, it signifies if the hypothesis can be confirmed or rejected for the observation made (Neyman & Pearson, 2023).

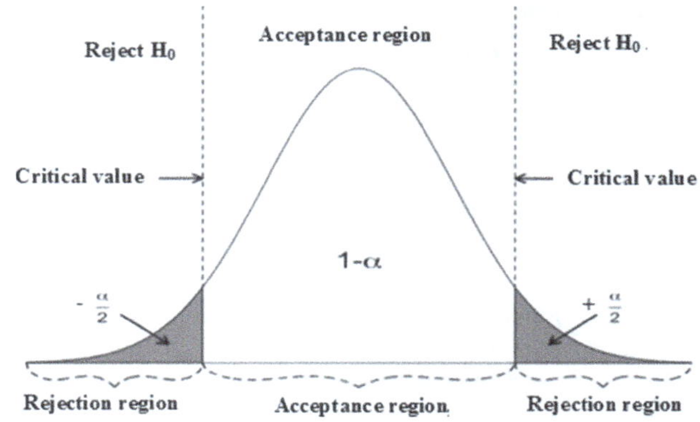

FIGURE 6.2 Hypothesis testing within the context of a normal distribution.

The assumption of hypothesis testing is called the null hypothesis. A null hypothesis is a type of assumption used in statistics indicating that there is no significant difference between the samples from the underlying population (Matusita et al., 1954). It is also known as the default hypothesis, represented by H_0. In contrast, there is a term 'Alternative Hypothesis', represented by H_1. For every null hypothesis, there is an alternative hypothesis that is opposite to what the null hypothesis states. In other words, if we reject the null hypothesis, we fail to reject the alternative hypothesis (Patriota, 2017). The decision to confirm or reject the null hypothesis is made by interpreting the result of the test (Raha, 2011). The result of hypothesis testing can be interpreted using p-values or critical values. The p-value is the probability of deducing the observed value, given the assumption (Figure 6.2). On the other hand, critical values are cut-off values that define regions where the test statistic is unlikely to lie (Suvorov et al., 2022).

Statistics is the science of analyzing huge amounts of data. In the real world, it is nearly impossible to deduce statistics about the entire population. This huge amount of data needs interpretation to draw meaningful conclusions. Hence, we take some random samples from the population, derive statistical measures (e.g. mean, standard deviation, variance), and draw conclusions about relationships from the data collected.

Data can be interpreted by assuming a specific outcome and using statistical methods to confirm or reject the assumption. This assumption is called a hypothesis, and the statistical test used for this purpose is called hypothesis testing.

In statistics, a hypothesis is a statement about a population that we want to verify based on information contained in the sample data.

Hypothesis testing quantifies an observation or outcome of an experiment under a given assumption. The result of the test enables us to interpret whether the assumption holds true or false. In other words, it signifies if the hypothesis can be confirmed or rejected for the observation made.

6.8 Perform Hypothesis Testing

Hypothesis testing can be performed using these three steps:

1. Design a test statistic.
2. Design a null hypothesis.
3. Compute the p-value or critical values.

How about looking at this coin-toss example to see how hypothesis testing is performed?

Given a coin, determine if the coin is biased towards heads or not.

What do you mean by saying a coin is biased toward heads? This means that the probability of getting heads is greater than 0.5, i.e. $(H_0)>0.5$.

6.9 Experimental Design

6.9.1 Flip a Coin 5 Times, and Count the Number of Heads

Consider the count of the number of heads as a random variable 'X'. In the theory of statistical hypothesis testing, it is known as a test statistic, i.e. 'X' = 5. The results are represented using the critical values in the same way they are interpreted using the p-value. While performing a statistical hypothesis test of whether the data sample is normally distributed, the test statistic was calculated and compared to the critical value at the 5% significance level. We can say that the hypothesis test found that the sample is normally distributed, failing to reject the null hypothesis at a 5% level of significance.

6.10 Errors in Hypothesis Testing

In hypothesis testing, relatively small samples are used to answer questions about population parameters. There is always a chance that the selected sample is not representative of the population; therefore, there is always a chance that the conclusion deduced is wrong. Hypothesis testing provides confidence in favor of a certain hypothesis. In other words, hypothesis testing refers to the use of statistical analysis to determine if observed differences between two or more data samples are due to random chance or true differences in the samples. Since we are computing the probability of an experiment, it can be deduced that the interpretation of the hypothesis test is purely probabilistic. This means that the outcome of the experiment can be misunderstood.

If the p-value is small, rejecting the null hypothesis indicates either of these two scenarios:

The H_0 hypothesis is false—we are right.

The H_0 hypothesis is true, and some rare and unlikely event occurred—we made a mistake.

This type of error is called a False Positive—Type I Error.

On the contrary, if the p-value is large, failing to reject the null hypothesis indicates either of these two scenarios:

The null hypothesis is true—we are right.

The H_0 is false, and some rare and unlikely event occurred—we made a mistake.

This type of error is called a False Negative—Type II Error.

There is always a possibility of making these kinds of errors while interpreting the results of hypothesis testing (Figure 6.3). Hence, we must be cautious of encountering such errors and verify the findings before concluding (Table 6.2).

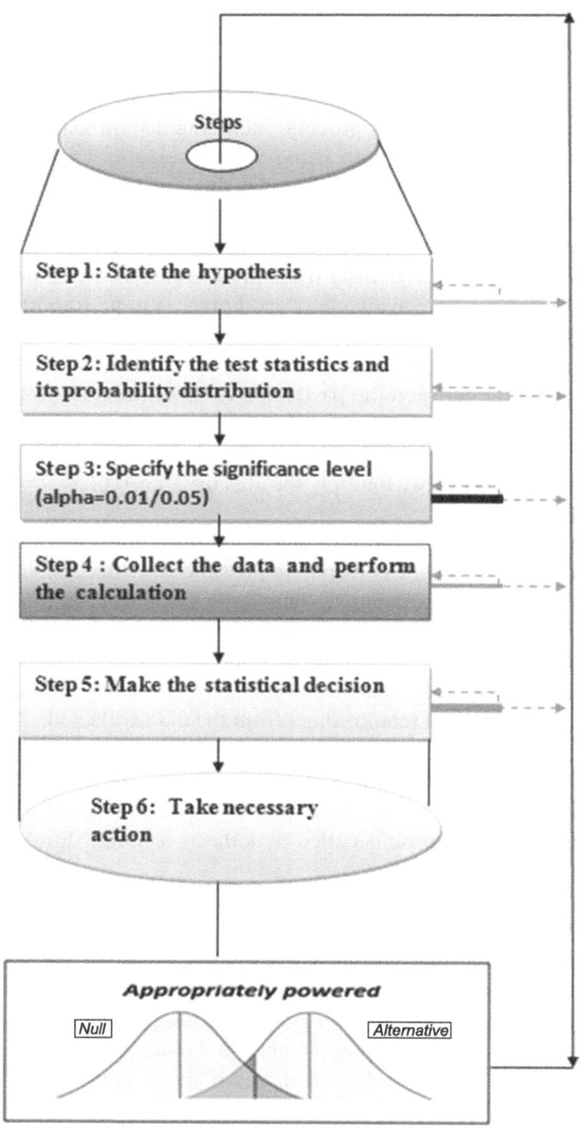

FIGURE 6.3 Flow chart detailing the various steps involved in hypothesis testing.

TABLE 6.2

Different Tests Used for Testing of Hypothesis

Tests	Diagnostic Tests	Criteria	Formula
Small Samples ($n \leq 30$)			
t-test	One sample test		$t_{\alpha=0.05/0.01} = \frac{\bar{X}-\mu}{\sigma/\sqrt{n}}$
	Two sample tests	Two variables considered (X,Y)	$t_{\alpha=0.05/0.01} = \frac{\lvert \bar{X}_1 - \bar{X}_2 \rvert}{\frac{\sigma_1^2}{n_1}+\frac{\sigma_2^2}{n_2}} \sigma_1, \sigma_2 = \text{Known}$ $t_{\alpha=0.05/0.01} = \frac{\lvert \bar{X}_1 - \bar{X}_2 \rvert}{Sp^2 \sqrt{\frac{1}{n_1}+\frac{1}{n_2}}}$ $Sp^2 = \frac{\sigma_1^2(n_1-1)\sigma_2^2(n_2-1)}{n_1+n_2-2}$ $\sigma_1, \sigma_2 = \text{Not Known}$
	Paired test	Two variables considered (X, Y) X: Before and Y: after experiments	$t_{\alpha=0.05/0.01} = \frac{\bar{d}}{\sigma/\sqrt{n}}$
Large Sample Test ($n \geq 30$)			
Z-test	One sample	Only one variable considered	$Z_{\alpha=0.05/0.01} = \frac{X-\mu}{\sigma/\sqrt{n}}$
	Two sample	Two variables considered (X, Y)	$t_{\alpha=0.05/0.01} = \frac{\bar{\bar{X}} - \bar{Y} - \Delta_0}{\frac{\sigma_1^2}{n_1}+\frac{\sigma_2^2}{n_2}}$ $H_0 = \mu_1 - \mu_2 = \Delta_0$ Upper tailed $H_A : \mu_1 - \mu_2 > \Delta_0 \;\; z \geq Z_\alpha$ Lower tailed $H_A : \mu_1 - \mu_2 < \Delta_0 \;\; z \leq Z_\alpha$ Both tailed (Two) $H_A : \mu_1 - \mu_2 \neq \Delta_0 \;\; z \geq Z_{\alpha/2}$ Or $z \leq Z_{-\alpha/2}$
	One proportion		$z = \frac{p - p_0}{\sqrt{\frac{p_0(1-p_0)}{n}}}$
	Two proportion	Two proportion (X,Y)	$z_{\text{cal}} = \frac{(\hat{p}_1 - \hat{p}_2)}{\sqrt{\bar{p}(1-\bar{p})\left(\frac{1}{n_1}+\frac{1}{n_2}\right)}}$ $\bar{p} = \frac{n_1\hat{p}_1 + n_2\hat{p}_2}{n_1+n_2}$ $z_{\text{cal}} = \frac{(\hat{p}_1 - \hat{p}_2) - (p_1 - p_2)}{\sqrt{\frac{\hat{p}_1(1-\hat{p}_1)}{n_1}+\frac{\hat{p}_2(1-\hat{p}_2)}{n_2}}}$
Chi square	Chi-square fit test	H_0: The data follow a specified distribution H_a: The data do not follow the specified distribution, chi-square for $(k-1)$ df for specific alpha value (0.01/0.05)	$\chi^2 = \sum_{i=1}^{n} \left(\frac{O_{ij} - E_{ij}}{E_{ij}}\right)^2$
	Chi-square test of independence	Test observed and theoretical frequencies	$\chi^2 = \sum_{i=1}^{n} \left(\frac{O_{ij} - E_{ij}}{E_{ij}}\right)^2$ where O_{ij}=Observed value of ith row and jth column, E_{ij} = Expected value of ith row and jth
F-test	Equality of variance and homogeneity	Hypothesis derived two ample, unbiased estimation of population variance of two samples	$F = \frac{\sigma_1^2}{\sigma_2^2} \sim n(\mu, \sigma^2)$
ANOVA	One way Two way	One parameter considered Two parameters considered	$y_{ij} = \mu + \alpha_i + \epsilon_{ij}$ $y_{ij} = \mu + \alpha_i + \beta_{ij} + \epsilon_{ij}$
Post hoc test	Tukeys, Dunnet, DMRT	Pairwise means comparison	–

6.11 One Sample *t*-Test

The one-sample *t*-test, also referred to as the single-sample *t-test*, is a univariate analytical tool for testing a hypothetical assumption to determine whether the mean calculated from sample data collected from a single group is different from a designated value specified by the researcher.This designated value does not come from the data itself; rather, it is an external value chosen for scientific reasons. The null hypothesis for a one-sample *t*-test can be stated as: The population mean equals the specified mean value. An alternative hypothesis can be stated as: The population mean differs from the specified value. The formula for a one-sample *t*-test is as mentioned in Equation (6.1).

$$t_{\alpha=0.05/0.01} = \frac{\overline{X} - \mu}{\sigma/\sqrt{n}} \tag{6.1}$$

where, $\mathrm{SD}(\sigma) = \frac{1}{n}\sum_{i=1}^{n}\sqrt{(x-\overline{x})^2}$

Imagine we have collected a random sample of 31 energy bars from a number of different stores to represent the population of energy bars available to the general consumer. The labels on the bars claim that each bar contains 20 g of protein (Table 6.3).

TABLE 6.3

Distribution of Grams of Protein in Random Sample of Energy Bars

20.70	27.46	22.15	19.85	21.29	24.75	–
20.075	22.91	25.34	20.33	21.54	21.08	–
22.14	19.56	21.10	18.04	24.12	19.95	–
19.72	18.28	16.26	17.46	20.53	22.12	–
25.06	22.44	19.08	19.88	21.39	22.33	25.79

6.12 Working Procedure

Statement of Hypothesis

Our null hypothesis (H_0) is that the underlying population mean is equal to 20. The null hypothesis is written as:

$$H_0 : \mu = 20 = 20$$

The alternative hypothesis (H_a) is that the underlying population mean is not equal to 20. The labels claiming 20 g of protein would be incorrect. This is written as:

$$H_a : \mu \neq 20$$

This is a two-sided test. We are testing if the population mean is different from 20 g in either direction. If we can reject the null hypothesis that the mean is equal to 20 g, then we make a practical conclusion that the labels for the bars are incorrect. If we cannot reject the null hypothesis, then we make a practical conclusion that the labels for the bars may be correct (Figure 6.4).

$$t_{\alpha=0.05/0.01} = \frac{21.37 - 20.0}{2.55/\sqrt{31}} = 300$$

$t = 3.008916$, $df = 30$; p-value (one-tailed) $= 0.002635$, p-value (two-tailed) $= 0.005270$

The result was found to be significant, so we reject the H_0. This p-value describes the likelihood of seeing a sample average as extreme as 21.4, or more extreme, when the underlying population mean is actually 20. In other words, it represents the probability of observing a sample mean as different, or even more different from 20, than the mean we observed in our sample. A p-value of 0.0046 means there are about 46 chances out of 10,000. We feel confident in rejecting the null hypothesis that the population mean is equal to 20.

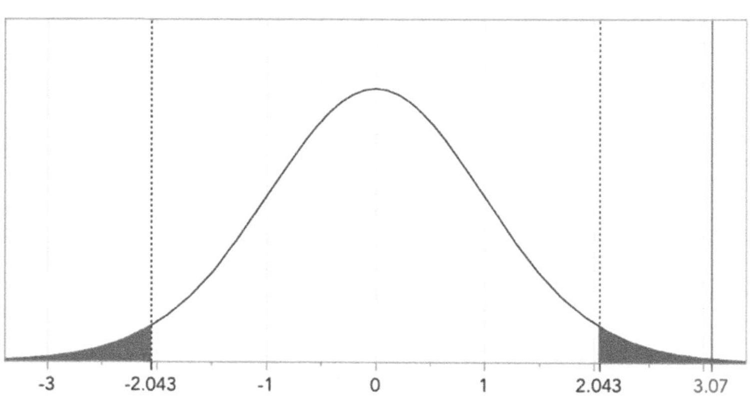

FIGURE 6.4 Displays results based on the *t*-distribution with 30 degrees of freedom.

TABLE 6.4

Body Fat Percentage Data Grouped by Gender

Male	13.30	6.0	20.0	8.0	14.0	19.0	18.0	25.0	16.0	24.0	15.0	1.0	15.0
Female	22.0	16.0	21.7	21.0	30.0	26.0	12.0	23.2	28.0	23.0	–	–	–

6.13 Two-Sample *t*-Test

Our sample data is from a group of men and women who did workouts at a gym three times a week for a year. Then, their trainer measured the body fat. The table below shows the data (Table 6.4).

Solution

6.13.1 Assumption

H_0: Our null hypothesis is that the underlying population means are the same. The null hypothesis is written as

$$H_0 : \mu_1 = \mu_2$$

H_a: The alternative hypothesis is that the means are not equal. This is written as

$$H_1 : \mu_1 \neq \mu_2$$

$$t_{\alpha=0.05/0.01} = \frac{\left|\bar{X}_1 - \bar{X}_2\right|}{Sp^2\sqrt{\frac{1}{n_1}+\frac{1}{n_2}}} \tag{6.2}$$

$$Sp^2 = \frac{\sigma_1^2(n_1-1)\sigma_2^2(n_2-1)}{n_1+n_2-2} \quad \sigma_1,\sigma_2 = \text{Not Known}$$

$$Sp^2 = \frac{5.32^2(13-1)6.84^2(10-1)}{13+10-2} = 38.88,$$

$$t_{\alpha=0.05/0.01} = \frac{|22.29-14.95|}{38.88\sqrt{\frac{1}{13}+\frac{1}{10}}} = 2.80$$

To find this value, we need the significance level ($\alpha=0.05$) and the *degrees of freedom*. The degrees of freedom (*df*) are based on the sample sizes of the two groups. For the body fat data, this is:$df = (n_1+n_2-2) = 13+10-2 = 21$. The *t*-value with $\alpha=0.05$ and 21 degrees of freedom is 2.080. We compare the value of our statistic (2.80) to the *t*-value. Since 2.80>2.080, we reject the null hypothesis that the mean body fat for males and females is equal, and conclude that we have evidence that body fat in the population is different between men and women (Table 6.5).

TABLE 6.5

Gender-Wise Distribution of Males and Females

Gender	Sample Size (*n*)	Average (*x*)	SD
Males (X_2)	10	14.95	6.84
Females (X_1)	13	22.29	5.32

6.14 Equal Variances

If the observations do not fall naturally into pairs, a different expression must be used for the variance of the mean (i.e. for the denominator in the expression for *t*). It may be noted that, unlike in the case of the paired*t*-test, the two samples need not necessarily have the same size. It should be firmly emphasized that it would be inefficient to test paired observations as though they were unpaired; unfortunately, this is often done. In general, this will lead to an overestimation of probability values, i.e. under-calling of significant differences. Returning to the case of unpaired observations, an estimate of the variance in the population (from which the two samples are supposed to have been drawn) is obtained by averaging the variances of the two samples. In fact, a weighted average is calculated, with the weights being (n_1-1) and (n_2-1), the degrees of freedom of S_1^2 and S_2^2. Thus, the estimate is given by:

6.14.1 Working Rule

$$t = \frac{\left|\bar{X}_i - \bar{X}_j\right|}{\left(\text{Pooled } S^2\right) * \sqrt{\frac{1}{n_i}+\frac{1}{n_j}}} \tag{6.3}$$

$$\text{Pooledvariance}\left(S^2\right) = \frac{(n_1-1)S_1^2 + (n_2-1)S_2^2}{n_1+n_2-2}$$

$$S_1 = \frac{1}{n_1-1}\sum_{i=1}^{n}\left(X_1-\bar{X}_1\right)^2 = S_1^2;\; S_2 = \frac{1}{n_2-1}\sum_{i=1}^{n}\left(X_2-\bar{X}_2\right)^2 = S_2^2$$

The two terms in the numerator are the sums of squares of deviations from the mean in samples 1 and 2, respectively. The above expression can therefore be rewritten as

$$S^2 = \frac{\left\{\sum X_1^2 - \left(\sum X_1\right)^2\right\}}{n_1} + \frac{\left\{\sum X_2^2 - \left(\sum X_2\right)^2\right\}}{n_1} \tag{6.3}$$

The variance of $\bar{X}_1 - \bar{X}_2$=(variance of $\bar{X}_1$+variance of $\bar{X}_2$)*

$$= s^2\left[\frac{1}{n_1}+\frac{1}{n_2}\right]$$

The test statistic then becomes:

$$t = \frac{\bar{X}_1 - \bar{X}_2}{\sqrt{SP^2\left(\frac{1}{n_1}+\frac{1}{n_2}\right)}} \tag{6.4}$$

which follows a t-distribution with ($df=n_1+n_2-2$). The underlying assumption for t-test are (i) the distributions of X_1 and X_2 are normal and (ii) the population variances of X_1 and X_2 are equal. However, minor deviations from these assumptions do not affect the validity of the test (Figure 6.5).

Illustration

Erythrocyte sedimentation rates (1-hour) were measured in an infected group of patients as well as in a control group (see below). Examine whether the infected group had a significantly higher ESR on average.

6.15 *P*-Value and Statistical Significance

The two-tailed P-value<0.001. By conventional criteria, this difference is considered to be extremely statistically significant. Confidence Interval: The mean ESR of the study group minus the ESR of the control group equals 22.3395% confidence interval of this difference ranges from 16.60 to 28.06, Intermediate values used in calculations'=7.9817, $df=28$, Standard error of the difference=2.798

The calculated value of t (9.98) is higher than the tabulated 0.1% value for t with 28 degrees of freedom, namely 43.850, indicating that the mean ESR in the infected group is significantly ($P<0.001$) higher than that of the control group (Table 6.6).

6.16 Unequal Variances

Student's t-test assumes that the sample means being compared for two populations are normally distributed and that the populations have equal variances. Welch's t-test is designed for unequal population variances, but the assumption of normality is maintained. Welch's t-test is an approximate solution to the problem where ($X_{1\ldots}$ X_n) and ($Y_1,\ldots, Y_m$) be *iid* samples from two populations that both come from the same location–scale family of distributions. Situations are sometimes encountered where the variances in the two samples differ considerably from one another for good reason, such as extreme chance fluctuations or differences in population variance. An example of this would be a situation where two technicians—one experienced (and therefore more consistent) and the other relatively inexperienced (and therefore more variable)—undertake a blood count or read a smear slide (Genin, 2022). Both technicians, in this instance, would be estimating the same population mean value, but the more experienced one would have smaller variability in their readings than the less experienced one. Another example is provided by comparisons between slow inactivators of isoniazid and rapid in activators. The former are, more or less, a homogeneous lot (two slow alleles), while the latter are a heterogeneous group and consist of homozygous rapid inactivators (two rapid alleles) and heterozygous rapid inactivators (one rapid and one slow allele). Consequently, the variability can be expected to be greater in the rapid inactivators than in the slow inactivators. It is difficult to suggest a definite course of action for all situations with unequal variances. Sometimes, a transformation of the readings to another scale (e.g. logarithmic) has the effect of equalizing the variances. When this is not possible, special methods are available (e.g. Cochran's modified t-test, Fisher-Behrens test), but it would be best to obtain expert guidance from a professional statistician.

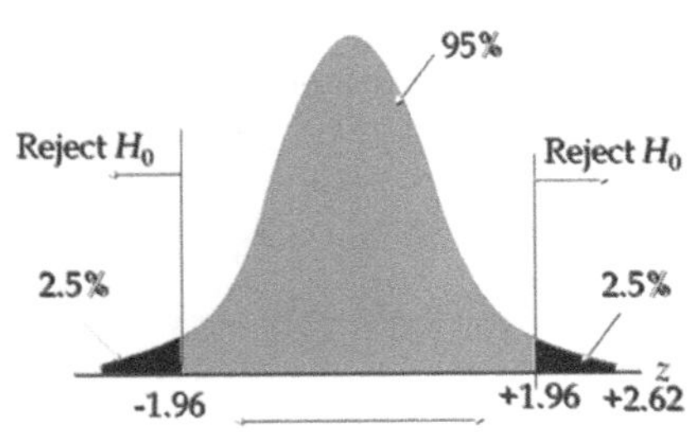

FIGURE 6.5 Acceptance and rejection of H_0 and H_a as derived from the normal distribution curve.

$$t=\frac{\overline{X_1}-\overline{X_2}}{\sqrt{\frac{S1^2}{N_1}+\frac{S2^2}{N_2}}} \tag{6.5}$$

where $\overline{X_1}$ & $\overline{X_2}$ are the two sample means, and S_1 and S_2 are their standard deviations. Fisher approximated the distribution of t, ignoring the random variation of the relative sizes of the SD (Table 6.7). The following formula is also used:

$$t=\frac{S_1/\sqrt{n_1}}{\sqrt{\frac{S1^2}{N_1}+\frac{S2^2}{N_2}}} \tag{6.6}$$

P-Value and Statistical Significance: The two-tailed P-value<0.0001. By conventional criteria, this difference is considered to be extremely statistically significant. Confidence Interval: The mean of the ESR study group minus the ESR control group equals 41.40. The 95% confidence interval of this difference ranges from 34.27 to 48.53. Intermediate values used in calculations- $t=12.4449$, $df=14$, Standard error of the difference=3.327 (Table 6.7).

TABLE 6.6

Mean-Wise Distribution of ESR on Study and Control Group

Group	ESR Study Group ($n=50$)	ESR CONTROL group ($n=25$)
Mean	45.80	23.47
SD	9.78	4.67
SEM	2.52	1.21

TABLE 6.7

ESR Data Frame for Control and Study Population

Group 1(ESR)	56	65	63	64	78	79	55	54	56	58
Group 2 (Control)	25	32	24	22	19	19	13	17	21	23

6.17 *t*-Test on Paired Observations

Situations often arise in which we wish to make comparisons based on paired observations on an individual, or on observations made on two individuals who have been 'matched'. In such situations, the difference between the two observations must, on average, be zero under the null hypothesis. The *t*-test described above, therefore, has frequent practical applications.

$$t_{\alpha=0.05/0.01} = \frac{\bar{d}}{\sigma/\sqrt{n}} \quad (6.7)$$

$$\text{where, } \sigma = \frac{1}{n-1}\sum_{i=1}^{n}\left(X_i - \bar{X}_i\right)^2, \bar{d} = \frac{\sum d}{n},$$

$$d = \text{Diff}\left(\text{Before} - \text{After experimenation absolute values}\right)$$

Testing of hypothesis

i. $t_{\text{Calvalue}} \geq t_{\alpha=0.05/0.01}$ Reject H_0
ii. $t_{\text{Calvalue}} \leq t_{\alpha=0.05/0.01}$ Accept H_0

Illustration

Erythrocyte sedimentation rates were measured before and after treatment of infection in ten patients. We examine whether the treatment conferred any significant benefit (Table 6.9).

TABLE 6.8

ESR Means of Two Group

Group	ESR Study Group ($n=50$)	ESR Control Group ($n=25$)
Mean	62.80	21.40
SD	9.15	5.19
SEM	2.89	1.64

TABLE 6.9

ESR Distribution before and after @ 1 Hour

Patient No	ESR @ 1 hour (mm) Before R_x	After R_x	Difference $d = (R_x - R_y)$	Square of difference $(d)^2$
1	25	8	17	289
2	43	10	33	1089
3	38	6	32	1024
4	20	7	13	169
5	41	10	31	961
6	48	5	43	1849
7	15	8	7	49
8	28	9	19	361
9	35	4	31	961
10	33	3	30	900
Total	$\sum R_{xa} = 326$	$\sum R_{xb} = 70$	$\sum d = 256$	$\sum d^2 = 7{,}652$

Consider the differences.

$\Sigma d=256$, $\Sigma d^2=7652$, $n=10$, Mean$=d=256/10=25.60$

Variance$=s^2=1/0\ [7{,}652-(256/10)^2]=1{,}098.4/9=122.04$

SD (s) $\sqrt{122.04} = 11.047$

$t=7.33$ With 9 degrees of freedom, the calculated value of t (7.33) is higher than the tabulated 0.1% value for two with ($df=9$) degrees of freedom, namely (4.781), indicating that the treatment had a significant ($P<0.001$) benefit in reducing the ESR.

Illustration

The paired *t*-test provides hypothesis testing for the difference between population means for a set of random samples whose variations are almost normally distributed. Subjects are often tested in a before-after situation or with subjects as alike as possible (Tables 6.10 and 6.11).

Working rule

$$t_\alpha = \frac{\bar{d}}{\frac{\text{SD}}{\sqrt{n}}} \quad (6.8)$$

TABLE 6.10

Data Frame of SAM and BMI

SAM (X_1) BMI kgm[3]	18.52	17.47	17.63	18.54	19.25	17.41	16.54	17.48	15.89	16.32
SAM (X_2) BMI kgm[3]	21.22	20.36	21.47	22.36	21.58	21.33	21.74	21.52	19.85	19.87

TABLE 6.11

Difference of Variable 1 and 2

SAM Cases Before (X_1) BMI kgm[3]	After (X_2) BMI kgm[3]	$d = X_1 - X_2$	d^2
18.52	21.22	−2.7	41.60
17.47	20.36	−2.89	29.029
17.63	21.47	−3.84	1.011
18.54	22.36	−3.82	0.72
19.25	21.58	−2.33	69.07
17.41	21.33	−3.92	2.64
16.54	21.74	−5.2	176.89
17.48	21.52	−4.04	6.66
15.89	19.85	−3.96	3.76
16.32	19.87	−3.55	1.29
Mean = 17.50	Mean = 21.13	−3.625	332.70
SD = 1.06	SD = 0.831	$\sum d = 0.817$	$\sum d^2 = 332.70$

$$\bar{d} = \frac{\sum d}{n} = \frac{137.42}{10} = 13.74,$$

$$\text{SD} = \frac{\sum \left(d - \bar{d}\right)^2}{n-1} = \frac{332.70}{10-1} = 36.96$$

$$\text{SEM} = \frac{\text{SD}}{\sqrt{n}} = \frac{36.96}{3.16} = 11.69,$$

$$t_\alpha = \frac{\bar{d}}{\frac{\text{SD}}{\sqrt{n}}} = \frac{13.74}{11.69} = 1.17$$

The calculated value of t (1.17) is higher than the tabulated 0.1% value ($df = 10$) degrees of freedom, indicating that the BMI had an insignificant ($p < 0.001$) benefit for SAM.

6.18 z-Tests

In hypothesis testing, the z-test follows the normal distribution. The z-test is used to determine whether the population means are different from each other when the variance values are known and the sample size (n) is relatively large. The z-score or z-statistic represents the number of standard deviations a data point is from the mean. The z-test and t-test are closely related, but the t-test produces better results when the sample size is small in the experiment.

$$Z_{\alpha=0.05/0.01} = \frac{\bar{X} - \mu}{\sigma / \sqrt{n}} \tag{6.9}$$

X: 85, 84, 74, 86, 96, 93, 85, 84, 74, 76, 96, 85, 77, 74, 68, 55, 58, 54, 57, 56, 53, 65, 56

X: Biomedical Training Post test score

Hypothesized mean difference = 6.48, Sample size ($n = 23$), SD = 14.39, Sample mean 73.52 Right-tailed test $H_0 : \mu = 6.48$ $H_a : \mu > 6.48$ statistic 22.34, $p = 0.00$

Z-score = 0.486

Probability of $x < 80$: 0.68 

Probability of $x > 80$: 0.33 

Probability of $73 < x < 80$: 0.18 

Steps

$$Z \text{ score} = \left(\frac{X - \mu}{\sigma}\right) = \frac{(80 - 73)}{14.39} = 0.486$$

P-value from Z-Table:

$P(x < 80) = 0.686$

$P(x > 80) = 1 - P(x < 80) = 0.313$ (31.33%)

$P(73 < x < 80) = P(x < 80) - 0.5 = 0.186$

6.19 Two-Sample z-Test

A two-sample z-test is used to test whether two population means are equal. This test assumes that the standard deviation of each population is known. For the results of a two-sample z-test to be valid, the following assumptions should be met: The data from each population are continuous (not discrete). Each sample is a simple random sample from the population of interest. Data in each population closely approximate a normal distribution. The population SD is known.

Illustration

Suppose the IQ levels among individuals in two different cities are known to be normally distributed, each with population standard deviations of 15.

A scientist wants to determine if the mean IQ levels between individuals in City A and City B are different. She selects a simple random sample of 20 individuals from each city and records their IQ levels.

To test this, she will perform a two-sample z-test at a significance level of $\alpha = 0.05$ using the following steps:

$\bar{X}_1$ (IQ)	$\bar{X}_2$ (IQ)
100.65	108.80
$n_1 = 20$	$n_2 = 20$
SD = 15	SD = 15

Hypothesis define

$H_0 : \mu_1 - \mu_2$ (the two population means are equal)

$H_a : \mu_1 \neq \mu_2$ (the two population means are not equal)

$$z_{\alpha=0.05/0.01} = \frac{\left|\bar{X}_1 - \bar{X}_2\right|}{\frac{\sigma_1^2}{n_1} + \frac{\sigma_2^2}{n_2}} \tag{6.10}$$

$$z_{\alpha=0.05/0.01} = \frac{|100.65 - 108.80|}{\frac{15^2}{20} + \frac{15^2}{20}}, \; z_{\alpha=0.05/0.01} = -1.78 \text{ is } 0.0858$$

Since the p-value (0.0858) is not less than the significance level (0.05), the scientist fails to reject the null hypothesis.

6.19.1 Proportion Tests

Single Proportion Test

A one-proportion z-test is used to compare an observed proportion to a theoretical one. The test statistic is calculated as

$$z = \frac{p - p_0}{\sqrt{\frac{p_0(1 - p_0)}{n}}} \tag{6.11}$$

where p = observed sample proportion, p_0 = hypothesized population proportion

n = sample size

Example: smoking rate in a town in the past was 21%. A sample of 100 individuals was taken,14 smokers were found. Has the smoking habit changed? use a 95% confidence interval to determine

$$z = \frac{014 - 0.21}{\sqrt{\frac{0.21(1-0.21)}{100}}} = -1.71,$$

P-value (one-tailed): 0.04284, p-value (two-tailed): 0.08569, 95% CI = [0.0720, 0.2080]

6.20 Two-Sample Test

A two-sample proportion test is a statistical method that compares the proportions of two independent groups. It can be used to determine if two samples come from the same population. It can also be used to determine if two populations differ significantly on specific characteristics.

Assumptions of the Two-Sample Z-Proportion Hypothesis Test

1. The data are simple random sample values from both populations
2. Both populations should follow a binomial distribution
3. The Samples are independent of each other
4. Test results are accurate when 'np' and $n(1-p)$ are greater than 5

$$z_{cal} = \frac{(\hat{p}_1 - \hat{p}_2)}{\sqrt{\bar{p}(1-\bar{p})\left(\frac{1}{n_1}+\frac{1}{n_2}\right)}} \quad \text{Pooled calculation} \tag{6.12}$$

$$\bar{p} = \frac{n_1\hat{p}_1 + n_2\hat{p}_2}{n_1 + n_2}$$

$$z_{cal} = \frac{(\hat{p}_1 - \hat{p}_2)-(p_1 - p_2)}{\sqrt{\frac{\hat{p}_1(1-\hat{p}_1)}{n_1}+\frac{\hat{p}_2(1-\hat{p}_2)}{n_2}}} \quad \text{Un pooled calculation} \tag{6.13}$$

6.20.1 Procedure to Execute Two-Sample Proportion Hypothesis Test

1. State the null hypothesis and alternative hypothesis.
2. State alpha (α); in other words, determine the significance level.
3. Compute the test statistic.
4. Determine the critical value (from the critical value table).
5. Define the rejection criteria.
6. Finally, interpret the result. If the test statistic falls in the critical region, reject the null hypothesis.

6.20.2 Hypothesis of Two-Sample Z-Proportion Test

H_0: The difference between population proportions is equal to the hypothesized difference.

H_A: The difference between population proportions is not equal to the hypothesized difference (two-tailed).

The difference between population proportions is greater than the hypothesized difference (right-tailed).

The difference between population proportions is less than the hypothesized difference (left-tailed).

Illustration

A car manufacturer aims to improve the quality of the products by reducing defects and increasing customer satisfaction. Therefore, he monitors the efficiency of two assembly lines on the shop floor. In Line A, there are 18 defects reported out of 200 samples (Figure 6.6), while Line B shows 25 defects out of 600 cars. At α 5%, is the difference between the two assembly procedures significant?

Define Null and Alternative Hypotheses

Null hypothesis: The two proportions are the same.

Alternative hypothesis: The two proportions are not the same (α = 0.05). State the decision rule.

Calculate Test Statistic

Line $A = \hat{p}_1 = 18/200 = 0.09 = 9\%$

Line $B = \hat{p}_2 = 25/600 = 0.0416 = 4.16\%$

$$\left(P_0 = \frac{X_1 + X_2}{n_1 + n_2}\right)$$

First compute $p_0 = 18+25/200+600 = 43/800 = 0.0537 = 5.37\%$

Now $\hat{p}_1 - \hat{p}_2 = 0.09 - 0.0416 = 0.0484$

$p_0 * (1-p_0) = 0.0537 * (1-0.0537) = 0.0537 * 0.9463 = 0.0508$

And $(1/n_1)+(1/n_2) = 0.006667$

$$z_{cal} = \left(\frac{(0.09-0.041)}{\sqrt{0.05(1-0.053)\left(\frac{1}{200}+\frac{1}{600}\right)}}\right)$$

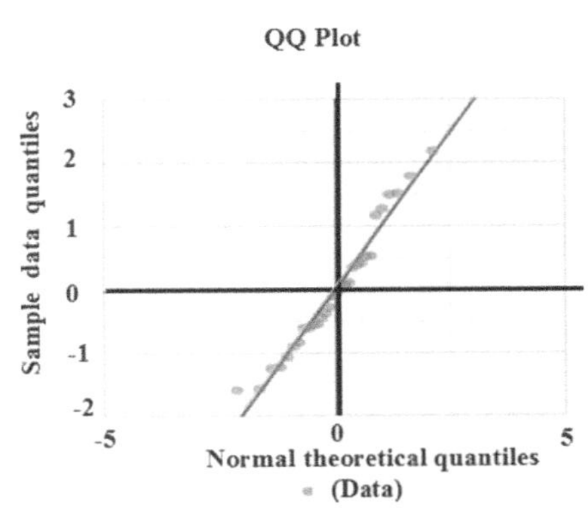

FIGURE 6.6 Normal distribution of observations in relation to replication and treatment.

So, $Z=(0.0484)/$ SQRT ((0.0508) * (0.006667))

$Z=(0.0484)/$SQRT (0.000339))=(0.0484)/(0.018406)=2.62

Inference

Compare Z_{calc} v/s C_{ritical}. In hypothesis testing, a critical value is a point on the test distribution that is compared to the test statistic to determine whether to reject the null hypothesis (H_0). The calculated test statistic value is 2.62, and it falls in the critical region. Hence, the null hypothesis (H_0) is rejected. Therefore, there is a significant difference between the two assembly line procedures.

6.21 Applications of Z-Proportion Test

The *z*-test is employed to assess whether there exists a statistically significant difference between the means of two independent samples. Additionally, it can be utilized to compare a population proportion against a hypothesized proportion or to evaluate the difference between the population proportions of two samples. The key applications of the Z-Proportion Test include the following:

1. Comparing levels of customer satisfaction
2. Assessing the effectiveness of treatments in clinical trials
3. Analyzing trends in voting behavior
4. Conducting market research on consumer habits
5. Implementing quality control measures in manufacturing.

6.22 The Chi-Square Test

Data collected in the field of medicine is often qualitative – for example, the presence or absence of a symptom, classification of a pregnancy as 'high risk' or 'non-high risk', or the degree of severity of a disease (mild, moderate, severe). The measure computed in each instance is a proportion, corresponding to the mean in the case of quantitative data such as height, weight, or serum cholesterol. Comparisons between two or more proportions are commonplace, and the test of significance employed for such purposes is called the 'Chi-square test.' In the case of comparison of two proportions, a test called the 'proportion test' can also be employed. However, it may be cumbersome as it may involve dealing with decimals. The Chi-square test is equivalent to the proportion test and is therefore preferable to use when comparing proportions. The Chi-square test is designed to examine whether a series of observed numbers in various categories of the data are consistent with the numbers expected in those categories based on a specific hypothesis (called The null hypothesis in statistical parlance). In practice, there will naturally be some differences between the observed (O) and the expected (E) numbers in each category due to sampling variation. The aim is to derive a single quantity to summarize all these variations. Since the sum of the deviations ($O-E$) will always be zero., the square of the deviations [$(O-E)^2$] (c.f. standard deviation). The larger the expected number in a category, the greater the spread of observed numbers due to chance variations. Consequently, a simple addition of all the values of $(O_{ij}-E_{ij})^2$ will tend to give too much emphasis to categories with large expected numbers and too little to categories with small expected numbers. Hence, each $O_{ij}-E_{ij})^2$ is divided by the corresponding expected value, and the sum of all these terms is obtained $\Sigma O_{ij}-E_{ij})^2$. This expression is always positive, except when the observed and expected values are identical, in which case it equals zero. The distribution followed by this quantity, if the null hypothesis is true, is known as the Chi-square distribution, and tables of it are found in most statistical textbooks. By comparing the computed value of $\Sigma O_{ij}-E_{ij})^2$, with the E_{ij} is tabulated value, the probability that the difference may have arisen from chance is obtained. This is called the chi-square test. This test is a very useful test of significance and can be used in a wide variety of situations, some of which are described below

$$\chi^2 = \sum_{i=1}^{n}\left(\frac{O_{ij}-E_{ij}}{E_{ij}}\right)^2 \tag{6.14}$$

where, O_{ij} =Observed value of *i*th row and *j*th column, E_{ij} = Expected value of *i*th row and *j*th column. The df=(row − 1)(column − 1)

Hypothesis testing: (i) $\chi^2_{\text{cal}} \geq \chi^2_{\text{cal}}$ @ df $(r-1)(c-1)$ Reject H_0 $(\alpha = 0.01/0.05)$

(ii) $\chi^2_{\text{cal}} \leq \chi^2_{\text{cal}}$ @df $(r-1)(c-1)$ Accept H_0 $(\alpha = 0.01/0.05)$

6.22.1 Chi-Square Test for Comparison of an Observed Proportion with a Hypothesized One

A pharmaceutical company claimed that a new product introduced by them can cure 80% of the patients with a particular disease in 7 days. In an experiment conducted to test this claim, it was observed that among 80 patients with the disease, only 56 (70%) were cured within the stipulated time (i.e. 7 days). Since 70% is appreciably less than 80%, can we conclude that the company's claim is exaggerated? To answer this question, we compute the chi-square under the null hypothesis that the real cure rate is indeed 80% as claimed by the company (Table 6.12).

Illustration

Steps to make the following table for easy calculation of the chi-square value. The chi-square value should be calculated for all the observed frequencies, in this case for both

TABLE 6.12

Vaccination of Lumpy Skin Disease in Cattle (2 × 2 Contingency Table)

Attributes	Vaccinated	Non-Vaccinated	Marginal Row Totals
Infected	450 (E_{11})	655 (E_{12})	1,105
Uninfected	324 (E_{21})	225 (E_{22})	549
Marginal Column Totals	774	880	1,654 (Grand Total)

Chi-square 49.24, $p<0.0001$.

categories: 'Cured' and 'Not cured.' On referring to the tabulated values of the Chi-square distribution with 1 degree of freedom, it is found that the probability corresponds to a value of 5 (or a greater value) <0.05. Hence, we reject the company's claim that the efficacy of the new product is 80%. The 'degrees of freedom' need some explanation. It may be intuitively obvious that the greater the number of terms of $(O-E)^2$, the greater will be the value of the computed chi-square. In assessing the statistical significance of a particular value of chi-square, the number of categories must be taken into account. The relevant number for this purpose is the number of 'degrees of freedom' of the configuration, that is, the number of categories to which the expected values can be assigned without restriction. In the above example, this number is 1, for if one of the two values (viz. cured or not cured) is known, the other can be obtained by subtraction from the total (Table 6.13).

6.22.2 Chi-Square Test for the Comparison of Percentages from Two Samples (2 × 2 Tables)

Consider a controlled clinical trial in which 90 (90%) of 100 patients who received Treatment A were cured compared with 105 (70%) of 150 who received Treatment B. This data can be set out in the form of a 2×2 table (see below) (Table 6.14).

In all, 195 of 250 patients, i.e. 78%, had a favorable response to treatment. Under the null hypothesis that the two treatments are equally effective, the expected number of patients with a favorable response to Treatment A is 78% of 100, i.e. 78; and the expected number for Treatment B is 78% of 150, i.e. 117. The corresponding numbers for unfavorable responses are, by subtraction, 22 for Treatment A and 33 for Treatment B.

$$X^2 = \Sigma\frac{(O-E)^2}{E} = \frac{(90-78)^2}{78} + \frac{(10-22)^2}{22} + \frac{(105-117)^2}{117} + \frac{(45-33)^2}{33}$$

$$= 1.85 + 6.55 + 1.23 + 4.36 = 13.99$$

On referring to tabulated values of the chi-square distribution with 1 degree of freedom (since if one of the four values in the core of the table is given, the others can be obtained by subtraction from the marginal totals), it is found that the probability corresponding to a value of 13.99 (or a greater value) is less than 0.001. Since this is a very small probability, we reject the null hypothesis that the two treatments are equally effective. The calculation of expected frequencies, though instructive, is not really necessary in routine practice . A simpler way of obtaining the value of chi-square is to compute the expression

$$X^2 = \frac{250\,[(90\times 45)-(10\times 105)^2}{195\times 55\times 100\times 150} = 13.99$$

The number is the square of the difference between the cross products multiplied by the total trial size and the denominator is the product of the four marginal totals (Table 6.15).

In general, in a 2×2 table, the value of

$$\chi^2 = \frac{N(ad-bc)}{(a+c)(b+d)(a+b)(c+d)} \quad (6.15)$$

where a, b, c, d, and N are as set out in the configuration below.

Since the number of observations in any group can only be a whole number, the calculated values of chi-square are essentially discontinuous and cannot exactly follow the continuous mathematical curve from which the probability levels are calculated. A correction for continuity is, therefore, prudent. In the formula above, the correction consists of subtracting half the total trial size from the absolute value of $(ad-bc)$.

$$\chi^2 = \frac{N\left[(ad-bc)-\frac{N}{2}\right]^2}{(a+c)(b+d)(a+b)(c+d)}$$

where $(ad-bc)$ denotes the absolute value of $(ad-bc)$. In the example above, the value of the corrected chi-square is

$$\chi^2 = \frac{250\left[(90\times 45-10\times 105)-\frac{250}{2}\right]^2}{195\times 55\times 100\times 150} = 12.84$$

TABLE 6.13

Working Procedure for Estimation of Chi-Square Value

Observed Value (O_{ij})	Expected Value $E_{ij} = \frac{\text{Row total * Column total}}{\text{Grand total}}$	$\chi^2 = \sum_{i=1}^{n}\left(\frac{O_{ij}-E_{ij}}{E_{ij}}\right)^2$
450	$E_{11} = \frac{450*1{,}105}{1{,}654} = 300.63$	$\left(\frac{450-300.63}{300.63}\right)^2 = 74.84$
655	$E_{12} = \frac{655*1{,}105}{1{,}654} = 437.59$	$\left(\frac{655-437.59}{437.59}\right)^2 = 108.11$
324	$E_{21} = \frac{324*549}{1{,}654} = 107.54$	$\left(\frac{324-107.54}{107.54}\right)^2 = 435.67$
225	$E_{22} = \frac{225*549}{1654} = 74.68$	$\left(\frac{225-74.68}{74.68}\right)^2 = 302.55$
	Total chi-square	$\chi^2 = 920.45$

Degrees of freedom (Row − 1) * (Column − 1) = 1*df*.

TABLE 6.14

Relative Comparison of Treatment A and B

Treatment	Response to Treatment: Cured	Response to Treatment: Not Cured	Total
Treatment A	90 (78)	10 (22)	100
Treatment B	105 (117)	45 (33)	150
Total	195	55	250

[a] Number in brackets is the expected frequency.

Source: Data Bowring and Lady Curzon hospital, BMC&RI, HIV ART annual report RCT data 2015.

TABLE 6.15

Distribution of Treatment with Respect to Response

	Response to Treatment: Cured	Response to Treatment: Not Cured	Total
Treatment A	a	b	$(a+b)$
Treatment B	c	d	$(c+d)$
Total	$(a+c)$	$(b+d)$	$a+b+c+d=N$

Note that the corrected value will always be smaller than the uncorrected value. This means that the uncorrected value of chi-square tends to exaggerate the significance of a difference.

6.22.3 Chi-Square Test for Proportion

In the test described in the preceding section, each patient received only one treatment, either A or B. There are situations, however, where a patient is administered both treatments, in a random sequence, and a comparison is made within each patient. (Alternatively, patients may be matched in pairs and one member, selected at random, is given treatment A and the other treatment B). Examples of this kind may be found in chronic conditions like asthma or bronchitis, where alleviating drugs are administered, and studies of acute toxicity to different doses of a drug. Consider a trial in which each of 100 patients receives two drugs, A and B, in a random sequence, and 15 manifest toxicities to A and five to B, including four to both A and B. It would be incorrect to compare 15 of 100 with five of 100 using the formula given in the preceding section, as the same 100 patients are tested twice. The proper method of analysis consists of identifying four groups of patients (see below) (Table 6.16).

For the comparison between the two drugs, it is clear that groups (1) and (4) can make no contribution, as both drugs are unsatisfactory (toxic) in group (1) and satisfactory (non-toxic) in group (4). Considering groups (2) and (3), the expected number of patients in each, under the null hypothesis that the two drugs have the same toxicity, is $11+1=6.2$

The chi-square value $\frac{(11-6)^2}{6}+\frac{(1-6)^2}{6}=\frac{25}{6}+\frac{25}{6}=8.33$.

This expression is the same as $\frac{(11-1)^2}{(11+1)}=\frac{100}{12}=8.33$. Applying Correction for continuity $\left(X_c^2\right)$

This has $df=1$. The probability value is read off as approximately 0.01, indicating that Drug A is significantly more toxic than Drug B. In general, if *a*, *b*, *c*, and *d* are the frequencies in the four groups cited above.

6.22.4 Trend Chi-Square (2 × 2) Table

The formula in the preceding section is applicable when there is no particular ordering of the strata – e.g. the strata may be Hindus, Muslims, Christians, and Others. However, in medical inquiries, and particularly in clinical trials, the strata often follow a ranking order – e.g. mild, moderate, severe, and very severe diseaseOne's interest is not in a general test of heterogeneity, but in a test for examining the significance of trends – in the example quoted, an increasing tendency for failure with increasing severity of the disease (Table 6.17). Consider the following example:

TABLE 6.16

Estimation of Likelihoods of Drug Effect

Group	Drug A	Drug B	No. of Patients	General
1	Toxic	Toxic	4	*a*
2	Toxic	Not toxic	11	*b*
3	Not toxic	Toxic	1	*c*
4	Not toxic	Not toxic	84	*d*

It will be noted that with the increasing extent of disease, there is a tendency for a higher proportion of patients with an unfavorable response, i.e. 10%, 15%, 20%, and 25% in the four strata. To examine the significance of this trend in proportions, we score the extent of disease (*X*) as *O* if mild, as 1 if moderate, as 2 if severe, and as 3 if very severe: and the response variable (*Y*) as *O* if favorable and as 1 if unfavorable. We will then have the data in the following form:

6.22.5 Chi-Square Test (3 × 3) Table

The example (hypothetical) given below is a 3×3 table, which examines the site and type of cerebral tumors (A, B, and C denote benign, malignant, and other types respectively, and I, II, and III denote frontal lobes, temporal lobes, and other cerebral areas, respectively) (Table 6.18).

The null hypothesis under study is that there is no association between site and type of tumor. The expected frequency for each cell is given in brackets. Considering the observed (*O*) and expected (*E*) values for each cell, with $(3-1)\times(3-1)=4$ degrees of freedom, the corresponding probability is 0.001, indicating that there is a strong association between site and type.

6.23 Precautions

1. The observations analyzed must be independent of one another.

   For example, if we are comparing the proportions of tuberculosis patients on two regimens. A and B, who are culture-positive at the end of treatment (2×2 table), it would be incorrect to pool duplicate specimens from the same patient and undertake a chi-square test on the amalgamated observations. This is because duplicate specimens from a patient

TABLE 6.17

Response of Treatment with Extent of Disease

Extent of Disease	Response to Treatment		Total ($N=400$)
	Favorable ($n=325$)	Unfavorable ($n=75$)	
Mild	45	05 (10%)	50
Moderate	85	15 (15%)	100
Severe	120	30 (20%)	150
Very severe	75	25 (25%)	100

TABLE 6.18

3×3 Chi-Square Analysis for Varied Sites and Treatment Effect

Site	Type			Total
	A	B	C	
I	15(10)	03(6)	02(4)	20
II	41(30)	12(18)	07(12)	60
III	44(60)	45(36)	31(24)	120
Total	**($n=100$)**	**($n=60$)**	**($n=40$)**	**($n=200$)**

cannot be regarded as independent; indeed, one would expect them to be highly correlated.
2. The calculations must be based on actual numbers of observations, not percentages.
3. Chi-square tests must never be undertaken on mean values or ratios; the observations must always be frequencies.
4. When the expected numbers are small, special steps must be taken. In the case of a 2×2 table, the use of a continuity correction was mentioned earlier. In the case of pxq tables, if there is only one or two cells with expected numbers less than five, no special action is indicated; if, however, there are several, it is better to amalgamate adjacent rows or columns in such a way as to get expected numbers of at least five (i.e. absorbing most of the cells with small expectations), and then apply the Chi-square test to the condensed table.
5. When comparing observed frequencies in a frequency distribution with frequencies expected on the basis of a mathematical distribution or some other specific hypothesis, a class with an expected number of less than five must be amalgamated with an adjacent class before applying the Chi-square test.

6.24 Comparison of Two Variances: *F*-Test

When undertaking a *t*-test on unpaired observations, the variances in the two samples should be compared. If there are good a priori reasons for them to be different, special tests may be employed (see above). If no such reasons are present, the likelihood that the difference observed is due to sampling fluctuation may be evaluated. For this purpose, we use a test called the 'variance ratio' test (*F*-test). We calculate the ratio of the larger variance to the smaller variance, i.e. $s_1^2/s_2^2 = F$. Under the null hypothesis that the population variances are equal, the distribution of F in samples of size n_1 and n_2 is known. The values of F corresponding to various probability levels are tabulated in all standard statistical textbooks in terms of the degrees of freedom of the two variances, $n_1 - 1(\vartheta_1)$ and $n_2 - 1\,(\vartheta_2)$. The *F*-test provides an exact treatment of the comparison of two variances under the following conditions:

a. The two samples must be independent (e.g. two series of patients and not the same series tested twice, before and after treatment).
b. Both samples must come from a normal distribution.

The assumption of normality is important, much more so than in the case of the *t*-test.

The *F*-test in statistics is a hypothesis-testing procedure that considers two variances from two samples. The *F*-test is used when the difference between two variances needs to be significantly assessed, i.e., when determining whether or not two samples can be taken as representative of the normal population with the same variance. This is the variance ratio test because it calculates the ratio of variances. The goal of the test is to determine whether the variance in two populations is equal. It was propounded by the British polymath R.A. Fisher and named in his honor. G.W. Snedecor later developed the test.

The following conditions are critical for using the *F*-test to compare the variances of two populations:

6.24.1 Normality

Observations are normally distributed (normalcy).

6.24.1.1 *Independent and Random Selection of Sample Items*

The selection of the samples' components should be independent and random.

6.24.2 More than Unity

The variance ratio must be one or larger than one; it cannot be less than one. When dividing variance estimates, smaller estimates divide the larger estimates of variances. The additive property states that the total of different variance components will equal the total variance, i.e. the total variance is the sum of the variance between samples and the variance within samples (Table 6.19).

H_0: Two samples are from the same group
H_a: The population's variances concerning both samples are equal.

Illustration: Consider the example of the population in a village:

Testing the equality of sample variances with a significance level of@ 5% with the above-given data.

$$\text{Variance sample for } S_1^2(\text{sample} 1) = \frac{1}{n_1 - 1}\sum_{i=1}^{n}\sqrt{\left(X_1 - \overline{x_1}\right)^2}$$
$$= 10.22$$

And

$$\text{Variance sample for } S_2^2(\text{sample } 2) = \frac{1}{n_2 - 1}\sum_{i=1}^{n}\sqrt{\left(X_2 - \overline{x_2}\right)^2}$$
$$= 10.0$$

$$F = S_1^2/S_2^2 \qquad (6.16)$$

$$10.22/10 = 1.022$$

TABLE 6.19
Methods for Estimation of Variance for Testing the Two Treatment Effect (*A* & *B*)

Village (Survey Site)	*A*	*B*
Sample size	10	12
Mean monthly income	150	140
Sample variance	92	110

The critical value for v_1 (10−1)=9 and v_2 (12−1)=11, and the table value of F at the 5% significance level=2.90.

Inference

In the example above, F's computed value (1.022) is less than its table value obtained from the F table (2.90). As a result, one can conclude that the null hypothesis is true and that the variance of the two samples is equal.

6.25 Analysis of Variance (ANOVA)

It is a unique statistical method used to analyze the differences between the means of two or more groups or treatments. It is often used to determine whether there are any statistically significant differences between the means of different groups. Developed by Ronald Fisher in 1918, this test extends the t-test and the z-test, which have the limitation of only allowing the nominal level variable to have two categories. This test is also called Fisher's analysis of variance. The use of ANOVA depends on the research design. Commonly, ANOVAs are used in three ways: one-way, two-way, and N-way ANOVA.

6.25.1 Assumptions

i. An ANOVA can only be conducted if there is no relationship between the subjects in each sample. This means that subjects in the first group cannot also be in the second group (e.g. independent samples/between groups).
ii. The different groups/levels must have equal sample sizes.
iii. An ANOVA can only be conducted if the dependent variable is normally distributed, so that the middle scores are the most frequent and the extreme scores are the least frequent.

Population variances must be equal (i.e., homoscedastic). Homogeneity of variance means that the deviation of scores (measured by the range or standard deviation, for example) is similar between populations.

A one-way ANOVA (analysis of variance) has one categorical independent variable (also known as a factor) and a normally distributed continuous (i.e., interval or ratio level) dependent variable.

Hypothesis

$$H_0: \mu_1 = \mu_2 = \mu_3$$

$$H_A: \mu_1 \neq \mu_2 \neq \mu_3$$

$$H_0: \sigma_1^2 = \sigma_2^2 = \sigma_3^2$$

$$H_A: \sigma_1^2 \neq \sigma_2^2 \neq \sigma_3^2$$

Illustration

1 The drug trial experiment was conducted at a tertiary care hospital. The following data expresses the efficacy of the drug over a period of 2 years (Table 6.20).

TABLE 6.20

Efficacy of Drug with Different Formulation

Group 1 (R_1)	Group 2 (R_2)	Group 3 (R_3)	Sum of Column
85	91	79	255
86	92	78	256
88	93	88	269
75	85	94	254
78	87	92	257
94	84	85	263
98	82	83	263
79	88	85	252
71	95	82	248
80	96	81	257
Sum R_1=834	Sum R_2=893	Sum R_3=847	GT= 2,574
Group Means 83.44	89.30	84.70	

Steps 1: Determine degrees of freedom (df)

Total df = ($r\,t$ − 1) = 3 × 10 = 30 − 1 = 29

Reps df (r − 1) = 3 − 1 = 2

Treatment df (t − 1) = 10 − 1 = 9

Error df = (r − 1) (t − 1) = 2 × 9 = 18

Step 2: Calculate Correction factor (CF)$=\dfrac{(\text{GT})^2}{r*t}=$

$$\frac{(2{,}574\ )^2}{30} = 220{,}849$$

Step 3: TSS (Total sum of square) $=\sum X_{ij}^2 - \text{CF}$

where X_{ij} = Observed value of ith row and jth column (Take square of all observation in the table matrices)

$$\text{TSS} = \sum 85^2 + 86^2 + 88^2 + 75^2 \ldots\ldots .81^2 - 220{,}849$$

$$\text{TSS} = \sum 85^2 + 86^2 + 88^2 + 75^2 \ldots\ldots .81^2 - 220{,}849$$

$$\text{TSS} = 222{,}142 - 220{,}849 = 1{,}292.80$$

Step 4: Treatment sum of square Trss$=\dfrac{\sum_{i=1}^{t} T_i^2}{r} - \text{CF}$

$$\text{Trss} = \frac{255^2 + 256^2 + 269^2 + 254^2 \ldots . .257^2}{3} - 220{,}849$$

$$= 192.20$$

Step 5: Replication sum of square RSS $=\dfrac{\sum_{i=1}^{t} R_i^2}{t} - CF$

$$\text{RSS} = \frac{834^2 + 893^2 + 847^2}{10} - 220{,}849$$

$$= 221{,}041.4 - 220{,}849 = 192.20$$

Step 6: Error sum of square with respect to treatment ESSt = TSS − Trss = 1,100.60

Step 7: Error sum of square with respect to replication ESSr = TSS − Trss − RSS

Step 8: Mean sum of square with resects to Treatment
$SSt/t-1=192.20/2=96.10$
Step 9: Mean sum of square with respect to error (One-way ANOVA)
$SSE/(R-1)(T-1) = 40.80$
Step 10: Mean sum of square with respect to error
$SSE/(R-1)(T-1)=989.14/18=54.95$ (Two way ANOVA)
Make a tabulation with respect to rows and columns in Table ANOVA (Table 6.21).

The *F*-test statistic for this one-way ANOVA is **2.358**. To determine if this is a statistically significant result, we must compare this to the *F*-critical value found in the *F* distribution table with the following values:

α (significance level) = 0.05
DF1 (numerator degrees of freedom) = *df* treatment = 2
DF2 (denominator degrees of freedom) = *df* error = 27

We find that the *F* critical value is **3.35**.

Since the *F*-test statistic in the ANOVA table is less than the *F*-critical value in the *F*-distribution table, we fail to reject the null hypothesis. This means we don't have sufficient evidence to say a statistically significant difference exists between the mean efficacy of the three groups (Figure 6.7) (Table 6.22).

Factor A

1. H_0 hypothesis
   Since the p-value > α, H_0 cannot be rejected.
   The averages of all groups are assumed to be equal.
   In other words, the difference between the sample averages of all groups is not large enough to be statistically significant.
   A non-significant result cannot prove that H_0 is correct, only that the null hypothesis cannot be rejected.
2. *P*-value
   The p-value equals 0.9863 ($P(x \leq 0.2254)=0.0137$). This means that the chance of a Type I error, rejecting a correct H_0, is too high: 0.9863 (98.63%). The larger the p-value, the more it supports H_0.
3. Test statistic
   The test statistic F_A equals **0**.2254, which is in the 95% region of acceptance: [0, 2.4563].
4. Effect size
   The observed effect size η^2 is medium (0.1). This indicates that the magnitude of the difference between the averages is moderate.

Factor B

1. H_0 hypothesis
   Since the p-value > α, H_0 cannot be rejected.

TABLE 6.21
One-Way ANOVA Model Output

Source of Variation	*df*	SS	MSS	F_{cal}	F_{table}
Treatment	$(k-1)$ $(3-1)=2$	192.20	96.10	2.358	
Error	$N-k$ $(30-3)=27$	1,100.60	40.80		
Total	$N-1$ $(30-1)=$ **29**	**1,292.80**			

TABLE 6.22
Two-Way ANOVA Model Output

Source	DF	Sum of Square (SS)	Mean Square (MS)	*F* Statistic (df_1, df_2)	*P*-value
Factor A - rows (A)	9	111.46	12.38	0.225 (9,18)	0.986
Factor B - columns (B)	2	192.20	96.1	1.748 (2,18)	0.202
Error	18	989.13	54.95		
Total	29	1292.8	44.57		

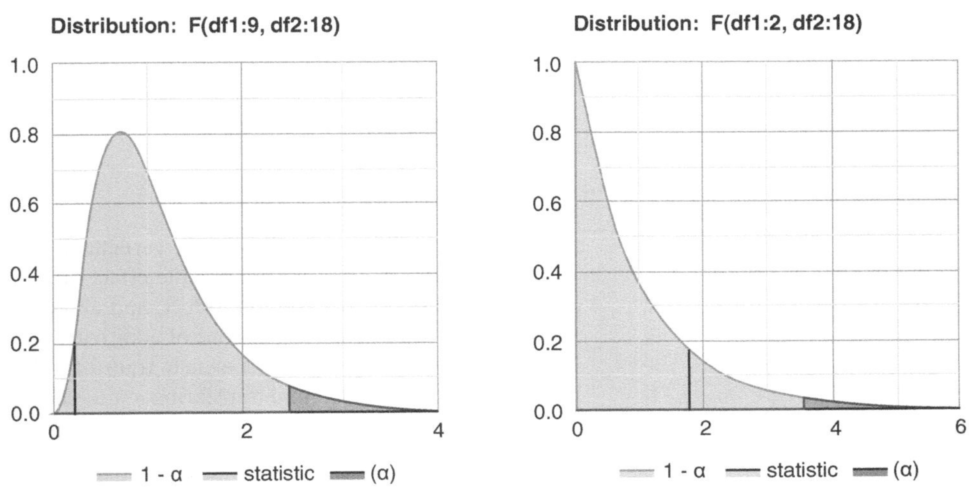

FIGURE 6.7 Critical value of the *F* distribution.

The averages of all groups are assumed to be equal.

In other words, the difference between the sample averages of all groups is not big enough to be statistically significant.

A non-significant result cannot prove that H_0 is correct, only that the null assumption cannot be rejected.

2. *P*-value

   The *p*-value equals 0.2023 ($P(x \leq 1.7488) = 0.7977$). This means that the chance of a Type I error, rejecting a correct H_0, is too high: 0.2023 (20.23%). The larger the p-value the more it supports H_0.
3. Test statistic

   The test statistic F_B equals 1.7488, which is in the 95% region of acceptance: [0, 3.5546].
4. Effect size

   The observed effect size η^2 is large, 0.16. This indicates that the magnitude of the difference between the averages is large.

6.26 ANOVA Data Preparation and Analysis (R-Software)

Data preparation is an art of the science. The researcher should arrange the data in a scientific manner with respect to treatment and replication. Without a proper data frame, the software will be unable to render the output (Table 6.23).

6.26.1 R-Code for ANOVA

```
# Create a data frame
dat<- data.frame ( group = c(1,2,3),
group1 = c(85,86,88,78,78,94,98,79,71,80),
group2 = c(91, 92, 93, 85, 87, 84, 82, 88, 95,
96),
group3 = c(1.45, 1.76, 2.08, 2.36, 2.49,
2.70),
Spring = c(79,78,88,94,92,85,83,85,82,81))
# However, to run an ANOVA, the data needs to
be in a long form
Library(tidyr)
dat.long <- gather (dat, key = "group",
value="Value", -rows)
```

TABLE 6.23

Data Frame for Analysis of ANOVA in *R*-Software

Group	Efficacy	Group	Efficacy	Group	Efficacy
1	85	2	91	3	79
1	86	2	92	3	78
1	88	2	93	3	88
1	75	2	85	3	94
1	78	2	87	3	92
1	94	2	84	3	85
1	98	2	82	3	83
1	79	2	88	3	85
1	71	2	95	3	82
1	80	2	96	3	81

6.27 Analysis of Covariance ANCOVA

The Analysis of Covariance (ANCOVA) procedure compares the means of one continuous dependent variable across two or more factor variables and determines the effects of covariates and covariate interactions with factors. The factor variables divide the population into groups.

6.27.1 Data

The dependent variable is quantitative. Factors are categorical and can have numeric or string values of up to eight characters. Covariates are quantitative variables that are related to the dependent variable.

6.27.2 Assumptions

1. The data are a random sample from a normal population, and in the population, all cell variances are the same. Analysis of variance is robust to departures from normality, although the data should be symmetric.
2. At least one categorical and at least one interval independent variable(s) may be categorical, but at least one must be a covariate (interval level).
3. Interval dependent: The dependent variable is continuous and at the interval level.
4. Low measurement error of the covariate
5. The covariate variables are continuous and at the interval level and are assumed to be measured without error.

The form of the relationship between the covariate and the dependent variable must be known. Most computer programs assume this relationship is linear, adjusting the dependent mean based on linear regression. Scatter plots of the covariate and the dependent variable for each of the '*k*' groups formed by the independent variables is one way to assess violations of this assumption. Homogeneity of covariate regression coefficients (*i.e.* 'parallel lines model'): The covariate coefficients (the slopes of the regression lines) are the same for each group formed by the categorical variables and measured on the dependent.

6.27.3 Additivity

The values of the dependent variable are an additive combination of its overall mean, the effect of the categorical independents, the covariate effect, and an error term. ANCOVA is robust against violations of additivity, but in case of severe violations, the researcher may transform the data, such as by using a logarithmic transformation to change a multiplicative model into an additive model. ANCOVA can automaticallyhandle interaction effects and, therefore, is not an additive procedure in the sense of regression models without interaction terms.

6.27.4 Independence of the Error Term

The error term is independent of the covariates and the categorical independents. Randomization in experimental designs ensures that this assumption will be met.

Independent Variables Orthogonal to Covariates

6.27.5 The Independents Orthogonal

If the covariate is influenced by the categorical independents, then the control adjustment ANCOVA makes on the dependent variable prior to assessing the effects of the categorical independents will be biased, as some indirect effects of the independents will be removed from the dependent.

6.27.6 Homogeneity of Variances

There is homogeneity of variances in the cells formed by the independent categorical variable. Heteroscedasticity refers to a lack of homogeneity of variances, which violates this assumption.

6.27.7 Multivariate Normality

For purposes of significance testing, variables follow multivariate normal distributions.

6.27.8 Compound Sphericity

The groups display sphericity (the variance of the difference between the estimated means for any two different groups is the same). A more restrictive assumption, called compound symmetry, is that the correlations between any two different groups are the same value. If compound symmetry exists, sphericity exists. Tests or adjustments for lack of sphericity are usually actually based on the possible lack of compound symmetry (Tables 6.24–6.26).

Adjusted means: $a_1=195$, $a_2=165.83$, $a_3=202.5$

Calculating sums of squares manually for an analysis of covariance.

$$\text{SS Total} = 582{,}800 - \left(\frac{3{,}200^2}{18}\right) = 13{,}911.11$$

$$\text{SS treat} = 990^2 + 995^2 + 1{,}215^2/6 - \frac{3{,}200^2}{18} = 5502.80$$

$$\text{SS pooled} = \frac{\left[632{,}800 - \left(3{,}530 * 3{,}200/18\right)^2\right]}{701{,}900 - \dfrac{3{,}200^2}{18}} = 2{,}856.80$$

TABLE 6.24

Treatment Response of Group A and B

Treat A		Treat B		Treat C	
Pre	**Post**	**Pre**	**Post**	**Pre**	**Post**
175	135	205	165	210	185
175	145	175	195	170	170
235	205	230	160	235	210
215	175	190	155	185	210
195	140	155	150	180	200
195	190	185	170	220	240

SS error = 1,3911.11 − 2,856.8 − 5,502.8 = 5,552.31

Calculation of Sum of Square

$SS_{Tot1} = 167{,}600 - 990^2/6 = 4{,}250$

$SS_{XY1} = 199{,}150 - (1{,}190 \times 990)/6 = 280$

$SS_{X1} = 238{,}750 - (1{,}190^2/6) = 2{,}733.33$

$SS_{Reg1} = 2{,}800^2/2{,}733.33 = 2{,}868.29$

$SS_{Error1} = 4{,}250 - 2{,}868.296 = 1{,}381.704$

$b_1 = 2{,}800/2{,}733.33 = 1.024$

$SS_{Total2} = 166{,}275 - 995^2/6 = 1{,}270.833$

$SS_{XY2} = 188{,}900 - (1{,}140 \times 995)/6 = -150$

$SS_{X2} = 219{,}900 - (1{,}140^2/6) = 3{,}300$

$SS_{Reg2} = -150^2/3{,}300 = 6.818$

$SS_{Error2} = 1{,}270.833 - 6.8182 = 1{,}264.015$

$\boldsymbol{b_2} = -150/3{,}300 = -0.04545$

$\mathbf{SS_{Total3} = 248{,}925 - 1{,}215^2/6 = 2{,}887.5}$

$\mathbf{SS_{XY3}} = 244{,}750 - (1{,}200 \times 1{,}215)/6 = 1{,}750$

$\mathbf{SS_{X3}} = 243{,}250 - (1{,}200^2/6) = 3{,}250$

$\mathbf{SS_{Reg3}} = 1{,}750^2/3{,}250 = 942.308$

$\mathbf{SS_{Error3}} = 2{,}887.5 - 942.308 = 19{,}451.192$

$\boldsymbol{b_3} = 0.5385 = 1{,}750/3{,}250$

TABLE 6.25

Calculation of Sums of Squares Table

Variables	Σ sums	Σ² products	ΣΣ² totals
X_1	1,190	238,750	
Y_1	990	167,600	
X_1Y_1		199,150	
$\Sigma X_1 \Sigma Y_1$			1,178,100
X_2	1,140	219,900	
Y_2	995	166,275	
X_2Y_2		188,900	
$\Sigma X_2 \Sigma Y_2$			1,134,300
X_3	1,200	243,250	
Y_3	1,215	248,925	
X_3Y_3		244,750	
$\Sigma X_3 \Sigma Y_3$			1,458,000
X overall	3,530	701,900	
Y overall	3,200	582,800	
XY overall		632,800	
$\Sigma X \Sigma Y$			11,296,000

TABLE 6.26

ANOVA Table for Regression Statistics on Each Treatment Separately

Source of Variation	*Df*	SS	MS	*F*-ratio	*P*
Diet	2	5,502.8			
Pre	1	2,085.5			
Heterogeneity of slopes	2	1,732.0	866.0	2.2635	0.1465
Residual	12	4,590.9	382.6		
Total	**17**	**13,911.11**			

$\mathbf{SS_{Total} = 4{,}250 + 1{,}270.833 + 2{,}887.5 = 8{,}408.333}$
$\mathbf{SS_{Xy}} = 2{,}800 - 150 + 1{,}750 = 4{,}400$
$\mathbf{SS_X} = 2{,}733.33 + 3{,}300 + 3{,}250 = 9{,}283.3$
$\mathbf{SS_{reg}} = 2{,}868.296 + 6.818 + 942.308 = 0.4740$
$\mathbf{SS_{Error}} = 1{,}381.704 + 1{,}264.015 + 1{,}945.192 = 4{,}590.911$
$\mathbf{b_{common}} = 4{,}400/9{,}283.3 = 3{,}817.422$

Summed regression statistics:

$SS_{Reg\ (common\ slope)} = 4{,}400^2/9{,}283.3 = 2{,}085.465$
$SS_{Heterogeneity\ of\ slopes} = 3{,}817.422 - 2{,}085.465 = 1{,}731.957$. I

6.28 Inference

At the $P = 0.05$ level the interaction is not significant—in other words, we can accept the lines as parallel. However, this is in direct contradiction to our visual assessment of the relationship between the response variable and the covariate, where we lack evidence of any regression relationship at all between 'post' and 'pre' for treatment B—let alone a similar slope to that found for treatments A and C. The apparent contradiction may result from the lack of power of tests for interaction, or it may result from random variation in treatment B. Which is the case is a matter of judgment (or guesswork!), but for the sake of the example, we will assume it results from random variation.

6.29 Data Transformation Methods

Many biological variables do not meet the assumptions of parametric statistical tests: they are not normally distributed, the standard deviations are not homogeneous, or both. Using a parametric statistical test (such as an ANOVA or linear regression) on such data may give a misleading result. In some cases, transforming the data will make it fit the assumptions better. You perform a mathematical operation on each observation to transform data, then use these transformed numbers in your statistical test (Figure 6.8).

How to do transformations in Stata

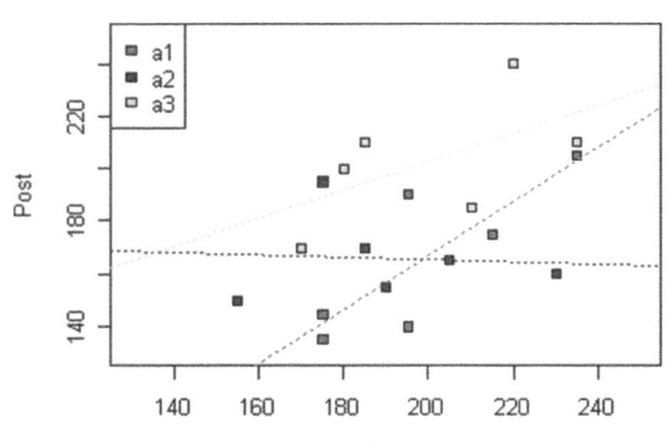

FIGURE 6.8 Adjusted means for pre- and post-treatment.

- Transformations for proportions and percents
- Transformations as a family
- Transformations for variables that are both positive and negative

6.29.1 Reasons for Using Transformations

There are many reasons for transformation. The list here is not comprehensive.

1. Convenience
2. Reducing skewness
3. Equal spreads
4. Linear relationships
5. Additive relationships

6.29.2 Convenience

A transformed scale may be as natural as the original scale and more convenient for a specific purpose (e.g. percentage rather than original data, sines rather than degrees). One important example is **standardization**, whereby values are adjusted for differing levels and spreads. In general, standardized values have a level of 0 and a spread of 1, and they have no units; hence, standardization is useful for comparing variables expressed in different units. Most commonly, a **standard score** is calculated using the mean and standard deviation (SD) of a variable:

$$\left(Z = \frac{X - \text{Mean}X}{\text{SD}(X)} \right) \tag{6.17}$$

The most useful transformations in introductory data analysis are the reciprocal, logarithm, cube root, square root, and square. In what follows, even when it is not emphasized, it is assumed that transformations are used only over ranges on which they yield (finite) real numbers as results.

6.29.3 Reciprocal

In this transformation, 'x' is replaced by the inverse of x ($1/x$). The reciprocal transformation will have little effect on the shape of the distribution. This transformation can only be used for non-zero values. The reciprocal, x to $1/x$, along with its sibling, the negative reciprocal, x to $-1/x$, is a very strong transformation with a drastic effect on the distribution shape. It cannot be applied to zero values. Although it can be applied to negative values, it is not useful unless all values are positive. The reciprocal of a ratio may often be interpreted as easily as the ratio itself: e.g. population density (people per unit area) becomes area per person; persons per doctor become doctors per person; rates of erosion become time to erode a unit depth.

6.29.4 Logarithmic

Numerical variables may have high skewness and non-normal distributions (Gaussian distribution) caused by outliers, highly exponential distributions, etc. Therefore, data transformation is needed. In a log transformation each variable of x will be replaced by log(x) with base 10, base 2, or natural log.

$$\text{New } X = \ln(X+1) \tag{6.18}$$

The logarithm, x to log base 10 of x, or x to log base e of x (ln x), or x to log base 2 of x, is a strong transformation with a major effect on the distribution shape. It is commonly used for reducing right skewness and is often appropriate for measured variables. It cannot be applied to zero or negative values. One unit on a logarithmic scale means a multiplication by the base of the logarithms being used. Exponential growth or decline

$$y = a\exp(bx) \tag{6.19}$$

is made linear by

$$\ln y = \ln a + \exp(bx) \tag{6.20}$$

so that the response variable y should be logged. (Here exp (4.19) means raising to the power e, approximately 2.71828, which is the base of natural logarithms.)

males/females;

Dependants/workers;

Downstream length/down valley length.

Then y is somewhere between 0 and infinity, or in the last case, between 1 and infinity. If $p=q$, then $y=1$. Such definitions often lead to skewed data, because there is a clear lower limit and no clear upper limit. The logarithm, however,

$\log y = \log p - \log q$ is somewhere between -infinity and infinity, and if $p=q$, then log $y=0$. Hence, the logarithm of such a ratio is likely to be more symmetrically distributed.

6.29.5 Cube Root

The cube root, x to $x^{(1/3)}$, is a fairly strong transformation with a substantial effect on distribution shape; it is weaker than the logarithm. It is also used for reducing right skewness and has the advantage that it can be applied to zero and negative values. Note that the cube root of a volume has the units of a length. It is commonly applied to rainfall data.

6.29.6 Square Root (Count Data)

This transformation has a moderate effect on the distribution. The main advantage of square root transformation is that it can be applied to zero values. Here, x is replaced by the square root(x). It is weaker than the log transformation.

$$X' = \sqrt{x + 3/8}$$

The square root, x to $x^{(1/2)}=\text{Sqrt}(x)$, is a transformation with a moderate effect on distribution shape; it is weaker than the logarithm and the cube root. It is also used for reducing right skewness and has the advantage that it can be applied to zero values. Note that the square root of an area has the units of a length. It is commonly applied to counted data, especially if the values are mostly rather small. This transformation has a moderate effect on the distribution shape and can also be used to reduce left skewness.

6.29.7 Angular Data Transformation

The arcsine transformation, or the angular transformation, is calculated as two times the arcsine of the square root of the proportion. To transform such data, use a combination of two functions:

$$X' = \text{ArcSin}\sqrt{X} \tag{6.21}$$

Used for proportion

Compare variance before and after transformation

SAS steps

```
Input x;
new x = log(x);
new x = sqrt (x+3/8);
new x = arc sin (sqrt (x));
Cards;
```

6.30 Measures of Association on Qualitative Data

The chi-square test is the usual method employed for examining the presence of an association between two qualitative variables; for instance, vaccination status and the incidence of disease. It does this by assessing whether the difference in the proportions with disease is (statistically) significantly different from zero. Although chi-square takes the value zero when there is no association, it increases without limit with increasing sample size, when there is an association. It is, therefore, not a very useful measure of association. To overcome this limitation, $C=\sqrt{(X^2)/(N+X^2)}$ (Pearson's coefficient of contingency), or Chuprov's coefficient (T), defined by $T=\sqrt{(X^2)/N}\sqrt{(p-1)(q-1)}$ (where p and q are the numbers of rows and columns), are employed (Widaman et al., 2023). An alternate approach to measuring an association consists of measuring the Relative Risk (RR), which is defined as the RATIO of the rate of disease among those exposed to the risk factor (e.g. to dust, oral contraceptives, cigarette smoking) to the rate among those not exposed. Consider a large population that has been classified as positive (+) or negative (–) according to some factor, and also as having a certain disease (+) or as not having it (–). The findings can then be represented in the following format, where a, b, c, and d are frequencies and N is the total population (Table 6.27).

TABLE 6.27

Distribution of Disease Diagnosis

	Disease		
Factor	Present (+)	Absent (–)	Total
Present (+)	(a)	(b)	($a+b$)
Absent (–)	(c)	(d)	($c+d$)
Total	($a+c$)	($b+d$)	N

6.30.1 Working Rule

By definition,

$$\text{Relative risk } = \left(\frac{a}{a+b} * \frac{c}{c+d}\right) \quad (6.22)$$

With the standard error of the log relative risk being calculated as:

$$\text{SE}\{l_n \text{ (RR)}\} = \sqrt{\frac{1}{a} + \frac{1}{c} - \frac{1}{a+b} - \frac{1}{c+d}} \quad (6.23)$$

And

$$\text{CI } 95\% = \exp\left(l_n \text{ (RR} - 1.96\text{*SE } \{l_n \text{ (RR)}\}\right) \text{ to } \exp\left(l_n \text{ (RR} + 1.96\text{*SE } \{l_n \text{ (RR)}\}\right)$$

The relative risk (RR), its standard error, and 95% confidence interval are calculated according to Altman, (1991). The relative risk or risk ratio is given by:

$$\text{RR} = \frac{a/(a+b)}{c/(c+d)} = \frac{26/58}{28/73} = 1.16\%$$

The standard error of the logarithm of the relative risk is given by:

$$\text{SE}\{l_n \text{ (RR)}\} = \sqrt{\frac{1}{a} + \frac{1}{c} - \frac{1}{a+b} - \frac{1}{c+d}} \quad (6.24)$$

$$\text{SE}\{l_n \text{ (RR)}\} = \sqrt{\frac{1}{26} + \frac{1}{28} - \frac{1}{58} - \frac{1}{73}}$$

$$\text{CI } 95\% = 0.77 \text{ to } 1.75$$

where zeros cause problems with the computation of the relative risk or its standard error, 0.05 is added to all cells (*a.b.c.d*). The number needed to treat (NNT) is the estimated number of patients who need to be treated with the new treatment rather than the standard treatment (or no treatment) for one additional patient to benefit. A standard normal deviate (z-value) is calculated as ln (RR)/SE {ln (RR)}, and the P-value is the area of the normal distribution that falls outside $\pm z$ (Table 6.28).

TABLE 6.28

Relative Risk Determination of Disease

Factor	Disease: Present (+)	Disease: Absent (–)	Total
Present (+)	26	32	$(a+b)=58$
Absent (–)	28	45	$(c+d)=73$
Total	$(a+c)=54$	$(b+d)=77$	$N=131$

NNT (Harm) 15.45395% CI4.273 (Harm) to ∞ to 9.558 (Benefit).

6.31 Measures of Agreement

One of the common problems medical researchers encounter is studying agreement between two or more observers. For example, the diagnosis of a disease (leprosy) by two paramedical workers/doctors, classification of an X-ray as normal or abnormal (e.g. tuberculosis) by two radiologists/doctors, or the labeling of a urine specimen as negative or positive (e.g. albumin) by two laboratory technicians. Consider the following data consisting of the radiographic assessment of the extent of cavitation in ($n=100$) patients by two readers (Table 6.29)

$$\text{Kappa 'K'} = \left(\frac{P_0 - P_e}{1 - P_e}\right) \quad (6.25)$$

$$P_e = \left(\frac{70}{100}\right) \times \left(\frac{75}{100}\right) + \left(\frac{30}{100}\right) \times \left(\frac{25}{100}\right)$$

$$P_e = 0.525 + 0.075 = 0.60$$

$$\text{Kappa 'K'} = \left(\frac{0.75 - 0.60}{1 - 0.60}\right)$$

$$\text{Kappa 'K'} = 0.375$$

'One way to interpret kappa is with this scale (1):

1. Kappa<0: No agreement
2. Kappa between 0.00 and 0.20: Slight agreement
3. Kappa between 0.21 and 0.40: Fair agreement
4. Kappa between 0.41 and 0.60: Moderate agreement
5. Kappa between 0.61 and 0.80: Substantial agreement
6. Kappa between 0.81 and 1.00: Almost perfect agreement'.

The column 'Total' (n_1+, n_2+,…, n_k+) indicates the sum of each row, known as row margins or marginal counts. Here, the total sum of a given row i is named n_i+ (Table 6.30).

The row 'Total' ($n+1$, $n+2$,…, $n+k$) indicates the sum of each column, known as column margins. Here, the total sum of a given column i is named $n+i$.

N is the total sum of all table cells.

TABLE 6.29

Measurement of Agreement Data Series

Response	+ve	–ve	Total
+ve	60	10	70
–ve	15	15	30
Total	75	25	100

TABLE 6.30

Comparison of Agreement with Respect To Two Raters

Rater-1		Rater-2: Level 1	Level 2	Level k	Total
	Level$_1$	n_{11}	n_{12}	n_{1k}	n_1+
	Level$_2$	n_{21}.	n_{22}	n_{2k}	n_2+
	Level$_3$	n_{31}	n_{32}	n_{3k}	n_3+
	Level$_4$	n_{41}	n_{42}	n_{4k}	n_4+
	Total	n_{1i}	n_{2i}	n_{ik}	GT

TABLE 6.31

Data Frame of Pane 1 and 2 to Assess the Sensory Evaluation

		Panel-1				
		A	B	C	D	Total
Panel-2	A	24	45	52	58	179
	B	26	32	40	39	137
	C	28	36	47	30	141
	D	25	21	20	19	85
	Total	103	134	159	146	542

For a given row/column, the marginal proportion is the row/column margin divided by *N*. This is also known as the marginal frequencies or probabilities. For row *i*, the marginal proportion is $Pi+=n_i+/N$. Similarly, for a given column *i*, the marginal proportion is $P+i=n+i/N$.

For each table cell, the proportion can be calculated as the cell count divided by *N*.

The proportion of observed agreement (P_0) is the sum of diagonal proportions, which corresponds to the proportion of cases in each category for which the two raters agreed on the assignment (Table 6.31).

$$P_0 = \sum \frac{(\text{Diagonal value})}{N} = \sum_{i=1}^{k} p_{ij} P_e = \sum_{j=1}^{k} P_{i+} + P_{+i} \tag{6.26}$$

$$\text{SE}(K) = \frac{P_e + P_e^2 - \sum P_{i+}P_{+i}\left(P_{i+} + P_{+i}\right)}{N(1-P_e)^2} \tag{6.27}$$

Once SE(*k*) is calculated, a 100(1 – alpha)% confidence interval for kappa may be computed using the standard normal distribution as follows:

$$k \pm \text{SE}(K)$$

Illustration 1: Two panels assessed the meat quality, and the agreement between the two panel scores on a 1–100 scale, including all sensory evaluation parameters (taste, smell, aroma, juiciness, tenderness, etc.)

Number of observed agreements: (n=122) 22.51% of the observations)

Number of agreements expected by chance: 132.1 (24.38% of the observations)

Kappa=–0.025, SE of kappa=0.023, 95% confidence interval 0.071 to 0.021

The calculations above only consider exact matches between observers. If the categories (A, B, C, etc.) are ordered, you may also wish to consider close matches. In other words, if one observer classifies a subject into group B and the other into group C, this is closer than if one classifies it into A and the other into D. The calculation of weighted kappa, below, assumes the categories are ordered and accounts for how far apart the two raters are. This calculation uses linear weights (Table 6.32).

TABLE 6.32

Fisher Exact Test

Attributes	Vaccinated	Not Vaccinated	Marginal Row Totals
Exposed to Covid	1,453 (*a*)	468 (*b*)	1,921 (*a*+*b*)
Unexposed Covid	733 (*c*)	112 (*d*)	845 (*c*+*d*)
Marginal Column Totals	2,186 (*a*+*c*)	580 (*b*+*d*)	2766 (*N*) (Grand Total)

Weighted Kappa=–0.074

$$\text{Fisher exact} = \frac{(a+b)!(a+c)!(c+d)!(b+d)!}{a!b!c!d!} \tag{6.28}$$

$$\text{Fisher exact} = \frac{(1921)!(2186)!(845)!(580)!}{1453!468!733!112!}$$

The Fisher exact test statistic value is<0.00001. The result is significant at $p<.05$.

6.32 Limitation of Significance Tests and Utility of Confidence Intervals

In the last three decades, statistical methods have made great inroads into the fields of medicine and public health. Significance tests and *P*-values have become commonplace in all leading medical journals. By and large, this has been a good development and investigators have learned to avoid attributing interpretations to differences/associations that could easily be due to chance. However, in recent years, coinciding with the easy availability of computers, statistical software packages, and programmed/programmable pocket calculators, there has been an upsurge in significance testing, sometimes bordering on indiscriminate. This, in turn, has led to a different type of problem, with investigators identifying statistically significant results, then working backward to set up a plausible hypothesis, but writing up the study as though it was undertaken to investigate this very hypothesis. Such practices have led M.J. Gardner and D.G. Altman to lament that 'Excessive use of hypothesis testing at the expense of other ways of assessing results has reached such a degree that levels of significance are often quoted alone in the main text and abstracts of the papers, with no mention of the actual concentrations, proportions etc., or other differences' (British Medical Journal, 1986, 292, 746).

6.33 Limitations of Significance Testing and *P*-Values

As mentioned in an earlier lecture (Concept of Test of Statistical Significance), some investigators think that demonstrating a difference to be statistically significant is tantamount to proving that it is biologically important, and conversely, that a non-significant difference is synonymous with 'no difference.' This has led to a great deal of overemphasis on significance testing in recent years, and the consequence is that we are fed an over-dose of $P<0.05$, P 0.05 or P=Non-significant, which convey little about the study findings and encourage oversimplistic and lazy thinking. Two examples will admirably illustrate the above statement. First, consider a situation in which the mean systolic BP of 200 diabetics is 146 mm Hg with an SD of 10 mm Hg. The difference is 3 mm Hg, and while the corresponding mean for 200 matched non-diabetics is 143 mm Hg with an SD of 10 mm Hg, the difference is 3 mm Hg, and its sampling error is 1 mm Hg, so that $t=3$. Although this is highly significant statistically, no clinician would attach any importance to the difference of 3 mm. Conversely, an appreciable observed difference in efficacy between two treatment regimens may turn out to be non-significant statistically because of inadequate trial size—for example, a 10% difference between two treatment regimens (60% v/s 70%) would be non-significant even if it is based on 100 patients in each regimen. The overemphasis on significance testing has unfortunately detracted from more useful approaches to interpreting study results, such as 'estimation.' It should be remembered that the investigator's aim is not just to say 'significant' or 'non-significant,' but to estimate the size of a treatment effect or the magnitude of an association. It is in this context that confidence intervals have a vital role.

6.34 Definition of Confidence Interval

In a representative sample of 100 observations, drawn at random from a large population, suppose the sample mean is found to be 68″ (and the SD is 10″). Can we make any statements about the Population Mean? We cannot automatically say that the Population Mean is 68″ because we are uncertain as to how much sampling fluctuation has occurred. What we do instead is determine a range of possible values for the Population Mean, with 95% degree of confidence. This range is called the 95% confidence interval and can be an important adjunct to a significance test. In general, the 95% confidence interval is given by Mean (or Proportion) ± 1.96 * sampling error of the mean (or Proportion). In the above example, ($n=100$), the sample mean (68) and its $SE=\left(\frac{10}{\sqrt{100}}\right)=1$; question of sample size, therefore, the 95% confidence interval ($68\pm1.96X_1$), approximately (66–70). Since the 95% confidence interval has been found to be 66″ to 70″, there may be a tendency on the part of the investigator to conclude that, the probability that the Population Mean lies between (66–70) is 95%. This is incorrect because the Population Mean either lies in this interval, or it does not, so the question of sample size being less than 60 would make it preferable to use the 5% level corresponding to a t with $(n-1)$ degrees of freedom. The correct meaning would be that if numerous random samples of size 100 are drawn and the 95% confidence interval is computed for each sample, the Population Mean will be within the computed interval in 95% of the instances. The essential difference between a significance test and a confidence interval is that while a non-significant result (with the former) indicates that only one previously chosen value (usually 0) is consistent with the data, a confidence interval indicates a range of underlying values that are consistent with the observed findings and therefore is more meaningful.

TABLE 6.33

Non-Significance in a Clinical Trial (N = 178) Patients

Treatment	Success	Failure	Total
A	76 (75%)	25	101
B	51 (66%)	26	77
Total	$n=127$	$n=51$	$N=178$

TABLE 6.34

Strong Significant Difference is a Clinical Trial in ($n=115$) Tuberculosis Patients

Treatment	Success	Failure	Total
A	49 (82%)	11	60
B	33 (60%)	22	55

6.35 Applications of the 95% Confidence Interval Method and Interpretation

The utility of the confidence interval method will now be illustrated with examples drawn from controlled clinical trials, the tuberculosis control program, a large-scale community study to evaluate an anti-leprosy vaccine, and laboratory studies of the ELISA test for HIV infection (Table 6.33).

Controlled Clinical Trials

The value of the chi-square (uncorrected) is 1.74, and this is not significant ($P>0.1$). This finding could lead some investigators to conclude that there is no difference in efficacy between the two treatments. However, the observed difference is 75% − 66% = 9%, and the 95% confidence interval is −4% to 22%. This indicates that compared to Treatment B, Treatment A has, at best an appreciable advantage (22%) and, at worst, a slight disadvantage (−4%). This inference is obviously more informative than just saying that the difference is not significant (Table 6.34).

The value of the chi-square (uncorrected) is 6.58, and this is highly significant ($P=0.01$). The observed difference in efficacy is 82% − 60% = 22%, and its 95% confidence interval is 6% to 38%. This indicates that changing from Treatment B to Treatment A can result in 6% to 38% more patients being cured. Again, this is more informative than just saying that the two treatments are significantly different (Table 6.35).

The value of the chi-square (uncorrected) is 3.61 ($P=0.06$). The observed difference in efficacy is 73% − 58% = 15%, and the confidence interval is - 1% to 31%. The latter finding suggests that Isoniazid 400 mg once a day is likely to be superior to 200 mg twice a day, even though the difference is, technically speaking, non-significant.

6.36 Disease Control Programs

Chi-square (uncorrected) = 4.32, which is significant ($P=0.04$). The impact of the special motivation program = 52% − 46% = 6^ in terms of improved case-holding. However, the 9% confidence interval is 0.5% to 11.6%, which indicates that the benefit from the special motivation program is not likely to be

TABLE 6.35

Border-Line' Significance in a Clinical Trial Based on (n=130) Patients

Isoniazid Dosage	Favorable Response (n_1)	Unfavorable Response (n_0)	Total ($N=n_1+n_0$)	Positive Agreement (PPA)	Negative Agreement (NPA)	Overall
400 mg (X_1)	47 (73%)	17	64	59.80 CI-95 (48.90–69.70%)	66.70. CI % (49.60–80.30%)	61.70 CI 95% (52.60–70.10%)
200 mg (X_2)	38 (58%)	28	66			

TABLE 6.36

Hypothetical Testing Pertaining to TB Subjects (Cases of National Tuberculosis Control Program).

Year	Motivation in Program	Completed Treatment (n)	Failed to Complete Treatment (n)	Total (N)	(PPA)	(NPA)	Overall
1987	Routine	276 (46%)	324	600	46.90%	47.10%	47.0%
1988	Special	312 (52%)	288	600	42.9%–51.0%	43.10%–51.0%	44.20–49.80

TABLE 6.37

Reported Cases of Large Community Anti-Leprosy Vaccine Trial in South India

Group	Cases (n)	No Cases (N)	Total (n) + (N)	Incidence per (1000)
Placebo	2,896	57,104	60,000	48.3
BCG vaccinated	4,896	115,445	120,000	38.0
Statistic	Value	95% CI		
Sensitivity	37.17%	36.09% to 38.25%		
Specificity	95.29%	95.17% to 95.41%		
Positive likelihood ratio	7.89	7.59 to 8.19		
Negative likelihood ratio	0.66	0.65 to 0.67		
Disease prevalence (*)	6.04%	5.91% to 6.17%		
Positive predictive value (*)	33.65%	32.80% to 34.51%		
Negative predictive value (*)	95.93%	95.86% to 96.00%		
Accuracy (*)	**91.77%**	**91.62% to 91.**		

(*) Significant at 1% level.

more than 11.6%. This knowledge may prompt the investigator to conclude that the special program was not really worthwhile, and that other strategies need to be explored to provide a greater magnitude of benefit (Table 6.36).

6.37 Field Trial of Vaccine

The contrast in the incidence, 48.3 versus 38.0, is highly significant ($P<0.001$), with 48.3 – 38.0. The vaccine efficacy is estimated at 38/48.3=21.4%. The 95% confidence interval is 17.7% to 24.9%. Now, let us suppose the trial had been undertaken on only 10% of the subjects (Table 6.37). The corresponding findings could be as follows:

The contrast in incidence is the same, 48.3 versus 38.0, and again the difference is highly significant ($P<0.001$) (Xia, 2017). The vaccine efficacy is also the same (21.4%), but the 95% confidence interval now is much broader (because the trial size is substantially smaller), i.e. 9.4% to 32.1%, indicating that the findings are consistent with a true efficacy that is as low as 9.4%, as compared to 17.7% above with the larger trial size. Once again, the confidence interval approach is more informative than the significance test, which merely indicates that the difference in incidences between unvaccinated and vaccinated subjects is very unlikely to be due to chance (Table 6.38).

6.38 Laboratory Studies

Suppose an 'ELISA' test was undertaken on 15 subjects who are known to be positive for 'HIV' infection by means of some other test, and all 15 are positive. The sensitivity of the 'ELISA' test is $\left(\frac{15}{15}\right)$ *ie* 100%. If similar findings had been obtained on 245 subjects, the sensitivity would again be $\left(\frac{245}{245}\right)$ *ie* 100%. Prima facie, there would seem to be little to choose between the two situations on the argument that both show the highest possible sensitivity, namely 100%. However, if the 95% confidence interval is computed, it will be found to be 78%–100% in the first case, indicating that the true sensitivity of 'ELISA' could be as low as 78%, which may well be unacceptable. In contrast, the 95% confidence interval in the second instance is 99%–100%, demonstrating that the test is highly acceptable.

TABLE 6.38

Correlation between BCG Vaccinated with Subjective Cases of Exposed and Unexposed

Group	Cases (*n*)	Non-cases (*N*)	Total (*n*) + (*N*)	Incidence per (1000)
Placebo	290	5,710	6,000	48.3
BCG vaccinated	456	11,544	12,000	38.0
Statistic	Value	95% CI		
Sensitivity	38.87%	35.36% to 42.48%		
Specificity	66.91%	66.20% to 67.61%		
Positive likelihood ratio	1.17	1.07 to 1.29		
Negative likelihood ratio	0.91	0.86 to 0.97		
Disease prevalence (*)	4.14%	3.86% to 4.45%		
Positive predictive value (*)	4.83%	4.43% to 5.28%		
Negative predictive value (*)	96.20%	95.98% to 96.41%		
Accuracy (*)	65.74%	65.05% to 66.44%		

(*) Significant at 1% level.

6.39 Concluding Remarks

The utility of the 95% confidence interval has become increasingly appreciated in medicine and public health since 1986, when the British Medical Journal declared that its editorial policy would be to encourage its use, wherever appropriate. This decision was followed the same year by the American Journal of Public Health and the Annals of Internal Medicine, and a year later by The Lancet. In 1988, the International Committee of Medical Journal Editors followed suit in their "Uniform Requirements for Manuscripts Submitted to Medical Journals," which is applicable to over ($n=300$) international journals, including three Indian journals: the Indian Journal of Dermatology, Venereology and Leprology, the Indian Journal of Gastroenterology, and the Indian Journal of Urology. It is merely a matter of time before other Indian journals will also insist on the use of confidence intervals, wherever appropriate.

REFERENCES

Brownlee, J. *15 Statistical Hypothesis Tests in Python (Cheat Sheet)* (Machine Learning Mastery, 2018).

Chen, K. S., Chen, S. C., Hsu, C. H. & Chen, W. Z. Statistical hypothesis testing for asymmetric tolerance index. *Applied Sciences* **11**, 6249 (2021).

Emmert-Streib, F. & Dehmer, M. Understanding statistical hypothesis testing: The logic of statistical inference. *Machine Learning and Knowledge Extraction* **1**, 945–962 (2019).

Genin, K. On falsifiable statistical hypotheses. *Philosophies* **7**, 40 (2022).

Hung, J. L., Chen, C. C. & Lai, C. M. Possibility measure of accepting statistical hypothesis. *Mathematics* **8**, 551 (2020).

J., N. L. & Lehmann, E. H. Testing statistical hypotheses. *The Incorporated Statistician* **10**, 98 (1960).

Kawamoto, Y., Sato, T. & Suenaga, K. Sound and relatively complete belief Hoare logic for statistical hypothesis testing programs. *Artificial Intelligence* **326**, 104045 (2024).

Lago, J. M. L. & Cruz, R. A. O. D. Linking to the real world: Contextual teaching and learning of statistical hypothesis testing. *LUMAT* **9**, 597–621 (2021).

Lehmann, E. L. Testing statistical hypotheses: The story of a book. *Statistical Science* **12**, 48–52 (1997).

Liu, C. C., Yu, C. H. & Chen, K. S. Using statistical test method to establish a decision model of performance evaluation matrix. *Applied Sciences* 13, 5139 (2023).

Matusita, K., Suzuki, Y. & Hudimoto, H. On testing statistical hypotheses. *Annals of the Institute of Statistical Mathematics* **6**, 133–141 (1954).

Murakami, Y. An introduction to medical statistics. *Journal of the Medical Society of Toho University* **61**, 238–243 (2014).

Neyman, J. & Pearson, E. S. in *Joint Statistical Papers*, 140–185 (University of California Press, 2023). https://doi.org/10.2307/jj.8501414.9

Patriota, A. G. On some assumptions of the null hypothesis statistical testing. *Educational and Psychological Measurement* **77**, 507–528 (2017).

Raha, S. A critique of statistical hypothesis testing in clinical research. *Journal of Ayurveda and Integrative Medicine* **2**, 105–114 (2011).

Sharma, N. Hypothesis statement and statistical testing: A tutorial. *BOHR International Journal of Operations Management Research and Practices* **1**, 52–58 (2022).

Suvorov, A. Y. et al. Statistical hypothesis testing: General approach in medical research. *Sechenov Medical Journal* **13**, 4–13 (2022).

Widaman, K. F. A note on statistical hypothesis testing: Probabilifying modus tollens invalidates its force? Not true! *Educational and Psychological Measurement* **83**, 1160–1172 (2023).

Xia, Y. & Sun, J. Hypothesis testing and statistical analysis of microbiome. *Genes and Diseases* **4**, 138–148 (2017).

Yamamoto, H. et al. Statistical hypothesis testing of factor loading in principal component analysis and its application to metabolite set enrichment analysis. *BMC Bioinformatics* **15**, 1–9 (2014).

7

Exploring Statistical Performance: Strategies to Study Molecular Genetics

7.1 Introduction

A gene or DNA sequence that can be used to identify individuals or species and has a known location on a chromosome is called a genetic marker. It can be defined as an observable variation that may result from a mutation or alteration in the genomic loci. A genetic marker can be a long DNA sequence, like minisatellites, or a short one, like the sequence surrounding a single base-pair change or single nucleotide polymorphism (SNP). For a long time, conventional phenotypic markers were the only means of identifying organisms through gene mapping (Civelek & Lusis, 2014). Among these were genes that encoded traits that could be readily observed, like blood types or seed shapes. The limited number of such traits present in a small number of organisms hindered mapping efforts. As a result, gene markers were developed that could identify genetic features in organisms that are difficult to observe directly (such as variations in proteins) (Drummond et al., 2012). DNA sequence variation's 'bread and butter' are single nucleotide polymorphisms. To ascertain the structure of a gene and determine which portions (the exons) make it into a functional mRNA molecule and which do not (the introns), they offer a wealth of information regarding the evolutionary history of human populations (Liu & Rattray, 2010). The significance of this comparative approach is increased by the high degree of alternative splicing found in vertebrates. Computational algorithms for gene-finding cannot accurately forecast the presence of alternative mRNA forms in the absence of experimental data; unfortunately, for rare mRNAs, this information is hard to obtain. For example, experiments may never identify an exon in an mRNA that is only used by a small percentage of brain cells (Zhang, 1998). That exon sequence would, however, most likely be present in the mouse genome. Because SNPs reflect past mutations that were mostly (but not exclusively) unique occurrences, two individuals who possess the same variant allele are recognized as sharing a common evolutionary history. Stated differently, by examining similar patterns of SNP variation, we can determine the identity of our ancestors. However, the abundance of SNPs is what really matters. Between every 1,000–2,000 nucleotides in two human DNA sequences, an estimated SNP is found. That may not seem like much until you realize that there are between 1.6 million and 3.2 million SNPs, or 3.2 billion nucleotides, in the human genome. And that is just from comparing two sequences; humans have a much larger total number of SNPs. Single nucleotide polymorphisms (SNPs) account for most genetically influenced variation in humans, especially in significant medically (and commercially) relevant traits like the propensity to develop a particular disease or the possible response to a particular drug, as Guillot & Frantz (2009) et al. address in their findings. Moreover, even in situations where an SNP is not the cause, the genes influencing these traits can be identified due to the large number of SNPs. A single nucleotide polymorphism (SNP) occurs when two sequences are compared position by position. A DNA sequence is made up of four linear combinations of nucleotides. Variations in the human genome's susceptibility to or resistance to a wide range of diseases are caused by single base differences. Thus, such information has enormous medical potential (Huang et al., 2006). SNP analysis in the human population has the potential to significantly improve human health by assisting with therapy selection and predicting disease susceptibility. This section provides a guide to selecting the optimal analytical approaches for SNPs model formulation, as well as an overview of new analytical tools for SNP for the early detection of emerging diseases and emerging technologies for large-scale SNP analysis (Kumar et al., 2004).

7.2 Molecular Markers Technology

7.2.1 DNA-Based Markers

To assess DNA polymorphism, a variety of molecular markers are used, which can be broadly categorized into three groups: (i) DNA based on hybridization; (ii) restriction fragment length polymorphisms (RFLPs); and (iii) oligonucleotide fingerprinting. The PCR-based DNA markers include random amplified polymorphic DNAs (RAPDs), simple sequence repeats or microsatellites (SSRs), and amplified fragment length polymorphisms (AFLPs), which entail the in vitro amplification of specific DNA sequences or loci with the aid of specifically or arbitrarily chosen oligonucleotide sequences (primers) and a thermostable DNA polymerase enzyme. Single nucleotide polymorphisms (SNPs) are DNA markers that are based on sequencing and DNA chips.

7.2.2 Properties of Ideal DNA Markers

The properties of ideal DNA markers are as follows:

a. Highly polymorphic nature
b. Co-dominant inheritance (determination of homozygous and heterozygous status of diploid organisms)

DOI: 10.1201/9781032713816-7

c. Frequent occurrence in the genome
d. Selectively neutral behavior (the DNA sequences of any organism are neutral to environmental conditions or management practices)
e. Easy access (availability)
f. Easy and fast assay
g. High reproducibility

7.2.3 Types of DNA Markers

Some commonly used types of genetic markers are

a. RFLP (Restriction Fragment Length Polymorphism)
b. SSLP (Simple Sequence Length Polymorphism)
c. AFLP (Amplified Fragment Length Polymorphism)
d. RAPD (Random Amplification of Polymorphic DNA)
e. VNTR (Variable Number Tandem Repeat)
f. SSR (Simple Sequence Repeat) Microsatellite Polymorphism
g. SNP (Single Nucleotide Polymorphism)
h. STR (Short Tandem Repeat)
i. SFP (Single Feature Polymorphism)
j. DArT (Diversity Arrays Technology)
k. RAD Markers (Restriction Site Associated DNA Markers)

7.3 Biochemical Markers

Biochemical markers are substances generated during the course of a disease, either at its onset or as it advances. These markers may serve as effect molecules, indicate joint damage, or fulfill both roles. They can include isozymes and different forms of other proteins, arising from alterations in the amino acid sequence of a protein. These markers can be distinguished through electrophoresis methods and histochemical staining techniques, which identify variations in proteins and amino acids, for example, at the gene product level.

7.4 Genetic Markers

Genetic markers identify nucleotide changes, such as deletion, duplication, inversion, and/or insertion, at the DNA level. Markers may display either co-dominant or dominant/recessive inheritance patterns. A marker is considered co-dominant if the genetic pattern of homozygotes can be differentiated from that of heterozygotes. Co-dominant markers typically provide more information than dominant markers (Krüger, 2018).

7.5 Polymorphism with Restricted Fragment Length (PCR-RFLP)

This method is based on amplifying variable regions of the target genome, followed by deriving the amplicon using one or more restriction enzymes specific to a particular sequence. Following electrophoresis, the various lengths of DNA fragments migrate in accordance with their weights, with the larger fragments migrating more slowly and the smaller fragments moving faster (Liang & Kelemen, 2004). Therefore, the term 'RFLP' generally refers to the variations in banding patterns that result from cleaving DNA fragments according to a specific sequence using restriction enzymes. Prior to the invention of PCR, restriction enzyme (RE) digestion of chromosomal DNA was usually required for RFLP analyses. This was followed by membrane blotting, electrophoretic separation, and hybridization with a labeled probe, which was typically radioactive (Figure 7.1).

7.6 Random Amplified Polymorphic DNA (RAPD)

Other names for random amplified polymorphic DNA include arbitrarily primed PCR (AP-PCR) and DNA amplification fingerprinting (DAF). This method, which detects amplified fragment length polymorphisms between organisms, is based on the use of short, arbitrary primers in PCR reactions. It can be used to create relatively complex and detailed DNA profiles. One short oligonucleotide, typically eight to ten nucleotides in length, is utilized in the most basic format. However, usually several primers are used, and if detection is combined with polyacrylamide gel electrophoresis, a range of five to 21 nucleotides has shown success. Multiple unspecific priming sites on opposing DNA strands are permissible under relaxed PCR conditions, even in cases where the match is not perfect. But a properly amplified template sequence can only extend from a priming site sequence to a complementary sequence that is close by. There is a range of less than 10 to over 100 amplified products, depending on the primer/template combination and ratios. This usually yields a spectrum of products characteristic of each template and primer combination, which can be effectively resolved and visualized with silver staining and polyacrylamide gel electrophoresis. While ethidium bromide staining and agarose gel electrophoresis can also be employed, they are limited to identifying major fragments.

7.7 Amplified Fragment Length Polymorphism (AFLP)

The DNA fingerprinting method known as amplified fragment length polymorphism is based on the identification of DNA fragments that have been exposed to restriction enzymes and then amplified via selective PCR amplification. Two restriction enzymes are used to cut the DNA, and the ends of the

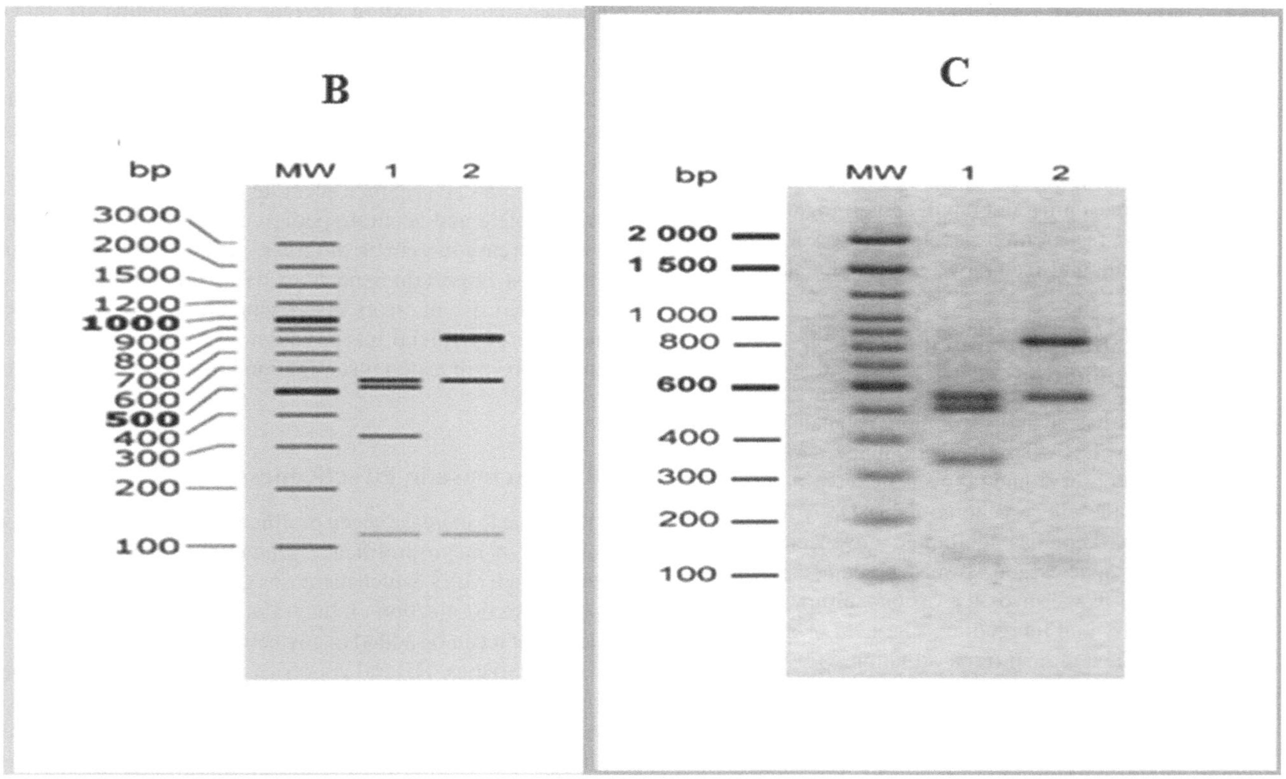

FIGURE 7.1 RFLP Polymorphism (restriction fragment length polymorphism RFLP).

DNA fragments are then ligated to create the template for PCR. The sequence of the AFLP primers is determined by the particular adapter that is ligated to the DNA fragment. As a result, these primers are actually somewhat arbitrary, and the assay uses DNA amplification to ascertain the distribution of DNA restriction sites across the relevant genomes.

7.8 Small-Scale Satellites

Both coding and non-coding DNA make up eukaryotic genomic DNA. Therefore, certain regions of the genome code for the translation of RNA, which is transcribed from the DNA template (coding DNA), and is then used to synthesize proteins. Nonetheless, non-coding DNA, which comprises a sizable fraction of eukaryotic genomes, frequently includes repetitive elements, such as multiple repeats, in addition to regulatory elements like enhancers and promoters. Satellite DNA, which has thousands of tandem repeats at a single site, as well as mini- and microsatellite DNA, depending on the number of repeats, are examples of these repetitions.

7.9 Microsatellites

Microsatellites, also known as 'sequence-tagged microsatellite sites' (STMS) or 'simple sequence repeats' (SSR), are made up of tandem repeats ranging from one to six base pairs (bp) that can be repeated up to 60 times. These domains were first introduced and demonstrated by Hamada and associates in the early 1980s. Repeats of (A), (TG), (CA)n, or (AAT) can make up repeat units. Microsatellites differ in the number of repeats, which makes them highly polymorphic. Finding up to 10 alleles per locus and 60% heterozygosity values in a comparatively small number of samples is not unusual. Microsatellites have a tendency to mutate at rates of up to 10–2 per generation.

7.10 SNPs or Single Nucleotide Polymorphisms

Single nucleotide polymorphisms (also referred to as 'snip') are the most recent contribution to studying DNA sequence variation. An SNP is found where different nucleotides occur at the same position in the DNA sequence. These markers are abundant and found in both coding and non-coding regions of the genome (Lasky et al., 2023). It is estimated that when comparing two human DNA sequences, there will be between 1.6 and 3.2 million SNPs. They are bi-allelic markers, indicating a specific polymorphism in only two alleles of a population. SNPs in coding regions can be directly associated with protein function, and as the inheritance pattern is more stable, they are more suitable markers for selection over time. SNP markers are significant because they are responsible for the majority of RFLP and AFLP markers. This is because SNPs occur in restriction enzyme recognition sites.

7.10.1 Using Genetic Markers

DNA markers hold promise for use in a wide range of genetics and animal breeding applications. Breeders of livestock can directly benefit from the technique in the form of individual identification, parentage verification, and the detection of specific genetic disorders. Mapping the genomes of different animals is an exciting and rapidly emerging use of genetic markers. In their research on genetic biodiversity and evolution, conservationists also employ a variety of genetic markers.

7.10.2 Applications in Animal Health and Production

Animal genetics has seen a revolutionary change as a result of the creation of DNA-based genetic markers. It is theoretically possible to observe and take advantage of genetic variation throughout the entire genome using DNA markers. The use of DNA markers has facilitated advancements in the identification of species and strains, parentage assignments, high-resolution linkage maps for various animal species, and genetic variability and inbreeding. The field of animal genetics has undergone a radical transformation with the introduction of DNA marker technology. Genomic research, broadly defined as the study of genes and their functions, is having a significant impact on a wide range of life aspects, from procreation and prolificacy to food safety and healthcare.

7.10.3 Enhancement of Milk Producing Characteristics

The main goal of selection in dairy animals has been to increase milk yield; notable genetic advancements in cattle have been achieved through the application of classical quantitative genetics. Due to milk's polygenic nature, it is challenging to identify every gene that affects yield; however, some likely candidate genes have been identified. An identified link between physiological or biochemical processes serves as the basis for selecting a potential quantitative trait locus (QTL) and analyzing the traits. Growth hormone (GH) released by the pituitary gland is well known to be essential for breast gland development, lactation, and growth. The main distinction between genetically modified animals and low-yielding ones was how the latter controlled the utilization of nutrients, a process in which growth hormone was an essential component. There have been reports relating differences in milk yield to higher levels of growth hormone in dairy cattle. Numerous mammalian species contain the GH gene, which has a similar nucleotide sequence consisting of five exons and four introns. In small ruminants, Ezon-4 and Ezon-5 correlate with performance metrics and accounted for greater diversity. The application of molecular marker-assisted selection (MAS) has started the process of increasing milk and meat production in milch animals. The alpha-lactalbumin-encoding gene is required for the mammary gland to produce lactose in dairy animals and may be a quantitative trait locus. It thus offers a fascinating candidate gene for MAS research. Studies have been conducted to explore the allelic variability of this gene and search for genetic polymorphism throughout the alpha-lactalbumin coding region. The Jamunapari breed of Indian goat is the breed that produces the most milk in the area. The four alpha-lactalbumin (ALA) exons of the Jamunapari goat breed were identified by PCR-SSCP to yield nine gel phenotypes. The only ALA gene polymorphism that has been documented worldwide is in exon-3, between positions 3E+5 and 3E+7. Since accurate genetic evaluation based on a complete and accurate pedigree is partially responsible for the increase in genetic progress, genetic evaluation is one of the most important aspects of dairy breeding. There has been a noted rise in errors in dairy breeding due to the widespread use of artificial insemination. For instance, the errors resulting from repeated insemination hinder the advancement of genetics.

7.10.4 Increase in Effectiveness

It takes time to increase sheep prolificacy through within-breed selections. A major prolificacy gene can be introduced into a flock through MAS, which increases genetic gain and increases the pressure of selection on the traits. The benefit of the main gene is that it can be added to any new breed and still maintain its characteristics. Several countries are presently conducting breeding programs to enhance the genetic characteristics of sheep's meat and wool, as well as their fecundity and disease resistance. MAS using the FecB (Borolo) gene is justified in order to increase sheep litter size, and it will be very profitable for mutton producers (Minamikawa, 2023). The FecB gene increases the rate of ovulation and the size of the litter. Furthermore, introducing the FecB gene may increase the fecundity of flocks with low levels of procreation.

7.10.5 Animal Welfare and Tameness

The rigorous selection practices used to raise farm animal productivity have an adverse effect on animal behavior. These kinds of side effects can be studied with the help of modern genetics and genomics. Taming silver foxes (Vulpes) is one of the most well-known instances of classical artificial selection for a behavioral trait. After more than 20 years of selective breeding, lines with human-friendly behavior have been developed. Dense marker maps for most farm animal species, including pigs, chickens, and cattle, have been created as a result of the identification and mapping of polymorphic DNA markers (Werhli & Husmeier, 2007). These dense marker maps identify chromosomal regions influencing polygenic traits, also known as quantitative trait loci (QTL), and the causal genes at these loci, making them indispensable tools for dissecting the underlying genetics of behavioral traits in animals. The temperament of cattle can have a significant impact on animal welfare as well as the farmer's safety. 'The behavioral reaction of an animal to human handling' is the definition of temperament. Another study on dairy cattle defined temperament as 'nervous, aggressive, or docile behavior during milking'. Both the number and locations of temperament QTLs varied significantly among the studies (Tibshirani et al., 2002). Subsequent investigations into gene expression patterns and stress-induced epigenetic modifications will likely be essential resources for understanding the genomics of animal welfare.

7.11 Conclusion

Animal genetic advancement is a basic, ongoing, and intricate process. Numerous techniques have been created and put to the test recently. Many markers have been made available by genetic polymorphism at the DNA sequence level, which has also shown promise for use in animal breeding. An almost infinite number of molecular markers have been identified since the first demonstration of DNA-level polymorphism, also known as restriction fragment length polymorphism (RFLP). Another significant development in this endeavor is the discovery of the polymerase chain reaction (PCR), which was made possible by the steadily growing body of DNA sequence data. Selecting the right marker system for a particular application is crucial to implementing marker-based data for genetic improvement. Certain factors—such as the degree of polymorphism (PIC), automation of the analysis, radioisotopes used, technique reproducibility, and cost—impact the selection of markers for various applications. Currently, molecular markers are undergoing enormous development, and this trend will likely continue. It is anticipated that geneticists and breeders will use molecular markers as a foundational tool to produce animals that society needs and wants (Figure 7.2) (Tables 7.1 and 7.2).

7.12 ANOVA Fixed Effect Model for Repeated Measures of SNPs

It is an extension of the independent *t*-test and is comparable to a one-way ANOVA, but for related groups rather than independent ones. Another name for repeated measures ANOVA is a within-subjects ANOVA or an ANOVA for correlated samples. All of these terms indicate the type of repeated measurements found in the above table. Standard software was used to ascertain the composition of the protein sequences (annotations were taken from the NCBS open virtual platform). The repeated sequences of SNPs are calculated. The matrix of the table shows that, many numerical values of composition were frequently repeated too many times. We can use the appropriate statistical techniques in this paradigm to determine whether there are any notable compositional differences between protein sequences. Experts in biochemistry and biology can use a range of standard models to explain the salient features of sequences. In order to describe the wellbeing of sequences and their compositions, we described and used special ANOVA repeated measures in this section. Additionally, this approach applies broadly to the study of genomic, proteomic, and metabolomic data sequences, among other things. Life scientists should apply this strategy in any further research interventions pertaining to life science research.

7.12.1 Assumptions of Repeated Measures

- The dependent variable is continuous (ratio or interval).
- One categorical independent variable (SNPs Sequences).
- Each subject appears in each group/treatment.
- The residual distribution is normal.
- Sphericity: Equality of variances of the differences between the groups.
- Required Sample Data: Sample data from all compared groups (Tables 7.3 and 7.4).

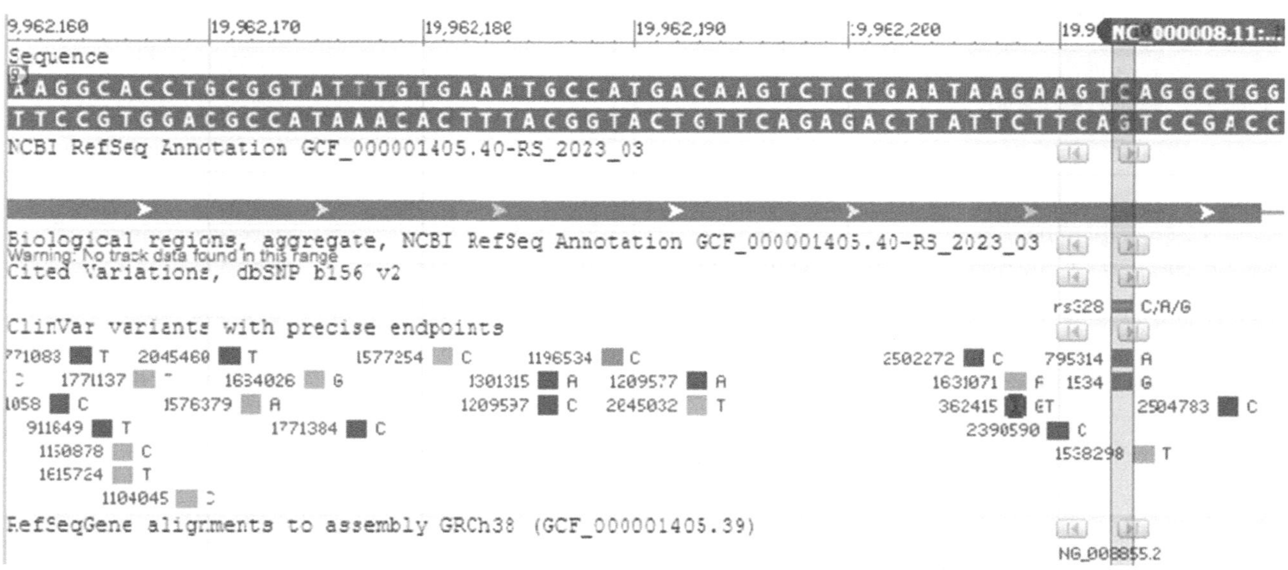

FIGURE 7.2 Identification of mutant genes in extensive sequences (high-throughput data) is illustrated through the alignment of sequences C/R/G with the NC0000084.1 mutant gene.

(Continued)

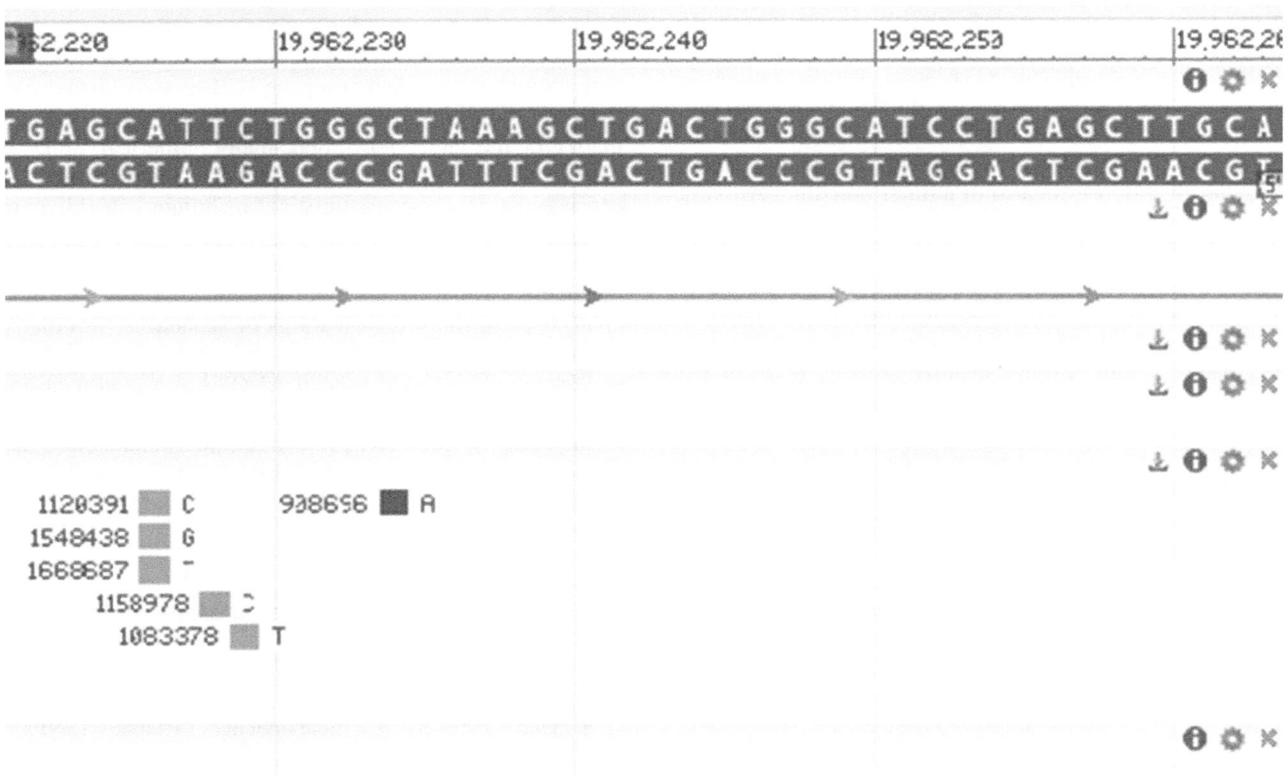

FIGURE 7.2 (*Continued*) Identification of mutant genes in extensive sequences (high-throughput data) is illustrated through the alignment of sequences C/R/G with the NC0000084.1 mutant gene.

TABLE 7.1

Proteomics Matching

Protein	Sequence ID	Length	No. of Matches	Sequence
Lipoprotein lipase [*Tupaia chinensis*]	**ELW52756.1**	479	1	KAPAVFVKCHDKSLNKKSGW
Hypothetical protein [*Ovis ammon polii* × *Ovis aries*]	**KAI4589084.1**	580	1	KAPAVFVKCHDKSLNKKSG
Lipoprotein lipase, partial [*Homo sapiens*]	**AAA59528.1**	135	1	KAPAVFVKCHDKSLNKKSG
Lipoprotein lipase [*Chrysochloris asiatica*]	**XP_006859966.1**	476	1	K P VFVKCHDKSLNKKSGW
Hypothetical protein [*Ovis ammon polii* × *Ovis aries*]	**KAI4573613.1**	447	1	KAPAVFVKCHDKSLNKKSGW K+P +FVKCHDKSLN+KSGW KSPVIFVKCHDKSLNRKSGW
LPL, partial [synthetic construct]	**AKI70063.1**	475	1	KAPAVFVKCHDKSLNKKSG
Unnamed protein product [*Homo sapiens*]	**BAF82979.1**	475	1	KAPAVFVKCHDKSLNKKSG
Lipoprotein lipase [*Gorilla gorilla*]	**XP_004046773.1**	475	1	KAPAVFVKCHDKSLNKKSG
Unnamed protein product [*Rangifer tarandus platyrhynchus*]	**CAI9705602.1**	479	1	KAPAVFVKCHDKSLNKKSGW K+P +FVKCHDKSLN+KSGW KSPVIFVKCHDKSLNRKSGW
Hypothetical protein [*Bos mutus*]	**MXQ89853.1**	541	1	KAPAVFVKCHDKSLNKKSGW K+P +FVKCHDKSLN+KSGW KSPVIFVKCHDKSLNRKSGW

Parameters

a: The number of subjects, number of rows.

b: The number of groups, number of columns.

n: Overall sample size, $n = a * b$.

$\bar{Y}_i$: An average of all the observations of subject i of variable A - subjects (row i).

$\bar{Y}_j$: An average of all the observations of group j of variable B - treatments/groups (column j).

$\bar{Y}$: Overall average ($\Sigma Y_{i,j}/n$, $i = 1$ to a, $j = 1$ to b).

Model

$$y_{ij} = \mu + \alpha_i + \epsilon_{ij} \quad (7.1)$$

where is the observed SNPs sequence at the *i*th row and *j*th position (bp 1,500)

μ = General mean of the position of SNPs (distance)
α_i = Effect of the *i*th row and *j*th SNPs type
ϵ_{ij} = Error associated with the *i*th row and *j*th SNPs type

TABLE 7.2

Protein Sequencing Compositions

Sequences	Composition
KAPAVFVKCHDKSLNKKSGW	C(102) H(161) N(29) O(**25**) S(1)
KAPAVFVKCHDKSLNKKSG	C(91) H(**151**) N(27) O(**24**) S(1)
KAPAVFVKCHDKSLNKKSG	C(91) H(**151**) N(27) O(**24**) S(1)
K P VFVKCHDKSLNKKSGW	C(96) H(**151**) N(27) O(**23**) S(1)
KAPAVFVKCHDKSLNKKSGW K+P +FVKCHDKSLN+KSGW KSPVIFVKCHDKSLNRKSGW	C(187) H(**291**) N(53) O(46) S(2)
KAPAVFVKCHDKSLNKKSG	C(91) H(**151**) N(27) O(**24**) S(1)
KAPAVFVKCHDKSLNKKSG	C(91) H(**151**) N(27) O(**24**) S(1)
KAPAVFVKCHDKSLNKKSG	C(91) H(**151**) N(27) O(**24**) S(1)
KAPAVFVKCHDKSLNKKSGW K+P +FVKCHDKSLN+KSGW KSPVIFVKCHDKSLNRKSGW	C(292) H(458) N(84) O(72) S(**3**)
KAPAVFVKCHDKSLNKKSGW K+P +FVKCHDKSLN+KSGW KSPVIFVKCHDKSLNRKSGW	C(292) H(458) N(84) O(72) S(**3**)

Analytical Steps

1. Determine the sum of the rows and columns to get the grand total
2. Calculate the grand average $\bar{y} = 1{,}246.16$
3. Calculate the SS subjects and MS subjects

$$\text{SSA} = \text{Column}^* \sum_i^a \left(Y_i - \bar{Y}\right)^2$$

$$\text{SSA} = 2^* \sum_i^a (3{,}083 - 1{,}246.16)^2 + (643 - 1{,}246.16)^2$$

$$+ (12.5 - 1{,}246.16)^2 = 10{,}519{,}400.33$$

$$\text{DFA} = \text{Rows} - 1 = 3 - 1 = 2$$

$$\text{MSA} = \frac{\text{SSA}}{\text{DFA}} = \frac{10{,}519{,}400.33}{2} = 5{,}259{,}700.16$$

4. Calculate the SS sequences (treatment) and MS treatments

$$\text{SSB} = \text{Row}^* \sum_i^a \left(Y_j - \bar{Y}\right)^2$$

$$\text{SSB} = 3^* \sum_i^a (1{,}028.33 - 1{,}246.16)^2 + (1{,}464 - 1{,}246.16)^2$$

$$= 284{,}708.16$$

$$\text{DFB} = \text{Column} - 1 = 2 - 1 = 1$$

$$\text{MSA} = \frac{\text{SSB}}{\text{DFA}} = \frac{284{,}708.16}{1} = 284{,}708.16$$

TABLE 7.3

ANOVA Table

Source of Variation	Degrees of Freedom (*df*)	Sum of Squares (SS)	Mean Square (MS)	*F*-Statistic	*P*-Value
Between the SNP_S	$(a-1)$	$SS_A = \Sigma_i^a b\,(\bar{Y}_i - \bar{Y})^2$	$MS_A = SS_A/df_A$	$F_A = MS_A/MS_E$	$P(x > F_A)$
Between the row	$(b-1)$	$SS_B = \Sigma_j^b a(\bar{Y}_j - \bar{Y})^2$	$MS_B = SS_B/df_B$	$F_B = MS_B/MS_E$	$P(x > F_B)$
Error within the cells	$(n-a-b+1)$	$SS_E = \Sigma_i^a \Sigma_j^b (Y_{i,j} - \bar{Y}_i - \bar{Y}_j + \bar{Y})^2$	$MS_E = SS_E/df_E$		
Total	$(\boldsymbol{n-1})$	$\mathbf{SS_T}$	$\mathbf{MS_E = S^2 = SS_T/(n-1)}$		

Total - All the deviations from the average
$SS_T = \Sigma_i^a \Sigma_j^b (Y_{i,j} - \bar{Y})^2$
SS_T = Sample Variance * $(n-1)$
$SS_T = SS_A + SS_B + SS_E$

TABLE 7.4

ANOVA Repeated Measures Model Output

	SNPs (Distance)					
Statistics	**Site 1 (n=16)**	**Site 2 (n=16)**	**Site 3 (n=16)**	**Site 4 (n=16)**	**Site 5 (n=16)**	**Total (n=16)**
ΣX	2372	3794	688	598	25	7,477
Mean	148.25	237.12	43	37.37	1.56	93.46
ΣX^2	47,017	117,766	39,164	29,190	51	1,716
Std. Dev.	88.89	136.14	25.27	21.35	0.8921	113.48

5. Calculate the SS total

$$SST = \sum_{i}^{a}\sum_{j}^{b}\left(y_{ij} - \bar{y}\right)^2$$

$$SST = (2{,}372 - 1{,}246.16)^2 + (3{,}794 - 1{,}246.16)^2$$
$$+(688 - 1{,}246.16)^2 + (598 - 1{,}246.16)^2$$
$$+(25 - 1{,}246.16)^2 + (0 - 1{,}246.16)^2 = 11{,}534{,}804.83$$

6. Calculate the SS error and MS error

$SSE = SST - SSA - SSB$

$DFE = n - a - b + 1 = 6 - 3 - 2 + 1 = 2$

$$MSE = \frac{SSE}{DFE} = \frac{730{,}696.33}{2} = 365{,}348.16$$

$$F(DFB, DFE) = \frac{MSB}{MSE} = \frac{284{,}708.16}{365{,}348.16} = 0.7793$$

7. Calculate the ETA squared effect size (Tables 7.5–7.7)

$$\eta^2 = \frac{SSB}{SST} = \frac{284{,}708.166}{11{,}534{,}804.83} = 0.024$$

TABLE 7.5

ANOVA Repeated Measures

Source	SS	df	MS	Fisher Value
Between-treatments	604,455.45	4	151,113.86	F=47.682*
Within-treatments	412,972.43	75	5,506.29	
Error	190,148.15	60	3,169.13	

* *Significant at 1% level.*

F_{cal} hypothetical value > F_{table} value, rejection of H_0 (null hypothesis).

TABLE 7.6

Partition of Sequences for Subgroup Analysis

Var *A*\Var *B*	*B*-Treat-1	*B*-Treat-2	Total
A-sub-1	2,372	3,794	3,083
A-sub-2	688	598	643
A-sub-3	25.0	0.00	12.50
Total	1,028.3333	1,464	1,246.1667

***R*-Software code**

```
set. seed (0)
#create data frame
data <- data. frame (program = rep (c ( "A",
"B", "C","D", "E"), each = 5),
                        Sequence = c (Seq if (102,
161, 25,1),
                                                Seq if
(91, 151, 24,1),
                                                Seq if
(292,458, 84, 72, 3)))
#view first six rows of data frame
head (dataset)
# Program Seq analysis
#1        A    102
#2        A    91
#3        A    96
#4        A    187
#5        A    91
#6        A    91
#7       A    91
#8       A    292
#9        A   292
#1        B   161
#2        B   151
#3        B   151
#4        B   151
#5        B   291
#6        B   151
#7       B    151
#8       B    151
#9       B    458
Box plot techniques
# view first 6 rows of "Seq protein" dataset
head (seq data)
# Seq data frame .R protein seq composition
              C    H         N      O   S
#1        102   161 29251
#2        9115127241
#3        91 15127 241
#4        9615127231
#5        18729153     462
#6        91               15127      041
#7        91               15127 241
#891      891      151     4      24       1
#9        292       84     72     23      3
#10       292       847    2       24     3
```

TABLE 7.7

Residual Calculation from the Observed Composition of Sequences

Variable *A*	Variable *B*	$Y_{i,j}$	Formula: $Y_{i,j} - \bar{Y}_i - \bar{Y}_j + \bar{Y}$	Residual
A-sub-1	*B*-Treat-1	2,372	2,372 – 3,083 – 1,028.3333 + 1,246.1667	–493.1667
A-sub-1	*B*-Treat-2	3,794	3,794 – 3,083 – 1,464 + 1,246.1667	493.1667
A-sub-2	*B*-Treat-1	688	688 – 643 – 1,028.3333 + 1,246.1667	262.8333
A-sub-2	*B*-Treat-2	598	598 – 643 – 1,464 + 1,246.1667	–262.8333
A-sub-3	*B*-Treat-1	25	25 – 12.5 – 1,028.3333 + 1,246.1667	230.3333
A-sub-3	*B*-Treat-2	0	0 – 12.5 – 1,464 + 1,246.1667	–230.3333

7.13 Box Plot Preparation for Proteomics Data

A box plot is one of the graphical representations of protein sequence data series (sometimes cal led a box-and-whisker plot). It shows the five-number summary of a dataset: the minimum (Min), first quartile (Q_1), median (Me), third quartile (Q_3), and maximum value (Max). A box plot allows us to easily visualize the distribution of values in a dataset using one simple plot (Figure 7.3).

Steps

Step 1: Extract the sequence composition and determine the max and min values

Step 2: Arrange the data in ascending or descending order of their magnitude

Step 3: Calculate the median values $\text{Me} = \left(\frac{(N+1)}{2}\right)^{\text{th}}$

Step 4: Calculate Q_1 and Q3 using the following formula

$$Q_1 = \left(\frac{N+1}{4}\right)^{\text{th}},\ Q_3 = \left(\frac{3\ N+1}{4}\right)^{\text{th}} \text{ items} \tag{7.2}$$

Box plots are useful because they help us visualize five important descriptive statistics of a sequence dataset: the minimum, lower quartile, median, upper quartile, and maximum. From the sequence data table (1) for carbon: 102, 91, 91, 96, 187, 91, 91, 91, 292, 292. The descriptive statistics of the series are as follows: Minimum: 91, First quartile: 91, Median: 93.5, Third quartile: 187, and Maximum: 292 (Tables 7.8 and 7.9).

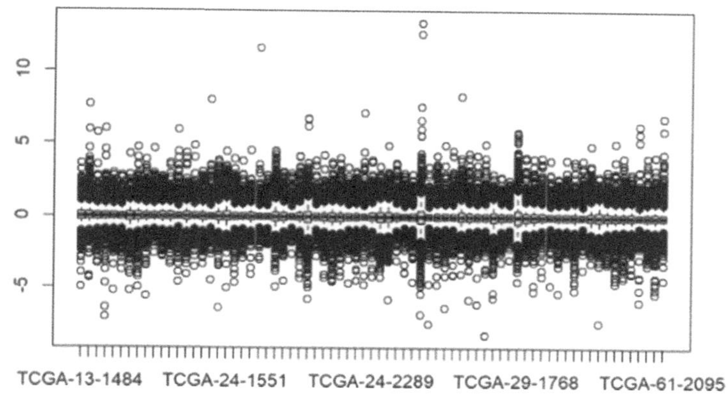

FIGURE 7.3 Box Plot for proteomic data.

TABLE 7.8

Regression Slope for Seq Data

b+	Seq No	AA		y+	b+	Seq No	AA	Seq No	y+
129.10	1	K	18		129.10	1	K	36	
226.15	2	P	17	1,973.04	200.13	2	A	35	3,970.07
325.22	3	V	16	1,875.98	297.19	3	P	34	3,899.03
472.29	4	F	15	1,776.92	368.22	4	A	33	3,801.98
571.36	5	V	14	1,629.85	467.29	5	V	32	3,730.94
699.45	6	K	13	1,530.78	614.36	6	F	31	3,631.87
802.46	7	C	12	1,402.68	713.43	7	V	30	3,484.80
939.52	8	H	11	1,299.68	841.52	8	K	29	3,385.74
1,054.55	9	D	10	1,162.62	944.53	9	C	28	3,257.64
1,182.64	10	K	9	1,047.59	1,081.59	10	H	27	3,154.63
1,269.67	11	S	8	919.49	1,196.62	11	D	26	3,017.57
1,382.76	12	L	7	832.46	1,324.71	12	K	25	2,902.55
1,496.80	13	N	6	719.38	1,411.75	13	S	24	2,774.45
1,624.89	14	K	5	605.34	1,524.83	14	L	23	2,687.42
1,752.99	15	K	4	477.24	1,638.87	15	N	22	2,574.33
1,840.02	16	S	3	349.15	1,766.97	16	K	21	2,460.29
1,897.04	17	G	2	262.11	1,895.06	17	K	20	2,332.20
–	18	W	1	205.09	1,982.10	18	S	19	2,204.10
b+	Seq No	AA	Seq No	y+	2,039.12	19	G	18	2,117.07

(Continued)

TABLE 7.8 (*Continued*)

Regression Slope for Seq Data

b⁺	Seq No	AA		y⁺	b⁺	Seq No	AA	Seq No	y⁺
129.10	1	K	19		2,225.20	20	W	17	2,060.05
200.13	2	A	18	1,929.03	2,353.29	21	K	16	1,873.97
297.19	3	P	17	1,858.00	2,450.34	22	P	15	1,745.87
368.22	4	A	16	1,760.94	2,597.41	23	F	14	1,648.82
467.29	5	V	15	1,689.91	2,696.48	24	V	13	1,501.75
614.36	6	F	14	1,590.84	2,824.58	25	K	12	1,402.68
713.43	7	V	13	1,443.77	2,927.58	26	C	11	1,274.59
841.52	8	K	12	1,344.70	3,064.64	27	H	10	1,171.58
944.53	9	C	11	1,216.61	3,179.67	28	D	9	1,034.52
1,081.59	10	H	10	1,113.60	3,307.77	29	K	8	919.49
1,196.62	11	D	9	976.54	3,394.80	30	S	7	791.40
1,324.71	12	K	8	861.51	3,507.88	31	L	6	704.37
1,411.75	13	S	7	733.42	3,621.92	32	N	5	591.28
1,524.83	14	L	6	646.38	3,750.02	33	K	4	477.24
1,638.87	15	N	5	533.30	3,837.05	34	S	3	349.15
1,766.97	16	K	4	419.26	3,894.07	35	G	2	262.11
1,895.06	17	K	3	291.16	–	36	W	1	205.09
1,982.10	18	S	2	163.07	–	–	–	–	–
–	19	G	1	76.03	–	–	–	–	–

TABLE 7.9

Regression Slope for Sequence Data (Arrays)

b⁺		AA		y⁺	b⁺		AA		y⁺
129.10	1	K	56		4,392.33	39	P	18	2,114.13
200.13	2	A	55	6,280.31	4,491.40	40	V	17	2,017.08
297.19	3	P	54	6,209.27	4,604.48	41	I	16	1,918.01
368.22	4	A	53	6,112.22	4,751.545	42	F	15	1,804.92
467.29	5	V	52	6,041.18	4,850.626	43	V	14	1,657.85
614.36	6	F	51	5,942.11	4,978.72	44	K	13	1,558.79
713.43	7	V	50	5,795.05	5,081.73	45	C	12	1,430.69
841.52	8	K	49	5,695.98	5,218.78	46	H	11	1,327.68
944.53	9	C	48	5,567.88	5,333.81	47	D	10	1,190.62
1,081.59	10	H	47	5,464.87	5,461.91	48	K	9	1,075.60
1,196.62	11	D	46	5,327.81	5,548.94	49	S	8	947.50
1,324.71	12	K	45	5,212.79	5,662.027	50	L	7	860.47
1,411.75	13	S	44	5,084.69	5,776.07	51	N	6	747.38
1,524.83	14	L	43	4,997.66	5,932.17	52	R	5	633.34
1,638.87	15	N	42	4,884.58	6,060.26	53	K	4	477.24
1,766.97	16	K	41	4,770.53	6,147.29	54	S	3	349.15
1,895.06	17	K	40	4,642.44	6,204.32	55	G	2	262.11
1,982.10	18	S	39	4,514.34		56	W	1	205.09
2,039.12	19	G	38	4,427.31	–	–	–	–	–
2,225.20	20	W	37	4,370.29	–	–	–	–	–
2,353.29	21	K	36	4,184.21	–	–	–	–	–
2,450.34	22	P	35	4,056.12	–	–	–	–	–
2,597.41	23	F	34	3,959.06	–	–	–	–	–
2,696.48	24	V	33	3,811.99	–	–	–	–	–
2,824.58	25	K	32	3,712.93	–	–	–	–	–
2,927.58	26	C	31	3,584.83	–	–	–	–	–
3,064.64	27	H	30	3,481.82	–	–	–	–	–
3,179.67	28	D	29	3,344.76	–	–	–	–	–
3,307.77	29	K	28	3,229.74	–	–	–	–	–

(*Continued*)

TABLE 7.9 (*Continued*)

Regression Slope for Sequence Data (Arrays)

b^+		AA		y^+	b^+		AA		y^+
3,394.80	30	S	27	3,101.64	–	–	–	–	–
3,507.88	31	L	26	3,014.61	–	–	–	–	–
3,621.92	32	N	25	2,901.53	–	–	–	–	–
3,750.02	33	K	24	2,787.48	–	–	–	–	–
3,837.05	34	S	23	2,659.39	–	–	–	–	–
3,894.07	35	G	22	2,572.36	–	–	–	–	–
4,080.15	36	W	21	2,515.33	–	–	–	–	–
4,208.25	37	K	20	2,329.25	–	–	–	–	–
4,295.28	38	S	19	2,201.16	–	–	–	–	–

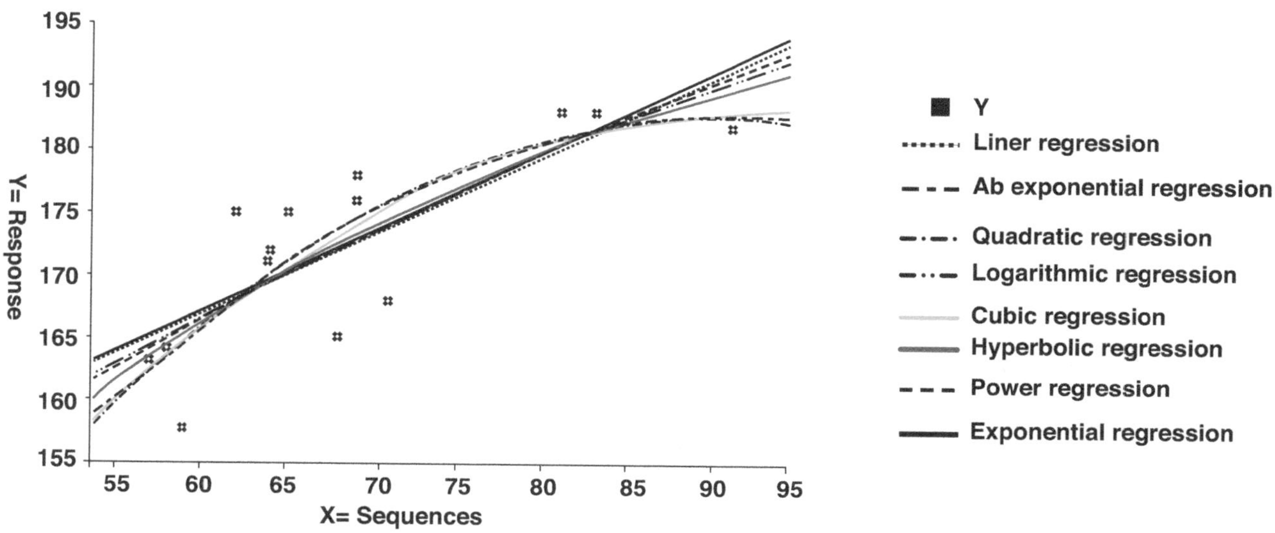

FIGURE 7.4 Different regression methods simulated based on least squares analysis.

7.14 Linear Least Square Analysis

The linear least squares method is one of the most widely used statistical methods to approximate the linear function of DNA and RNA sequences. In biological and life science experiments, this method is the standard approach in regression analysis to approximate the solution of genetic sequences and omic data series. There involves a set of equations where there are more equations than unknowns, and it minimizes the sum of the squares of the residuals in the results of each set of equations. Suppose we have a set of sequence data points (slope and estimator values, Figures 7.4 and 7.5). Let's consider a random sequence of proteins $(x_1, y_1)....(x_m, y_m)$. We will approximate the function as a linear combination of parameters modeled by the following equation (7.3):

$$y(x_i, a_1, a_2 \ldots a_n) = a_1 + a_2 x + a_3 . \ln(x) \ldots a_n x^{10} \quad (7.3)$$

$$\begin{bmatrix} \hat{y}_1 \\ . \\ \bar{y}_m \end{bmatrix} = \begin{bmatrix} 1 & x_1 & l_n x & x_1^{10} \\ 1 & x_m & l_n x_m & x_m^{10} \end{bmatrix} \begin{bmatrix} a_1 \\ . \\ a_m \end{bmatrix} \quad (7.4)$$

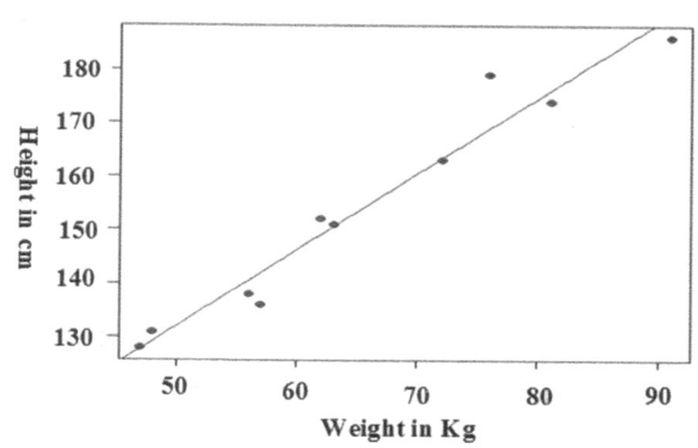

FIGURE 7.5 Scatter plot of linear regression.

$$\hat{y} = X_a \text{ using for least square approximation} \quad (7.5)$$

$$f(a) = \sum_{i=1}^{m} \left[y_i(x_i; a) - y_i \right]^2 \text{ or in matrix of} \quad (7.6)$$

$$f(a) = |X_a - y|^2 \quad (7.7)$$

TABLE 7.10
Least Square Analysis of Proteomics Data (Model Output)

Regression Methods	Equation	Correlation	R^2 (%)	Average Relative Error (%)
Linear	$y = 0.6233x + 129.57$	0.788	0.62	2.44
Quadratic	$y = -0.0168x^2 + 3.086x + 41.37$	0.816	0.66	2.00
Cubic	$y = 0.001x^3 - 0.039x^2 + 4.69x + 3.19$	0.816	0.66	2.01
Power	$y = 56.48x^{0.264}$	0.798	0.63	2.34
ab exponential	$y = 134.44 - 1.0036^x$	0.784	0.61	2.47
Logarithmic	$y = -20.17 + 45.62.lnx$	0.801	0.64	2.30
Hyperbolic	$y = 220.77 - \frac{3264.22}{x}$	0.810	0.65	2.16
Exponential	$y = e^{4.90+0.003x}$	0.784	0.61	2.47

This value is the distance between vector y and vector X_a. To minimize this distance, X_a should be projected onto the x column space, and X ay vectors should be orthogonal to that space.

$$(XV)(X_a - y) = V^T\left(X^T X_a - X^T Y\right) = 0 \tag{7.8}$$

$$X^T X_a - X^T Y = 0 \tag{7.9}$$

$$X^T X_a = X^T Y \text{ Hence, } a = \left(X^T X\right)^{-1} X^T Y \tag{7.10}$$

The curve fit must pass through particular sequential points, since the slope of the curve at these points must be multiplied by the Lagrangian function (Table 7.10). The equation becomes

$$f(a) = |X_a - y|^2 \tag{7.11}$$

$$\begin{bmatrix} \hat{y}_{c1} \\ . \\ \overline{y}_{ck} \end{bmatrix} = \begin{bmatrix} 1 & x_{c1} & l_n xc & x_{c1}^{10} \\ 1 & x_{ck} & l_n x_{ck} & x_{ck}^{10} \end{bmatrix} \begin{bmatrix} a_1 \\ . \\ a_k \end{bmatrix} \tag{7.12}$$

$$b = CA \tag{7.13}$$

$$F(a,\lambda) = f(a) + \lambda\psi(a) \tag{7.14}$$

$$F(a,\lambda) = |X_a - y|^2 + \lambda(Ca - b) \tag{7.15}$$

$$\begin{bmatrix} a \\ \lambda \end{bmatrix} = \begin{bmatrix} 2X^T X & C^T \\ C & 0 \end{bmatrix}^{-1} \begin{bmatrix} 2X^T Y \\ b \end{bmatrix} \tag{7.16}$$

7.15 Linear Regression Model

Regression analysis is a powerful statistical method, Back in 1900, Prof. Francis Galton used it to study the hereditary traits of father and son's height. It is also known as a switch-back design and is a widely used statistical tool to establish a relationship model between two variables. One of these variables is called the predictor variable (X), whose value is gathered through biological experiments. The other variable is called the response variable (Y), whose value is derived from the predictor variable. In linear regression, these two variables are related through an equation, where the exponent (power) of both variables is 1. Mathematically, a linear relationship represents a straight line when plotted as a graph. A non-linear relationship where the exponent of any variable is not equal to 1, creates a curve. The general mathematical equation for linear regression is as following (equation). This is the description of the parameters used: 'Y' is the response variable (Seq Protein), 'X' (charge of composition) is the predictor variable, and 'a' and 'b' are constants, called the coefficients.

$$\hat{Y} = a + bX \tag{7.17}$$

$$a = \frac{\sum X_i \sum Y_i - n\sum_{i=1}^{n} X_i Y_i}{\left(\sum X_i\right)^2 - n\sum X_i^2},$$

$$b = \frac{\sum X_i \sum X_i Y_i - n\sum_{i=1}^{n} X_i^2 \sum Y_i}{\left(\sum X_i\right)^2 - n\sum X_i^2}$$

$b+$	129	226	325	472	571	699	802	940	1,055	1,183	1,270	1,383	1,497	1,625	1,753	1,840	1,897
$y+$	**1,973**	**1,876**	**1,777**	**1,630**	**1,531**	**1,403**	**1,300**	**1,163**	**1,048**	**919**	**832**	**719**	**605**	**477**	**349**	**262**	**205**

Linear Correlation Coefficient

$$r = \frac{n\sum X_iY_i - \sum X_i \sum Y_i}{\sqrt{\left(n\sum X_i^2 - \left(\sum X_i\right)^2\right)\left(n\sum Y_i^2 - \left(\sum Y_i\right)^2\right)}} \quad (7.18)$$

Coefficient of Determination R^2 (%) $= r_{xy}^2$. Standard error of the regression is given by the equation (10.18).

$$\bar{A} = \frac{1}{n}\sum\left|\frac{Y_i - \hat{Y}_i}{Y_i}\right| * 100 \quad (7.19)$$

Case 1: R-Code for calculation

```
x <-c (129,226,325,472,571,699,802,940,1055,11
83,1270,1383,1497,1625,1753,1840,1897)
y <- c (1973,1876,1777,1630,1531,1403,1300,116
3,1048,919,832,719,605,477,349,262,205)
# Apply the lm () function.
relation <- lm (y ~ x)
print (relation)
# The predictor vector.
x <- c (129,226,325,472,571,699,802,940,1055,1
183,1270,1383,1497,1625,1753,1840,1897)
# The response vector.
y <- c (1973,1876,1777,1630,1531,1403,1300,116
3,1048,919,832,719,605,477,349,262,205)
# Apply the lm () function.
relation <- lm(y ~ x)
# Find seq protein energy by 472.
a <- data. frame (x =472)
Result <- predict (relation,a)
print (result)
```

Case 2. Illustration: Fit a regression line for the following datasets on height and weight by using R software.

R-Code

```
# Values of height
151, 174, 138, 186, 128, 136, 179, 163, 152,
131
# Values of weight.
63, 81, 56, 91, 47, 57, 76, 72, 62, 48
x <-c (151,174,138,186,128,136,179,163,152,13
1)
y <-c (63, 81, 56, 91, 47, 57, 76, 72, 62, 48)
# Apply the lm() function.
Relation <- lm (y ~ x)
Print (relation)
Call: lm (formula = y ~ x)
# Apply the lm () function.
Relation <- lm(y ~ x)
#  Find weight of a person with height 170.
a <- data. Frame (x =170)
result < -   predict (relation, a )
print (result)
```

```
#  Create the predictor and response variable.
x <-c (151,174,138,186,128,136,179,163,152,131)
y <-c (63, 81, 56, 91, 47, 57, 76, 72, 62, 48)
relation < - lm(y ~ x)
#  Give the chart file a name.
Png (file = "linearregression.png")
# Plot the chart.
Plot (y, x, col = "blue", main = "Height &
Weight Regression",
abline (lm( x~ y)), cex = 1.3,pch = 16,xlab =
"Weight in Kg", y lab = "Height in cm")
# Save the file.
dev. Off ()
```

Model output

```
Coefficients:
(Intercept)            x
   -38.4551       0.6746
Call:
lm(formula = y ~ x)

Residuals:
    Min       1Q  Median      3Q     Max
-6.3002 -1.6629  0.0412  1.8944  3.9775

Coefficients:
             Estimate Std. Error t value
Pr(>|t|)
(Intercept) -38.45509    8.04901  -4.778
0.00139 **
x             0.67461    0.05191  12.997
1.16e-06 ***
---
Signif.codes:  0 ,***' 0.001 ,**' 0.01 ,*'
0.05 ,.' 0.1 ' ' 1

Residual standard error: 3.253 on 8 degrees of
freedom
Multiple R-squared:  0.9548,Adjusted
R-squared:  0.9491
F-statistic: 168.9 on 1 and 8 DF,  p-value:
1.164e-06
       1
76.22869
Call:
lm(formula = y ~ x)

Residuals:
    Min       1Q  Median      3Q     Max
-6.3002 -1.6629  0.0412  1.8944  3.9775

Coefficients:
             Estimate Std. Error t value
Pr(>|t|)
(Intercept) -38.45509    8.04901  -4.778
0.00139 **
x             0.67461    0.05191  12.997
1.16e-06 ***
---
```

```
Signif.codes:  0 ,***' 0.001 ,**' 0.01 ,*'
0.05 ,.' 0.1 ' ' 1
Residual standard error: 3.253 on 8 degrees of
freedom
Multiple R-squared:   0.9548, Adjusted
R-squared:  0.9491
F-statistic: 168.9 on 1 and 8 DF, p-value:
1.164e-06
```

7.16 Quadratic Regression

Quadratic regression is the process of finding the equation of a parabola that best fits the data series of Seq DNA, RNA, and Protein data (Figure 7.6). The equation can be defined in the following equation (7.20):

$$\hat{y} = aX^2 + bx + c \tag{7.20}$$

The quadratic regression is an extension of the simple linear regression model. While linear regression can be performed with as few as two points, quadratic regression requires more data points to ensure the series falls into a "U" shape.

System of equations to find a, b, and c

$$\begin{cases} a\sum X_i^2 + b\sum X_i + nc = \sum y_i \\ a\sum X_i^3 + b\sum X_i^2 + c\sum X_i = \sum x_i y_i \\ a\sum X_i^4 + b\sum X_i^3 + c\sum X_i^3 = \sum x_i^2 y_i \end{cases} \tag{7.21}$$

Correlation coefficient 'r'

$$R(\%) = \sqrt{1 - \frac{\sum\left(Y_i - \hat{Y}_i\right)^2}{\sum\left(Y_i - \bar{Y}_i\right)^2}} \tag{7.22}$$

FIGURE 7.6 Plot illustrating quadratic regression.

$$\bar{A} = \frac{1}{n}\sum\left|\frac{Y_i - \hat{Y}_i}{Y_i}\right| * 100 \tag{7.23}$$

7.17 Cubic Regression

Cubic regression allows us to model the relationship between two biological variables by finding an accurate approximation of the curve that best fits the observed samples of scoring in genetic arrays. If this curve corresponds to a polynomial, we are dealing with polynomial regression. The following equations show the mathematical formulation of the cubic model (Figures 7.7–7.9).

$$\hat{Y} = ax^3 + bx^2 + cx + d \tag{7.24}$$

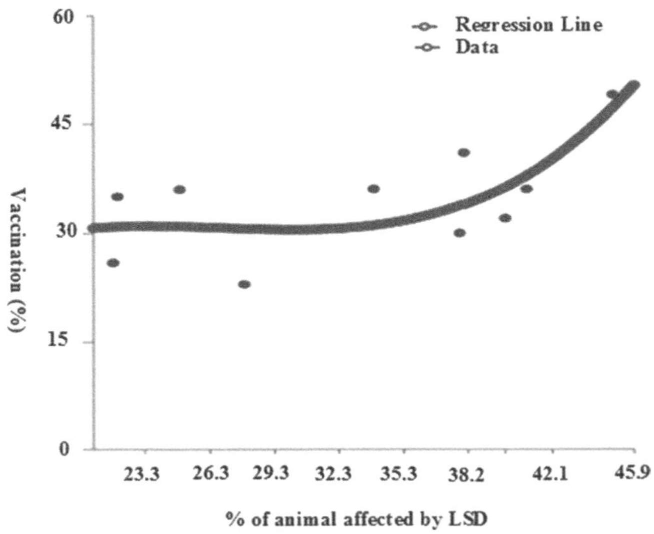

FIGURE 7.7 Scatter plot of power regression.

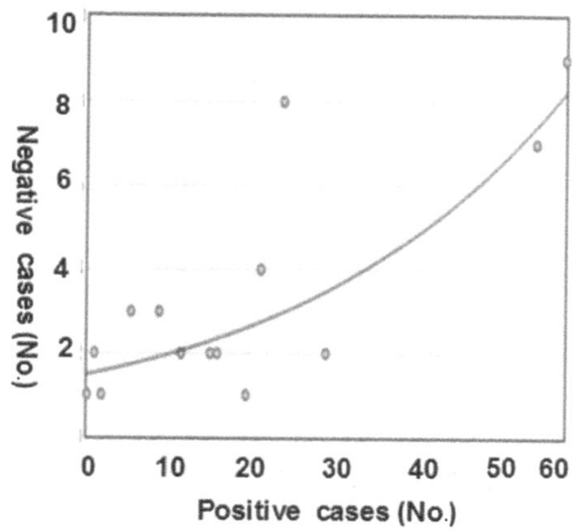

FIGURE 7.8 Lumpy skin disease in cattle (Abs exponential distribution).

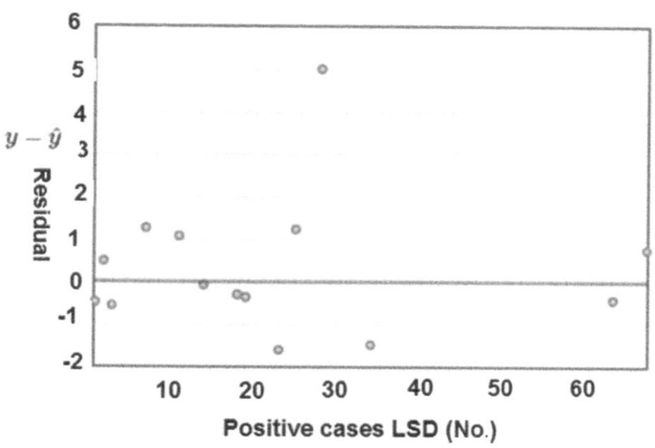

FIGURE 7.9 Continues with lumpy skin disease in cattle (Abs exponential distribution).

System of equations to find a, b, c, and d

$$\begin{cases} a\sum X_i^3 + b\sum X_i^2 + c\sum x_i + nd = \sum y_i \\ a\sum X_i^4 + b\sum X_i^3 + c\sum X_i^2 + d\sum x_i = \sum x_i y_i \\ a\sum X_i^5 + b\sum X_i^4 + c\sum X_i^3 + d\sum X_i^2 = \sum x_i^2 y_i \\ a\sum X_i^6 + b\sum X_i^5 + c\sum X_i^4 + d\sum X_i^3 = \sum x_i^3 y_i \end{cases} \quad (7.25)$$

Power regression

$$\hat{y} = a.x^b \quad (7.26)$$

$$b = \frac{n\sum \ln x_i . \ln y_i - \sum \ln x_i \sum \ln y_i}{n\sum \ln^2 x_i - \left(\sum \ln x_i\right)^2} \quad (7.27)$$

'a' is the coefficient

$$a = \exp\left(\frac{1}{n}\sum \ln y_i - \frac{b}{n}\sum \ln x_i\right) \quad (7.28)$$

$$R(\%) = \sqrt{1 - \frac{\sum\left(Y_i - \hat{Y}_i\right)^2}{\sum\left(Y_i - \bar{Y}_i\right)^2}} \quad (7.29)$$

$$\bar{A} = \frac{1}{n}\sum\left|\frac{Y_i - \hat{Y}_i}{Y_i}\right| * 100 \quad (7.30)$$

Fitted coefficients:

a=-28.42, b=6.809, c=-0.256, d=0.0032

Model: $y = -28.429 + 6.809x - 0.2568x^2 + 0.0032x^3$

- Goodness of fit: Standard Error of Regression: 6.3332, Coefficient of determination R^2: 0.5093

Ab-Exponential Regression

$$\hat{y} = a.b^x \quad (7.31)$$

$$b = \exp\frac{n\sum x_i \ln y_i - \sum x_i . \sum \ln y_i}{n\sum x_i^2 - \left(\sum x_i\right)^2} \quad (7.32)$$

$$a = \exp\left(\frac{1}{n}\sum \ln y_i - \frac{\ln b}{n}\sum x_i\right) \quad (7.33)$$

$$R(\%) = \sqrt{1 - \frac{\sum\left(Y_i - \hat{Y}_i\right)^2}{\sum\left(Y_i - \bar{Y}_i\right)^2}} \quad (7.34)$$

$$\bar{A} = \frac{1}{n}\sum\left|\frac{Y_i - \hat{Y}_i}{Y_i}\right| * 100 \quad (7.35)$$

Negative cases of lumpy skin disease (LSD) in milching dairy cows: LSD (1.39 – 1.02); Positive cases: LSD negative cases (1.39 – 1.02), positive cases (r=0.72, r^2=0.727, r^2=0.52).

7.18 Hyperbolic Regression (Figure 7.10)

$$\hat{y} = a + \frac{b}{x} \quad (7.36)$$

$$b = \frac{n\sum \frac{y_i}{x_i} - \sum \frac{1}{x_i}\sum y_i}{n\sum \frac{1}{x_i^2} - \left(\sum \frac{1}{x_i}\right)^2}, \quad a = \frac{1}{n}\sum y_i - \frac{b}{n}\sum \frac{1}{x_i} \quad (7.37)$$

$$R(\%) = \sqrt{1 - \frac{\sum\left(Y_i - \hat{Y}_i\right)^2}{\sum\left(Y_i - \bar{Y}_i\right)^2}} \quad (7.38)$$

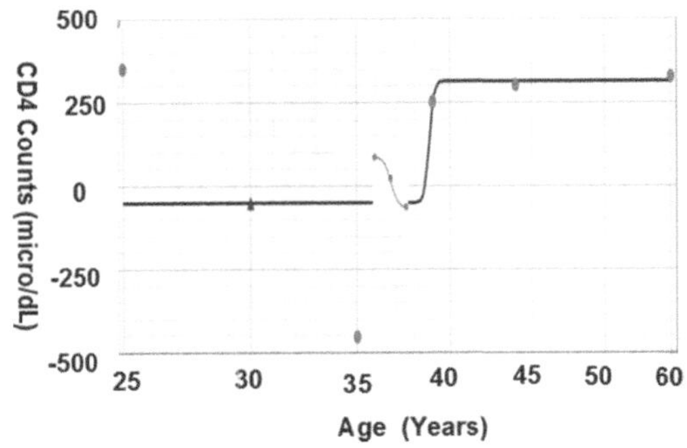

FIGURE 7.10 Scatter distribution of Hyperbolic regression.

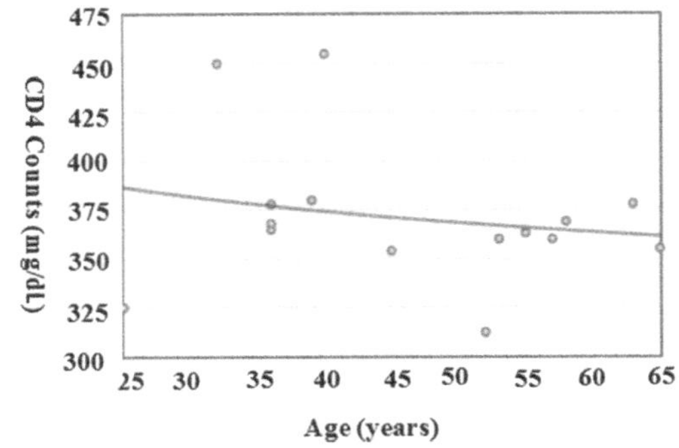

FIGURE 7.11 Scatter distribution of Logarithmic regression.

TABLE 7.11

The Hyperbolic Regression Model Output

Statistics	R^2	aR^2	p	SE	SSE	F	AIC	BIC	df
Goodness measures	0.308	−2.46	0.958	569.5	3243	0.12	77.59	76.03	1
Co efficient	a	b	c	d	–	–	–	–	–
Values	312.5±402.7	−295.81±336,300	38.79±235,100	−52.50±569.50		–	–	–	–

$$\bar{A}=\frac{1}{n}\sum\left|\frac{Y_i-\hat{Y}_i}{Y_i}\right|*100 \tag{7.39}$$

Model equation: $\boldsymbol{Y=-52.5+(312.50-(-52.50)/(X/38.76)\ \wedge -295.81}$

7.19 Logarithmic Regression (Figure 7.11)

$$\hat{y}=a+b\ln x \tag{7.40}$$

$$b=\frac{n\sum(y_i\ \ln x_i)-\sum\ln x_i.\sum y_i}{n\sum\ln^2 x_i-\left(\sum\ln x_i\right)^2},\quad a=\frac{1}{n}\sum y_i-\frac{b}{n}\sum\ln x_i \tag{7.41}$$

CD_4 counts (mg/dL) = 471.95 − 26.46 * ln (age (years)) CD_4 counts, 471.95 − 26.46 * ln (age (years)), r^2=0.038 (Table 7.11).

7.20 Exponential Regression

$$\hat{y}=e^{a+bx} \tag{7.42}$$

$$b=\frac{n\sum x_i\ \ln y_i-\sum x_i\sum\ln y_i}{n\sum x_i^2-\left(\sum x_i\right)^2},\quad a=\frac{1}{n}\sum\ln y_i-\frac{b}{n}\sum x_i \tag{7.43}$$

CD_4 Counts (mg/dL) = 399.84 − 0.99 age (years) CD_4 counts (mg/dL) = 399.85 − 0.99 age (years), r^2=0.045 (Figure 7.12).

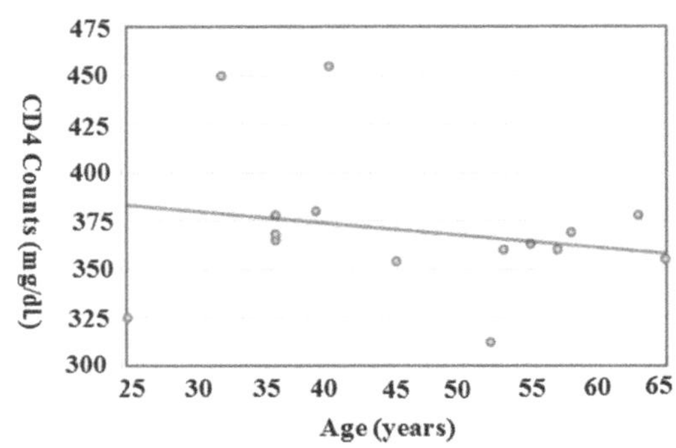

FIGURE 7.12 Exponential distribution of age and CD_4 counts.

Matrix: e DNA full Gap penalty: 2.0 Extend penalty: 2.0 Score: 4,482.0 Sequence 1 length: 900 Sequence 2 length: 900 Alignment length: 901 Identity: 898/901 (99.67%) Similarity: 898/901 (99.67%) Gaps: 2/901 (0.22%) (Figures 7.13 and 7.14)

```
# Python program to convert
# altered DNA to protein
inputfile ="DNA_sequence_original.txt"
f =open (inputfile, "r")
seq =f.read()
seq =seq.replace("\n", "")
seq =seq.replace("\r", "")
 deftranslate(seq):
       table ={
         'ATA':'I', 'ATC':'I', 'ATT':'I',
'ATG':'M',
         'ACA':'T', 'ACC':'T', 'ACG':'T',
'ACT':'T',
         'AAC':'N', 'AAT':'N', 'AAA':'K',
'AAG':'K',
         'AGC':'S', 'AGT':'S', 'AGA':'R',
'AGG':'R',
         'CTA':'L', 'CTC':'L', 'CTG':'L',
'CTT':'L',
```

```
        'CCA':'P', 'CCC':'P', 'CCG':'P',
'CCT':'P',
        'CAC':'H', 'CAT':'H', 'CAA':'Q',
'CAG':'Q',
        'CGA':'R', 'CGC':'R', 'CGG':'R',
'CGT':'R',
        'GTA':'V', 'GTC':'V', 'GTG':'V',
'GTT':'V',
        'GCA':'A', 'GCC':'A', 'GCG':'A',
'GCT':'A',
        'GAC':'D', 'GAT':'D', 'GAA':'E',
'GAG':'E',
```

FIGURE 7.13 Lumpy skin disease within the cattle population.

```
        'GGA':'G', 'GGC':'G', 'GGG':'G',
'GGT':'G',
        'TCA':'S', 'TCC':'S', 'TCG':'S',
'TCT':'S',
        'TTC':'F', 'TTT':'F', 'TTA':'L',
'TTG':'L',
        'TAC':'Y', 'TAT':'Y', 'TAA':'_',
'TAG':'_',
        'TGC':'C', 'TGT':'C', 'TGA':'_',
'TGG':'W',
    }
    protein =""
    iflen (seq)%3  =0:
        for i in range (0,len (seq ) ,  3) :
            codon = seq [ I : i +3 ]
            protein + = table[codon]
    return protein
defread_seq ( inputfile ) :
    With open (input file, "r") as f:
        seq = f.read()
    seq = seq.replace("\n", "")
    seq = seq.replace("\r", "")
    return seq
 prt = read_ seq ("amino_acid_sequence_
original.txt ")
dna = read_ seq ("DNA_sequence_original.txt ")
 p = translate (dna [20:935])
p = prt
```

DNA_ sequence

Attatatacgtaaataacatacctgctaaaaaccatagtaatttagaattcaaatcaa aaattatcattattataataaataaaataataagtgctcctattatactaatatcaaa-tataccaaaaaatgaaaccaatggatgggatacatagtaagaaaaatcaggaaatc-tatgagccatccattttccaactctattccatataccgtttttcatttcgttagctcttttttt-gacaaatacttagtattgtgaagtatatacttagatgagtcatttaaaacttgtctatt-gagttttaattctttttccaacttatacatttcaaaacttaaactggtagaaatacctttat-tttttataatttcgtttattatttttatggcagacgataatctaattacatatgcggataagg-taaaatcatatccccctgtgtacgaatacataaaattattattttcttgattgaacaatattc-

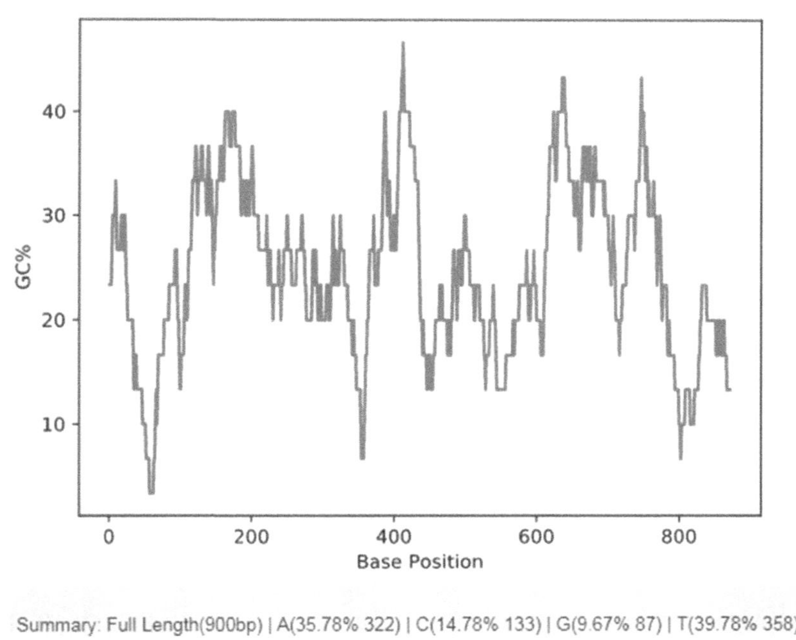

FIGURE 7.14 Base position and GC% (900 bp).

tagaattactattatgaaaagtttcacgtaattggaaaatgtctatattttttcttg-catttcaattataatgttatgtattgttattaaatcttttaatgtgttatcatcttctataa-caacaataaacttattctcagaatttttatatccgcatcggcatacgatttccataaac-taaagtgtttcgcaatagaactttttgcctcttttgtacaaagagcattacataatc-cagaaaaatattctgtaaaattttcaacacctcctgatttttctaccttttcccatata-aggaacttatatgataaactgatatctttttatctttaaaaaaaaaatttacatctgaatttt-taaaatcttttactgtgtcaactttttataaaatatatcattgtcacttttaat.

Protein concentration

$$PC = \left(\frac{\text{Absorbance at } \lambda_{\max}}{\text{Extinction Co effcient*Path length}} \right)$$

$$*(\text{Molecular weight})*\text{Dilution factor}$$

$$= \left(\frac{5}{15{,}960 \text{ M}^{-1}/\text{cm}^{-1} \text{ *3 cm}} \right) *(77{,}780)*7 = 56.85 \text{ mg/mL}$$

D DNA Concentration

$$\text{d DNA} = \left(\frac{\text{Absorbance at } \lambda_{\max}}{\text{Path length}} \right)$$

$$*(\text{Conversion factor})*\text{Dilution factor}$$

$$= \left(\frac{18}{3\text{cm}} \right) *50 \ \mu\text{g/mL}*5 = 1{,}500 \ \mu\text{g/mL}$$

Ss DNA Concentration

$$\left(\frac{18}{3\text{cm}} \right) *(\text{Conversion factor})*\text{Dilution factor}$$

$$= \left(\frac{18}{3 \text{ cm}} \right) *33 \ \mu\text{g/mL}*5 = 990 \ \mu\text{g/mL}$$

 # 5′3′ Frame 1

att ata tac gta aat aac ata cct gct aaa aac cat agt aat tta gaa ttc aaa tca aaa

I I Y V N N I P A K N H S N L E F K S K

att atc att att ata ata aat aaa ata ata agt gct cct att ata cta ata tca aat ata

I I I I I I N K I I S A P I I L I S N I

cca aaa aat gaa acc aat gga tgg gat aca tag taa gaa aaa tca gga aat cta tga gcc

P K N E T N G W D T - - E K S G N L - A

atc cat ttt cca act cta ttc cat ata ccg ttt ttc att tcg tta gct ctt ttt ttt gac

I H F P T L F H I P F F I S L A L F F D

aaa tac tta gta ttg tga agt ata tac tta gat gag tca ttt aaa act tgt cta ttg agt

K Y L V L - S I Y L D E S F K T C L L S

ttt aat tct ttt tcc aac tta tac att tca aaa ctt aaa ctg gta gaa ata cct tta ttt

F N S F S N L Y I S K L K L V E I P L F

ttt ata att tcg ttt att att ttt atg gca gac gat aat cta att aca tat gcg gat aag

F I I S F I I F M A D D N L I T Y A D K

gta aaa tca tat ccc cct gtg tac gaa tac ata aaa tta tta ttt tct tga ttg aac aat

V K S Y P P V Y E Y I K L L F S - L N N

att cta gaa tta cta tta tga aaa gtt tca cgt aat tgg aaa atg tct ata ttt ttt tct

I L E L L L - K V S R N W K M S I F F S

tgc att tca att ata atg tta tgt att gtt att aaa tct ttt aat gtg tta tca tct tct

C I S I I M L C I V I K S F N V L S S S

ata aca aca ata aac tta ttc tca gaa ttt ttt ata tcc gca tcg gca tac gat ttc cat

I T T I N L F S E F F I S A S A Y D F H

aaa cta aag tgt ttc gca ata gaa ctt ttt gcc tct ttt gta caa aga gca tta cat aat

K L K C F A I E L F A S F V Q R A L H N

cca gaa aaa tat tct gta aaa ttt tca aca cct cct gat ttt tct acc ttt tcc cat ata

P E K Y S V K F S T P P D F S T F S H I

agg aac tta tat gat aaa ctg ata tct ttt tta tct tta aaa aaa aaa ttt aca tct gaa

R N L Y D K L I S F L S L K K K F T S E

ttt tta aaa tct ttt act gtg tca act ttt tta taa aat ata tca ttg tca ctt ttt aat

F L K S F T V S T F L - N I S L S L F N

 # 5′3′ Frame 2

atta tat acg taa ata aca tac ctg cta aaa acc ata gta att tag aat tca aat caa aaa

L Y T - I T Y L L K T I V I - N S N Q K

tta tca tta tta taa taa ata aaa taa taa gtg ctc cta tta tac taa tat caa ata tac

L S L L - - I K - - V L L L Y - Y Q I Y

caa aaa atg aaa cca atg gat ggg ata cat agt aag aaa aat cag gaa atc tat gag cca

Q K M K P M D G I H S K K N Q E I Y E P

tcc att ttc caa ctc tat tcc ata tac cgt ttt tca ttt cgt tag ctc ttt ttt ttg aca

S I F Q L Y S I Y R F S F R - L F F L T

aat act tag tat tgt gaa gta tat act tag atg agt cat tta aaa ctt gtc tat tga gtt

N T - Y C E V Y T - M S H L K L V Y - V

tta att ctt ttt cca act tat aca ttt caa aac tta aac tgg tag aaa tac ctt tat ttt

L I L F P T Y T F Q N L N W - K Y L Y F

tta taa ttt cgt tta tta ttt tta tgg cag acg ata atc taa tta cat atg cgg ata agg

L - F R L L F L W Q T I I - L H M R I R

taa aat cat atc ccc ctg tgt acg aat aca taa aat tat tat ttt ctt gat tga aca ata

- N H I P L C T N T - N Y Y F L D - T I

ttc tag aat tac tat tat gaa aag ttt cac gta att gga aaa tgt cta tat ttt ttt ctt

F - N Y Y Y E K F H V I G K C L Y F F L

gca ttt caa tta taa tgt tat gta ttg tta tta aat ctt tta atg tgt tat cat ctt cta

A F Q L - C Y V L L L N L L M C Y H L L

taa caa caa taa act tat tct cag aat ttt tta tat ccg cat cgg cat acg att tcc ata

- Q Q - T Y S Q N F L Y P H R H T I S I

aac taa agt gtt tcg caa tag aac ttt ttg cct ctt ttg tac aaa gag cat tac ata atc

N - S V S Q - N F L P L L Y K E H Y I I

cag aaa aat att ctg taa aat ttt caa cac ctc ctg att ttt cta cct ttt ccc ata taa

Q K N I L - N F Q H L L I F L P F P I -

gga act tat atg ata aac tga tat ctt ttt tat ctt taa aaa aaa aat tta cat ctg aat

G T Y M I N - Y L F Y L - K K N L H L N

ttt taa aat ctt tta ctg tgt caa ctt ttt tat aaa ata tat cat tgt cac ttt tta at

F - N L L L C Q L F Y K I Y H C H F L

DNA sequence

**ATTATATACGTAAATAACATACCTGCTAAAAACCATAGTAATTTAGAATTCAAATCAAAAATTATCATTATTATAATAAATAAAATAATAA
GTGCTCCTATTATACTAATATCAAATATACCAAAAAATGAAACCAATGGATGGGATACATAGTAAGAAAAATCAGGAAATCTATGAGC
CATCCATTTTCCAACTCTATTCCATATACCGTTTTTCATTTCGTTAGCTCTTTTTTTTGACAAATACTTAGTATTGTGAAGTATATACT
TAGATGAGTCATTTAAAACTTGTCTATTGAGTTTTAATTCTTTTTCCAACTTATACATTTCAAAACTTAAACTGGTAGAAATACCTTT
ATTTTTTATAATTTCGTTTATTATTTTTATGGCAGACGATAATCTAATTACATATGCGGATAAGGTAAAATCATATCCCCCTGTGTACG
AATACATAAAATTATTATTTTCTTGATTGAACAATATTCTAGAATTACTATTATGAAAAGTTTCACGTAATTGGAAAATGTCTATATTT
TTTTCTTGCATTTCAATTATAATGTTATGTATTGTTATTAAATCTTTTAATGTGTTATCATCTTCTATAACAACAATAAACTTATTCTCA
GAATTTTTTATATCCGCATCGGCATACGATTTCCATAAACTAAAGTGTTTCGCAATAGAACTTTTTGCCTCTTTTGTACAAAGAGCA
TTACATAATCCAGAAAAATATTCTGTAAAATTTTCAACACCTCCTGATTTTTCTACCTTTTCCCATATAAGGAACTTATATGATAAACT
GATATCTTTTTTATCTTTAAAAAAAAAAATTTACATCTGAATTTTTAAAATCTTTTACTGTGTCAACTTTTTTATAAAATATATCATTGT
CACTTTTTAAT**

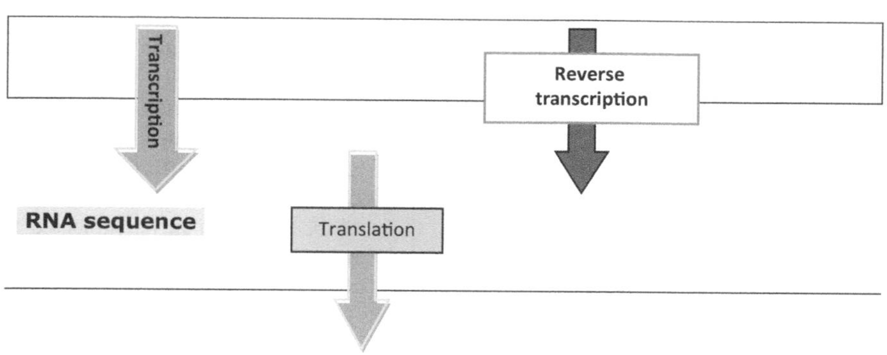

**AUUAUAUACGUAAAUAACAUACCUGCUAAAAACCAUAGUAAUUUAGAAUUCAAAUCAAAAAUUAUCAUUAUUAUAAUAAAUAAAAUA
AUAAGUGCUCCUAUUAUACUAAUAUCAAAUAUACCAAAAAAUGAAACCAAUGGAUGGGAUACAUAGUAAGAAAAAUCAGGAAAUC
UAUGAGCCAUCCAUUUUCCAACUCUAUUCCAUAUACCGUUUUUCAUUUCGUUAGCUCUUUUUUUUGACAAAUACUUAGUAUUGUG
AAGUAUAUACUUAGAUGAGUCAUUUAAAACUUGUCUAUUGAGUUUUAAUUCUUUUUCCAACUUAUACAUUUCAAAACUUAAACUG
GUAGAAAUACCUUUAUUUUUUAUAAUUUCGUUUAUUAUUUUUAUGGCAGACGAUAAUCUAAUUACAUAUGCGGAUAAGGUAAAAU
CAUAUCCCCCUGUGUACGAAUACAUAAAAUUAUUAUUUUCUUGAUUGAACAAUAUUCUAGAAUUACUAUUAUGAAAAGUUUCACG
UAAUUGGAAAAUGUCUAUAUUUUUUUCUUGCAUUUCAAUUAUAAUGUUAUGUAUUGUUAUUAAAUCUUUUAAUGUGUUAUCAUCU
UCUAUAACAACAAUAAACUUAUUCUCAGAAUUUUUUAUAUCCGCAUCGGCAUACGAUUUCCAUAAACUAAAGUGUUUCGCAAUAG
AACUUUUUGCCUCUUUUGUACAAAGAGCAUUACAUAAUCCAGAAAAAUAUUCUGUAAAAUUUUCAACACCUCCUGAUUUUUCUA
CCUUUUCCCAUAUAAGGAACUUAUAUGAUAAACUGAUAUCUUUUUUAUCUUUAAAAAAAAAAAUUUACAUCUGAAUUUUUAAAAUC
UUUUACUGUGUCAACUUUUUUAUAAAAUAUAUCAUUGUCACUUUUUAAU**

Translation (Start /stop codon)

Protein sequence

IIYVNNIPAKNHSNLEFKSKIIIIIINKIISAPIILISNIPKNETNGWDT..EKSGNL.AIHFPTLFHIPFFISLALFFDKYLVL.SIYLDESFKTC LLSFNSFSNLYISKLKLVEIPLFFIISFIIFMADDNLITYADKVKSYPPVYEYIKLLFS.LNNILELLL.KVSRNWKMSIFFSCISIIMLCIVI KSFNVLSSSITTINLFSEFFISASAYDFHKLKCFAIELFASFVQRALHNPEKYSVKFSTPPDFSTFSHIRNLYDKLISFLSLKKKFTSEF LKSFTVSTFL.NISLSLFN

A-ala, C-Cys D-Asp E-Glu F-Phe G-Gly H-His I –Ile K-Lys L-Leu M-Met N-Asn P-Pro Q –Gln R-Arg S –Ser T- Thr V-Val W-Trp Y-Tyr. Unused letters: BJOUXZ

7.21 Next-Generation Sequencing Analysis

Deoxyribonucleic acid (DNA) is the carrier of genetic material for all living organisms and many viruses. It is the most essential component of chromosomes and plays an important role in the development and functioning of organisms. It consists of four kinds of nucleotides: Adenine (A), Cytosine (C), Guanine (G), and Thymine (T) (Figure 7.15). The important role of DNA has led people to explore and research it. This has given rise to the rapid development of DNA sequencing technology. DNA sequencing is a process for determining the exact type and order of nucleotides in a fragment of the genome or the whole genome. Evolving from the traditional sequencing technology, the Sanger method to the currently widely used next-generation technologies (NGS) (Metzker, 2010; Mardis, 2013), and the next next-generation sequencing technologies (next-NGS), DNA sequencing technologies are rapidly developing and moving towards the direction of low-cost, high-speed, and high-accuracy methods. With the help of DNA sequencing, researchers can illuminate genetic information from a biological system. Deciphering DNA sequences is necessary for almost all branches of life sciences, and its understanding has grown exponentially in the past decades. The first-generation sequencing method, called Sanger sequencing, was developed by Edward Sanger in 1975. Sanger sequencing was considered the gold standard for DNA sequencing for around three decades. The first major breakthrough of first-generation sequencing was the Human Genome Project (HGP), a 13-year-long, $3 million project, completed in 2003. Due to inherent limitations in throughput, speed, scalability, and resolution of first-generation Sanger sequencing, the second-generation sequencing method, or Next-Generation Sequencing (NGS), was developed to meet the high demand for cheaper and faster sequencing technology. NGS, also known as Massive Parallel Sequencing, allows the complete genome (of humans) to be sequenced in less than a day. More recently, third-generation sequencing (TGS) has evolved (Figure 7.16) (Table 7.12).

Step 1: Sample extraction

Any sample (such as cell cultures, fresh-frozen tissues, formalin-fixed paraffin-embedded (FFPE) tissues, blood, saliva, and bone marrow) that yields DNA or RNA can be subjected to NGS. There are several extraction protocols that vary depending on

FIGURE 7.15 Distribution of GC content.

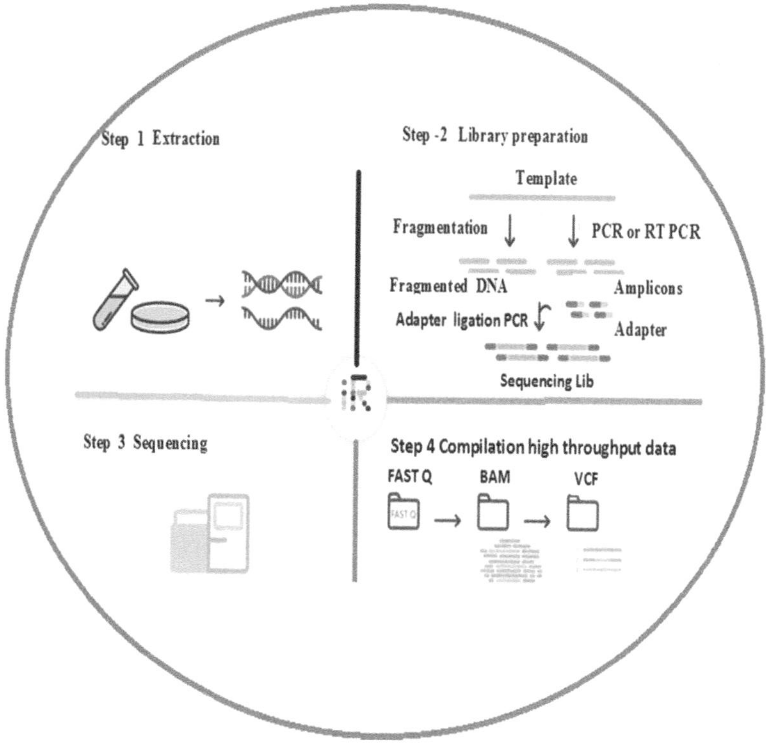

FIGURE 7.16 Flowchart of Next Generation Sequencing (NGS).

TABLE 7.12
Total and Relative Usage of Codons

Codons	Amino Acid	Count	Total Usage	Relative Usage	Codons	Amino Acid	Count	Total Usage	Relative Usage
TTT	F	28	9.33	82.35	GCT	A	3	1	27.27
TTC	F	6	2.00	17.65	GCC	A	2	0.67	18.18
TTA	L	19	6.33	51.35	GCA	A	5	1.67	45.45
TTG	L	4	1.33	10.81	GCG	A	1	0.33	9.09
CTT	L	4	1.33	10.81	TAT	Y	4	1.33	36.36
CTC	L	0	0.00	0.00	TAC	Y	7	2.33	63.64
CTA	L	8	2.67	21.62	TAA	*	2	0.67	28.57
CTG	L	2	0.67	5.41	TAG	*	1	0.33	14.29
ATT	I	16	5.33	39.02	CAT	H	6	2.00	100
ATC	I	2	0.67	4.88	CAC	H	0	0.00	0.00
ATA	I	23	7.67	56.1	CAA	Q	1	0.33	100
ATG	M	3	1.00	100	CAG	Q	0	0.00	0.00
GTT	V	2	0.67	18.18	AAT	N	15	5.00	71.43
GTC	V	0	0.00	0.00	AAC	N	6	2.00	28.57
GTA	V	6	2.00	54.55	AAA	K	23	7.67	92.00
GTG	V	3	1.00	27.27	AAG	K	2	0.67	8.00
TCT	S	14	4.67	36.84	GAT	D	7	2.33	77.78
TCC	S	3	1.00	7.89	GAC	D	2	0.67	22.22
TCA	S	14	4.67	36.84	GAA	E	10	3.33	90.91
TCG	S	3	1.00	7.89	GAG	E	1	0.33	9.09
CCT	P	6	2.00	54.55	TGT	C	3	1.00	75.00
CCC	P	1	0.33	9.09	TGC	C	1	0.33	25
CCA	P	3	1.00	27.27	TGA	*	4	1.33	57.14
CCG	P	1	0.33	9.09	TGG	W	2	0.67	100
ACT	T	4	1.33	33.33	CGT	R	1	0.33	33.33

(Continued)

TABLE 7.12 (*Continued*)

Total and Relative Usage of Codons

Codons	Amino Acid	Count	Total Usage	Relative Usage	Codons	Amino Acid	Count	Total Usage	Relative Usage
ACC	T	2	0.67	16.67	CGC	R	0	0.00	0.00
ACA	T	6	2.00	50.00	CGA	R	0	0.00	0.00
ACG	T	0	0.00	0.00	CGG	R	0	0.00	0.00
AGT	S	4	1.33	10.53	AGG	R	1	0.33	33.33
AGC	S	0	0.00	0.00	GGT	G	0	0.00	0.00
AGA	R	1	0.33	33.33	GGC	G	0	0.00	0.00
GGA	G	2	0.67	100	GGG	G	0	0.00	0.00

* STOP.

the starting material. In general, each extraction technique has been refined to produce the greatest quantity and best quality of nucleic acid from the appropriate sample type. Following extraction, the amount and quality of DNA or RNA should be determined, as high-quality starting material is critical for successful sequencing. The purity of a nucleic acid sample is typically expressed as an A260/280 value, which can be determined using a spectrophotometer such as Thermo Scientific's Nanodrop (Cat. No. ND-8000-GL). 'Pure' DNA generally reports an A260/280 reading of 1.8, whereas RNA is closer to 2.0.

Step 2: Library preparation

There are two main steps involved in creating a sequencing library from your RNA or DNA sample. The two steps involved are (i) amplification to produce a pool of target sequences of the right size, and (ii) the addition of sequencing adapters that will subsequently communicate with the NGS platform. If RNA serves as the initial template, reverse transcription must be performed to convert the RNA to cDNA.

PCR amplification yields a collection of specifically sized DNA fragments called a library that are compatible with the sequencing system to be used. The primers used in library preparation are designed based on the sequences of interest, which range from a whole genome to particular RNA transcripts.

The amplified DNA or cDNA fragments, known as amplicons, are effectively bookended by the adapter ligation step, which adds specific oligonucleotide sequences that will interact with a sequencing flow cell's surface. An additional barcode is ligated to the amplicon if more than one sample needs to be sequenced in a single sequencing run. It is possible to combine the finished libraries into a single sequencing run, which is subsequently 'demultiplexed' for data analysis.

Step 3: Sequencing DNA

Parallel sequencing is performed using an NGS platform. The library is loaded onto the sequencer which then 'reads' the nucleotides one by one. The number of reads produced will vary depending on the sequencing platform and kit used. Several methods of NGS have been developed, including pyrosequencing, sequencing by ligation (SOLiD), sequencing by synthesis (SBS – Illumina), and Ion Torrent sequencing. Ninety percent of the world's sequencing data comes from Illumina sequencing, which is by far the most popular (data per Illumina's website).

7.22 Illumina Platforms

There are various sequencing technologies, but all NGS platforms are capable of sequencing millions of tiny DNA fragments or cDNA. Some platforms produce more reads or varying read lengths than others. Illumina pioneered the most widely used and successful sequencing technology. Benchtop (iSeq, MiniSeq, MiSeq, NextSeq) and production-scale (NextSeq, HiSeq, and NovaSeq) sequencing platforms are among the instruments offered by Illumina. Illumina sequencers use a glass flowcell coated with millions of oligonucleotides that are complementary to the sequencing adaptors. Each library fragment hybridizes with the primers and is further amplified to generate millions to billions of clonal clusters. Then, fluorescently labeled nucleotides are used to synthesize a complementary strand for each fragment. After the addition of each tagged nucleotide, the flow cell is imaged, and the emission from each cluster is recorded. The fluorescent emission wavelength and intensity are used to identify the sequence of the templates (Figure 7.17).

Step 4: Alignment and data analysis

Following sequencing, specialized software is used to interpret the massive amount of data generated. First, the reads need to be filtered for quality, amplicon size, and agreement between paired ends. Next, the reads need to be assembled and aligned to a reference genome. finally, the reads, whether assembled or raw, can be compared to a reference sequence or to reads from another sample to identify variants based on disease state, etc. If reads are aligned with a reference genome, variant annotation can be used to associate variants with known genes or regulatory sequences.

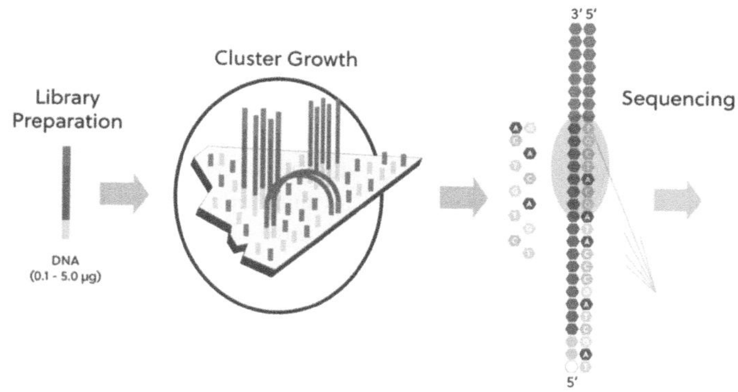

FIGURE 7.17 Flowchart of DNA sequencing.

7.23 Sequencing in Diagnostic/Clinical Applications

The application of sequencing to diagnostic and clinical settings has become more common as sequencing costs decrease. The data generated by NGS can be used to elucidate the mechanisms that cause diseases, support clinical diagnoses, guide therapeutic choices, and inform prognoses. Compared to other analytical methods, such as protein staining-based approaches, sequencing is more informative as it provides an array of details about the genome. A significant focus area for NGS is cancer treatment, as the genetic composition of a particular tumor can help oncologists predict which treatments are likely to be effective. NGS can identify tumor-specific variants such as copy number alterations, mutations, and gene expression changes. Sequencing panels, which include primers that directly target specific transcripts or genomic regions of interest, are often used in oncology to reduce costs and efficiently initiate genotype-directed targeted therapies. Based on sequencing results, patients may then undergo targeted therapy or immunotherapy.

7.24 The NGS's Capacities

Among the many biological study fields that have benefited from the strength, versatility, indispensability, and universality of NGS are the following:

High-throughput sequencing (HTS) of the human genome: HTS allows the identification of genes and regulatory pathways linked to disease.

Targeted sequencing: Sequencing of particular genes or genomic regions aids in the discovery of mutations that cause disease.

RNA-Seq (NGS of RNA): RNA-Seq provides all transcriptomic information of a sample without prior knowledge of an organism's genetic sequence. This capability facilitates quicker diagnoses and potential benefits for improved therapeutic decision-making for various genetic diseases, including many cancers.

Variant studies: By comparing DNA sequence data with a reference sequence to catalog differences, variant studies are common in medical genetics. The differences can range from single nucleotide polymorphisms (SNPs) to complex chromosomal rearrangements. RNA-Seq offers a robust alternative to microarrays for gene expression studies, enabling researchers to visualize RNA expression as sequences.

The size of data files is often in the magnitude of megabytes or gigabytes (up to 1,000 gigabase pairs, or Gb, in a single sequencing run) and is expected to increase further in the coming years. While sequencing costs have dropped precipitously and sequencing speed and efficiency have increased exponentially, the development of computational tools for preliminary analysis of these enormous datasets has lagged compared to data generation. As a result, there is an increasing demand for effective and user-friendly tools for high throughput next-generation sequencing (NGS) technologies.

7.25 The Generations of NGS Are Classified as Follows

1. First generation sequencing technology: (i) Sanger Sequencing
2. Second generation sequencing technology: (i) Roche -454, (ii) Illumina – GA II, (iii) SOLiD
3. Third generation sequencing technology: (i) Helicose, (ii) PacBio, (iii) Illumina = HiSeQ, MiSeq, (iv) Ion Torrent, (v) Oxford Nanopore

Next-generation sequencing technology is primarily offered by four companies in the commercial market: Roche/454, Illumina/Solexa, AB/SOLiD, and ABI/Ion Torrent. Roche 454 produces the longest reads and also the fastest method with high accuracy. Illumina GA offers a wide range of read lengths from 50 bp to 250 bp and high throughput with median running speed, while AB SOLiD utilizes the shortest read. A comparison of these methods is discussed by Mardis (2013). These massively parallel DNA sequencing technologies have been applied to transcriptome sequencing (RNA-Seq), de novo genome sequencing, and genome re-sequencing. RNA-Seq is a widely used approach for transcriptomic profiling. The Next-NGS of third-generation sequencing is a recent development. Helicos single-molecule sequencing technologies, SMRT technology from Pacific Biosciences and single-molecule nanopore sequencing technology from Oxford Nanopore Technologies are examples. This sequencing technology is moving towards high throughput, low cost, and long read length. That being said, this paper solely focuses on Illumina, the company that produces the most well-known databases.

7.26 NGS Read Types

DNA sequencing technology uses three different types of reads: mate-pair, paired-end, and single-end read. Single-end reads are the outcome of sequencing only one end of the DNA fragments, whereas paired-end reads and mate-pair reads obtain both ends of the fragments during sequencing. Paired-end sequencing differs from mate-pair sequencing in the techniques utilized to build the sequencing library and sequence the DNA fragment (Figure 7.18).

FASTA File Format: FASTA is a standard text-based format for sequencing. Each sequence contains two lines. The first line starts with a '>' character and is followed by the sequence identifier and/or description. The second line is the sequence containing A, C, G, T, or N (unknown base)

@sequence_id GATTCCTGTAAGCTTAAAGCTCCATTGT ACCCG ATATACGCCTTT

FASTQ File Format: The final sequence output is in another popular file format called FASTQ, even though the nucleotide is determined by gathering the fluorescence signal. The file format used by FASTQ is text-based. It includes every nucleotide sequence along with the quality scores that go with it. An illustration of the FASTQ format is presented next. A sequence is represented by four lines in the FASTQ format. The first line starts with the '@' character and contains the sequence identifier. The second line consists of the nucleotide sequence. The third line starts with the "+" character and may include an optional description. The fourth line provides the quality score (Figure 7.2). Each score corresponds to the quality of its matching base. As a result, the number of scores and the number of letters in the sequence must match. The following formula can be used to determine the quality score for the Illumina GA platform:

$$q_{\text{Solexa}} = -10 \log 10\left(\frac{P}{1-P}\right) \tag{7.44}$$

@sequence_id GATTCCTGTAAGCTTAAAGCTCCATT GTACCCG ATATACGCCTTT+&??#55CCFF%%>>>>>661 5%%+++***09@??=><<=++@@AB

NGS Assembly In sequence assembly, shorter DNA sequences with overlapping bases are combined to recreate the original DNA structure. Since the entire genome cannot be sequenced at once using current sequencing technologies, this process is essential. It is necessary to divide the entire genome into manageable chunks before sequencing it. De-novo assembly and mapping assembly are the two distinct forms of assembly: De-novo assembly assembles short reads to create longer sequences without relying on a reference genome. Mapping assembly assembles reads to an existing backbone sequence template and builds a sequence similar to the backbone. De-novo Assembly Example: There are four short sequence reads that overlap: S_1, S_2, S_3, and S_4. Together, these four reads can yield the longer sequence:

S_1: ACCTGTTA
S_2: TGTTACCA
S_3: ACCAGATA
S_4: ATACGCGG Counting ACCTGTTACCAGATACG CGG

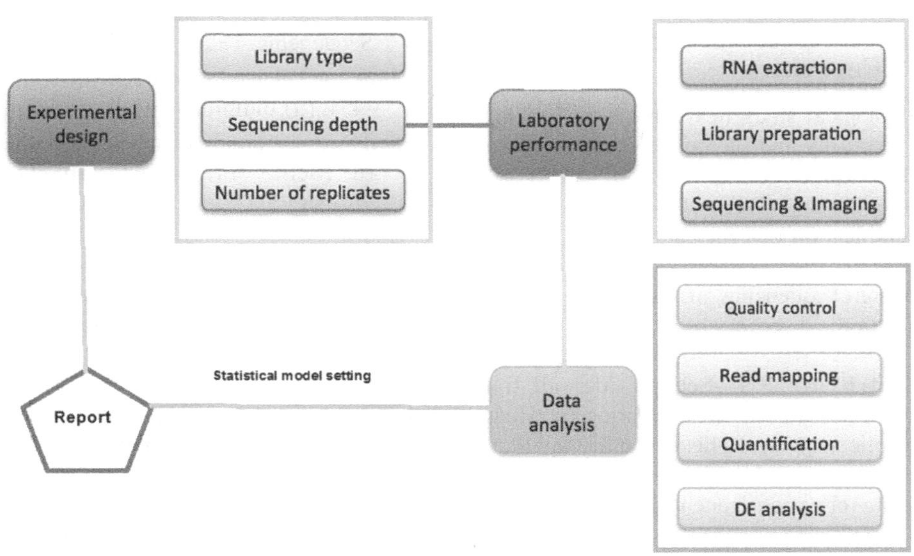

FIGURE 7.18 Workflow for RNA-seq.

Assembly Tools: There are several freely available assembly tools. Some of the important free assembly software are MIRA, SOAP de novo, ABySS, EULER, RAY and Commercial software includes: CLC, Newbler, etc.

7.27 NGS Sequence Alignment

Sequence alignment is used to compare the similarity of two sequences. Sequence alignment is theoretically based on Darwin's theory of evolution. It is hypothesized that two sequences with a high degree of similarity evolved from a common ancestor via nucleotide substitution, sequence fragmentation, deletion, and genetic variations. Sequence alignment is the process of joining two or more sequences so that the same nucleotide bases appear in the same column. Periodically, gaps are added to the sequence to maximize the alignment outcome. There are several alignment tools available, most of which make use of the Smith-Waterman and Needleman-Wunsch algorithms. The Needleman-Wunsch algorithm is a global alignment technique, while the Smith-Waterman algorithm is a local alignment method. Both of these algorithms are based on dynamic programming. Widely used alignment software includes BLAST, BWA, MOSAIK, BFAST, Bowtie, SOAP, and SSAHA. Prior to the thorough analysis of NGS deep sequencing data, which includes the investigation of alternative splicing and differential gene expression in RNA-Seq studies, it is essential to examine structural variants. This process involves handling a substantial volume of raw data, employing assembly and alignment algorithms to derive the sequence, and performing comparative analyses to address specific biological inquiries..

7.28 NGS Data Analysis

Figure 7.19 shows the classification of NGS genomic data analysis, which is based on data from studies related to genomics, transcriptomics, and epigenomics. Using NGS data at the genomic level, we search for structural variation, gene fusion, point mutations, and smallindels. Using NGS data at the transcriptomics level, we search for differential gene expression, RNA editing, and gene fusion. Using NGS data at the epigenomics level, we search for methylation, histone modification, and transcription factor binding. Figure 7.20 shows the general flow of analysis. The massive volume of raw data must be pre-processed, assembly/alignment algorithms should be used to obtain the sequence, and comparisons should be made to answer the question.

NGS Data Analysis The classification of NGS genomic data analysis is illustrated in Figure 7.4. This analysis incorporates data derived from genomics, transcriptomics, and epigenomics studies. At the genomic level, the focus of NGS data includes the identification of point mutations, small insertions and deletions (indels), copy number variations, and structural variations. In the realm of transcriptomics, we investigate gene fusions, alternative splicing events, RNA editing, and variations in gene expression. The epigenomics aspect of NGS data is employed to examine methylation patterns, histone modifications, and the binding of transcription factors. Figure 7.20 presents the overall workflow of the analysis. The sequence can only be derived after preprocessing the vast amounts of raw data, utilizing assembly and alignment algorithms to generate the sequence, and performing comparisons to address specific biological inquiries.

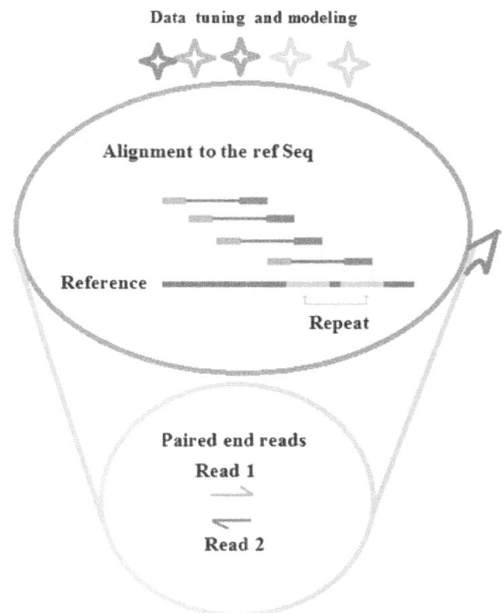

FIGURE 7.19 Paired-end sequencing and alignment.

Statistical models: Since NGS data are basically count data, only discrete models are used for analysis. The most important models are the Poisson, Inflated Poisson, Generalized Poisson, and Negative Binomial. Hidden Markov Models (HMM) and hierarchical HMM are also commonly used. However, NGS data presents significant challenges for data storage, analysis, and solutions. A typical NGS experiment produces gigabytes to terabytes of raw data, which presents challenges for data management and analysis. The full benefits of NGS sequencing will not become apparent until very high-performance computing and extensive bioinformatics support are available. Efficiently aligning short reads to a reference genome can be challenging, especially particularly challenging, especially when creating new algorithms to handle ambiguities. De novo assembly is even more difficult than reference-based assembly. Developing software that can discriminate between de novo or extremely rare mutations and low-frequency alleles derived from ancient ancestors is a complex task. Numerous programs are freely available, but custom programming might be required to adapt them to the specific requirements of an NGS project. For example, constructing ancestral genomes or analyzing rare mutations often requires additional programming effort. Since many programs are open source, modifications may be necessary to suit the needs of a particular project.

7.29 RNA Sequencing

RNA sequencing (RNA-seq) is the application of next-generation sequencing technologies to cDNA molecules. Reverse transcription is the method used to obtain information about

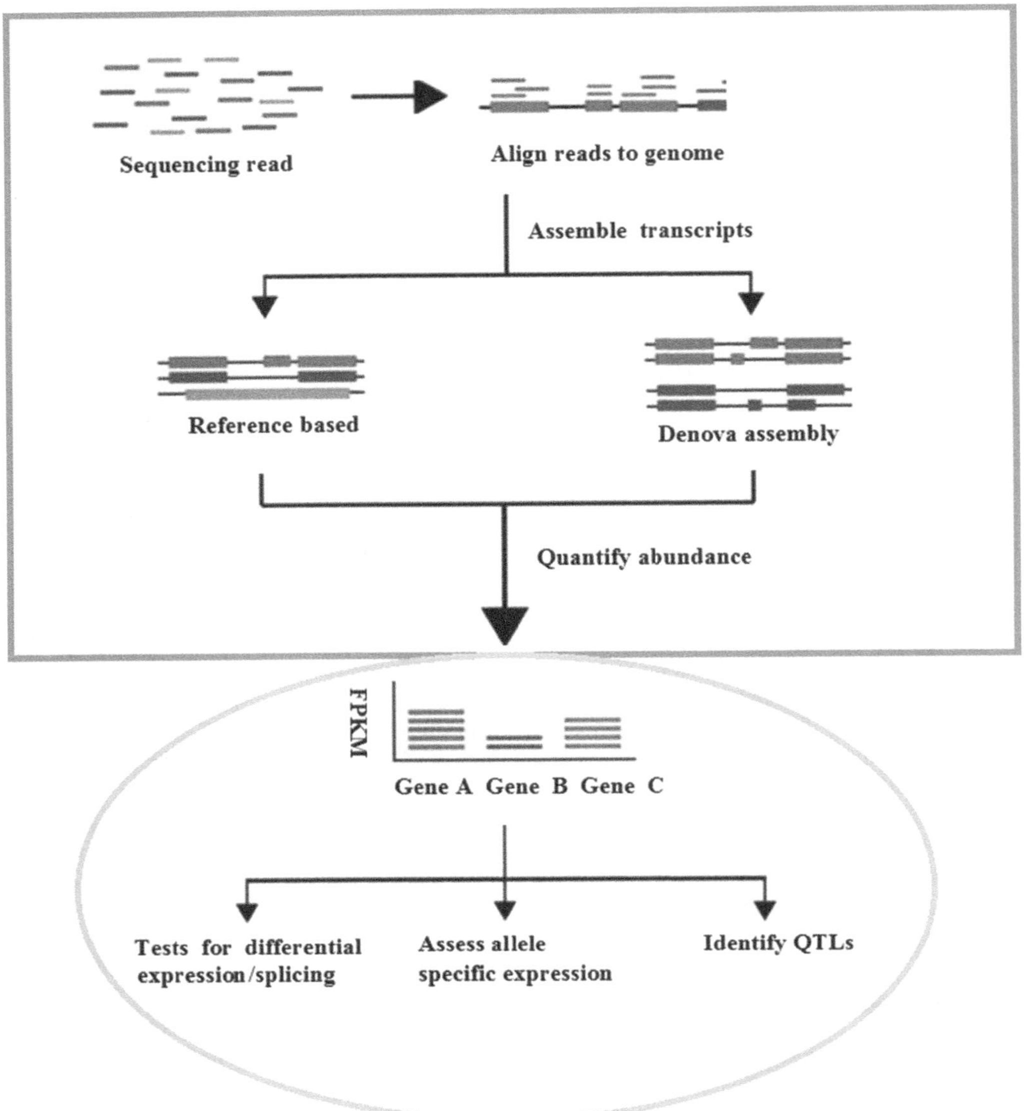

FIGURE 7.20 RNA-Sequencing Data Analysis Workflow.

a sample's RNA content. For this reason, cDNA molecules are created from RNA molecules through the RNA-seq series of experiments. The next steps are massively parallel deep sequencing and sequencing library construction (Figure 7.18).

The following factors need to be taken into account when designing an RNA-seq experiment (9):

- Building a library
- Depth of sequencing or library size
- Quantity of duplicates
- Getting the Library Ready

Library preparation methods vary depending on the type of RNA being used, whether strand specificity is necessary, and the type of reads. Selecting the population of RNA to be transformed into a cDNA library is a crucial step. Total RNA can be used to detect both mRNAs and non-coding RNAs, but additional enrichment steps (like ribosomal RNA depletion) might be necessary to detect low-abundance transcripts. Poly A+ RNA enrichment is helpful in eukaryotic organism studies and can be used to purify mRNAs (Figure 7.19) (Table 7.13).

While RNA-seq is a commonly employed method for transcriptome profiling, concerns regarding the effectiveness of various platforms and protocols are brought up by the rapid advancement of sequencing technologies and techniques. A multitude of factors can be used to explain variation in RNA-seq data, including the NGS platform utilized, the quality of the input RNA, and the experimenter themselves. To account for these sources of technical variability, many labs use positive controls, also known as "spike-ins," for sequencing libraries. A collection of common RNA synthetic spike-in standards for RNA-seq and microarray research was created by the External RNA Controls Consortium (ERCC). Approximately 2,800 bp of a vector containing 96 DNA

TABLE 7.13

RNA-Seq Library Protocols

RNA-Seq Library	Usage	Description
Poly-A selection	Sequencing mRNA	Selects for RNA species with a poly-A tail and enriches for mRNA
Ribo-depletion	Sequencing mRNA, pre-mRNA, ncRNA	Removes ribosomal RNA and enriches for mRNA, pre-mRNA, and ncRNA
Size selection	Sequencing miRNA	Selects RNA species using size fractionation by gel electrophoresis
Duplex-specific nuclease	Reduce highly abundant transcripts	Cleaves highly abundant transcripts, including rRNA and other highly expressed genes
Strand-specific	De novo transcriptome assembly	Preserves strand information of the transcript
Multiplexed	Sequencing multiple samples together	Uses genetic barcoding to enable sequencing of multiple samples together
Short-read	Higher coverage	Produces 50–100 bp reads, generally with higher read coverage and reduced error rate compared to long-read sequencing
Long-read	De novo transcriptome assembly	Produces >1,000 bp reads, advantageous for resolving splice junctions and repetitive regions

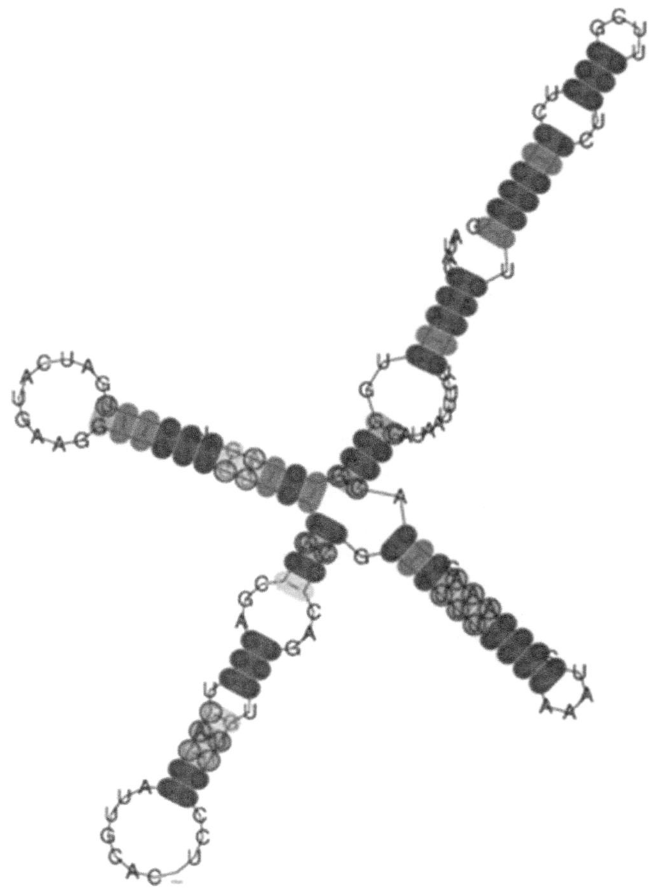

FIGURE 7.21 Biochemical structure of LSDV.

plasmids with standard sequences ranging from 273 to 2,022 bp make up the spike-ins. At varying concentrations, the spike-in standard sequences are incorporated into sequencing libraries to evaluate sensitivity, quantification, and coverage. When it comes to distinguishing biological variability from technical variability found in differential transcriptome profiling studies, these RNA standards are a useful quality control tool (Figures 7.20–7.22) (Tables 7.14–7.18).

7.29.1 Assumed Hypotheses

Factor *A*: H_0: $\mu_1 = .. = \mu_a$

There is no difference in the means of variable *A* categories.

Factor *B*: H_0: $\mu_1 = .. = \mu_b$

There is no difference in the means of variable *B* categories.

H_0: Interaction $(A_iB_j) = 0$ ($\forall$ i = 1 to a, $j = 1$ to b)

There is no interaction between variable *A* and variable *B*, i.e. for all the cells, the effect of variable *A* on the cells' means does not depend on the effect of variable *B*, and vice versa.

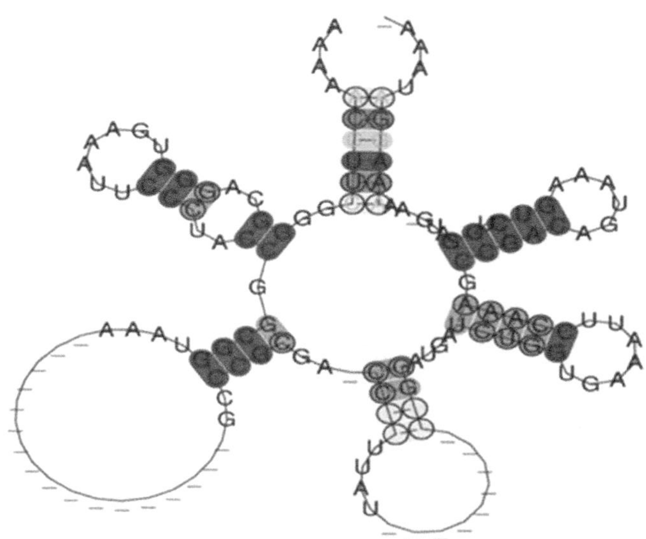

FIGURE 7.22 Biochemical structure of LSDV in relation to a long-standing mutant gene.

7.29.2 Assumptions

The following assumptions were made to test the hypothetical values: (i) The selected dependent variables are continuous (ratio or interval). (ii) Two or more categorical independent variables are considered. (iii) All independent observations are not repeated (no repeated measures). (iv) The residuals distribution is normal. (v) Lastly, there is homogeneity of variances, meaning a similar variance for each cell.

TABLE 7.14

UlsnRNA *Asellus aquaticus*—RNAz Statistics for Result Alignment

Attributes	*Asellus aquaticus*	*Bacillus cereus*
Mean pairwise identity	88.15	67.89
Shannon entropy	0.22	0.54
GC content	0.55	0.48
Mean single sequence minimum free energy	−61.72	−37.45
Consensus minimum free energy	−56.73	−24.16
Energy contribution	−56.13	−21.30
Covariance contribution	−0.6	−2.85
Combinations pair	1.25	1.69
Mean z-score	−1.11	−1.06
Structure conservation index	0.92	0.64
Background model	dinucleotide	dinucleotide
Decision model	structural RNA alignment quality	structural RNA alignment quality
SVM decision value	0.35	0.29
SVM class probability	0.709	0.671
Prediction	RNA	RNA

TABLE 7.15

Seq RNA Descriptive

Attributes	*Asellus aquaticus*	*Bacillus cereus*
Sequence Number	267	74
Effective Sequences	3.22	1.46
Consensus Length	164	103
Expected Maximum Hit-length	184	200
Base Pairs	45	24
Bifurcations	3	4
Model Type	"cm"	"cm"
Relative Entropy CM	0.59	0.59
Relative Entropy HMM	0.364	0.389

TABLE 7.16

RNA Code Statistics for Result Alignment: Asellus aquaticus

Hss	Frame	Length	From	To	Start	End	Score	p-value
0	2.0	18	25	42	74	127	17.515	0.323
1	1.0	10	8	17	22	51	13.821	0.659
2	−2.0	16	19	34	56	103	11.263	0.882
3	−3.0	8	38	45	114	137	8.437	0.991
4	−2.0	3	47	49	140	148	8.328	0.991
5	−1.0	6	9	14	25	42	8.257	0.992
6	−3.0	8	15	22	45	68	7.619	0.997
7	3.0	7	30	36	90	110	7.397	0.998
8	2.0	8	1	8	2	25	5.697	1.000
9	1.0	3	38	40	112	120	4.099	1.000

TABLE 7.17

RNA Code Statistics for Result Alignment *Bacillus cereus*

Hss	Frame	Length	From	To	Start	End	Score	p-value
0	1.0	15	5	19	13	57	7.995	0.818
1	−1.0	5	14	18	40	54	7.723	0.85
2	−3.0	9	5	13	15	41	6.298	0.965
3	3.0	5	35	39	105	119	5.292	0.994
4	−1.0	4	42	45	124	135	4.94	0.997
5	2.0	4	30	33	89	100	4.699	0.998
6	−3.0	9	39	47	117	143	4.227	1.0
7	−3.0	3	25	27	75	83	3.558	1.0
8	2.0	3	25	27	74	82	3.447	1.0
9	2.0	3	11	13	32	40	3.006	1.0
10	−2.0	3	36	38	107	115	2.972	1.0
11	−1.0	6	6	11	16	33	2.376	1.0

TABLE 7.18

Gene Distribution with Respect to Replications (Number of Genes)

Var *A*\Var *B*	A-1	A-2	A-3	B-1	B-2	B-3	Total
Gene 1	5.70	11.70	9.90	5.70	4.50	3.20	6.78
Gene 2	9.43	10.67	9.39	3.40	2.50	2.30	6.28
Gene 3	10.59	9.89	8.50	4.20	5.75	2.50	6.90
Gene 4	10.38	10.20	8.50	5.10	2.80	2.40	6.56
Gene 5	5.75	10.85	10.9	3.90	4.20	2.80	6.40
Gene 6	9.82	10.45	8.50	2.50	3.40	3.50	6.36
Gene 7	8.90	11.02	10.33	5.75	2.40	5.75	7.35
Gene 8	10.59	10.67	9.94	6.55	5.75	5.55	8.17
Gene 9	2.50	5.20	3.50	9.21	9.76	11.47	6.94
Gene 10	3.50	4.20	2.50	9.21	11.78	5.75	6.15
Gene 11	4.50	3.20	2.70	9.89	10.72	8.81	6.63
Gene 12	3.50	2.20	1.90	10.85	9.76	9.73	6.32
Gene 13	3.50	4.20	3.50	11.33	9.76	10.49	7.13
Gene 14	5.10	2.20	5.75	11.22	9.76	8.81	7.14
Total	6.69	7.61	6.84	7.05	6.63	5.93	6.79

7.29.3 Parameters

- a: The number of categories in variable A: number of rows.
- b: The number of categories in variable B; number of columns.
- n_i : Sample size of category i of variable A (row i).
- n_j : Sample size of category j of variable B (column j).
- $n_{i,j}$: Sample size of cell (i,j) (row i, column j). In balance: $n_{i,j}=n/(a*b)$
- n: Overall sample size, includes all the groups ($\Sigma n_{i,j}$, $i=1$ to a, $j=1$ to b).
- $\bar{Y}_i$: Average of all the observations of category i of variable A (row i).

$\bar{Y}_j$: Average of all the observations of category j of variable B (column j).

$\bar{Y}$: Overall average ($\Sigma Y_{i,j,k}/n$, $i=1$ to a, $j=1$ to b, $k=1$ to $n_{i,j}$).

Sum of squares

The sum of squares accumulates the squared differences related to the effect we try to estimate.

SS_A: The squared differences related to the effect of variable A. Compare the average of every category to the total average. The same value as the sum of squares between groups in one-way ANOVA.

SS_B: The same as SS_A, but for variable B.

SS_{AB}: The squared differences related to the effect of the combination of variable A and variable B in each cell. To understand the influence of the interaction AB, we consider the interaction of the specific value of variable A and the specific value of variable B. Take the average of each cell, remove the influence of variables A and B, and compare it to the total average.

A effect of (gene) $=\bar{Y}_i - \bar{Y}$

B effect (replications) $=\bar{Y}_j - \bar{Y}$

Interaction effect (AB effect) = Cell average - A effect - B effect - Total average.

$$=\bar{Y}_{i,j}-\left(\bar{Y}_i-\bar{Y}\right)-\left(\bar{Y}_j-\bar{Y}\right)-\bar{Y}.$$

$$=\bar{Y}_{i,j}-\bar{Y}_i-\bar{Y}_j+\bar{Y}.$$

Take the square of each difference

$$\bar{Y}_{i,j}-\bar{Y}_i-\bar{Y}_j+\bar{Y})^2.$$

Count the squared differences of each value in the cell; hence multiply by the sample size of each cell ($n_{i,j}$).

$$SS_{AB}=\Sigma_i^a\Sigma_i^b n_{i,j}\left(\bar{Y}_{i,j}-\bar{Y}_i-\bar{Y}_j+\bar{Y}\right)^2$$

Fixed and Random Effects (The fixed and random effects are related to the independent variables)

Fixed Effect (The effect is constant across individual genetic sequences)

The categories of the variable contain the entire category' list:

a. The effect of this variable is interesting; the difference between the categories is important.
b. There is no known pattern in the difference between the categories.
c. Random Effect (The effect varies across individual genetic sequences).
d. The category' list is only a sample from the entire category' list.
e. The effect of this variable is not interesting by itself: the difference between the categories is not important.
f. There is no known pattern in the difference between the categories (Tables 7.19 and 7.20).

TABLE 7.19

Marginal and Residual Differences of Seq RNA Groups

Variable *A*	Variable *B*	$Y_{i,j,k}$	Formula $n_{i,j}$ $(Y_{i,j,k}-\bar{Y}_i-\bar{Y}_j+\bar{Y})$	Residual
Gene 1	A-1	5.70	5.70−6.73−6.69+6.79	−0.98
Gene 1	A-2	11.70	11.70−6.78−7.61+6.79	4.09
Gene 1	A-3	9.90	9.90−6.7−6.84+6.79	3.06
Gene 1	B-1	5.70	5.70−6.78−7.05+6.79	−1.34
Gene 1	B-2	4.50	4.50−6.7−6.63+6.79	−2.11
Gene 1	B-3	3.20	3.20−6.78−5.93+6.79	−2.71
Gene 2	A-1	9.43	9.43−6.28−6.69+6.79	3.24
Gene 2	A-2	10.67	10.67−6.28−7.61+6.79	3.56
Gene 2	A-3	9.39	9.39−6.27−6.84+6.79	3.06
Gene 2	B-1	3.40	3.40−6.28−7.05+6.79	−3.14
Gene 2	B-2	2.50	2.50−6.28−6.63+6.79	−3.61
Gene 2	B-3	2.30	2.30−6.28−5.93+6.79	−3.11
Gene 3	A-1	10.59	10.59−6.90−6.69+6.79	3.78
Gene 3	A-2	9.89	9.89−6.90−7.61+6.79	2.16
Gene 3	A-3	8.50	8.50−6.90−6.84+6.79	1.54
Gene 3	B-1	4.20	4.20−6.90−7.05+6.79	−2.96
Gene 3	B-2	5.75	5.75−6.90−6.63+6.79	−0.98
Gene 3	B-3	2.50	2.50−6.90−5.93+6.79	−3.54
Gene 4	A-1	10.38	10.38−6.53−6.69+6.79	3.91
Gene 4	A-2	10.2	10.2−6.56−7.617+6.79	2.81
Gene 4	A-3	8.50	8.50−6.56−6.84+6.79	1.88

(Continued)

TABLE 7.19 (*Continued*)

Marginal and Residual Differences of Seq RNA Groups

Variable *A*	Variable *B*	$Y_{i,j,k}$	Formula $n_{i,j}$ $(Y_{i,j,k}-\bar{Y}_i-\bar{Y}_j+\bar{Y})$	Residual
Gene 4	B-1	5.10	5.10−6.53−7.05+6.79	−1.72
Gene 4	B-2	2.80	2.80−6.56−6.63+6.79	−3.59
Gene 4	B-3	2.40	2.40−6.56−5.93+6.79	−3.29
Gene 5	A-1	5.75	5.75−6.40−6.69+6.79	−0.55
Gene 5	A-2	10.85	10.85−6.40−7.61+6.79	3.62
Gene 5	A-3	10.90	10.9−6.40−6.84+6.79	4.45
Gene 5	B-1	3.90	3.90−6.40−7.05+6.79	−2.76
Gene 5	B-2	4.2	4.20−6.4−6.63+6.79	−2.03
Gene 5	B-3	2.8	2.80−6.4−5.93+6.79	−2.73
Gene 6	A-1	9.82	9.82−6.36−6.69+6.79	3.55
Gene 6	A-2	10.45	10.45−6.36−7.61+6.79	3.26
Gene 6	A-3	8.5	8.50−6.36−6.84+6.79	2.09
Gene 6	B-1	2.5	2.50−6.36−7.05+ 6.79	−4.12
Gene 6	B-2	3.4	3.40−6.36−6.63+6.79	−2.79
Gene 6	B-3	3.50	3.50−6.36−5.93+6.79	−1.99
Gene 7	A-1	8.90	8.9-7.35−6.69+6.79	1.64
Gene 7	A-2	11.02	11.02−7.33−7.61+6.79	2.84
Gene 7	A-3	10.33	10.33−7.35−6.84+6.79	2.92
Gene 7	B-1	5.75	5.75−7.35−7.05+6.79	−1.86
Gene 7	B-2	2.40	2.40−7.35−6.63+6.79	−4.79
Gene 7	B-3	5.75	5.75−7.35−5.93+6.79	−0.74
Gene 8	A-1	10.59	10.59−8.17−6.69+6.79	2.51
Gene 8	A-2	10.67	10.67−8.17−7.61+6.79	1.67
Gene 8	A-3	9.94	9.94−8.17−6.84+6.79	1.71
Gene 8	B-1	6.55	6.55−8.17−7.05+6.79	−1.88
Gene 8	B-2	5.75	5.75−8.17−6.63+6.79	−2.25
Gene 8	B-3	5.55	5.55−8.17−5.93+6.79	−1.76
Gene 9	A-1	2.50	2.50−6.94−6.69+6.79	−4.34
Gene 9	A-2	5.20	5.20−6.94−7.61+6.79	−2.56
Gene 9	A-3	3.50	3.50−6.94−6.84+6.79	−3.48
Gene 9	B-1	9.21	9.21−6.94−7.05+6.79	2.00
Gene 9	B-2	9.76	9.76-6.94−6.63+6.79	2.98
Gene 9	B-3	11.47	11.47−6.94−5.93+6.79	5.39
Gene 10	A-1	3.50	3.50−6.15−6.69+6.79	−2.55
Gene 10	A-2	4.20	4.20−6.15−7.61+6.79	−2.77
Gene 10	A-3	2.50	2.50−6.15−6.84+6.79	−3.70
Gene 10	B-1	9.21	9.21−6.15−7.05+6.79	2.79
Gene 10	B-2	11.78	11.78−6.15−6.63+6.79	5.78
Gene 10	B-3	5.75	5.75−6.15−5.93+6.79	0.45
Gene 11	A-1	4.50	4.50−6.63−6.69+6.79	−2.03
Gene 11	A-2	3.20	3.20−6.63−7.617+6.79	−4.25
Gene 11	A-3	2.70	2.70−6.63−6.84+6.79	−3.98
Gene 11	B-1	9.89	9.89−6.63−7.057+6.79	2.99
Gene 11	B-2	10.72	10.72−6.63−6.63+6.79	4.24
Gene 11	B-3	8.81	8.81−6.63−5.93+6.79	3.03
Gene 12	A-1	3.50	3.5−6.32−6.69+6.79	−2.72
Gene 12	A-2	2.20	2.2−6.32−7.61+6.79	−4.94
Gene 12	A-3	1.90	1.9−6.32−6.84+6.79	−4.47
Gene 12	B-1	10.85	10.85−6.32−7.05+6.79	4.26
Gene 12	B-2	9.76	9.76−6.32−6.63+6.79	3.60
Gene 12	B-3	9.73	9.73−6.32−5.93+6.79	4.27
Gene 13	A-1	3.50	3.50−7.13−6.69+6.79	−3.53
Gene 13	A-2	4.20	4.20−7.13−7.61+6.79	−3.75

(*Continued*)

TABLE 7.19 (*Continued*)

Marginal and Residual Differences of Seq RNA Groups

Variable *A*	Variable *B*	$Y_{i,j,k}$	Formula $n_{i,j}$ $(Y_{i,j,k}-\bar{Y}_i-\bar{Y}_j+\bar{Y})$	Residual
Gene 13	A-3	3.50	3.50−7.13−6.84+6.79	−3.67
Gene 13	B-1	11.33	11.33−7.13−7.05+6.79	3.93
Gene 13	B-2	9.76	9.76−7.13−6.63+6.79	2.79
Gene 13	B-3	10.49	10.49−7.13−5.93+6.79	4.22
Gene 14	A-1	5.10	5.10−7.14−6.69+6.79	−1.94
Gene 14	A-2	2.20	2.20−7.14−7.61+6.79	−5.76
Gene 14	A-3	5.75	5.75−7.14−6.84+6.79	−1.43
Gene 14	B-1	11.22	11.22−7.14−7.05+6.79	3.81
Gene 14	B-2	9.76	9.76−7.14−6.63+6.79	2.78
Gene 14	B-3	8.81	8.81−7.14–5.93+6.79	2.53

TABLE 7.20

ANOVA Table for Factor *A* and *B* Significance

Source	*df*	Sum of Square (SS)	Mean Square (MS)	*F* Statistic (df_1,df_2)	*P*-value
Factor *A* - rows (Gene)	13	22.813	1.7549	0.1332 (13,65)	0.99^{ns}
Factor *B* - columns (Reps)	5	21.394	4.2788	0.3247 (5,65)	0.89^{ns}
Error	65	856.507	13.177		
Total	**83**	**900.714**	**10.852**		

7.30 4 R-Code

A <- c(“Gene 1”,”Gene 2”,”Gene 3”,”Gene 4”,”Gene 5”,”Gene 6”,”Gene 7”,”Gene 8”,”Gene 9”,”Gene 10”,”Gene 11”,”Gene 12”,”Gene 13”,”Gene 14”,”Gene 1”,”Gene 2”,”Gene 3”,”Gene 4”,”Gene 5”,”Gene 6”,”Gene 7”,”Gene 8”,”Gene 9”,”Gene 10”,”Gene 11”,”Gene 12”,”Gene 13”,”Gene 14”,”Gene 1”,”Gene 2”,”Gene 3”,”Gene 4”,”Gene 5”,”Gene 6”,”Gene 7”,”Gene 8”,”Gene 9”,”Gene 10”,”Gene 11”,”Gene 12”,”Gene 13”,”Gene 14”,”Gene 1”,”Gene 2”,”Gene 3”,”Gene 4”,”Gene 5”,”Gene 6”,”Gene 7”,”Gene 8”,”Gene 9”,”Gene 10”,”Gene 11”,”Gene 12”,”Gene 13”,”Gene 14”,”Gene 1”,”Gene 2”,”Gene 3”,”Gene 4”,”Gene 5”,”Gene 6”,”Gene 7”,”Gene 8”,”Gene 9”,”Gene 10”,”Gene 11”,”Gene 12”,”Gene 13”,”Gene 14”,”Gene 1”,”Gene 2”,”Gene 3”,”Gene 4”,”Gene 5”,”Gene 6”,”Gene 7”,”Gene 8”,”Gene 9”,”Gene 10”,”Gene11”,”Gene 12”,”Gene 13”,”Gene 14”)

B <- c(“A-1”,”A-1”,”A-1”,”A-1”,”A-1”,”A-1”,”A-1”,”A-1”,”A-1”,”A-1”,”A-1”,”A-1”,”A-1”,”A-1”,”A-2”,”A-2”,”A-2”,”A-2”,”A-2”,”A-2”,”A-2”,”A-2”,”A-2”,”A-2”,”A-2”,”A-2”,”A-2”,”A-2”,”A-3”,”A-3”,”A-3”,”A-3”,”A-3”,”A-3”,”A-3”,”A-3”,”A-3”,”A-3”,”A-3”,”A-3”,”A-3”,”A-3”,”B-1”,”B-1”,”B-1”,”B-1”,”B-1”,”B-1”,”B-1”,”B-1”,”B-1”,”B-1”,”B-1”,”B-1”,”B-1”,”B-1”,”B-2”,”B-2”,”B-2”,”B-2”,”B-2”,”B-2”,”B-2”,”B-2”,”B-2”,”B-2”,”B-2”,”B-2”,”B-2”,”B-2”,”B-3”,”B-3”,”B-3”,”B-3”,”B-3”,”B-3”,”B-3”,”B-3”,”B-3”,”B-3”,”B-3”,”B-3”,”B-3”,”B-3”)

DV <- c(5.7,9.43,10.59,10.38,5.75,9.82,8.9,10.59,2.5,3.5,4.5,3.5,3.5,5.1,11.7,10.67,9.89,10.2,10.85,10.45,11.02,10.67,5.2,4.2,3.2,2.2,4.2,2.2,9.9,9.39,8.5,8.5,10.9,8.5,10.33,9.94,3.5,2.5,2.7,1.9,3.5,5.75,5.7,3.4,4.2,5.1,3.9,2.5,5.75,6.55,9.21,9.21,9.89,10.85,11.33,11.22,4.5,2.5,5.75,2.8,4.2,3.4,2.4,5.75,9.76,11.78,10.72,9.76,9.76,9.76,3.2,2.3,2.5,2.4,2.8,3.5,5.75,5.55,11.47,5.75,8.81,9.73,10.49,8.81)

ID<c(0,1,2,3,4,5,6,7,8,9,10,11,12,13,0,1,2,3,4,5,6,7,8,9,10,11,12,13,0,1,2,3,4,5,6,7,8,9,10,11,12,13,0,1,2,3,4,5,6,7,8,9,10,11,12,13,0,1,2,3,4,5,6,7,8,9,10,11,12,13,0,1,2,3,4,5,6,7,8,9,10,11,12,13)

df1 <- data. Frame(A, B, DV,ID)

Microarray and RNA-seq experiments can both benefit from the use of many techniques for visualizing and interpreting gene expression data. The most popular techniques are covered in the section below. Data on gene expression is frequently visualized using a heatmap (Figure 7.22). The heatmap can also be used in conjunction with clustering techniques, which group genes and/or samples according to how similar their patterns of gene expression are. Identifying frequently regulated genes or biological markers linked to a specific condition (such as an illness or an environmental factor) can be aided by this. Heatmaps show the data as a grid, with each row denoting a gene and each column denoting a sample. Gene expression variations, not absolute values, are represented by the color and intensity of the boxes. Red indicates up-regulated genes, and blue indicates down-regulated genes. A black expression indicates no change (Figure 7.23).

Gene set analysis, which is based on the functional annotation of differentially expressed genes, is a popular method for interpreting gene expression data (Figure 7.23). This helps determine whether the genes in the A-1 and A-2 groups that are differentially expressed are linked to a particular biological process or molecular function. For this, the Gene Ontology, which provides standardized annotations of gene products, is frequently used. For the gene list x-axis to work, every annotation (such as differentially expressed genes) must have its frequency compared to a reference list, which usually includes all the genes identified in the genome or on microarrays (Figure 7.24).

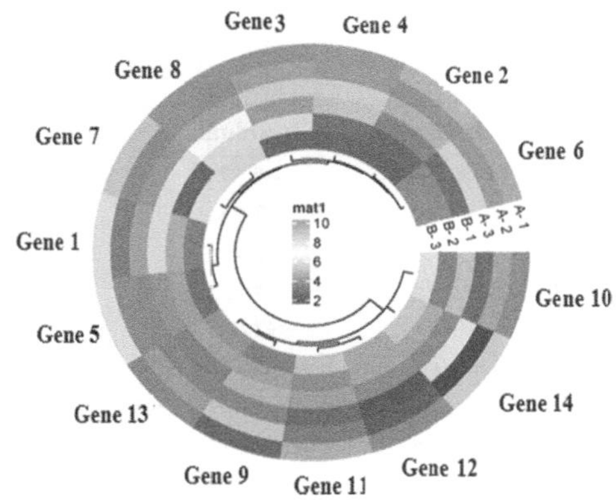

FIGURE 7.23 Heat maps where data is organized in a grid format.

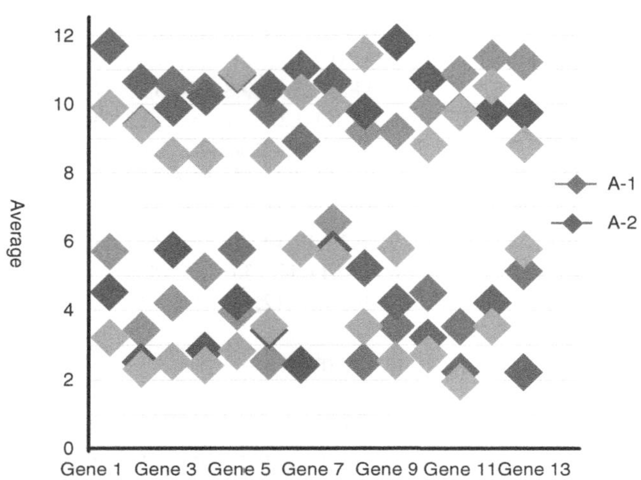

FIGURE 7.24 Average distribution of gene sequences.

To visually verify that your sequence data satisfies the homoscedasticity and normality assumptions of linear regression, utilize residual plots and Q-Q plots. If your data seem homoscedastic, you can observe this with a residual plot. This shows that the residuals, or the difference between the observed and predicted values of gene expression, are equal for every value of your predictor variable (Figure 7.23). If your data are homoscedastic, they will appear to be randomly distributed around the x-axis. If they are not (for instance, if they form a bowtie, fan, curve, etc.), the conclusions from your data are not supported. We should check if the data meet the normalcy assumption by using 'Q plots' and compare the distribution of data to a normal distribution by plotting the quartiles of the data against the normal distribution (Figure 7.25).

1. P-value

   The p-value equals **0.9999**, $(P\ (x \leq 0.1332) = 0.00014)$. This means that the chance of a type-I error, rejecting a correct H_0, is too high: 0.9999 (99.99%). The larger the p-value, the more it supports H_0.

m. Test statistic

   The test statistic F_A equals **0.1332**, which is in the 95% region of acceptance: [0, 1.8739].

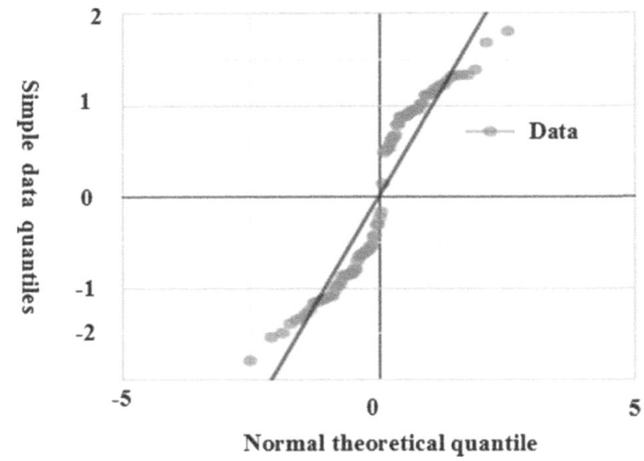

FIGURE 7.25 Q-Q plots for the residuals, allowing for a comparison with the normal distribution.

n. Effect size

   The observed effect size η^2 is small, 0.026. This indicates that the magnitude of the difference between the averages is small.

Factor - *B*

H_0 hypothesis

Since the p-value $> \alpha$, H_0 cannot be rejected.

The averages of all groups are assumed to be equal.

In other words, the difference between the sample averages of all groups is not big enough to be statistically significant.

A non-significant result cannot prove that H_0 is correct, only that the null assumption cannot be rejected.

1. P-value

   The p-value equals 0.8963, $(P(x \leq 0.3247) = 0.1037)$. This means that the chance of a type-I error, rejecting a correct H_0, is too high: 0.8963 (89.63%). The larger the p-value, the more it supports H_0.

o. Test statistic

   The test statistic F_B equals 0.3247, which is in the 95% region of acceptance: [0, 2.356].

p. Effect size

   The observed effect size η^2 is small, 0.024. This indicates that the magnitude of the difference between the averages is small (Tables 7.21–7.23).

7.31 Applications of RNA-Seq

Gene fusions to comprehend, for example, disease-causative variants in cancer, disease-associated single nucleotide polymorphisms (SNPs), and gene expression profiling across samples are just a few of the scientific questions that can be addressed with RNA-seq. It can also be used to investigate disease-related alternative splicing events, defined as the differential inclusion or exclusion of exons in the processed RNA product following the splicing of a precursor RNA segment.

TABLE 7.21

Compares the Mean Absolute Deviations (MAD), Similar to Comparing the Standard Deviations (Leven's Test)

Pair	Difference 'Δ'	SE	*Q*	CI-95% (Lower-Upper	Critical Mean	*P*-value
X_1–X_2	0.59	0.58	1.03	[−1.80–2.99]	2.39	0.98
X_1–X_3	0.14	0.58	0.24	[−2.25–2.53]	2.39	1.00
X_1–X_4	0.04	0.58	0.07	[−2.35–2.43]	2.39	1.00
X_1–X_5	0.33	0.58	0.57	[−2.06–2.72]	2.39	1.00
X_1–X_6	0.10	0.58	0.17	[−2.29–2.49]	2.39	1.00
X_2–X_3	0.45	0.58	0.78	[−1.94–2.85]	2.39	0.99
X_2–X_4	0.55	0.58	0.96	[−1.84–2.95]	2.39	0.98
X_2–X_5	0.27	0.58	0.46	[−2.13–2.66]	2.39	1.00
X_2–X_6	0.49	0.58	0.85	[−1.90–2.89]	2.39	0.99
X_3–X_4	0.10	0.58	0.17	[−2.29–2.49]	2.39	1.00
X_3–X_5	0.19	0.58	0.32	[−2.20–2.58]	2.39	1.00
X_3–X_6	0.04	0.58	0.07	[−2.35–2.43]	2.39	1.00
X_4–X_5	0.29	0.58	0.50	[−2.10–2.68]	2.39	1.00
X_4–X_6	0.06	0.58	0.10	[−2.33–2.45]	2.39	1.00
X_5–X_6	0.23	0.58	0.39	[−2.16–2.62]	2.39	0.90

TABLE 7.22

Group Correlation Matrix

Group	X_2	X_3	X_4	X_5	X_6
X_1	0.59	0.14	0.039	0.33	0.1
X_2	0	0.45	0.55	0.27	0.49
X_3	0.45	0	0.1	0.19	0.039
X_4	0.55	0.1	0	0.29	0.061
X_5	0.27	0.19	0.29	0	0.23

TABLE 7.23

Absolute Differences from the Group's Center

Group-1	Group-2	Group-3	Group-4	Group-5	Group-6
3.22	8.845	6.60	3.65	3.35	3.35
2.22	7.845	6.00	2.75	3.25	3.25
2.22	6.845	5.80	2.25	2.95	3.15
2.22	5.845	5.00	1.95	2.35	2.85
1.22	5.845	5.00	1.05	1.55	2.45
0.63	4.845	2.75	0.45	1.25	2.15
0.02	0.155	0.00	0.40	0.00	0.10
0.03	0.155	0.00	0.40	0.00	0.10
3.18	0.405	0.00	3.06	4.01	0.10
3.71	0.625	0.89	3.06	4.01	3.16
4.10	0.625	1.40	3.74	4.01	3.16
4.66	0.805	1.44	4.70	4.01	4.08
4.87	0.975	1.83	5.07	4.97	4.84
4.87	1.655	2.40	5.18	6.03	5.82

Furthermore, single-cell RNA-seq has gained popularity as a tool for researching cellular heterogeneity, diversity, and intricate biological processes in the fields of neuroscience and stem cell biology (Figure 7.25).

7.32 Advantages of RNA-Seq over Hybridization-Based Techniques

RNA-seq provides several advantages over hybridization-based approaches: A greater dynamic range of expression levels over which transcripts can be detected (>8,000-fold range) and higher sensitivity for genes expressed at low or very high levels are features of RNA-seq. It is also more reproducible and has less technical variation. RNA-seq is not constrained by prior understanding of the organism's genome. Moreover, RNA-seq is especially appealing for non-model organisms because it can be carried out in species for which genomes are not yet available. When it comes to transcriptional characteristics like newly transcribed regions, alternative splicing, and allele-specific expression, RNA-seq provides previously unheard-of detail (down to single-base resolution). Lastly, although it is commonly known that cross-hybridization bias affects microarrays, RNA-seq is thought to be impartial. However, a bias in the guanine-cytosine content of RNA-seq data has been noted in multiple studies, and RNA-seq mapping ambiguity can occur for paralogous sequences.

7.33 Conducting an RNA-Seq Investigation

There is no universal pipeline for all the different types of RNA-seq. Both the experimental design and the analysis procedures will differ depending on the organism being studied and the research goals. We describe a potential workflow for RNA-seq-based differential expression detection, as comparing gene expression across samples is a common goal of biological studies.

7.34 Application of Proteomics Knowledge to Statistics

The term proteome, which describes the entire set of proteins expressed in a given cell at a given time, was first used in 1994 by Australian postdoctoral fellow Marc Wilkins. Wilkins demonstrates this in the flowchart. The modification made to protein research is also highlighted. Proteomics is, broadly speaking, the study of the makeup, arrangement, dynamics, and function of the proteins that control the activity of individual cells. It is mainly divided into two categories: (i) interactomes, which encompass protein complexes and binary protein interactions, and (ii) the quantification of proteins via advanced proteomic techniques. The former pertains to the investigation of interactions between individual proteins. Quantifying proteins involves the rate of gene transcription, the rate of protein degradation, and the effectiveness of mRNA translation into proteins all influence the quantity of a particular protein in the cell at any given time.

7.34.1 Applications of Proteomics

- Manufacturing and production of potential new drugs for the treatment of diseases, e.g., diabetes, cancer.
- This depends on genomic information in the form of protein expression.
- Manufactured drugs to inactivate proteins involved in disease.
- Computer software used as a target for new drug discovery.
- Competitive molecules that inactivate the enzymes.
- Forms the basic new drug discovery tool (Figures 7.26 and 7.27).

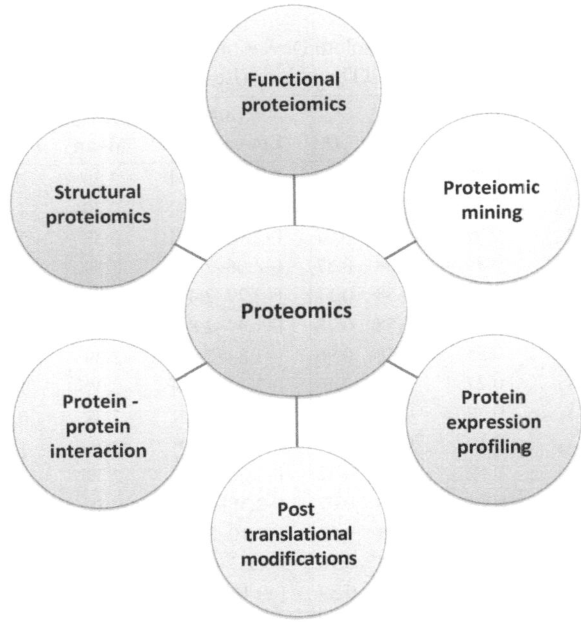

FIGURE 7.27 Classification of proteomics.

7.35 Subclassifications

- Proteomic mining: (i) Drug discovery, (ii) target identification/validation, (iii) differential display
- Protein profiling: (i) Medical microbiology, (ii) signal transduction, and (iii) disease mechanisms
- Post-translational modifications: (i) Glycosylation, (ii) phosphorylation, and (iii) proteolysis
- Protein interaction: (i) Yeast two-hybrids, (ii) coprecipitation, and (iii) phage display
- Structural proteomics: (i) Organelle composition, (ii) sub-proteome isolation, and (iii) protein complex
- Functional proteomics: (i) Yeast genomics, (ii) affinity-purified protein complex, and (iii) mouse knockouts

7.36 Biological Applications of Proteomics

- Proteome maps: Starting point for major studies in genomics.
- Questions of interest: How much of the genome is transcribed and translated in living organisms?
- What effect do different growth conditions have on the proteome?
- Studies are ongoing in eukaryotes like humans.
- Eukaryotes extensively modify their proteins by N- or C-terminal cleavages.
- They decorate them with sugars and/or phosphates, sulfates, and other PTMs.
- Yeast Saccharomyces cerevisiae.

7.37 Tracking Complexity

- Host-pathogen or host interactions—nitrogen fixation in legumes through association with bacteria to form nodules, infection of flax by flax rust.

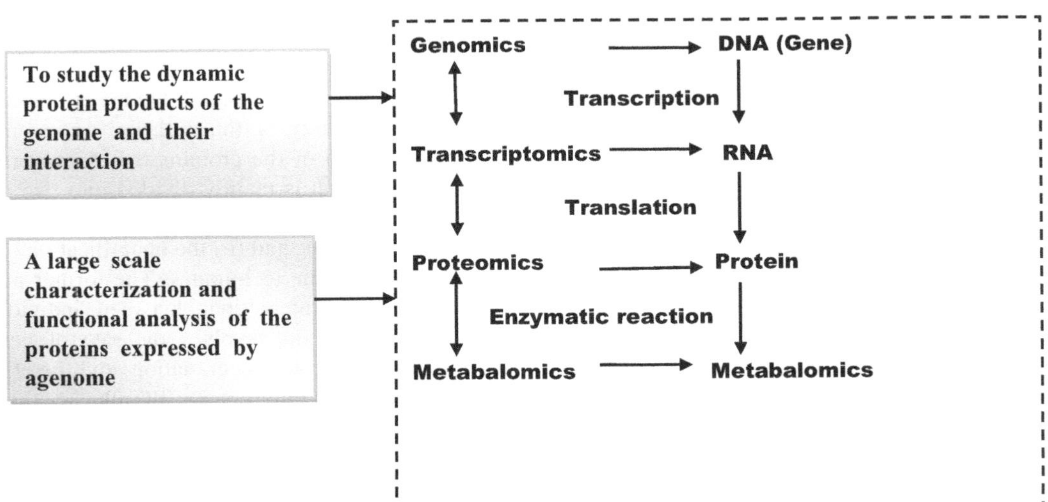

FIGURE 7.26 Flow shows the working steps of proteomics.

7.38 Immunogenic Proteins

- Identifying proteins from infectious disease agents recognized by the immune system.

7.39 Vaccine Candidate for Microbial Pathogens

- Allergy research to identify which grass pollens are immunogenic agents.
- Identification of allergens (proteins) in latex (gloves) through a 2D PAGE run using latex as a sample.

7.40 Improvements in Agricultural Products

- Engineering resistance to pathogens/parasites into various plants; most of these resistance mechanisms involve the expression of toxic or protective proteins
- Discovery of new toxic or protective proteins
- Wool proteome project to investigate economically important characteristics like color and fiber strength
- Remanufacturing low-value products—proteinaceous whey as a by-product of cheese manufacture, and also investigating whether this whey can be used to grow recombinant bacteria for biotechnological production

7.41 Microarrays

Microarrays consist of a set of DNA probes that are typically attached to a solid surface, like a glass slide, at specific locations so that sample DNA fragments can be hybridized with them. Typically, the probes consist of oligonucleotides that are either synthesized in situ (Affymetrix) or "ink-jet printed" onto slides (Agilent). Labeled single-stranded DNA or antisense RNA fragments from a sample of interest are hybridized to the DNA microarray under high stringency conditions. The number of nucleic acid fragments in the sample correlates with the amount of hybridization detected for a given probe.

7.42 One-Color and Two-Color Arrays

Whether to measure the expression levels from each sample on separate microarrays (one-color array) or compare the relative expression levels between a pair of samples on a single microarray (two-color array) is a crucial design decision in microarray experiments (Figures 7.27 and 7.29). One-color and two-color arrays perform similarly overall.

In two-color microarrays, two biological samples (experimental/test sample and control/reference sample) are labeled with different fluorescent dyes, usually Cyanine 3 (Cy3) and Cyanine 5 (Cy5). Equal amounts of labeled cDNA are then simultaneously hybridized to the same microarray chip. Following this competitive hybridization, the fluorescence measurements are performed independently for each dye and show the relative

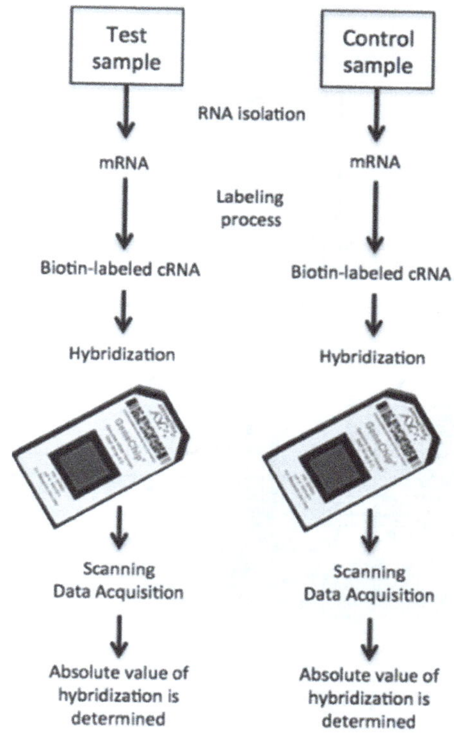

FIGURE 7.28 Process of RNA isolation.

abundance of each gene in the test sample (Cy5) and control sample (Cy3). The ratio of the fluorescent signals at each probe, Cy5/Cy3, is used to report the hybridization data. In contrast, each sample is labeled and hybridized to a unique microarray in single-color microarrays, producing an absolute fluorescence value for each probe (Figures 7.28 and 7.29).

7.43 Drawbacks with Microarrays

Although hybridization-based methods have high throughput and are reasonably priced, they have a number of drawbacks, including

- Dependence on currently available genome sequence knowledge
- Elevated background concentrations as a result of hybridization
- The dynamic range of detection constrained by background and saturation signals

*Comparing expression levels across different experiments is often difficult and can require complicated normalization methods.

7.44 Statistical Modeling for Testing SAM (Significance Analysis of Microarray)

$$d_i = \frac{r_i}{s_r + s_0}; i = 1, 2 \ldots p \quad (7.45)$$

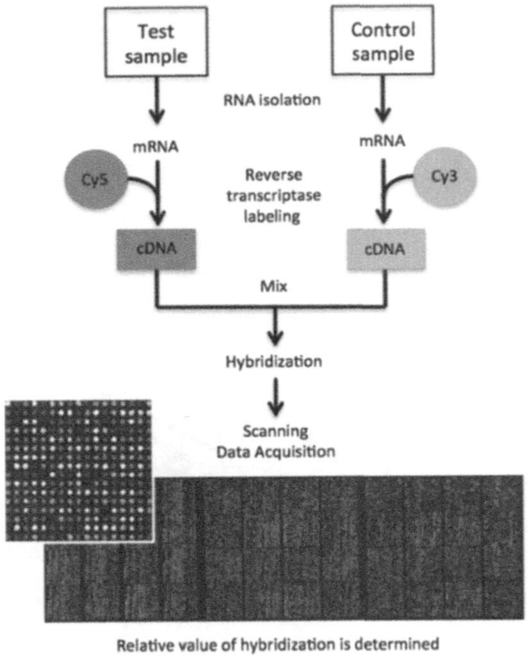

FIGURE 7.29 Liquid-phase separation technique designed to extract pure RNA.

where d_i = Test statistic calculating the relative difference in gene expression

r = is the linear regression coefficient of gene ith, s_r = SE of r, $s_0 = s$ is computed as a percentile based on alpha

$$r_i = \frac{\sum_j y_j \left(x_{ij} - x_i\right)}{\sum_j \left(y_j - \hat{y}_i\right)^2} \quad x_{ij} = \sum_j x_{ij}/n \tag{7.46}$$

$$s_r = \frac{\widehat{\sigma_i}}{\left[\sum_j \left(y_j - \hat{y}_i\right)^2\right]^{1/2}} \tag{7.47}$$

$$\widehat{\sigma_i} = \sqrt{\frac{\sum_j \left(x_{ij} - x_{ij}\right)^2}{n-2}} \tag{7.48}$$

The s_0o constant is chosen to minimize the coefficient of variation of d_i. r_i is equal to the expression levels (x) for gene i under y experimental conditions.

$$\text{False discovery rate} = \left(\frac{\begin{array}{c}\text{Median}(\text{or } 90 \text{ th prcentile})\text{of}\\ \#\text{ of falsely called genes}\end{array}}{\text{Number of genes called significant}}\right) \tag{7.49}$$

Fold changes (t) are specified to guarantee that genes called significant change by at least a pre-specified amount. This means that the absolute value of the average expression levels of a gene under each of two conditions must be greater than the fold change (t) to be called positive and less than the inverse of the fold change (t) to be called negative.

7.45 The SAM Algorithms

1. Order test statistics according to magnitude
2. Compute the ordered null (unaffected) scores for each permutation.
3. Plot the expected null scores against the ordered test statistic.
4. If the difference between the mean test statistic and the absolute value of the test statistic for each gene exceeds a specified threshold, consider each gene to be significant.
5. Determine the false discovery rate by comparing the observed and expected values.

7.46 Significant Gene Sets

Positive gene set: Greater values of the phenotype y are correlated with higher expression of the majority of the genes in the gene set.

Negative gene set: Most genes in the gene set are expressed less frequently when the phenotype 'y' is higher (Figure 7.30).

7.47 Bioinformatics Pairwise Protein Algorithms

A pairwise algorithm is an algorithmic technique with its origins in dynamic programming. A protein profile, which is a residue scoring matrix for one or more aligned sequences, can be compared against a DNA strand's three translation frames using pairwise algorithms, which enables frameshifting. Pairwise is the most unique protein-DNA alignment tool available because it allows frameshifting to occur during alignment.

7.48 Frameshifting

It is the phenomenon where multiple translation frames exist in a single DNA strand. Using one of the three frames to convert the DNA into a protein sequence, standard Protein-DNA alignment tools compare the resultant sequence to the target protein. The assumption behind this type of alignment is that the translation frame of the entire DNA strand stays intact. A variation of the Smith–Waterman best local alignment algorithm is the Pair-Wise algorithm. All of these algorithms are members of the minimal string edit algorithm class. The primary distinction between Pairwise and other alignment

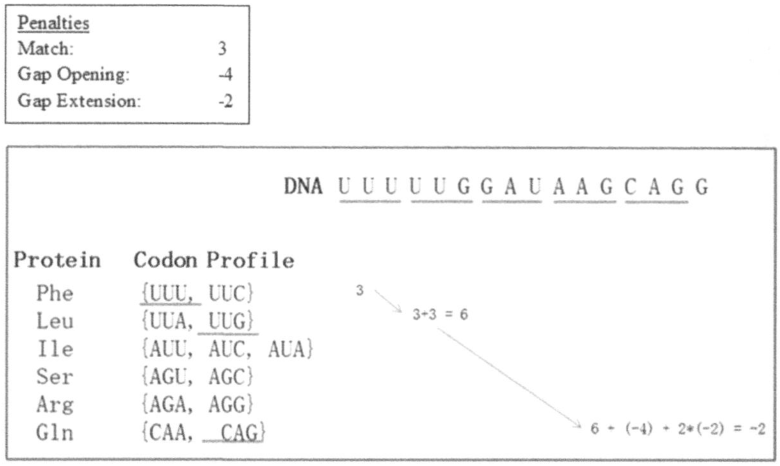

FIGURE 7.30 Alignment algorithms for proteins.

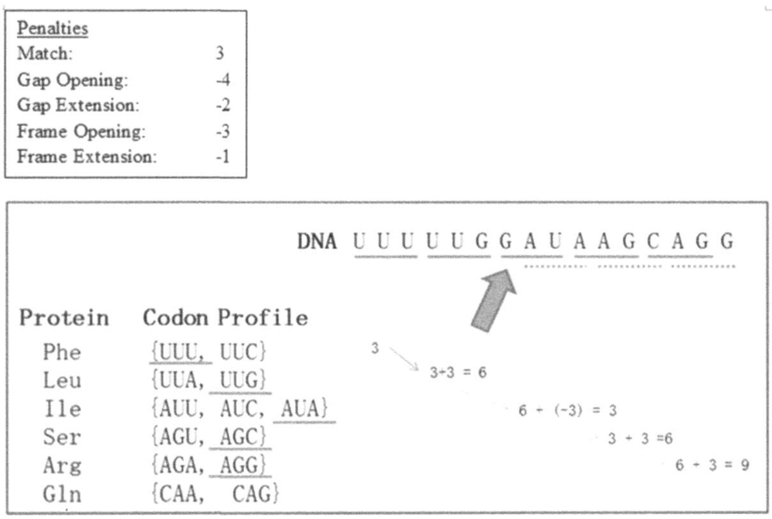

FIGURE 7.31 Shift in the translation frame from the first frame to the second.

algorithms is that Pairwise introduced two new penalties, Frame Opening Penalty (FOP) and Frame Extension Penalty (FEP), in addition to the standard penalties: Match, Gap Opening Penalty (GOP), and Gap Extension Penalty (GEP).

Illustration 1

The alignment result of one protein sequence and one DNA sequence using the standard protein-DNA alignment algorithm is shown in the figure. Frame 1 of the DNA sequence was used. As the image illustrates, two amino acids, or six nucleic acids, were not aligned, which resulted in a low overall score of −2 Figure 7.30.

Illustration 2

Figure 7.2 illustrates the aligned result using Pairwise. The penalties have been adjusted as follows while maintaining the same DNA and protein sequence. The location of the frameshifting event is indicated by the arrow. The translation frame was moved from frame one to frame two at nucleotide (G) (dotted line). This modification produced a much improved alignment with a score of (Figure 7.31).

7.49 Protein SWIS Modeling of Lumpy Skin Disease

Homology modeling, also known as comparative modeling of proteins, refers to constructing an atomic-resolution model of the ‘target’ protein from its amino acid sequence and an experimental three-dimensional structure of a related homologous protein (Figures 7.32 and 7.33, Tables 7.24–7.26).

Oligo-State Monomer: (matching prediction)
GMQE: 0.04
Q Mean D is Co Global: 0.43 ± 0.12
Q Mean D is Co Local

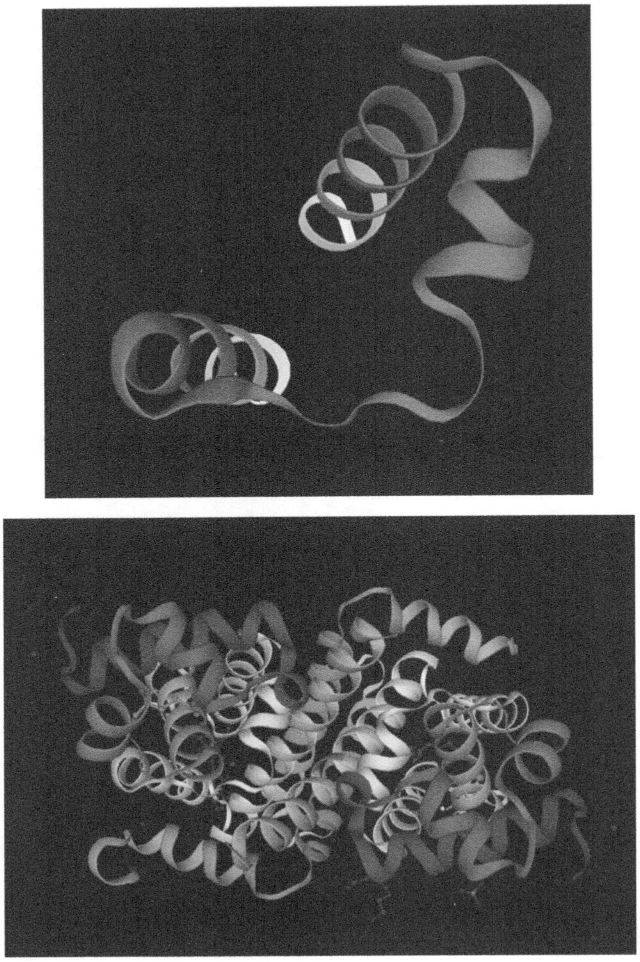

FIGURE 7.32 PDF related to Lumpy Skin Disease in affected animals.

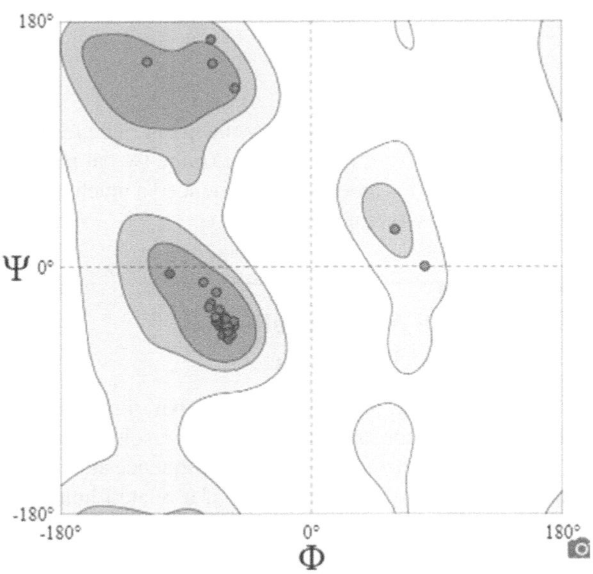

FIGURE 7.33 Ramachandran plots associated with Lumpy Skin Disease.

TABLE 7.24

Mol Probity Score

Mol Probity Score	1.16
Clash Score	1.29
Ramachandran Favored	95.56%
Ramachandran Outliers	0.00%
Rotamer Outliers	0.00%
C-Beta Deviations	1
B237 LYS	
Bad Bonds	2/402
B253 HIS, B233 HIS	
Bad Angles	4/543
B253 HIS, (B231 ALA-B232 LEU), B233 HIS	

TABLE 7.25

Ligand Model Results

Biounit Oligo State	Homo-Dimer
QSQE	0.05
Method	X-ray, 1.75Å
Seq Similarity	0.29
Coverage	0.16
Range	217–263

TABLE 7.26

Heuristic Model

Template	Sequence Identity	Biounit Oligo State	Description
7sqk.1	16.98	Hetero-octamer	Haus augmin-like complex subunit 2, Cryo-EM structure of the human augmin complex
5fkk.1	17.02	homo-dimer	Tetracycline Repressor, Class D, TetR(D) N82A mutant in complex with anhydrotetracycline and magnesium
5fkk.1	17.02	Homo-dimer	Tetracycline Repressor, Class D, TetR(D) N82A mutant in complex with anhydrotetracycline and magnesium
4ac0.1	24.00	Homo-dimer	Tetracycline repressor protein, class b from transposon TN1 0 TETR(B) in complex with minocycline and magnesium
7q93.1	20.00	Homo-dimer	NADQ transcription factor: Crystal Structure of Agrobacterium tumefaciens NADQ, NAD complex

Q Mean Z-Scores
Template
5fkk.1.B Tetracycline repressor, Class D
TetR(D) N82A mutant in complex with anhydrotetracycline and magnesium
Seq Identity: 17.02% Coverage (Figures 7.34–7.36)

Ligand	Added to Model	Description
CL	× - Not biologically relevant.	CHLORIDE ION
CL	× - Not biologically relevant.	CHLORIDE ION
CL	× - Not biologically relevant.	CHLORIDE ION
CL	× - Not biologically relevant.	CHLORIDE ION
CL	× - Not biologically relevant.	CHLORIDE ION
CL	× - Not biologically relevant.	CHLORIDE ION
CL	× - Not biologically relevant.	CHLORIDE ION
CL	× - Not biologically relevant.	CHLORIDE ION
CL	× - Not biologically relevant.	CHLORIDE ION
MG	× - Not in contact with model.	MAGNESIUM ION
PG4	× - Not biologically relevant.	TETRAETHYLENE GLYCOL
PGE	× - Not biologically relevant.	TRIETHYLENE GLYCOL
SO4	× - Not biologically relevant.	SULFATE ION
TDC	× - Binding site not conserved.	5A,6-ANHYDRO TETRACYCLINE

Model-Template Alignment
A total of 26 templates were found to match the target sequence. This list was filtered by a heuristic down to 18. The top templates are:

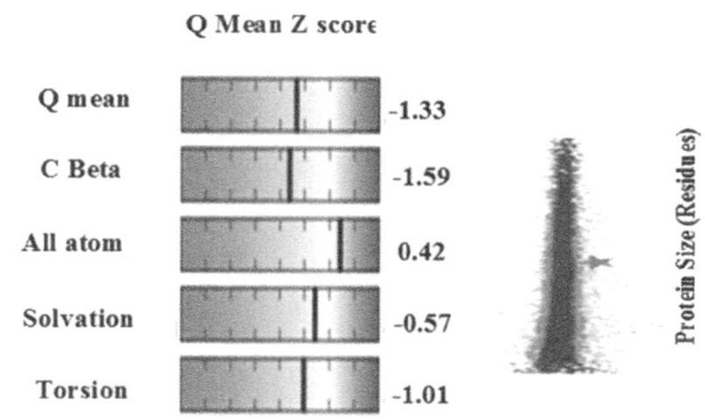

FIGURE 7.34 Q mean Z-scores, where the Q mean Z-score or normalized Q mean score.

7.50 Statistical Applications

Statistical models represent a widely accepted method for clustering noisy gene expression data, which is essential for investigating gene function. Given that gene expression data derived from microarray experiments frequently encounter substantial technical and biological noise, researchers often rely on replicated experiments to address data variability. Internal replication, such as utilizing multiple probes per gene within a single experiment, offers critical insights into the technical noise present. Nevertheless, existing implementations of mixture models either fail to account for the correlation among replicated measurements under the same experimental conditions or neglect the measurement error at the probe level, thereby missing out on valuable information regarding technical noise. The proposed descriptive tools

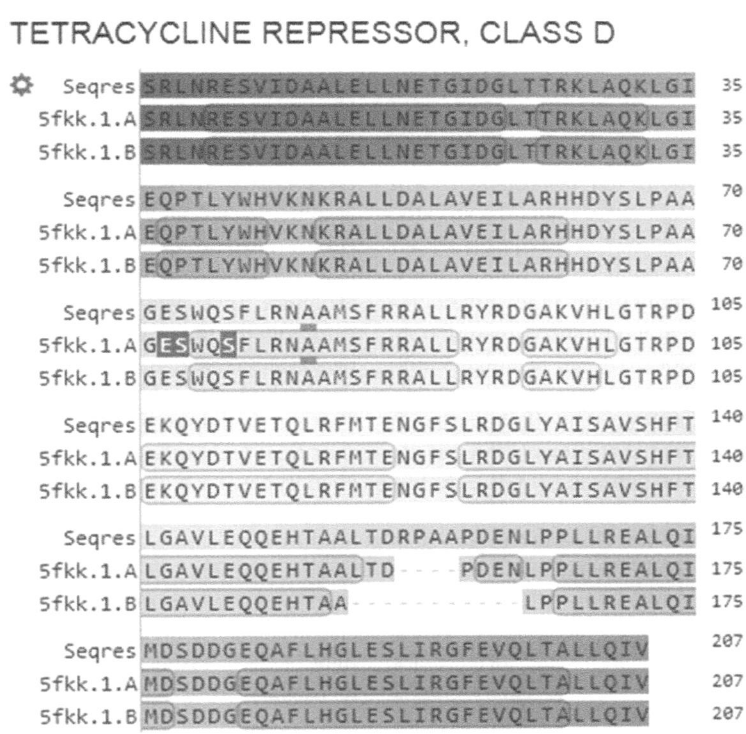

FIGURE 7.35 Protein sequences of the Tetracycline Repressor Class D (TetR).

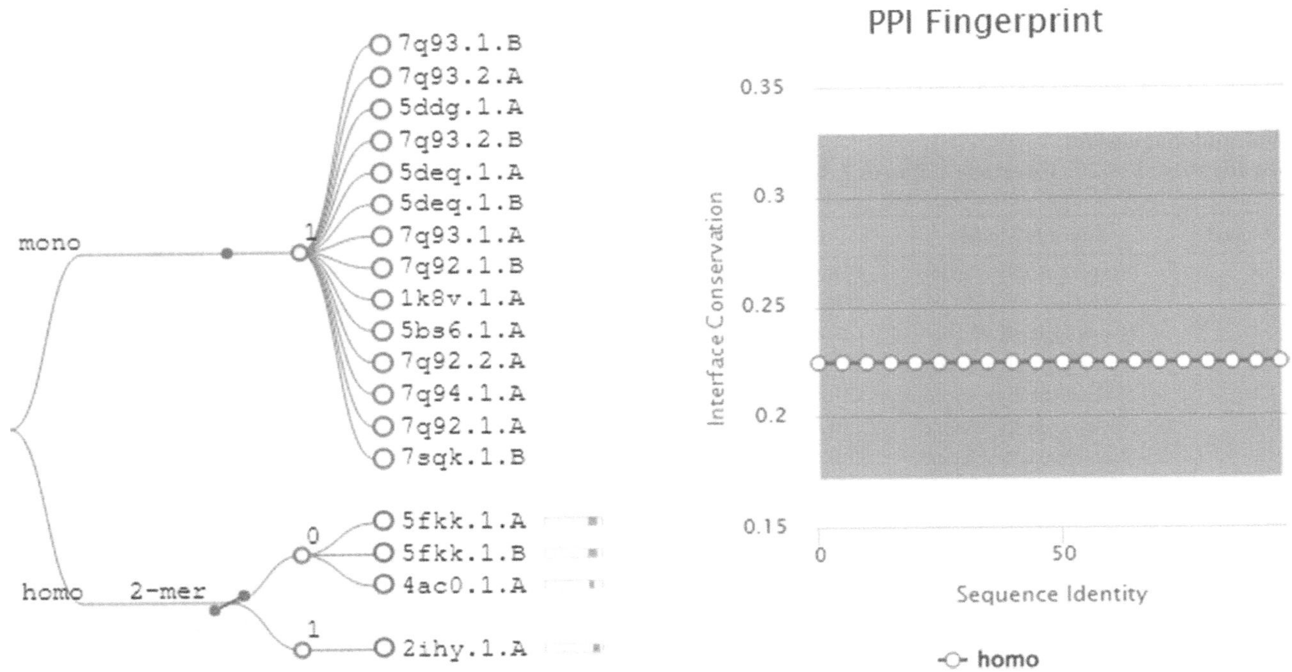

FIGURE 7.36 Oligomeric templates (7sqk.1.B HAUS).

are expected to demonstrate robust performance on large genomic datasets characterized by realistic noise, particularly when compared to standard Gaussian mixture models. Additionally, our framework incorporates a Bayesian latent model for observations. We will draw inferences using a contemporary Markov Chain Monte Carlo (MCMC) method for Bayesian model selection. Through these techniques, we will evaluate the latent effects in extensive genomic analyses, with model simulations indicating that our approach is preferable to conventional statistical methods, including *k*-means, hierarchical clustering, and various machine learning projection models. Furthermore, our fitted models will extend to the examination of pre-genomic data, enabling the identification of adaptive alleles—a task that requires meticulous integration of genetics, molecular biology, physiology, and ecology. The advent of genomics and innovative computational methodologies is poised to enhance this research. Genotype-environment associations (GEAs) leverage statistical correlations between allele frequencies and their environmental origins to test the hypothesis that allelic variation at specific genes is adapted to local conditions (Lasky et al., 2023). Researchers can explore the genome for GEAs to formulate hypotheses regarding adaptive genetic traits. Newer methods were proposed to explore the Genetic Environment Associations (GEAs) in the foundational hypotheses of genetic architecture, thereby enhancing the testing of hypotheses generated by GEAs through the lenses of genetics and ecophysiology. Recommendations are offered for novice users interested in exploring the molecular foundations of adaptation. When integrated with a robust hypothesis testing framework, GEAs have the potential to deepen our comprehension of the molecular mechanisms driving climate adaptation, which is essential for plant improvement. The study conducted by Drummond et al. (2012) highlights that computational evolutionary biology, statistical phylogenetics, and coalescent-based population genetics are increasingly pivotal in analyzing and interpreting molecular sequence data. We introduce version 1.7 of the Bayesian Evolutionary Analysis by Sampling Trees (BEAST) software package, which employs a suite of MCMC algorithms for Bayesian phylogenetic inference, divergence time estimation, coalescent analysis, phylogeography, and related molecular evolutionary studies. This package features an upgraded graphical user interface known as Bayesian Evolutionary Analysis Utility (BEAUti), which provides access to sophisticated models for molecular sequence and phenotypic trait evolution that were previously restricted to developers. Additionally, the research by Civelek and Lusis (2014) examines systems genetics, an approach aimed at elucidating the flow of biological information that underpins complex traits. This methodology employs a variety of experimental and statistical techniques to quantify and integrate intermediate phenotypes, such as transcript, protein, or metabolite levels, within populations exhibiting variation in traits of interest. Studies in systems genetics have yielded the first comprehensive insights into the molecular architecture of complex traits and are instrumental in identifying genes, pathways, and networks associated with prevalent human diseases. In light of the pressing need to comprehend how the numerous loci identified in genome-wide association studies influence disease susceptibility, systems genetics is poised to gain prominence. In microarray research, a significant challenge lies in comparing disease outcome predictors derived from gene expression levels with traditional clinical predictors. Conducting this comparison on the same dataset utilized to develop the microarray predictor may result in a substantial bias favoring the microarray predictor. To address this issue, we introduce a novel approach termed 'pre-validation', which aims to facilitate a more equitable comparison between the two predictor sets. We analyze this method and investigate its application in a recent breast cancer study (Tibshirani & Efron, 2002). The field of Statistical Human Genetics has been established for over a century, during which the meanings of many terms have evolved, primarily due to advancements in molecular discoveries. This evolution has led to challenges in communication between molecular and statistical geneticists. Consequently, as molecular genetics increasingly transitions into an in silico statistical science, it is essential to establish well-defined and shared terminology (Elston et al., 2012). Inferring large-scale covariance matrices from sparse genomic data presents a common challenge in bioinformatics. The conventional covariance and correlation estimators are evidently inadequate for this task. As a statistically efficient and computationally rapid alternative, we propose a new shrinkage covariance estimator that utilizes the Ledoit-Wolf lemma for the analytical determination of optimal shrinkage intensity. We subsequently apply this enhanced covariance estimator, which guarantees minimum mean squared error, is well-conditioned, and remains positive definite even with small sample sizes, to the task of inferring large-scale gene association networks. This chapter demonstrates that our approach exhibits significant advantages over competing methods, both in simulation scenarios and when applied to actual expression data (Schafer & Strimmer, 2005; Liang & Kelemen, 2004). Several critical challenges must be addressed in the analysis of temporal patterns in gene expression: firstly, the correlation structure inherent in multidimensional temporal data; secondly, the multitude of variation sources compounded by existing high-level noise; and finally, the fact that gene expression often encompasses diverse and dynamic patterns. To tackle these issues, we propose a Hierarchical Bayesian Neural Network model designed to accommodate the input correlations present in time-course gene array data. This hierarchical Bayesian framework allows for the estimation of variations in absolute gene expression levels and the associated noise. The optimization of both network parameters and hyperparameters was conducted simultaneously using Monte Carlo Markov Chain simulation techniques. The results indicate that the proposed model and algorithm effectively capture the dynamic characteristics of gene expression temporal patterns, even in the presence of high noise levels, strong input correlations, complex interactions, and other intricate features commonly found in microarray data. We validate and illustrate the efficacy of the proposed models using temporal datasets from yeast cell cycle studies. The performance of the Hierarchical Bayesian Neural Network model was benchmarked against other widely used machine learning techniques, including Nearest Neighbor, Support Vector Machine, and Self-Organized Map.

REFERENCES

Civelek, M. & Lusis, A. J. Systems genetics approaches to understand complex traits. *Nature Reviews Genetics* **15**, 34–48 (2014).

Drummond, A. J., Suchard, M. A., Xie, D. & Rambaut, A. Bayesian phylogenetics with BEAUti and the BEAST 1.7. *Molecular Biology and Evolution* **29**, 1969–1973 (2012).

Elston, R. C., Satagopan, J. M. & Sun, S. Genetic terminology. *Methods in Molecular Biology* **850**, 1–9 (2012).

Guillot, G. & Frantz, A. C. Supplementary material to molecular ecology review article statistical methods in spatial genetics. *Gestion* **18**, 10–15 (2009).

Huang, Y., Peruggia, M., Hsu, J. C. & Scott, A. A. Statistical selection of maintenance genes for normalization of gene expressions. *Statistical Applications in Genetics and Molecular Biology* **5**, 4 (2006).

Krüger, T. Editorial change at statistical applications in genetics and molecular biology. *Statistical Applications in Genetics and Molecular Biology* **17**, 20180046 (2018)

Kumar, S., Tamura, K. & Nei, M. MEGA3: Integrated software for molecular evolutionary genetics analysis and sequence alignment. *Briefings in Bioinformatics* **5**, 150–163 (2004).

Lasky, J. R., Josephs, E. B. & Morris, G. P. Genotype-environment associations to reveal the molecular basis of environmental adaptation. *Plant Cell* **35**, 125–138 (2023).

Liang, Y. & Kelemen, A. G. Hierarchical Bayesian neural network for gene expression temporal patterns. *Statistical Applications in Genetics and Molecular Biology* **3**, 20 (2004).

Liu, X. & Rattray, M. Including probe-level measurement error in robust mixture clustering of replicated microarray gene expression. *Statistical Applications in Genetics and Molecular Biology* **9**, 42 (2010).

Minamikawa, M. Development of an efficient breeding platform for fruit trees using data science and statistical and molecular genetics. *Breeding Research* **25**, 181–186 (2023).

Schafer, J. & Strimmer, K. A shrinkage approach to large-scale covariance matrix estimation and implications for functional genomics. *Statistical Applications in Genetics and Molecular Biology* **4**, 32 (2005).

Tibshirani, R. J. & Efron, B. Pre-validation and inference in microarrays. *Statistical Applications in Genetics and Molecular Biology* **1**, 1 (2002).

Werhli, A. V. & Husmeier, D. Reconstructing gene regulatory networks with Bayesian networks by combining expression data with multiple sources of prior knowledge. *Statistical Applications in Genetics and Molecular Biology* **6**, 15 (2007).

Zhang, M. Q. Statistical features of human exons and their flanking regions. *Human Molecular Genetics* **7**, 919–932 (1998).

8

Unleash the Potential of Proteomics Data with Statistical Application in Biochemistry

8.1 Introduction

The *Bos taurus* milk constituents (80%) casein and 20% whey protein (WP), are a group of soluble proteins comprising approximately 20% of the protein content of cow's milk. They are usually isolated from the liquid portion of milk during cheese making, followed by filtration and drying. The WP fraction is made up of several proteins, including *lactoglobulin, lactalbumin, immunoglobulins,* and *glycomacropeptides,* which have a range of varied properties. They are high-quality proteins containing all the essential and non-essential amino acids (nAA) and are relatively rich in the branched-chain amino acids (BCAA) *leucine, isoleucine,* and *valine.* A number of different classes of WP products are commercially available, including the filtrates, whey protein concentrates (WPC), protein isolates (WPI), WP hydrolysate (WPH), which are filtered and then enzymatically digested. These differ in terms of nutrient profile, with WPI and WPH containing the most protein in commercially available products, typically 85–95 g of protein per 100 g of product, and the highest concentration of BCAA among commercially available WP products and indeed among all milk proteins (Almeida et al., 2016). The consumption of whey protein is commonly associated with bodybuilding and sports nutrition, where it is used as a supplemental protein source to enhance muscle protein synthesis. Biochemically, it is embedded in an acyclic network of heavy and light chain *immunoglobulins* (*IGs*), *bovine serum albumin* (*BSA*), *lactoferrin* (LF), *and lactoperoxidase glycomacropeptide* (GMP) with low molecular weights (Alu'Datt et al., 2012; Abrahao, 2012). Given the emerging evidence suggesting a potential role of dairy proteins in improved vascular function (VF), it was examined in several studies to determine its short- and long-term effects on the vascular system. According to biochemical reactions, the potent intracellular antioxidant glutathione and one-carbon metabolism contain 3–4 times more bioavailable cysteine than other proteins (Amalfitano et al., 2019; Barac et al., 2011). Cysteine is an essential amino acid for the biosynthesis of glutathione (a tripeptide with antioxidant, anticarcinogenic, and immune-stimulatory properties (Boye et al., 2012; Brodkorb et al., 2016). A major health benefit of WP is to develop strength and body composition, promote weight loss by boosting metabolism, suppressing appetite, maintaining a good metabolism, nourishing bones, reducing inflammation, promoting rapid healing of wounds, lowering blood pressure, reducing hunger, repairing, and preserving muscle tissues, *etc.* (Brodkorb et al., 2016; Chen & Barkley, 1998). Nowadays, WP is rarely used for degenerative diseases like cancer and various surgical operations. It provides rehabilitative benefits for autoimmune disorders and develops immunity for immunocompromised (AIDS/HIV) patients. It can also show undesirable effects that can be combated with increased intake of WP. Regarding realistic insight, it is one of the important bioavailability sources for maintaining a better gut health system (Croy et al., 1980; Corrochano et al., 2019). Currently, numerous micro and macro research studies on WP have been conducted worldwide, and they have also explored the binding of specific amino acids (peptides) and the retention time of WP binding to form 'helix and sheath'. In this paradigm, very limited studies have developed algorithms (separate) and formulated the bindings of amino acid acyclic networks and their binding capacity. This driven algorithm will help food and dairy technologists explore whey protein by-products in the form of capsules and oral administration (nutraceuticals) for the geriatric population and growing children. It can also be used it for treating patients with terminal illnesses (cancer), HIV/AIDS, and other immunocompromised diseases. During the pandemic caused by SARS-CoV-2, many medical and nutritional specialists recommended WP for the development of gut microbiota (immunity). As of this calendar year 2025, numerous studies have been conducted on WP with amino acid binding (AAb) and its usage. However, in the case of human health intervention, a large number of studies will be necessary to describe the overall well-being of whey protein for the eradication of different diseases and ailments in humans. In this paradigm, the present study attempts to address the importance of whey protein by reviewing published articles worldwide. This basic systematic analysis review will help us describe the most important health benefits and also know the retention time of WP binding with different peptide sequences. This newer research (evidence-based approach) will help researchers extend basic and advanced (nutritional) translational health studies.

8.2 Methods

8.2.1 Study Intervention

Whey protein isolates (WPI) are liquid by-product that can be further processed into spray-dried products like whey protein concentrates (WPC), WPI, or whey protein hydrolysate (WPH). The whey isolates have had their base component (water) removed and are generally considered almost lactose- and cholesterol-free, typically containing at least 90% protein. These were isolated by two methods: (i) ion exchange and (ii) membrane filtration. The WPC was prepared by three

DOI: 10.1201/9781032713816-8

various methods common to the dairy manufacturing industry and concentrated to 80% total solids. Whey protein concentrate lactic acid (WPC-L) was produced from milk using LAB cultures to convert lactose to lactic acid, lowering the pH to 4.6 to precipitate the casein. WPC-LR was produced using a combination of rennet and LAB. The rennet enzyme mixture is inactivated during processing, and whey protein concentrate mineral acid (WPC-MA) was produced using mineral acid, either sulfuric or hydrochloric, to lower the pH (~4.6) to precipitate the casein. The study was conducted based on meta-analysis reports (effective population size was determined). Most cited *Bos taurus* protein accessions were downloaded from the NCBI virtual platform (website).

Accession	bp	Software	Linear DNA/miRNA
NG_028350.2; GI: 345110543	1,502	FASTA	Linear DNA
AY267466.1; GI: 34398413	4,058	BLAST/ FASTA	Linear DNA
GJ062448.1; GI:270038557	5,008,217	FASTA	Linear DNA
GK000016.2; GI: 257481329	81,724,687	FASTA	Linear DNA
NM 001080219.1; GI: 122692296	1,749	FASTA	Linear DNA

Source: https://www.ncbi.nlm.nih.gov/nuccore.

8.3 Model Formulation

8.3.1 Markova Random Two-State Models of Peptide Sequence

The Markova (two-state) random model was used to derive probabilities for the most cited peptide sequences (S_1, S_2, S_3,… S_n), and each of the peptide sequences was assigned numerical values to form an accession number. These accession numbers are called 'tags' or alphabetic symbols (represented). The Markov chain model (MCM) was formulated using Monte Carlo simulation techniques to quantify the retention time and binding capacity of different WP peptides. We have predicted clear binding sequences with particular amino acids (determining the current state). The state of the sequences was taken before defining the current state of random variables (peptide sequences). This postulate (model) does not affect the future action of hidden state variables driven by MCM. All possible state variables were coded as [1, 2, … *n*], where numeral one indicates the lower binding of amino acids (AA), and numeral two (coded) indicates the strong binding of AA for the formation of WP. The probability of binding of a particular AA (P=0.20) for the formation of WP was extracted from the regression model using bootstrapping techniques (model building).

8.3.2 Model Optimization

The MCM model was superimposed on the artificial neural network (ANN) to obtain promising transient probabilities of retention time (peptide sequences) and the nature of the Seq peptide. However, during the process of binding WP, enormous amounts of retrieving energy were released (energy pathway) for the synthesis of different essential amino acids. Regarding the ambient pathway on WP, we need better optimization techniques for accurate extrapolation. During the first stage of model formulation, we defined the state variables using the substitution method of MCM. The binding capacity and likelihood were efficiently determined by iteration techniques (bootstrap), and we also derived the specific amino acids to estimate the marginal difference between the retention time and the binding capacity of peptides. Furthermore, we formulated a machine learning superimposed (MLS) model using the Thompson iteration method to predict accurate retention time and label various amino acids (AA) for the creation of the Seq Peptide Library (model estimation purpose). Based on this above intervention, the MLS algorithms are essential components for predicting unobservable amino acids from the labeled data. From previous literature, researchers estimated the retention time of WP using traditional methods like stochastic gradient descent with different momentum, AdaGrad, RMS Prop, and Adam Optimizer, among others. The above-formulated models show various analytical gaps because they are based on simple extrapolation techniques for noisy data. In this research, we reduced the noisy data by smoothing the MLS through different iterations. The sequence peptide data were derived closer to the original function and connected to the different kinds of probability distributions. Simultaneously, we assigned the weighted averages to each Seq peptide based on the different eclectic charges, and protein index models were estimated using standard software and correlated with the Protein Index (defined the sequences in 'μ')

$$\mu_t = \beta\mu_{t-1} + (1-\beta)S_t \tag{8.1}$$

$$\mu_{t-1} = \beta\mu_{t-2} + (1-\beta)S_{t-1} \tag{8.2}$$

$$\mu_{t-2} = \beta\mu_{t-3} + (1-\beta)S_{t-2} \tag{8.3}$$

μ = Peptide sequence, time taken for biding, β = parameter constant, S = noisy

$$\mu_t = \frac{\mu_t}{1-\beta^t} \tag{8.4}$$

The proteomics of WP has frequently been described as protein expression resulting from gene transcription and translation (it is the broad study of proteins). This cutting-edge technology was used to identify the pathway of nucleotides and AAb, as well as the diversity of AA and the effect of interactions between different biochemical pathways. These underlying biological processes El Hiani et al., 2019; Dalsgaard et al., 2007) are very useful for researchers to formulate products. Many food technology industries (FTI) have used rationalized bioinformatics based on proteomic research (knowing the pathways) to formulate dairy food products by utilizing different biochemical methods. So far, numerous studies have conveyed different pathways of WP to identify the most promising AA for completing biochemical processes. Many R&D institutes have developed

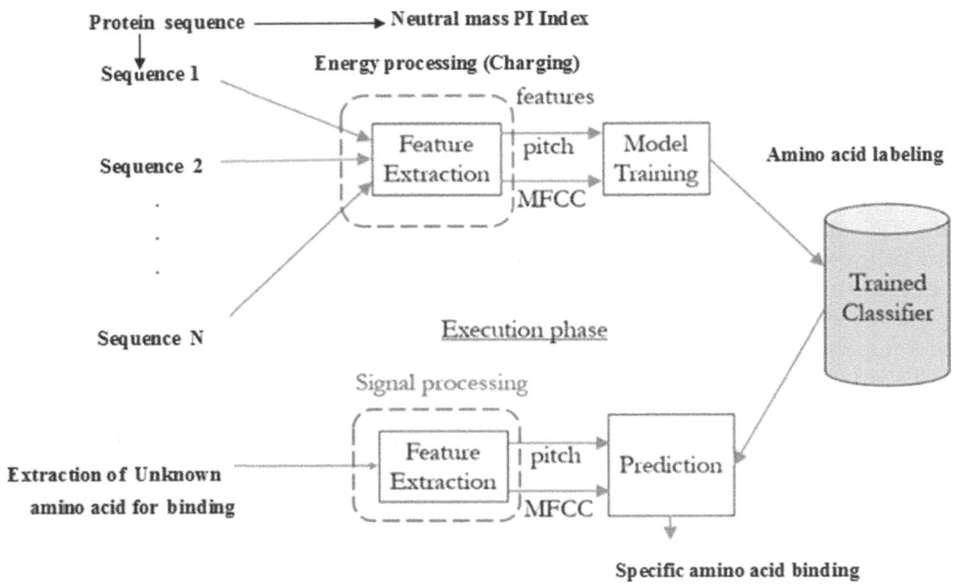

FIGURE 8.1 Flow chart designed for the identification of unknown amino acid.

health quotient scores for dairy products and correlated algorithms for biochemical pathways. Since the binding of amino acids in whey protein is an essential component used to investigate homeostatic effects in humans, WP will support studies of the quality of life (QOL) on individuals and formulate nutritional studies around the world. In fact, proteomics is a more direct way to study biochemical processes than genomics and transcriptomics. Proteomics will be widely used to estimate milk components and their protein profile. The pathway of unknown amino acid binding of whey protein is described in (Figure 8.1). All Seq peptide accession analysis was carried out usig 'FASTA' software based on parent proteins and their composition. The protein index was developed as per the SOP.

We formulated a stochastic matrix using regression estimator with a P-value (P=0.20) and a latent p-value is $(1-p)=0.80$, assuming that a 0.50 (50%) random state variable converges for Seq peptide completed binding.

$$P = \begin{bmatrix} \text{State} & 1 & 2 \\ 1 & 0.50 & 0.50 \\ 2 & 0.80 & 0.20 \end{bmatrix} \quad (8.5)$$

For binding state, we solved the equation (1.1) $X * P = X$; and $x_1 + x_2 = 1$

$$X = [x_1,\ x_2]P = \begin{bmatrix} 0.50 & 0.50 \\ 0.80 & 0.20 \end{bmatrix}$$

$$ie \Rightarrow [x_1,\ x_2]\begin{bmatrix} 0.50 & 0.50 \\ 0.80 & 0.20 \end{bmatrix} = [x_1,\ x_2]$$

$$x_1(0.50) + x_2(0.80) = x_1; \quad x_1(0.50) + x_2(0.80) = x_2$$

$$x_1 + x_2 = 1 \text{ again we simplified; } x_1(-0.50) + x_2(0.80) = 0$$

$$x_1(0.50) + x_2(-0.80) = 0;\ x_1 + x_2 = 1$$

$$[x_1,\ x_2] = [0.615,\ 0.384] \text{from stochastic matrix eqn} \quad (8.6)$$

An initial state $I = [0,\ 0.20]$. The value of (P=0.20) was determined by the MLS model (X=neutral mass of each peptide sequence and Y=PI); $\hat{Y} = 8.66 - 0.0025\ X'\ P = 0.20$

Probability of the state after 'I' process is as follows:

$$I * P = [0,\ 0.20] * \begin{bmatrix} 0.50 & 0.50 \\ 0.80 & 0.20 \end{bmatrix} = [16.0, 4.0]$$

$$I * P^2 = 1 * P^1 * P = [16.0, 4.0] * \begin{bmatrix} 0.50 & 0.50 \\ 0.80 & 0.20 \end{bmatrix} = [11.2, 8.8]$$

$$I * P^3 = 1 * P^2 * P = [11.2, 8.8]\begin{bmatrix} 0.50 & 0.50 \\ 0.80 & 0.20 \end{bmatrix} = [12.64, 7.36]$$

8.3.3 Denaturation of Whey Protein (Kinetic Analysis) Model

In order to determine the thermal denaturation of WP, it was determined using the following mathematical equation (rate of change of time).

$$\frac{dC_t}{dt} = K_n C^n{}_t \quad (8.7)$$

For $n \neq 1$, this equation will yield:

$$\left(\frac{C_t}{C_o}\right)^{1-n} = 1 + (n-1) K_n C^{n-1}{}_0 t \quad (8.8)$$

where n=1 this equation yields

$$l_n\left(\frac{C_t}{C_o}\right) = -K_n C_0{}^{n-1} t \quad (8.9)$$

where n=reaction order, $C_0(\text{g/L})$ = concentration of protein before heat treatment, and $C_t(\text{g/L})$ = Concentration of native protein at time t(s) and $K_n\left(\text{g}^{1-n}/\text{L/s}\right)$ = rate constant. Equations (1.8 and 1.9) were used to calculate the reaction order 'n' at different temperatures of WP. The rate constant K_n was calculated $\left(\frac{C_t}{C_o}\right)^{1-n}$ and plotted against time 't'. The temperature dependence rate with constant K_n was modeled using the Arrhenius equation (1.10)

$$l_n(K_n) = l_n(K_{n0}) - \frac{E_\alpha}{RT} \tag{8.10}$$

where K_n frequency factor $\left(\text{g}^{1-n}/\text{L/s}\right) = E_\alpha$ = activation energy (J/mol), R=universal gas constant, T=Absolute temperature.

8.4 Model Results

Despite the complete exploration of the protein sequences of whey protein, the Seq peptide had a higher neutral mass index and protein index (PI) with alphabetic expression (EKTKIPAVF, parent protein *β-lactoglobulin*, neutral mass index 1031.60, PI 8.63; similarly 'KILDKVGIN' - *α-lactalbumin*, neutral mass index 998.61; and KTKIPAVF - *β-lactoglobulin*, neutral mass index 902.55, and PI 10.0). The present results show that the above Seq peptide has relatively higher expression and binding capacity and is found to be excellent, with retention times ranging from 5 to 11.50 minutes (cross-linked) and 10 to 13.50 minutes (unlinked). It was found to be statistically significant, with R^2 (%)=0.88 (Table 8.1).

Figure 8.2 shows the *X*-axis (peptide length) and *Y*-axis (retention time) to express the binding of peptides with different

TABLE 8.1
Whey Protein Bio-Informatics (Seq Peptide Sequencing)

Peptide Sequence	Parent Protein	Composition	Neutral Mass	PI
Low Bindings Peptide Sequences				
VIESPPEIN	*K-casein*	C (44) H(70) N (10) O (15) S (0)	996.51	3.79
VEELKPTPEGDLEIL	*β-lactoglobulin*	C (75) H (122) N (16) O (26) S (0)	1,680.88	3.91
TPVVVPPFLQPEVM	*β-casein*	C (74) H(115) N(15) O (18) S(1)	1,551.53	4.00
ELKPTPEGDLEIL	*β-lactoglobulin*	C(65) H(106) N(14) O (22) S(0)	1,452.77	4.00
LVRTPEVDDE	*β-lactoglobulin*	C(49) H(79) N(13) O (19) S(0)	1,171.57	4.16
VRTPEVDDE	*β-lactoglobulin*	C(43) H(68) N(12) O (18) S(0)	1,058.48	4.16
TPVVVPPLQPE	*β-casein*	C(55) H (88) N (12) O (15) S(0)	1,174.65	4.59
Medium Binding Sequences				
HHKAPGPEDSLHEQ	*Folate receptor-α*	C (67) H (98) N (22) O (22) S (0)	1,580.83	5.75
DTDYKKY	*β-lactoglobulin*	C (42) H (59) N (9) O (14) S (0)	931.42	5.96
SWMHQPHQPLPPTVM	*β-casein*	C (81) H (118) N (22) O (19) S (2)	1,784.84	6.66
WMHQPHQPLPPTVM	*β-casein*	C (78) H (113) N (21) O (17) S (2)	1,697.81	6.92
DHKSEEDKHLKIR	*Osteopontin-k*	C (69) H (113) N (23) O (22) S (0)	1,633.85	6.93*
Strong Binding Sequences				
YWLAHK	*α-lactalbumin*	C (41) H (54) N (10) O (7) S (0)	816.42	8.60*
EKTKIPAVF	*β-lactoglobulin*	C (49) H (79) N (11) O (12) S (0)	1,031.6	8.63*
KILDKVGIN	*α-lactalbumin*	C (45) H (80) N (12) O (12) S (0)	998.61	8.59*
KTKIPAVF	*β-lactoglobulin*	C (44) H (72) N (10) O (9) S (0)	902.55	10.0**

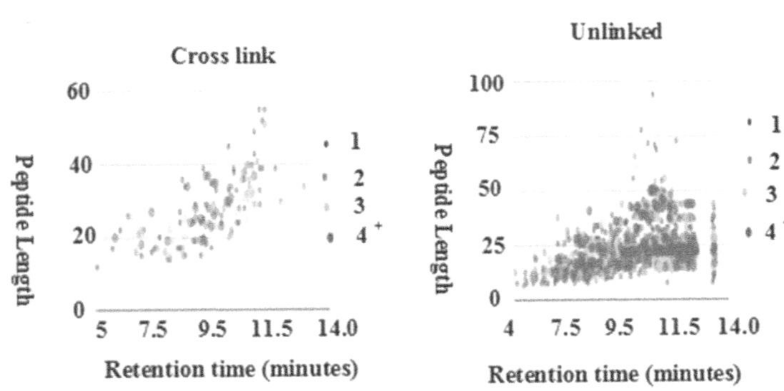

FIGURE 8.2 Scatter plot for relationship between peptide length and retention time.

PI charges (represented by different colors 1 to 4$^+$). As per the scattered graph, all observed values (peptide length) have an ordinate between 5.0 and 11.75 minutes (retention time). The peptide length (unlinked) versus retention time was correlated based on the machine learning superimposed (MLS) scatter plot. The results show more retention time in unlinked Seq peptides (ranging between 4.0 and 13.75 minutes). In a similar concordance study conducted by Kersey et al. (2004), he opined that the genomic sequences of several eukaryotes and their proteomes remain relatively poorly defined. Information about proteins identified by different experimental and computational methods is stored in different databases, meaning no single resource offers full coverage of known and predicted proteins. The IPI (International Protein Index) has addressed these issues and offers complete, non-redundant datasets representing the human, mouse, and rat proteomes, built from the Swiss-Prot, TrEMBL, Ensembl, and RefSeq databases (Pegolo et al., 2020). Potential functional relationships among milk protein fractions in dairy cattle were formulated in a structural equation model (SEM) (Fairley, 1996). GWAS provides a decomposition of total SNP effects into direct effects and those mediated by traits that are upstream in phenotypic networks (fitted to a mixed Bayesian multitrait genomic model). This model infers the genomic correlations among six milk nitrogen fractions (CN), namely κ-, β-, $\alpha\ S_1$-, and $\alpha\ S_2$-CN, and two whey proteins, namely *β-lactoglobulin (β-LG) and α-lactalbumin (α-LA),* in a population of 989 Italian Brown Swiss cows. The information acquired might be leveraged for the setting of optimal management and selection strategies aimed at improving milk quality and technological characteristics in dairy cattle (Amalfitano et al., 2019). Together with whey proteins, CNs are important sources of biologically active peptides characterized by different physiological properties and nutritional implications (Silva and Malcata, 2005). Whey protein concentrate (WPC) and hydrolysate (WPH) are protein ingredients used in sports and medical formulations (Dalziel et al., 2016). Whey is generally considered a dietary protein supplement that can provide antimicrobial activity, immune modulation, improved muscle strength and body composition, and protection against cardiovascular disease and osteoporosis (Figure 8.3).

The observed PI and neutral mass index were tested using regression smoothing techniques, and the results show that both the observed and predicted PI are strongly correlated with binding capacity ($R^2=0.82$). However, in the case of neutral mass with PI, binding capacity shows a very low correlation because of the non-homology of amino acids in whey ($R^2=0.50$) (Table 8.2).

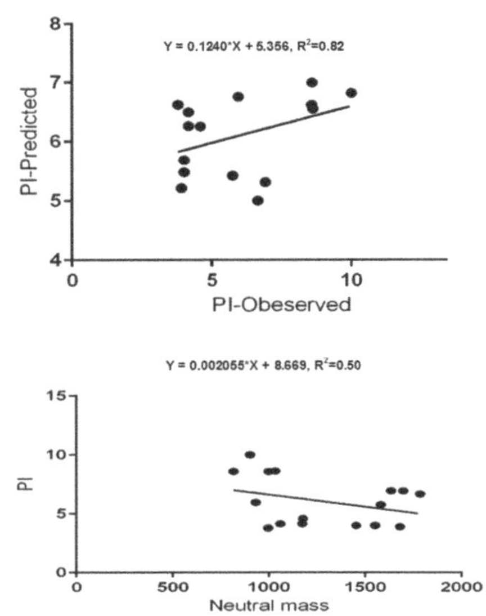

FIGURE 8.3 Scatter plot for comparing the observed and predicted values of the protein index for whey protein.

TABLE 8.2

Whey Protein Peptide Sequencing with Different Charge Correlation

	Charge									
Peptide Sequence	**+1**	**+2**	**+3**	**+4**	**+5**	**+6**	**+7**	**+8**	**+9**	**+10**
HHKAPGPEDSLHEQ	1,581.74	791.37	527.92	396.19	317.15	264.46	226.83	198.60	176.64	159.08
TPVVVPPFLQPEVM	1,552.84	776.93	518.29	388.97	311.37	259.65	222.70	194.99	173.43	156.19
VEELKPTPEGDLEIL	1,681.89	841.45	561.30	421.23	337.18	281.15	241.13	211.12	187.77	169.10
LVRTPEVDDE	1,172.58	586.79	391.53	293.90	235.32	196.27	168.37	147.45	131.18	118.16
ELKPTPEGDLEIL	1,453.78	727.39	485.26	364.20	291.56	243.14	208.55	182.60	162.43	146.28
DHKSEEDKHLKIR	1,634.86	817.93	545.63	409.47	327.78	273.32	234.41	205.24	182.55	164.39
EKTKIPAVF	1,032.61	516.81	344.87	258.91	207.33	172.94	148.38	129.96	115.63	104.17
TPVVVPPLQPE	1,175.67	588.34	392.56	294.67	235.94	196.78	168.82	147.84	131.22	118.47
KILDKVGIN	999.62	500.31	333.88	250.66	200.73	167.44	143.67	125.83	111.96	100.87
VIESPPEIN	997.52	499.26	333.18	250.14	200.31	167.09	143.37	125.57	111.73	100.66
YWLAHK	817.44	409.22	273.15	205.11	164.29	137.08	117.64	103.06	91.72	82.65
KTKIPAVF	903.57	452.29	301.86	226.65	181.52	151.43	129.94	113.83	101.29	91.26
SWMHQPHQPLPPTVM	1,785.86	893.43	595.96	447.22	357.98	298.48	255.99	224.11	199.32	179.49
DTDYKKY	932.44	466.72	311.48	233.86	187.29	156.25	134.07	117.44	104.50	94.15
VRTPEVDDE	1,059.50	530.25	353.84	265.63	212.70	177.42	152.22	133.32	118.62	106.86
WMHQPHQPLPPTVM	1,698.82	849.92	566.95	425.46	340.57	283.98	243.55	213.23	189.65	170.79

Between and within the sequences were tested by one way -ANOVA, *df* (9,159), $\text{Fisher}_{\text{Cal}} = 86.63^{**}$, $p<0.001$.

8.5 Discussion

The present study discusses the salient findings of the binding capacity of WP and its retention times (low, medium, and strong) for amino acids (Goyal et al., 2021). This is a new area of research on WP to explore the bioactive linkage of peptides (BAP) that are encrypted in the acyclic structure of whey protein, including the protein index (PI) of each peptide array. All dairy proteins have derived different peptides with semi-structured bonding to form food proteins, commonly referred to as bioactive peptides (BAP), i.e., protein fragmentation, that have a positive influence on physiological and metabolic functions or conditions of the body and may have ultimately beneficial effects on human health. The bioactive peptide typically contains 2–20 amino acids. In our study, we extrapolated peptide sequences of whey protein, and peptide arrays like 'EKTKIPAVF' (*β-lactoglobulin*), 'KILDKVGIN' (*α-lactalbumin*), and 'KTKIPAVF'(*β-lactoglobulin*) were found to have the highest protein index (PI), with an IQR range between 7 and 8.63. The above peptide array shows the strong binding capacity of WP in *Bos taurus* milk. The retention time of bindings is 50–137.50 seconds, and there is a strong association for both cross-linked and unlinked peptides (R^2=0.82). Many studies suggest the diverse peptide array and the potential role of peptide link and PI in enhancing good binding between hydrophobic ligands (i.e. retinol and long-chain fatty acids), enzyme regulation, and passive immunity in pediatric and geriatric populations. However, the bioactive peptide (BAP) will support studies of the gut (human physiology) and signal cell pathways in cell division (Madureira et al., 2010). The present study is not intended to indicate that health benefits have been convincingly demonstrated in humans, but rather to maintain consistency with its use in the scientific literature, based on evidence derived largely from isolated cell systems and animal studies. To continue to show that consumption of dairy foods is linked to beneficial or neutral associations with the risk of hypertension, cardiovascular disease (CVD), and type-2 diabetes, an examination of the research on dairy protein-derived peptides (cross- and unlinked) in relation to these three conditions of public health concern is of interest (Liu et al., 2019). Pace et al. (2011) opined that the three BCAAs *leucine*, isoleucine, and valine are the essential amino acids found in both whey protein and casein, with whey protein being the richer source at approximately 17 g per 100 g of total protein. The administration of BCAA has been demonstrated to alter cell signaling, particularly the mechanistic target of rapamycin (*mTOR*), which plays a key role in nutrient sensing and cell growth and proliferation (El Hiani et al., 2019). Circulating BCAA concentrations have been positively correlated with CVD risk, including hypertension, in a number of cohorts (Tobias et al., 2020; Yamaguchi et al., 2017). In addition, while BCAA intake was positively associated with hypertension in an Iranian cohort (Teymoori et al., 2017), twin studies in a UK cohort demonstrated an association between BCAA dietary intake and reduced twin risk of hypertension, reduced risk of insulin resistance, and metabolites such as alpha-hydroxyisovalerate, also known as 2-hydroxyisovaleric acid, is a metabolite that can be found in urine, which are highly associated with adiposity (Jennings et al., 2016). Plasma BCAA and phenylalanine concentrations are the main indicators for ingestion of WPH and whey, which increased plasma valine for 60 minutes, leucine for 90 minutes, *isoleucine* for 60 minutes, and *phenylalanine* for 45 minutes post-ingestion (p 0.05). Plasma amino acid concentrations did not differ between the lineages. Only valine tended to be slightly higher at 15 minutes WPH than whey ingestion (p 0.001). In our study, we did not test ingestion because of analytical limitations (Dalsgaard et al., 2007). WPC is the least processed and is 35%–80% protein by weight, whereas WPI is further processed to remove fat and carbohydrates to produce a product that is 90% protein by weight. WPH is enzymatically and acid-pretreated to reduce particle size and is the most digestible protein of the three variants. Whey protein typically comprises an abundance of amino acids (the building blocks of protein). WP has gained extensive use for anabolic recovery, generating maximum response with the least total protein and calories, which are important considerations for weight management, conditions of limited food intake such as illness or aging, or post-exercise recovery, including muscle-building use for athletes. The importance of the *mTOR* signal and the high concentration of leucine in whey protein (WP) can only be appreciated in the context of daily protein turnover. Every day, adults need to make 250–300 g of new protein to repair and replace existing proteins (FitzGerald, 2020). This process occurs with continuous cycling between the synthesis and breakdown of proteins. While skeletal muscle accounts for approximately 50% of total body protein, it receives only about 25% of the newly formed protein. The model-based study was conducted by (De Carvalho-Silva et al., 2012). The whey protein derived peptides have been proposed to have anti-inflammatory properties, and Swiss mouse models have demonstrated the ability of WPH, when dosed at 300 mg/kg body weight, to reduce markers of inflammation as measured by examining oedema levels and *leucocyte* activity.

8.6 Conclusion

An overall summing of the results concludes that whey protein is a commonly consumed protein irrespective of age group and a significant source of high-quality protein. The retention time and sequences of peptides for binding are essential, unobservable components for dairy specialists. In the above study, we derived the WP retention time and categorization of peptides in a sequential manner. As observed from the study, peptide sequences had relatively higher expression and excellent binding capacity, with retention times ranging from 5 to 11.50 minutes in cross-linked and unlinked cases. Moreover, as WP is an especially rich source of BCAAs, including leucine, future areas of study should include the impact of this higher usage of whey on different diseases and clarify the effect of the risk of unlinked amino acids. The machine learning superimposed (MLS) model (used for the estimation of likelihood calculation) is most robust for determining sequence peptide looping.

This model is very useful for bioinformatics, dairy chemists, and life scientists for the accurate estimation of likelihoods and transient probabilities.

Conflict of Interest

There is no conflict of interest with any extramural or institutional funding agencies.

Ethical Clearance

Only secondary Seq Peptide data are used in the current investigation. There are no ethical clearance concerns.

Acknowledgment

We express our heartfelt gratitude to the responsible officers of KVAFSU (B) for their moral support as well as to the NCBI for displaying Seq Data for the researchers.

REFERENCES

Abrahao V. Nourishing the dysfunctional gut and whey protein. *Current Opinion in Clinical Nutrition & Metabolic Car* **15**, 480–484 (2012).

Almeida, A, et al. Exploring the gating mechanisms of aquaporin-3: new clues for the design of inhibitors? *Mol Biosyst* **12**(5), 1564–1573 (2016).

Alu'Datt, M. H., Alli, I. & Nagadi, M. Preparation, characterization and properties of whey-soy proteins co-precipitates. *Food Chemistry* **134**(1), 294–300 (2012).

Amalfitano, N. et al. Milk protein fractions strongly affect the patterns of coagulation, curd firming, and syneresis. *Journal of Dairy Science* **102**(4), 2903–2917 (2019).

Barac, M. et al. Functional properties of pea (*Pisum sativum* L) protein isolates modified with chymosin. *International Journal of Molecular Sciences* **12**(12), 8372–8387 (2011). https://doi.org/10.3390/ijms12128372

Boye, J. I., Ma, C. Y. & Ismail, A. Thermal stability of β-lactoglobulins A and B: Effect of SDS, urea, cysteine and N-ethylmaleimide. *Journal of Dairy Research* **71**(2), 207–215 (2012). https://doi.org/10.1017/S0022029904000184

Brodkorb, A., Croguennec, T., Bouhallab, S. & Kehoe, J. J. Heat-induced denaturation, aggregation and gelation of whey proteins, in *Advanced Dairy Chemistry: Volume 1B: Proteins: Applied Aspects*, 155–178 (2016).

Chen, Y. & Barkley, M. D. Understanding tryptophan fluorescence in proteins. *Biochemistry* **37**(28), 9976–9982 (1998).

Chihi, M. L., Mession, J. L., Sok, N. & Saurel, R. Heat-induced soluble protein aggregates from mixed pea globulins and β-lactoglobulin. *Journal of Agricultural and Food Chemistry* **64**(13), 2780–2791 (2016). https://doi.org/10.1021/acs.jafc.6b00087

Corrochano, A. R., Sariçay, Y., Arranz, E., Kelly, P. M., Buckin, V. & Giblin, L. Comparison of antioxidant activities of bovine whey proteins before and after simulated gastrointestinal digestion. *Journal of Dairy Science* **102**, 54–67 (2019). https://doi.org/10.3168/jds.2018-14581

Croy, R. R., Gatehouse, J. A., Evans, I. M. & Boulter, D. Characterisation of the storage protein subunits synthesised *invitro* by polyribosomes and RNA from (developing pea *Pisum sativum*) L. *Planta* **148**(1), 57–63 (1980). https://doi.org/10.1007/ BF00385442

Dalsgaard, T. K., Holm Nielsen, J. & Bach Larsen, L. Proteolysis of milk proteins lactosylated in model systems. *Molecular Nutrition &* Food Research **51**(4), 404–414 (2007).

Dalsgaard, T. K., Otzen, D., Nielsen, J. H. & Larsen, L. B. Changes in structures of milk proteins upon photo-oxidation. *Journal of Agricultural and Food Chemistry* **55**(26), 10968–10976 (2007). https://doi.org/10.1021/jf071948g

Dalziel, J. E. et al. Influence of bovine whey protein concentrate and hydrolysate preparation methods on motility in the isolated rat distal colon. *Nutrients* **8**, 809 (2016). https://doi.org/10.3390/nu8120809

De Carvalho-Silva, L. B. et al. Anti-inflammatory activities of enzymatic (alcalase) hydrolysate of awhey protein concentrate. *African Journal of Biotechnology* **11**, 2993–2999 (2012). https://doi.org/10.5897/AJB11.2330

El Hiani, Y, Egom, E. E. A. & Dong, X. P. mTOR signalling: Jack of all-trades. *Biochemistry and Cell Biology* **97**, 58–67 (2019). https://doi.org/10.1139/bcb-2018-0004

Fairley, P., Monahan, F. J., German, J. B. & Krochta, J. M. Mechanical properties and water vapor permeability of edible films from whey protein isolate and n-ethylmaleimide or cysteine. *Journal of Agricultural and Food Chemistry* **44**(12), 3789–3792 (1996). https://doi.org/10.1021/jf9601731

FitzGerald, R. J. et al. Application of in silico approaches for the generation of milk protein–derived bioactive peptides. *Journal of Functional Foods* **64**, 103636 (2020).

Goyal, A. et al. *ACE Inhibitors. Stat Pearls* (StatPearls Publishing, 2020). Available from: https://www.ncbi.nlm.nih.gov/Books /NBK430896. Accessed May 26, 2021.

Jennings, A., MacGregor, A., Pallister, T., Spector, T. & Cassidy, A. Associations between branched chain amino acid intake and biomarkers of adiposity and cardiometabolic health independent of genetic factors: A twin study. International Journal of Cardiology 223, 992–998 (2016). https://doi.org/10.1016/j.ijcard.2016.08.307

Kersey, P. J. et al. The International Protein Index: An integrated database for proteomics experiments. Proteomics **4**(7), 1985–1988 (2004). https://doi.org/10.1002/pmic.200300721

Liu, R., Cheng, J. & Wu, H. Discovery of food-derived dipeptidyl peptidase IV inhibitory peptides: A review. *International Journal of Molecular Sciences* **20**, 463 (2019).

Madureira, A. R. et al. Invited review: Physiological properties of bioactive peptides obtained from whey proteins. *Journal of Dairy Science* **93**, 437–455 (2010).

Pace, C. N. et al. Contribution of hydrophobic interactions to protein stability. *Journal of Molecular Biology* **408**(3), 514–528 (2011).

Pegolo, S. et al. Structural equation modeling GWAS of milk proteins. *Journal of Dairy Science* **104**, 5705–5718 (2020).

Silva, S. V. & Malcata, F. X. Caseins as source of bioactive peptides. *International Dairy Journal* **15**, 1–15 (2005). https://doi.org/10.1016/j.idairyj.2004.04.009

Teymoori, F., Asghari, G., Mirmiran, P. & Azizi, F. Dietaryamino acids and incidence of hypertension: A principle component analysis approach. *Scientific Reports* **7**, 16838 (2017). https://doi.org/10.1038/s41598-017-17047-0

Tobias, D. K. et al. Circulating branched-chainaminoacids and long-term risk of obesity-related cancers in women. *Scientific Reports* **10**, 16534 (2020). https://doi.org/10.1038/s41598-020-73499-x

Yamaguchi, N. et al. Plasma free amino acid profiles evaluate risk of metabolic syndrome, diabetes, dyslipidemia and hypertension in a large Asian population. *Environmental Health and Preventive Medicine* **22**, 1–8 (2017).

9

Statistical Tools for Clinical Trials in Medical Research

9.1 Background and Rationale of Randomized Controlled Trials

Historically, the method of assessing potential advancements in medical care has been ineffective. It has only been generally acknowledged in the last few years that well-conducted clinical trials, which adhere to the rules of scientific experimentation, offer the only trustworthy foundation for assessing the effectiveness and safety of novel medicines (Bernadt et al., 1990).

9.2 Types of Clinical Trial

Any planned experiment with patients that aims to determine the best course of therapy for patients with a certain medical problem in the future is referred to as a 'clinical trial'. Essential to a clinical trial is the ability to conclude how best to treat patients in the general population who will need treatment in the future based on the outcomes of a small sample of patients. Drug trials are commonly categorized into four experimental phases: Phase I to IV (Bartlett & Cornell, 2007).

9.2.1 Phase I Trial

Since the main goal of these early human trials was safety rather than effectiveness, healthy human volunteers were typically used. The initial goal is to use dose-escalation tests to establish an appropriate single drug dosage, or how much medication may be administered without producing significant negative effects. This category will encompass studies pertaining to drug metabolism, bioavailability, and various dosages (Figure 9.1).

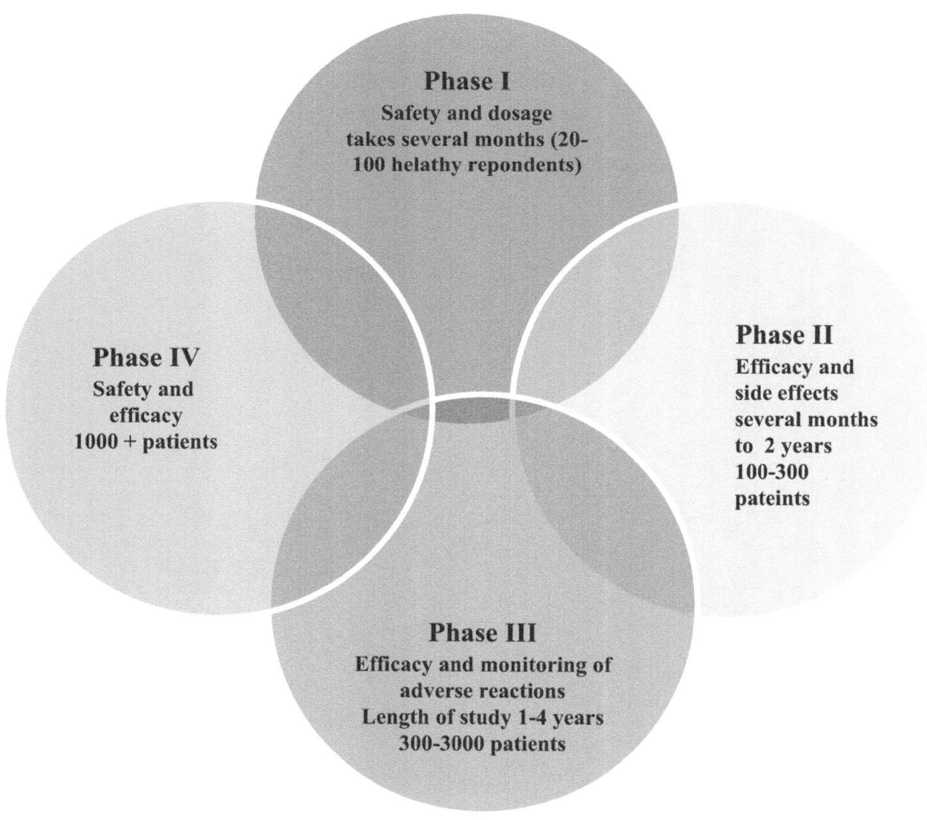

FIGURE 9.1 Flowchart of a clinical trial encompassing various phases.

DOI: 10.1201/9781032713816-9

9.2.2 Phase II Trial

These are fairly small-scale investigations into the effectiveness and safety of a drug and require close monitoring of each patient.

9.2.3 Phase III Trial

After a drug is shown to be reasonably effective, it is essential to compare it with the current standard treatment(s) for the same condition in a large trial. This is the most rigorous and extensive type of scientific clinical investigation of a new treatment.

9.2.4 Phase IV Trial

These are post-marketing surveillance studies designed to track side effects, mortality, and long-term morbidity.

9.3 Randomized Controlled Trials (RCTs)

Let us now examine the full-scale (Phase II) controlled clinical trials' scientific justification. Developing a good idea to improve treatment is the top priority for clinical research. Any new therapy's supporters will, of course, naturally be excited about its potential; preclinical research and phase I and II trials may show a great deal of promise. Unfortunately, when new treatments are put to the rigorous test of a well-designed phase III trial, they often prove to be less effective than anticipated. Phase III trials must adhere to a basic rule: they must be comparable. That is, it is necessary to contrast the experience of a cohort of patients receiving the novel treatment with that of a control cohort comprising comparable patients undergoing a conventional course of care. It is usually appropriate to have a control group of patients who are not receiving treatment (placebo group) if there is no standard treatment. Furthermore, it is typically necessary to randomly assign each patient to receive either the new treatment or the standard treatment in order to get an objective assessment of the new treatment's worth. As a result, the randomized controlled trial is now recognized as the most dependable technique for carrying out clinical research (Pocock et al., 2015a).

9.4 RCT Steps

9.4.1 Describe the Trial's Objectives and Provide a Specific Hypothesis

One must develop a more detailed definition of trial aims in terms of specific hypotheses regarding treatment efficacy and safety from an initial concept about a potential therapy improvement. In other words, precise definitions of the patient type, treatments to be compared, and techniques for assessing each patient's response to treatment are required.

9.4.2 Create a Documented Protocol for the Trial

The next steps are to draft a comprehensive design for a randomized trial and document one's approach in a research protocol. The design of the trial needs to follow scientific, ethical, and organizational guidelines for it to be conducted successfully. There are two important matters to consider: Dimension (a): The trial needs to include a sufficient number of patients to assess each treatment's response fairly accurately. To avoid irrelevant factors from influencing treatment comparisons, bias must be prevented by ensuring that patient selection, care, and evaluations are consistent across treatments.

9.4.3 Execute the Trial with Proper Planning

The trial must be carried out word for word in strict accordance with the protocol. Generally speaking, deviating from the protocol during the trial is prohibited. Unexpected deviations, however, might be implemented after the procedure has been adjusted appropriately. Examine the time.

9.5 Hypothesis Testing and Descriptive Statistics

The results should be subjected to statistical techniques to test the predetermined study hypotheses. The strength of the evidence supporting a real treatment-related difference can be specifically assessed using significance tests.

9.5.1 Make a Determination: Report Results

The final say on the relative benefits of the treatment must be made, and the results must be published for use by other medical professionals in their clinical practice.

9.5.2 Examples of RCTs

1. New treatment vs. standard treatment
   Disease: Anal fistula
   New treatment: Kshaarasootra (Ayurvedic medicated thread)
   Standard treatment: Surgery
   Reference: Indian Council of Medical Research et al. Multi-centric randomized controlled clinical trial of Kshaarasootra (Ayurvedic medicated thread) in the management of fistula-in-ano. *Indian J Med Res* 1991: 94: 188-85.
2. New treatment vs. placebo
   Disease: Acute myocardial infarction
   New treatment: Anturan
   Placebo: Inactive tablets resembling Antura
   Reference: Anturane Reinfarction Trial Research Group. Sulfinpyrazone in the prevention of sudden death after myocardial infarction. *New Eng J Med* 1980: 392: 250–6.
3. Two active treatments and a placebo
   Disease: Mild hypertension
   Active treatment: 1. Bendrofluazide
   Active treatment: 2. Propranolol
   Placebo: Inactive substance
   Reference: Medical Research Council Working Party. Randomized controlled trial of treatment for mild hypertension: design and pilot trial, *B M J* 1988: i: 1438–40.

4. Patient management: Special vs. General
   Disease: Acute stroke
   New management: Special stroke unit
   Standard management: General medical ward
   Reference: Garraway, W.M., Akhtar., A.J., Presscott, R.J., et al. Management of acute stroke in the elderly: preliminary results of a controlled trial. B M J 1980: i: 1040–3.
5. Two standard treatments
   Disease: Operable breast cancer
   Standard treatment 1: Radical mastectomy
   Standard treatment 2: Simple mastectomy+ radiotherapy
   Reference: Langlands, A.O., Presscott, R.J., and Hamilton, T. A clinical trial in the management of operable cancer of the breast. *Brit J Surg* 1980: 68: 180–4.

Any clinical trial's goal has to be to discover the real solution to a pertinent medical problem. This indicates that the results must be based on an objective assessment of objective evidence rather than a subjective compilation of clinical opinion. Therefore, randomized controlled trials are the most reliable method of determining the true efficacy of a medication or treatment for any given medical issue.

9.6 Essential Design Features of RCT

In the past, there have been many haphazard and frequently inconclusive attempts to evaluate the use of therapeutic procedures. Over the past three decades, the effectiveness and safety of novel treatments have been assessed through the application of scientific experimentation principles. This section aims to clarify the key scientific and statistical concepts that are crucial for determining the efficacy of a treatment at the conclusion of a clinical trial (Pocock, 2023).

9.7 Research Design Features

Selecting the test and control treatments: The treatments used in the test and the control must differ from one another (Porkess, 2023). For investigators to assign patients to either treatment, both treatments must be medically justified. The test and control treatments need to meet the following general requirements:

- They have to be able to be identified from one another.
- *They need to be supported by medicine.
- The use of either treatment requires an ethical justification.
- The treatments used must be appropriate for the study participants' medical needs.
- For both the patients being studied and the doctors giving them care, the treatment must be acceptable.
- The effectiveness of the test treatment must be reasonably questioned.
- There ought to be good reason to think that the advantages of treatment will exceed the hazards.

*The mode of treatment delivery must be in line with the trial's design requirements (e.g. in a double-masked trial, the mode of delivery must be the same for each treatment) and as close to actual use as is reasonably possible.

The size of the expected treatment difference is significantly influenced by the control treatment selection. It will be harder to prove that the test treatment is superior the more successful the control treatment is. Depending on the type of disease, a placebo, a standard course of medication therapy, or no treatment at all may be used as the control treatment in a drug trial. Even if it would be unethical to deny control patients treatment, this cannot happen. If a patient needs treatment for a condition, they must get some kind of medical attention.

1. The classical trial design comparing two treatments A and B can be depicted as:
   Treatment **A**

   Study sample Outcome

   Treatment **B**
   The asterisk indicates the randomization or other mode of treatment assignment.
2. Outcome measure
   The outcome measure used for treatment comparisons will be a clinical event (e.g. death, myocardial infarction, significant loss of vision, recurrence of a disease) or a surrogate outcome measure (e.g. blood pressure change, serum lipid level). The desired characteristics for the primary outcome measure are:

   - Easy to diagnose or observe
   - Free of measurement or ascertainment errors
   - Capable of being observed independent of treatment assignment
   - Clinically relevant
   - Chosen before the start of data collection

   The investigator has the usual difficult choice between 'hard' (objective, easily measurable) and 'soft' (subjective, not so easily measurable) outcomes. The latter may lead to differential information bias in favor of a new experimental treatment, and suitable blinding of both study subjects and observers of the outcome is necessary to protect against such bias.

9.8 Establishing Comparable Study Groups

For there to be a meaningful basis for comparison, the test and control groups' baseline characteristics need to be roughly comparable (Rodda et al., 1992). When comparing two treatments, the best experimentation mode is one in which the two study groups' baseline characteristics are exact replicas of each other. To do this, a uniform set of patients must be randomly assigned to receive the test and control treatments. One patient in each pair is assigned to the test treatment and the other to the control treatment in an alternative trial design wherein pairs of patients are enrolled, matched on all relevant baseline characteristics. In real life, matching is not practical. By first grouping patients into subgroups based on these characteristics and then allocating members of each subgroup to the test or control treatment in the same proportion as for all other subgroups, it is possible to assume the comparability of the study groups for a few important baseline characteristics. But this method, called stratification, can only manage the distribution of a small number of variables at most. Appropriate patient selection can partially satisfy the need for comparability. Most trials have eligibility and exclusion criteria aimed at minimizing study population variability. However, when a highly homogeneous population is studied, the desire for patient homogeneity and the resulting improvement in study precision must be balanced against decreased opportunities for generalizations (Pocock et al., 2015b). To prevent selection biases in the formation of study groups, it is imperative that treatment assignments be made impartially, independent of the patient and clinic staff, once the eligible patient has consented to be enrolled. For an allocation scheme to be considered sound, the following general conditions must be met:

- Assignment remains masked to the patient, physician, and all other clinic personnel until it is needed for initiation of treatment.
- Future assignments cannot be predicted from past assignments.
- The order of allocations is reproducible.
- Methods for generation and administration of the schedule are documented.
- The process used for generation has known mathematical properties.
- The process provides a clear audit trail.
- Departures from the established sequence of assignments can be detected.

The accepted standard for creating treatment groups is randomization. There is still a good deal of misunderstanding regarding the reasons for randomizing. While the process does provide a basis for certain types of statistical analysis, it is far more useful as a method of making bias-free treatment assignment. The term random is often misused in medical circles by investigators who equate a haphazard and random process (as in referring to a random blood sugar determination, or in characterizing a group of arbitrarily selected individuals as a random sample). The details of the randomization procedure will be separately dealt with.

9.9 Parallel vs Crossover Design

In the clinical trial designs, we have considered, a group of patients receiving a test treatment is compared with a control treatment. Such a design is called a parallel design because the study groups receive their respective treatments simultaneously, i.e., in parallel. In a crossover design, however, each study patient receives each treatment in series by "crossing over" in sequence from one to the other. For example, patients with asthma might each receive in sequence two different treatment regimens to see which of the two treatments is more efficacious among the group as a whole. Crossover trials have a major advantage over parallel trials in statistical efficiency, in that a given treatment difference is demonstrable with fewer patients. There are two reasons for this: (i) each patient receives both treatments (in a two-treatment comparison) and thus counts twice, and (ii) variability in treatment response due to individual patient characteristics is eliminated. Proper conduct of crossover trials, however requires randomization of treatment sequences (A, B vs. B, A), time-dependent (rather than outcome-dependent) crossover treatments, and elimination of (or control for) carry-over effects from the first treatment. More about crossover trials will be dealt with separately (Nauta, 2011).

9.10 Blinding and Bias Control

The aim of any trial should be to collect data that are free of bias, particularly treatment-related bias. This has the potential to obscure a treatment difference or create the impression that one exists when, in fact, it does not. The usual practice to protect against treatment-related bias is blinding or masking. Blinding of study patients is needed to protect against contamination (confounding) of the true treatment effect by the so-called placebo effect. The placebo effect is the nonspecific effect that any treatment can have on the outcome, especially when the patient believes it to be efficacious. The main reason for providing look-alike, feel-alike, smell-alike, taste-alike placebo treatments when comparing an experimental treatment with no treatment is to facilitate patient blinding. Blinding of observers is also necessary to prevent the information bias that would occur if the outcomes are determined by observers who are aware of the treatment received. Such awareness can influence the outcome assessment, either consciously or unconsciously, especially if the outcome is subjective. It is conventional to refer to the trials as un-blinded, single-blinded, or double-blinded (Kestin, 2018). A double-blinded trial is one in which neither the patient nor the physician responsible for treatment is informed of the patient's treatment assignment; a single-blinded trial is one in which the patient is not informed of the treatment assignment, but the treating physician is; and an un-blinded trial is one in which both the patient and the physician are informed of the treatment assignment. The logistics of blinding are not easy. They are discussed separately. Unfortunately, blinding may be infeasible for some treatments. This is obviously true for most surgical procedures but also pertains to many behavioral and health-care interventions. Furthermore, un-blinding can arise due to differences inside effects that occur with different treatments, even if those treatments originally seem indistinguishable. A double-blinded trial, as defined above, is characterized by masked data collection. However, even single-blinded or un-blinded trials may be designed so that data collection is done in a masked fashion, that is, treatment information is withheld from personnel responsible for data collection. This requires one set of personnel to administer the study treatments and another to collect the data needed for assessing the study treatments. Laboratory tests should be performed, recorded, and reported by personnel who are masked to treatment assignment, regardless of the level of treatment masking in the rest of the trial (Kestin, 2006). The only exceptions are cases in which treatment assignment is needed to determine the tests to be performed. Also, data processing should be performed by personnel who are masked.

9.11 Efficacy vs Effectiveness in Clinical Trials

One of the major problems likely to arise during a clinical trial is that patients may either not have received or complied with the treatment to which they were assigned, or, having received it initially, may have switched to another. Another problem is that subjects may withdraw from participation in the trial before treatment or follow-up is complete. Such problems have led to two different ways of analyzing the results of a clinical trial and, consequently, to two different interpretations. Which of the two approaches is taken depends on whether one is interested in treatment efficacy or treatment effectiveness. Efficacy refers to the potential effect of treatment under optimal circumstances, i.e., whether treatment can have an effect on the outcome. Effectiveness refers to the actual effect of treatment in the "real world" of people who comply poorly, change treatment, or become lost to follow-up, i.e. whether treatment does have an effect on the outcome. Both efficacy and effectiveness may be important (Jokura et al., 2016). Efficacy is of primary interest to those who are interested in the biological potency and mechanism of action. Clinical trials remain the 'gold standard' for evaluating treatments. They have contributed enormously to the evolution of more effective therapeutic and preventive measures for a variety of clinical conditions and are represented by an increasing proportion of articles published in leading medical journals. Although the use of a clinical trial design, even with randomized treatment assignment, does not offer certainty on the conclusions of the study, the RCT is the research design of choice whenever a randomized assignment is feasible and ethical.

9.12 Randomization

In a controlled trial, the mode of deciding the regimen for any individual patient must not only be free of bias but also appear to be free of bias. One can readily see the danger in entrusting the choice of treatment for individual patients to the clinician. To take a simple example, if patients were to be treated at home or in a sanatorium at the clinician's discretion, it is almost certain that the more ill patients would tend to be admitted to the sanatorium while the less ill patients would be treated at home. Another undesirable procedure is the method of alternation, whereby the first patient is prescribed treatment A, the second treatment B, the third treatment A, the fourth treatment B, and so on. A variant of this is to admit all patients on odd days to treatment A and those on even days to treatment B. Such procedures, however, are capable of bias because the order in which patients are admitted to a study can be manipulated without much difficulty. The best protection against all accusations of bias is random allocation. Apart from validating statistical tests of significance, the purpose of random allocation is to avoid personal preferences in the choice of treatment for individual patients. It must be emphasized that these personal preferences can be conscious or, more often, subconscious. The great advantage of random allocation is that it is highly likely to result in the construction of two groups that are similar in all aspects. The baseline characteristics of the test-treated and control-treated groups must be more or less similar to provide a valid basis for comparison. The ideal experimental model for comparing two treatments is one in which the baseline characteristics of the two study groups are identical in all aspects (Jaykaran et al., 2010). This requires a homogeneous group of patients who are arbitrarily assigned to the test and control treatments. Once an eligible patient has agreed to be enrolled, it is essential that the treatment assignment is made free of influence from both the patient and clinic personnel to avoid selection bias. The general conditions that should be satisfied to have a sound allocation scheme are detailed below:

i. Assignment remains masked to the patient, physician, and all other clinic personnel until it is needed for initiation of treatment.
ii. Future assignments cannot be predicted from past assignments.
iii. The order of allocations is reproducible
iv. Methods of generation and administration of the schedule are documented
v. The process used for generation has known mathematical properties.
vi. Departures from the established sequence of assignments can be detected.

9.13 Patient Registration

In practice, it is advisable to adopt a more formal, well-defined routine for registering and randomizing patients. For each patient who might be considered suitable for inclusion in a clinical trial, the following formal sequence of events should take place:

Patient registration typically involves the following steps: scheduling an appointment, checking in at the facility, providing personal information (name, contact details, insurance), verifying identity, completing medical history forms, gathering necessary documentation, and confirming insurance coverage; essentially, collecting all necessary data to create a patient record for billing and treatment purposes.

- Appointment scheduling
- Check-in at the facility
- Patient Information Gathering
- Identity Verification
- Medical History Collection
- Insurance Eligibility Check
- Form Completion
- Data privacy and Entry

9.14 Patient Recruitment

The identification of appropriate patients may seem like an obvious step, but it is important to ensure that the patients entered in the trial are representative of all patients with the disease under investigation so that the trial's conclusions can be readily applied to the entire population of such patients. The clinician should avoid being unduly selective in their choice of patients.

9.15 Checking Eligibility

The eligibility of each potential patient should be checked. The clinician should review the list of eligibility criteria in the protocol and exclude the patient automatically if any criterion is not fulfilled. Of course, some ineligible patients may slip through the net only to be identified later, and such patients may be excluded from the analysis. A more serious problem arises when ineligible patients go undetected due to poor supervision.

9.16 Agreement to Randomize

It is vital to ensure that the clinician agrees to accept any of the random treatment assignments before formally entering the patient into the trial. Otherwise, there may be a problem where, after the treatment is assigned the clinician opts to use another treatment instead. Such action is entirely unacceptable and could seriously invalidate the trial. Ideally, each clinician should be willing to enter and randomize all eligible patients; however, in practice, one must allow a clinician to exclude the occasional patient they consider unsuitable.

9.17 Patient Consent

In general, there is an ethical need to inform each patient about their entry into a clinical trial and obtain their agreement before proceeding further. In the United States, it is a legal requirement that fully informed patient consent be obtained in writing prior to randomization, whereas in some countries, there is a more informal attitude toward patient consent.

9.18 Formal Entry to Trial

It is essential that each patient entering the trial be formally identified before the random assignment of treatment is made. This can be achieved by recording the patient's name and possibly a few other details, such as date of birth and date of entry. There is value in keeping a separate log of eligible patients who are not entered into the trial, as this provides insight into the representativeness of the patients who are in the trial.

9.19 Random Treatment Assignment

Only after the above steps have been successfully completed does the doctor learn which treatment the patient has been assigned. In most clinical trials, a randomization list of consecutive random treatment assignments is prepared in advance. One essential feature is that the clinician does not know the order of this list and is unable to predict the next assignment. To achieve this aim the list should be prepared by an independent person, possibly a statistician.

The randomization list could be transferred to a sequence of sealed envelopes, each containing the name of the allotted treatment on a card. The clinician opens the envelopes in sequence when the patient has formally entered the trial. Sealed envelopes are a well-tested method of randomization but are not totally foolproof.

If the trial is a double-blind evaluation of drug therapy, the pharmacist preparing the drugs should be involved. They need to be given the randomization list to produce a corresponding sequence of drug packets containing the appropriate treatments, but identical in appearance, etc. These packets are then presented to the clinician, and the treatment assignment proceeds as for sealed envelope randomization, except that the clinician still does not know which treatment the patient is receiving, even after randomization.

An unsuitable alternative is to supply each clinician with sufficient quantities of each drug, coded A or B, so that they do not know which drug is which, and then to use one of the other methods for randomization to assign A or B to each patient. This procedure is ill-advised since, if the clinician "breaks the code" for any single patient—either by necessity or chance—the actual treatment is revealed for all their patients, and blindness is destroyed.

9.20 Preparing the Randomization List

First, we shall consider a simple unrestricted method of randomization, which is essentially equivalent to repeated coin tossing. Then, various methods are described for restricting randomization to ensure approximately equal numbers of patients on each treatment. There are also various stratified randomization methods, which take into account a few patient characteristics to ensure that the treatment groups are not dissimilar (Gilmore, 2008).

9.21 Simple Randomization

For a randomized trial with two treatments (say A and B) the basic concept of tossing a coin (head=A, tail=B) repeatedly is quite reasonable but rather clumsy and time-consuming. Thus, people find it more convenient to use a table of random numbers instead. For instance, Table 9.1 shows such a table of random digits from 0 to 9. Each digit occurs, on average, the same number of times. The table presents digits in pairs

TABLE 9.1

Remarkable Examples of Entirely Inaccurate Conclusions that an Investigator Might Draw in the Absence of a Control Group of Patients

Treatment	Group	Percentage of Cure
Cured or improved	Anti-histamine	68% of 589
(Common cold)	Placebo	65% of 588
Relapse rate	3 years treatment	1% of 88
(Pulmonary tuberculosis)	2 years treatment	0% of 84
Incidence of giddiness	(i) Streptomycin + INH	35% of 88
	(ii) PAS + INH	11% of 80

merely to help the user scan across the page (Evans, 2010). A randomization list may be generated by using the digits, one per treatment assignment, starting with the top row and working downwards:

For two treatments, assign A for digits 0–4
and B for digits 5–9
Hence, the numbers in the top row of Table – 1
0 5 2 8 8 4 3 8 4 1 6 etc.
Produce a list starting **A B A B B A A B A A** B etc.
For three treatments, say A, B, and C,
assign A for digits 1–3
B for digits 4–6
C for digits 8–9
and ignore 0
so that numbers 0 5 2 8 8 4 3 8 4 1 6 etc.
Produce a list starting **A B A B B A A B A A** B etc..
For four treatments, say A, B, C, and D,
assign A for digits 1–2
B for digits 3–4
C for digits 5–6
D for digits 8–8
and ignore 0 & 9
so that numbers 0 5 2 8 8 4 3 8 4 1 etc.
Produce a list starting **C A D D B B D B A** etc..

Although the process is straightforward, it is advisable to prepare the list ahead of time and ensure that it is long enough to complete. These randomization lists can be created for any duration needed.

The advantage of employing such a simple strategy is that probability theory guarantees that each treatment assignment is completely random and that, over time, the number of patients assigned to each treatment will not differ significantly (Day et al., 2000). However, because clinical trials can only include a certain number of participants, it is necessary to account for the possibility that the final treatment numbers may differ. For example, using the above list with two treatments and 20 patients may result in 8 patients on treatment A and 12 on treatment B. This can be inconvenient and inefficient, especially if treatment A is the new treatment and treatment B is the standard. It would be particularly problematic to have 4 patients on one treatment and 16 on the other in a trial with 20 participants. The likelihood of a severe imbalance becomes negligible only in large trials (say, more than 200 patients), at which point simple randomization may be advised. However, this could still present a challenge if one plans to analyze preliminary data while the trial is still ongoing. To guarantee consistent treatment numbers throughout a trial, it is frequently preferable to restrict randomization.

9.21.1 Replacement Randomization

After preparing a randomization list as described above, one can check if there is any serious inequality in treatment numbers. If it is unsatisfactory, then one can generate an entirely new simple randomization list to replace the first one. It is unlikely that one will need to repeat this more than once to achieve reasonably equal-sized treatment groups throughout the trial. The decision to replace a randomization list should be based on objective criteria. For instance, one could stipulate that, in a trial involving two treatments and about 100 patients, a straightforward randomization with an inequality of 10 or more at any point should be used instead (Cleophas et al., 2009).

9.21.2 Random Permuted Blocks

A more conventional method of restricted randomization involves ensuring that there are precisely equal treatment numbers at certain equally spaced points in the patient assignment sequence. A table of random numbers (Table 9.1) can be used to generate the randomization list for blocks that are relatively small in size.

For two treatments and blocks of two patients,
assign AB for digits 0–4
and BA for digits 5–9
C for digits 5–6
Then the numbers 0 5 2 8 8 4 3 8 etc.
Produce a list starting **AB BA AB BA BA AB AB BA** etc.
For three treatments and blocks of three patients,
assign **ABC** for digit 1
ACB for digit 2
BAC for digit 3
BCA for digit 4
CAB for digit 5
CBA for digit 6
and ignore 0 & 8–9
Then the numbers 0 5 2 8 8 4 3 etc
produce a list starting **- CAB ACB - - BCA BAC** etc.
For two treatments and blocks of four patients,
assign **AABB** for digit 1
ABAB for digit 2
ABBA for digit 3
BBAA for digit 4
BABA for digit 5
BAAB for digit 6
and ignore 0 & 8–9
Then the numbers 0 5 2 8 8 4 etc
Produce a list starting **- BABA ABAB - - BBAA etc.**

One issue with this is that a clinician who records previous assignments could figure out what the next course of treatment would be at the end of each block. It appears that the likelihood of randomization becoming predictable increases with block size, so a block size of two should be avoided in particular. In order to decrease predictability, a trial should, in general, have a fairly large block size; however, if interim analysis is to be conducted, the block size should not be so large as to raise the possibility of a significant imbalance. A trial with, let us say, 100 patients or more might, for instance, have a block size of 20. A list of random permutations of the numbers 0 to 19 is provided below. You can use it as needed.

For two treatments, and blocks of ($n=20$) patients,

Assign A for digits	0–9
and B for digits	0–19

Then, starting with the top permutation and working downwards, we obtain the following list:

Block-1

11	19	15	5	9	0...............	8	3	4
B	B	B	A	A	A...............	A	A	A

Block-2

14	12	0	1	19	8...............	6	13	3
B	B	A	A	B	A...............	A	B	A

For three treatments, and blocks of 15 patients,

Assign A for digits	1 –59
B for digits	6–10
C for digits	11–15
and ignore 0 & 16–19	

Block 1

11	19	15	5	9	0...............	8	3	4
C	-	C	A	B	-...............	B	A	A

Block 2

14	12	0	1	19	8...............	6	13	3
C	C	-	A	-	B...............	C	C	A

If one prefers to use random permuted blocks of size 10 or less, one can still use Table 9.2 by simply ignoring numbers 10–19 in each block. To help reduce the predictability of random permuted blocks, one could vary the block size at random from one block to the next.

TABLE 9.2

Factor and Risk-Assigned Treatment

Factor	**Sex -1**		**Risk (2)**			
Level	**1 Male**	**2 Female**	**1(H)**	**2(M)**	**3(L)**	**Total**
Patient	1	16	10	13	9	26
Group	2	14	10	12	6	24
Total		30	20	25	15	50

9.22 Stratified Randomization

In any randomized trial, it is desirable that the treatment groups be similar in regard to certain relevant patient characteristics. For instance, in a breast cancer trial, it would be unfortunate if the proportion of premenopausal patients was very different between treatments. It would affect the credibility of any treatment comparisons. Although there are methods of statistical analysis to allow for such a lack of comparability, readers are more likely to be convinced when valid conclusions can be reached from a simple presentation of results for comparable treatment groups. Stratified randomization is like an insurance policy in that its primary aim is to guard against the unlikely event of the treatment groups ending up with some major difference in patient characteristics. The first issue is deciding which patient factors one should stratify by. This is best achieved by studying the outcomes of previous patients, preferably in earlier trials. One should be quite convinced about a factor's potential to have an impact on the outcome before it is included in stratification. If investigators have only a vague idea about which factor may affect the outcome, then it may be wise to proceed with unstratified randomization (Ashby, 2006). Take a breast cancer trial, for instance, where the two patient characteristics of age and nodal status (number of positive auxiliary nodes) were thought to be of primary prognostic significance. Regarding the former, patients with more positive nodes are generally expected to have a worse prognosis; age plays a significant role in this regard as older patients typically have shorter survival times. The next step is to classify each of the pertinent patient factors into two or more levels after selecting them. Age was therefore divided into under 50 and over 50, and nodal status into 1–3 and ≥ 4 positive nodes. Although the split decision is unavoidably somewhat arbitrary, it should consider how many patients are most likely to fall into each category. Furthermore, in this instance, the age of 50 offers a rough distinction between premenopausal and postmenopausal status.

The purpose of categorizing each factor is to end up with several patient types or strata. In this case the four ($=2\times2$) strata are:

Age <50 and 1–3 positive nodes
Age ≥ 50 and 1–3 positive nodes
Age <50 and ≥ 4 positive nodes
Age ≥ 50 and ≥ 4 positive nodes

Then, before the trial begins, a separate restricted randomization list is prepared for each of the patient strata, with randomly permuted blocks being the usual approach.

9.22.1 Need for Control

What is a 'control' treatment?

A 'control treatment' is a medication, device, test, or procedure used in a clinical trial that acts as the benchmark by which 'test treatments' are assessed. Depending on how the study is designed, the control treatment could be a placebo, a sham operation, a standard treatment plan, or no treatment at all.

Should we need a control treatment?

Considering the first example, which deals with the common cold, a British Medical Research Council trial showed that 68% of patients receiving an anti-histamine preparation had a complete cure or improvement in their condition within 2 days, a highly encouraging finding for the use of anti-histamines. However, practically the same proportion, namely 65% of the placebo patients, also had a cure or improvement within 2 days. In reality, therefore, the anti-histamine had no effect, and the improvement noted was probably the natural course of the disease (Au-Yong et al., 2020). The next example is from the Tuberculosis Research Centre, Madras, and refers to relapse rates in patients with bacteriologically quiescent tuberculosis. In patients who received chemotherapy for 3 years, the total relapse rate in the third, fourth, and fifth years was only 1%. This low proportion could have led to a recommendation that 3 years of chemotherapy is necessary to keep the relapse rate low (in fact, similar recommendations have been made in the literature on the basis of findings like those above). Such a recommendation is, however, totally unwarranted since, in the patients who received only 2 years of chemotherapy, the relapse rate was 0%. The third example is also from the Tuberculosis Research Centre, Madras, and gives the incidence of giddiness among patients while on treatment. The incidence of giddiness was 35% in 88 patients who were treated with streptomycin and isoniazid, compared to 11% in 80 patients who were treated with PAS and isoniazid. In the absence of the second group (i.e., PAS+INH), one may have concluded that the incidence due to streptomycin was 35%, but by considering the finding in the second group, the incidence attributable to streptomycin is more accurately estimated as (35 − 11= 24%). After accepting the need for a control group receiving standard treatment, many researchers are still reluctant to assign patients randomly to new or standard treatment. Such reluctance often stems from the investigator's desire to enroll all future patients in the new treatment because of his wish to gain as much experience of it as possible and his inclination to believe it is better anyway. The most common way of avoiding randomization is to compare retrospectively one's current patients on the new treatment with previous patients who had received standard treatment; this latter group of patients is commonly known as historical controls. Such an approach has one major flaw: how can one ensure that the comparison is fair to treatment? That is, if the treatment and control groups differ with respect to any feature other than the treatment itself, how can one guarantee that any apparent improvement in patient response is actually due to the new treatment? The 'potential incompatibility' can be divided into two broad areas: patient selection and the experimental environment, each of which may give rise to several sources of bias.

9.23 Patient Selection

1. A historical control group is less likely to have clearly defined criteria for patient inclusion, since such patients on standard treatment were not known to be in the clinical trial when their treatment began.
2. Since historical controls were recruited earlier and possibly from a different source, there may be a change in the type of patient available for selection.
3. There is a danger that the investigator may be more restrictive, either deliberately or subconsciously, in his choice of patients for a new treatment (Pocock et al., 2015c).

9.23.1 Experimental Environment

1. The quality of recorded data for historical controls may be inferior.
2. The criteria of response may differ between the two groups of patients.
3. Ancillary patient care may improve with the new treatment.

The result of all these problems is that studies with historical controls tend to exaggerate the value of a new treatment. Having a concurrent control group—one that is based on data gathered during the same time period as that used to generate all other data in the study—is the most satisfactory way to mitigate these risks. There might still be disagreements regarding whether future patient assignments to the different treatments should be made randomly, even if the investigators have consented to a prospective clinical trial in which those assignments will be made (McDermid et al., 2021). Alternatively, a degree of judgment by the investigator and/or patient may be used, or worse, a predetermined systematic method (assigning patients based on date of birth, date of presentation, etc.) may be chosen. These two approaches have the potential to add significant bias. In one trial, for example, patients who were admitted on odd days of the month were given anticoagulants, whereas patients who were admitted on even days were not. The trial involved myocardial infarction patients. There were 589 treated patients and 442 control patients, a significant disparity that suggests that anticoagulant admission is preferred. In general, the most important method for determining the effectiveness of new therapeutic developments is to conduct randomized controlled trials with concurrent controls. When randomization is used appropriately, it not only greatly lowers the chance of variations in the experimental environment but also ensures that patients are not selected unfairly for different treatments. Compared to other forms of treatment allocation, randomized allocation is easier to implement and makes trial conclusions more credible.

9.23.2 Preparation of Protocol

A protocol may be described as a document that sets out, in great detail, the aims and objectives of the investigation, the design of the experiment including the sample size, and the procedures involved in the execution (Singh et al., 2013). It serves many purposes. First of all, it compels the investigator to set out in writing all the relevant facts and procedures. There is, fortunately, a healthy fear in most minds of putting things down in writing and consequently, the process of writing will usually ensure that the investigator has spent a reasonable amount of time planning the study, in consultation with other experts, including a statistician. Secondly, in situations where many persons are involved in investigations—for example, a co-operative study involving a number of centers—a written protocol is the most satisfactory way of ensuring that all the co-investigators understand the purpose of the study and the method of conducting it. Finally, for individuals or institutions seeking financial support for their investigations, a good protocol can go a long way in influencing the decision of technical committees that allot funds for various projects.

9.24 Comprehensive and Satisfactory Protocol

9.24.1 Background or Preamble

To begin with, a brief explanation regarding the background of the proposed investigation should be given. This must make appropriate references to previous work on the subject by the investigator as well as by others, and might also indicate the value of the proposed investigation.

9.24.2 Clear Statement of the Aims

Often, investigations have more than one aim. It is therefore important to mention clearly the various aims, and also indicate their relative importance as there can sometimes be a clash of interests. For instance, if the proposed investigation is a controlled clinical trial to investigate a new regimen, the main aim of the investigation could be to determine the therapeutic efficacy of the new regimen in comparison with a standard regimen, and the evaluation of toxicity and acceptability could be subsidiary objectives (Seto et al., 2010). To take another example, the main aim of a laboratory study could be to determine the effect of storing sputum specimens at room temperature on the viability of the tubercle bacilli in them; a subsidiary objective could be to examine the implications of the susceptibility of the bacilli to anti-tuberculosis drugs.

9.24.3 Population

Next, the population for study must be clearly defined. This could be (i) all residents in a specified geographic area, (ii) all patients with a particular disease in a specified group of hospitals, (iii) all women in the child-bearing age in a specified urban community, (iv) all diagnostic specimens of sputum received in a laboratory over a specified period of time, etc. In order to avoid ambiguity and with a view to making meaningful generalizations, terms like residents, disease, child-bearing age, and diagnostic should be defined. One of the studies at our Centre is an investigation of the efficacy of isoniazid chemoprophylaxis in close family contacts of tuberculosis patients. The words 'close family contacts' and 'tuberculosis patients' may mean different things to different people. To us, the former meant persons related to the patient by blood or marriage and who have been living in the same house and using the same kitchen as the patient for at least the 3 months immediately preceding the start of treatment for the patient, and the latter meant persons with radiographic and bacteriological evidence of pulmonary tuberculosis.

9.24.4 Sampling Procedure

For economy in time, labor, and money, investigations are usually undertaken on a sample of the total defined population. The technique adopted for selecting the sample is extremely important and usually has a great bearing on the confidence that can be placed on the investigation findings. For one thing, the procedure must be free of bias. There are several methods of sampling from a population – for instance, purposive sampling, simple random sampling, stratified random sampling, multi-stage sampling, and cluster sampling. These methods vary considerably in their complexity and efficiency, and much thought must therefore be given to the subject of selecting the optimal sampling design.

9.24.5 Number of Subjects

The number of subjects to be included in an investigation must always be determined in consultation with a statistician and depends on various assumptions. For instance, in a study of a vaccine, the assumptions made could be that the incidence of disease in non-vaccinated subjects would be 10% and that the vaccine would be at least 50% effective. While fixing the sample size, allowances must be made for loss of information due to various causes – for instance, absconding by patients in clinical trials and non-response by subjects in surveys. It must be emphasized that if the proportion of such losses is large, the sample may no longer be representative.

9.24.6 Specification of Investigations to Be Undertaken

The nature of investigations that are to be undertaken must be specified clearly so that some degree of uniformity is preserved during the study. The techniques employed for laboratory tests or clinical assessments must be described extensively, unless they are standard methods.

9.24.7 Brief Review of Investigation

For investigations that involve one-time assessments, the main problem is one of non-response. The procedures to be employed for tackling this problem should be included in the protocol. For instance, a revisit may be made at a time convenient to the respondent. For investigations that involve repeated observations on subjects (e.g., guinea pigs, patients, contacts) over a period of time, the assessments and the periods at which they are to be undertaken must be specified (Ulm & Nelson, 2021). For example, in a diabetes trial blood sugar and HbA_{1c} estimations may be made monthly and once every 3 months, respectively. Also, the full details of management and methods for coping with non-co-operation during the course of the study must be spelled out. Finally, the circumstances under which a subject can be withdrawn from the study should also be stated (e.g. a diabetes trial patient may be withdrawn from the trial if the postprandial blood sugar is above 300 mg%, at any time during the trial treatment period, and confirmed by another assessment made 1 week later).

9.24.8 Forms

A copy of the forms that are proposed to be used in the study must be included in the protocol since the mode of recording the data is as important as that of collecting it. In the case of a clinical trial, a copy of the consent form should also be enclosed.

9.24.9 Planned Analyses

Although it may not always be necessary, it would nevertheless be valuable to state the various analyses contemplated for the data to be collected (Wang et al., 2007). This would have the advantage that analyses that are considered important at the time of planning the study are not forgotten at a later date, especially in studies of chronic diseases that are spread over several years. In the diabetes trial, for instance, the criteria for the control of diabetes may be stated as 'fasting blood sugar less than 120 mg%, postprandial blood sugar less than 180 mg%, and HbA_{1c} less than 8.5%.

9.24.10 Concluding Remarks

A protocol must be treated as a sacrosanct document, departures from which are not permissible. This being the case, it would be prudent to consider initially all aspects of the study, including possible complications that may arise later. Such a step would usually ensure that the study is completed strictly in accordance with the initial protocol. Despite the greatest care, however, situations may sometimes arise in which a departure from the protocol becomes necessary, probably due to ethical considerations or some new facts that have come to light since the time the protocol was prepared (Wong et al., 2019). In such situations, the original protocol should be amended, and copies of the amendment circulated to all those connected with the study.

9.25 Design of Proforma

The design of a proforma is the translation of the protocol into the form of well-defined, unambiguous questions, acceptable to all those associated with the conduct of the study. If the questions are incomplete, ill-conceived, or inadequately answered, no statistical analysis can compensate for the resulting defects and produce the answers that the investigator had hoped to get. Hence, it is worth spending considerable time on this aspect of proforma design, as the better the design, the easier and more efficient the subsequent analysis will be (Wu & Sargent, 2011).

9.25.1 Steps in Proforma Design

- Keep the study objectives in mind.
- Make a detailed list of information needed and the study variables.
- Decide on the main sub-divisions of information needed and select a logical order of types of information.
- For each piece of information or variable in each sub-division as decided earlier, write a question as it will be conversed.
- Assess each and every question to ensure that it is simple, clear, short, unambiguous, and that it can be answered well.
- Do pre-testing and revise if needed.

Assessment of each question to verify whether it is clear, short, unambiguous, and simple, or whether it could be answered well, must be done rigorously. This can be done by posing the following questions to each and every question.

a. What will I discover from the answer?
b. Will the answer be meaningful?
c. Will the answer be relevant?
d. Can I leave it out completely?

These questions ensure that we will fetch the required answer and address the content part of the question.

The following two questions, namely

a. Is the question unambiguous?
b. Are the words simple?

Will help us in framing the questions properly, i.e., the wording aspect of the question. Then, we must decide whether to have an open-ended question or a closed (pre-coded) question (i.e., form). This depends on the situation and also on the nature of the variables. Finally, the logical positioning of the questions in the proforma is to be determined, i.e., the ordering of the questions must be done.

9.25.2 Question Content

In our anxiety to include as many questions as possible, we should not add irrelevant questions to our proforma, as an irrelevant question is costly. It consumes precious time and may reduce efficiency. Responders may become tired and bored. Additionally, investigators may sense the irrelevance of certain questions and ask them carelessly. This can have a carry-over effect on subsequent, more important questions. The next aspect is whether it is possible to collect the required information precisely, even if the question is relevant. The responder may not remember the answer, as memory can fade over time. In some instances, the responder may not reveal the correct answer due to fear or concerns about prestige. This is especially true in the case of sensitive issues like income or history of abortions, as questions on these topics may arouse fear, emotions, or discomfort.

9.25.3 Question Wording

The wording should be sufficiently specific and the language should be simple, i.e., very specific and to the point ('crystal clear'). We must avoid ambiguity and vagueness. Finally, we must not lead the respondent. This is most important. Therefore, we must avoid putting leading questions that may imply spontaneous approval or disapproval of the issue being investigated.

9.25.4 Open-Ended Questions

Open-ended questions are characterized by having no predefined answer options, encouraging respondents to provide detailed, descriptive responses in their own words, often starting with "what," "how," or "why," and aiming to explore opinions, feelings, and motivations rather than just collecting factual data; they are considered qualitative in nature and require analysis through interpretation of the text responses provided.
Key parameters of open-ended questions:

- No set answer choices
- Prompts for elaboration
- Focus on qualitative data
- Requires interpretation
- Can be used to explore complex topics

Important considerations when crafting open-ended questions is the (a) clarity and conciseness; (b) avoid leading questions; (c) be mindful of length; and (d) follow-up questions, etc.

9.25.4.1 Advantages

a. Allows the respondent to give a more complete answer.
b. Does not constrain the respondent to precise responses, enabling us to obtain answers that were not anticipated or studied earlier.
c. Most suitable for exploratory studies.

9.25.4.2 Disadvantages

a. Interviews may be too long and difficult to keep on point.
b. Responses may be more difficult to code and analyze due to the wide variety of answers.
c. The respondent may forget to mention important points.

9.25.5 Closed Questions

9.25.5.1 Advantages

a. Responses can be obtained and recorded quickly.
b. Respondents are helped to concentrate on important issues/points.
c. Since the response is structured, data processing and analysis are made easier.
d. Grading of subjective responses can be done.

9.25.5.2 Disadvantages

a. Options are limited by previous knowledge.
b. There is a tendency for the respondent to select the middle option/alternative from a list of options/alternatives provided.
c. The interviewer may lead the respondent with the alternatives provided in the proforma.
d. May be less useful for exploratory studies.

9.25.6 Question Order

It is important that opening questions should be easy to gain the confidence of the respondent, and highly sensitive questions should be left as late as possible in case they may cause refusals. The sequence of questions with broad to specific answers often seems to be appropriate. We must set a logical sequence in the ordering of the questions.

9.25.7 Number of Questions

Much attention should be given to this aspect. Obviously, the number of questions that need to be asked varies widely with circumstances. But the designer of the proforma should ask himself the following three questions for each question to be included in the proforma:

a. Is this question essential?
b. Can I obtain a useful answer to it?
c. Can I analyze it usefully at the end?

Such self-discipline is likely to reduce the size of any proforma. However, in our desire to reduce the size of the proforma, we should not arbitrarily cut short the number of questions. We must strike a balance.

9.25.8 Number of Proforma

Consider a clinical trial in which patients are to be treated for a period of 1 year with monthly follow-ups. Can we just have a single form to collect the data for the entire 1 year period? 'No,' as it is operationally convenient and also efficient to have several forms for each patient. Additionally, multiple forms per patient ensure the compliance of patients. Errors can be detected by periodic scrutiny of the data collected, and necessary mid-course corrections, if needed, can be made. Furthermore, we can have an up-to-date database, which will facilitate quick interim analysis.

9.25.9 Practical Aspects

a. The Document—Layout.
b. The items should not be crowded.
c. No more than one question per line.
d. Adequate space should be provided for open-ended questions.
e. Provide boxes for coding open-ended questions.
f. The unit of measurement should be specified along with each item.
g. In the case of questions with many alternatives, provide a box next to each alternative to mark the answer with a tick (✓).
h. Skipping instructions should depend only on the answer to the question immediately preceding the instruction.
i. Provide a separate box for coding each numeral.
j. The size of each box for recording numbers should be sufficiently large.
k. Boxes should be arranged serially and vertically.
l. The layout should permit the data entry personnel to proceed in an orderly fashion.
m. The use of capitals, letters of different types, italics, and lines of varying thickness that appeal to the eye makes the schedule agreeable as well as easily understood.
n. Allot a running serial number to each item to be filled in.

9.26 Physical Characteristics

- Use good-quality and standard-size paper.
- Use clear fonts in printing.
- Print on one side only.
- use as few sheets as possible.

9.26.1 Patient Management

Before approaching the subject matter of this lecture, it is necessary to provide working definitions of three different processes.

9.26.2 Patient Follow-Up

A process involving periodic contact with the patient after enrollment into the trial for the purpose of administering the assigned treatment, observing the effects of treatment, modifying the course of treatment, and collecting data to evaluate the treatment.

9.26.3 Patient Close-Out

A process carried out to separate a patient from the trial, involving the cessation of treatment and **termination of regular follow-up.**

9.26.4 Patient Post-Trial Follow-Up

A process that involves patient follow-up after completion of the close-out stage of the trial and that is designed to yield information on the primary or a secondary outcome measure. This lecture deals with the various steps involved in carrying out these three processes.

9.27 Management of Investigator and Patient Interest during Follow-Up

The follow-up process requires a dedicated and committed staff to schedule and carry out the required examinations and a willing patient population. Both are needed for the trial to succeed.

9.28 Investigator Interest

Investigator commitment to the trial and interest in its activities must be high throughout if it is to succeed. Interest will be easy to maintain in a short-term trial, where the initial enthusiasm that usually accompanies the start of any new activity is enough to carry it through to completion. However, even in such cases, spirits can sag before the data analyses are done and the final paper has been written. They can sag long before that point in long-term trials. The following list describes some aids that can be used to maintain investigator interest:

a. Periodic meetings of all study personnel.
b. Distribution of periodic progress reports on patient recruitment and follow-up, data collection, and other performance characteristics of the trial for review by all members of the investigative group.
c. Periodic newsletters distributed to study personnel designed to inform them of study progress, protocol changes, and so forth.
d. Investigator participation in the analyses of results and in writing or presenting papers concerning the trial.
e. Preparation of reports and papers during the course of the trial summarizing the design, organizational and operating features of the trial.
f. Execution of concurrent ancillary studies.
g. Certificates of appreciation from the sponsor.

The above list is written with long-term multi-center trials in mind. However, morale problems are not unique to multi-center trials. They can be just as significant in single-center trials as well.

9.28.1 Patient Interest

A patient's interest in the trial and willingness to continue in it can be expected to diminish with the passage of time. The longer the period of follow-up, the greater the need for measures to counter act the loss of interest and improve participation levels. Factors and approaches that can help sustain patient interest in the trial are given below:

The most important factor is the attitude of the clinic staff. Clinic staff should treat patients with courtesy and dignity and should take an interest in meeting their needs.

The clinic should be located in pleasant physical surroundings and in a secure environment.

a. Payment of clinic registration fees and costs for procedures required in the trial.
b. Special clinics in which patients are able to avoid the confusion and turmoil of a regular outpatient clinic.
c. Scheduled appointments designed to minimize waiting time.
d. Clinic hours designed for patients' convenience.
e. Written or telephone contacts between clinic visits.

9.29 Losses to Follow-Up

A loss to follow-up occurs whenever an item of information required as part of a scheduled follow-up examination is not obtained within the permissible time window. The loss may be due to

- Failure of the clinic staff to complete an item on an otherwise properly completed data form.
- Failure of the patient to agree to certain procedures during an examination.
- Failure of the patient to return to the clinic for an examination within the time window specified for it.

Losses due to missed examinations or examinations that are not done within the specified time window, and hence are counted as missed, are more worrisome than losses resulting from failure to complete specific items or procedures during an examination. Furthermore, an occasional missed examination for a patient has different implications than a sequence of missed examinations. The longer the sequence, the greater the uncertainty regarding the outcome status of the patient. Patients who are no longer able or willing to return to the clinic for scheduled follow-up examinations are drop-outs. This declaration may be made by the patient (e.g., by announcing an intent to leave the study because of a lack of interest or because of a forthcoming move to another place) or by clinic staff. A clinic declaration should not be made until: (i) Clinic staff has made a concerted effort to locate the patient if they have disappeared and tried to convince them to return to the clinic for a follow-up examination. (ii) The patient has missed a specified number of follow-up clinic visits. The date of the patient's last completed follow-up examination should be used as the date of drop-out. The patient should remain classified as a drop-out until or unless they return to the clinic for a follow-up examination.

- The willingness of a patient to remain under active follow-up will depend on a variety of factors including the following:
- The amount of time and inconvenience involved in making follow-up visits to the clinic.
- The importance of the procedures performed at follow-up visits from a health maintenance point of view.
- The potential health benefits associated with treatment versus potential risks.
- The amount of trauma and discomfort produced by the study treatment or procedures performed.
- The number and type of side effects associated with treatment.

A certain number of drop-outs in long-term trials will occur simply because of patient relocations. Drop-outs should be contacted at periodic intervals. The contacts may be made via home visits or by mail.

9.30 Close-Out of Patient Follow-Up

The process of disengaging patients from a trial may require as much skill and care as the enrollment process. The following is a summary list of considerations that should be addressed in planning for close-out:

- Time schedule.
- Information to be collected.
- Phased treatment disengagement (applicable to drug trials)
- Nature of recommendations given to patients regarding subsequent treatment.
- Method of ensuring proper transfer of patient care responsibilities to an alternate clinic or physician when appropriate
- Ensuring patients have ample opportunity to make alternative arrangements for care and to have any questions answered regarding the trial and its outcome before separation.
- Method of summarizing baseline and follow-up data for subsequent use by the patient's private physician.
- The separation can be an emotional experience for both patients and clinic staff. It should be based on a detailed plan that has been constructed and reviewed before the start of close-out. The details of the separation should be discussed with patients well before separation occurs Clinic staff must spend whatever time is necessary to answer questions and help find suitable alternative sources of care. This is essential in any trial that has been providing with patients routine medical care.
- Close-out provides an opportunity to assess the adequacy of the mask in masked trials. Theoretically, such checks could be made at various time points

throughout the trial. However, they are not usually carried out due to a desire to discourage speculation concerning the treatment assignments since the assessment involves asking the masked individual(s) to state a guess regarding treatment assignment.

- A key consideration at close-out has to do with whether to carry out added data collection on patients as they are separated from the trial. The wisdom of making such provisions depends on the importance of the data generated in relation to the aims of the trial. Results obtained for tests or procedures for which there are no corresponding baseline values will be of limited use in making treatment comparisons.

The method of terminating therapy in masked drug trials must be given special consideration. A dosage step-down scheme may be necessary if an abrupt cessation of one or more of the drugs is considered unsafe. In addition, a patient will want to know the treatment they were on. Hence, study physicians must be supplied with treatment codes well in advance of close-out visits.

Patients should be told at close-out if the clinic plans to keep in touch with them and, if so, the reason for doing so and the way in which contact will be maintained (e.g. via mail, telephone, or home visits). They may be asked to sign a consent authorizing the contacts and to provide updated location information if contacts are planned. In fact, it is a good idea to alert patients to the possibility of future contacts and obtain consent for them even if subsequent contacts are not planned.

9.31 Post-Trial Patient Follow-Up

There should be a sound rationale for any post-trial follow-up involving direct patient contact. The prime motivation for most post-trial follow-ups stems from a desire to extend the period of observation for death or some other non-fatal events. Another reason may be to observe patients for a disease or condition that may be caused or aggravated by treatments administered during the trial. The usefulness of the information obtained will depend on the completeness of the follow-up and the nature of intervening treatments administered after close-out. Interpretation of the results will be easiest if patients have not been exposed to any additional treatment after separation from the trial. It will be problematic if they have been exposed.

9.31.1 Data Collection and Quality Assurance

Decisions regarding the data collection schedule and related forms are among the most important in the trial. They will determine both the amount and quality of data generated in the trial. First of all, one must decide on the schedule at which patients are seen, both before and after entry into the trial, and the specific items of information to be collected at each time point. Once this is decided, the process of developing data forms should be started. This should be done by personnel who are experienced in form construction and are familiar with methods of data collection and data processing in prospective studies.

A review of sample forms used in other trials, especially those from trials with design and operating features similar to the one in question, will be useful in the development of study forms. The sample forms may be obtained through published literature or by writing directly to the investigators. The data forms thus developed should be tested in the clinic, and suggestions, if any, received from the clinic personnel should be taken into account before finalizing them.

9.31.2 Baseline Visit

Every clinical trial must provide for data collection at a minimum of two time points: at or just before randomization and the initiation of treatment to provide baseline data, and at least once after randomization for the collection of follow-up data.

9.31.3 Follow-Up Visits of Baseline Investigations

- Determine a patient's eligibility for enrollment.
- Provide baseline data for assessing changes occurring after the treatment.
- Explain the purpose of the study to the patient and obtain consent for participation in the trial.
- Issue the treatment assignment.
- It is useful to have the patient make at least two visits to the clinic before enrollment. These repeat visits make it possible to replicate certain key baseline measurements. A time separation (days/weeks/months) between visits may be needed to:
    a. Perform the necessary screening and diagnostic procedures for determining the patient's eligibility.
    b. Allow adequate time for clinic staff to evaluate the data collected on the patient before enrollment.

Required diagnostic and data collection procedures should be designed to minimize patient inconvenience and exposure to unnecessarily risky procedures. Hence, whenever feasible, the simplest procedures with the least risk should be performed first so that patients who then prove to be ineligible can be spared risky, more complex, and time-consuming procedures.

9.31.4 Baseline Adaptive Randomization

Assume two treatments (1 & 2) with two prognostic factors (1 & 2) – gender and risk group

Factor 1–2 levels (M&F)

Factor 2–3 levels (M&F) – (High, Medium, and Low Risk)

$$\text{let } B(t)=\sum_{i=1}^{2} W_i \text{ Rage } \left(X_{i,j}^{t}, X_{i,2}^{t}\right) \; W_i = \text{Weight for each factor}$$

For example, $W_1 = 3, W_2 = 2, \; W_1 / W_2 = 1.5$

$X_{i,j}$ =Number of patients in *i*th factor and *j*th treatment
$X'_{i,j}$ =Change in $X_{i,j}$ if next patient is assigned *Rx* at *t*

$$p=\frac{2}{3}\text{ for smallest } B(t)=\left(\frac{2}{3},\frac{1}{3}\right)$$

Now we can determine *b*(1) and *b*(2) for patients #51. If we assigned treatment 1 (*t*=1)
9a), calculate *B*(*t*) and assign pt #51 to treatment 1 (*t*=1).

1. Factor 1, Level 1 (Male)

	Treatment Group	Now	Proposed
	K	K_1K	$K_1'K$
Treat Group	1	16	18
	2	14	14

Range=18−4=3
Calculate *B*(*t*) and assign pt #51 to Trt 1) *t*=1

2. Factor 2 Level 3 (Low risk)

	Treatment Group	Now	Proposed
	K	K_1K	$K_1'K$
Treat Group	1	4	5
	2	6	6

Range=5−6 =1

*B(1)=3(3)+2(1)=11
b) Calculate *B*(2) and assign pt #51to Trt 2 *t*=2

1. Factor 1 Level 1

	Treatment Group	Now	Proposed
	K	K_1K	$K_1'K$
Treat Group	1	16	16
	2	14	15

Range=16−15 =1

2. Factor 2, Level 3 (Low Risk)

	Treatment Group	Now	Proposed
	K	K_1K	$K_1'K$
Treat Group	1	4	4
	2	6	8

Range=4−8=3
B(2) =3 (1)+2 (3)=9
(c) Rank *B*(1) and *B*(2) measures of imbalance

t	*B*(*t*)	Assign *t* with Probability
2	9	2/3
1	11	1/3

TABLE 9.3
Correlation of Sex and Risk

Factor	Sex -1		Risk (2)			
Level	1 Male	2 Female	1(H)	2(M)	3(L)	Total
Patient	1	16*	10	13	4*	26
Group	2	14	10	12	6	24
Total		30	20	25	15	50

* Significant at 1% level.

9.31.5 Follow-Up Visit

A follow-up visit is any visit, either required or non-required, to the study clinic by a patient who has been enrolled in the trial that takes place after the randomization visit. Required visits should be scheduled to occur at specified time points after the randomization visit (Table 9.3).

- Follow-up visits are needed to:
- Carry out procedures specified in the study protocol, including treatment administration.
- Evaluate the patient's response to treatment.
- Assess patient adherence to the assigned treatment.
- Collect data needed for the evaluation of the treatments.

The timetable for required follow-up visits will be dictated by various factors, including

a. Requirements for treatment administration and for assessing adherence to treatment.
b. Rate of occurrence of the outcome(s) of interest.
c. Patient health care needs.
d. Cost of patient visits.
e. Patient convenience considerations.

The schedule for required follow-up visits may be designed to allow for more frequent visits immediately after enrollment (e.g. weekly) to permit clinic personnel to initiate and administer the assigned treatment. The interval between visits may then be increased to some maximum (e.g. monthly) and held constant thereafter. Follow-up visits that are made on an ad hoc basis because of special problems experienced by the study patients after enrollment are referred to as non-required follow-up visits. Provisions should be made in the data forms to distinguish between required and non-required follow-up visits. Differences among the treatment groups in the number of non-required follow-up visits can lead to biases in the diagnoses and reports of clinical events used to evaluate the study treatment.

9.31.6 Visit Time Limits

Ideally, the entire set of baseline and follow-up visits for a patient should occur at precise time points relative to the time of randomization. However, such precision is generally not possible, nor is it necessary for most of the observations required in a typical clinical trial. The usual approach is to consider a visit and the related data collection as valid if the visit took place within a defined interval (e.g. a week) on either side of the desired time point.

9.31.7 Quality Assurance

Quality assurance, as applied to clinical trials, refers to any method or procedure for collecting, processing, or analyzing study data that is aimed at maintaining or enhancing their reliability or validity. Examples include:

1. Visual checks of completed data forms for illegible responses and for unanswered or incorrectly answered items.
2. Rekeying data as a check for errors in the data entry process.
3. Edit procedures to verify the accuracy of items on completed data forms.
4. Repetition of a laboratory determination to check reproducibility.

9.31.8 Visual Check

A visual check is the process in which a member of the research team (e.g. statistical assistant) scrutinizes the completed form for completeness and correctness. It is good practice to conduct a visual check immediately after a form is completed, as far as possible, so that corrections, if any, can be carried out promptly. This process allows illegibly answered items, unanswered items, or incorrectly answered items to be identified and corrected.

9.32 Data Entry and Verification

Data that are modified and entered on the day of origination, or very soon after, are the best candidates for data intake systems. However, this is typically not feasible. Completing the research form while the patient is being examined is one workable strategy. The data should then be entered as quickly as possible. Completed forms from multi-centric trials may be delivered to the data center on a regular basis (e.g. once a month) for data entry and storage. Data verification is the process of having a different individual rekey the information on a completed form after it has been input into a computer to identify flaws in the data entry process. If there are any mistakes, they should be fixed immediately.

9.32.1 Data Edit

The process of identifying, querying, and, when necessary, correcting inaccurate values in a data set is referred to as data editing. A number of edit inquiries and edit checks are part of the standard editing procedure.

An edit check is a process applied to a single item or set of items on a completed data form to identify potential problems. For example, it may compare data from one part (e.g. anemia) to another (e.g. Hb) on the same form to detect discrepancies or identify items whose values fall outside designated ranges (e.g. age range: 15–45).

An edit query is a form-generated inquiry that questions the sufficiency or correctness of a particular piece of information on a completed form. It requires someone at the generation site to review the information and respond to the query.

9.32.2 Replication

Replication of an observation or reading is often used as a check on the quality of the data obtained. Examples are:

1. Comparison of two independent measurements, such as a lab test, to determine if the difference observed is outside a specified range.
2. Use of two independent readings of an ECG to identify items of disagreement for adjudication by a third reader.
3. Averaging two or more consecutive blood pressure readings made on a patient during a given clinic visit in order to have a more reliable estimate of the patient's 'true' BP.

9.32.3 Quality Control of Lab Data

It's crucial to design systems where bias cannot possibly exist as prejudice can infiltrate laboratory studies. For example, it should be set up so that the lab workers are not aware of the source of individual specimens when smears are checked or urine tests are conducted. Additionally, standards in the laboratory should be checked by unexpectedly adding controls.

9.32.4 Site Visits

A site visit can be described as an outside staff member's visit to a trial center to evaluate how well it is performing. One person or several people may carry it out. Activities for the visiting team could include: (i) meeting with the clinic director; (ii) meeting with clinic staff; (iii) watching clinic staff perform specific procedures; (iv) checking forms, etc., at the clinic to make sure they are current; (v) speaking with study participants to verify the informed consent process, etc. In conclusion, every attempt should be made to gather high-quality data so that, upon completion of a clinical trial, meaningful conclusions can be reached.

9.32.5 Open and Double-Blind Trials

Regardless of how well a medication works in a test tube or on experimental animals, it must eventually be evaluated in humans. In other word, it needs to go through some kind of clinical testing. A clinical trial is a planned study intended to evaluate a treatment's effectiveness in humans by contrasting the results of a test-treated patient group with those of a similar patient group receiving a control treatment. Both patient groups are enrolled, treated, and monitored for the same amount of time and under similar conditions. One way to create the groupings is by randomization. A laboratory test, a non-fatal clinical episode, or death could be the outcome measure. Depending on the outcome measure, the observation time may be short or long. Therefore, the impartial comparison of a test treatment (such as a new therapy) with a control treatment (such as a standard drug, placebo, or both) forms the foundation requirement of clinical trials.

9.32.6 Open Trial

In most clinical studies, the patient and the physician are aware of the specific treatment each patient is getting. We refer to these trials as 'open trials'. However, open trials are not without flaws, particularly when the assessment criterion is subjective (e.g., nausea). Let's say, for instance, that a researcher wishes to determine the prevalence of giddiness in patients with tuberculosis receiving streptomycin treatment. Streptomycin is known to make people giddy. Therefore, the evaluating doctor would be on the lookout for giddiness symptoms and would diagnose a patient with giddiness as soon as the patient complained of them if he knew that the patient in question was receiving streptomycin treatment. His evaluation would thus be skewed, albeit subconsciously.

9.32.7 Double-Blind Trials

A double-blind trial is a clinical trial in which the patient and the physician are not aware of the treatment the patient is receiving. Double-blind studies are a challenging task to carry out. The medications being tested (e.g., the control and test drugs) ought to be created in the same way, with respect to size, shape, color, taste, and other aspects. For both groups, the same dosage, timing of administration, and length of treatment should apply. While extremely challenging, this is not insurmountable. It takes a lot of preparation and creative thinking to carry out a double-blind study that is truly double-blind. I'll list a few of the issues we ran into while using traditional remedies in double-blind trials and describe the solutions we came up with.

9.33 Some Special Problems of Double-Blind Trials

The test and control medications may differ in color (for example, black and white), dosage (for example, 500 and 250 mg), length of treatment (for example, 40 and 12 days), or timing of administration (for example, before and after food). Let's examine how to solve these issues when creating a double-blind experiment.

9.33.1 Color Matching

Assume the control drug is white and the test drug is black in color. To make the white control medicine and the test drug capsules identical, we may add a coloring agent (like charcoal) to the test drug capsule.

9.33.2 Dosage Matching

The fact that the dosages of the control and test medications vary may also be an issue. For instance, the dosage of the control medicine might be 250 mg, whereas the dosage of the test drug might be 500 mg. 500 mg unfilled capsules can be used to encapsulate the test medication. We could combine 250 mg of an inert material (like lactose) with 250 mg of the control medication, then encapsulate the combination in similarly sized empty capsules.

9.33.3 Duration Matching

A further issue that could arise while organizing a double-blind trial is the differing treatment durations of the control and test medications. Let the treatment period for the test drug be 40 days and the treatment period for the control drug be 12 days. One way to solve this issue is by putting the test medication in a pink bottle for the first 12 days and a white bottle for the final 28 days. Likewise, place a placebo in a white bottle for 28 days and the control medication in a pink bottle for 12 days. Give all research participants instructions to finish the contents of the pink bottle before moving on to the white bottle.

9.33.4 Time of Administration Matching

Let the test drug be taken before food and the control drug after food. To overcome this problem, one may follow the method given below:

Capsulate	**Test** drug in **Green** capsules
	Control drug in **RED** capsules
	Placebo in **Green** & in **Red** capsules
Prescribe	**Green** capsules before food &
	Red capsules after food.

This will ensure that the test drug is taken before food and the control drug is consumed after food, as shown below:

	Test Series	Control Series
Before food	Test drug (Green)	Placebo (Green)
After food	Placebo (Red)	Control drug (Red)

9.34 Conclusion

An open trial may be conducted if the evaluation parameter (blood sugar, for example) is objective. On the other hand, a double-blind experiment is recommended if it is subjective (such as nausea). Double-blind studies are more easily completed.

9.35 Crossover Trials

The effectiveness of any treatment created in a lab or evaluated on test subjects must eventually be evaluated on human subjects. A clinical trial is a research project that involves human subjects. Nowadays, a well-respected technique for determining the relative effectiveness of various illness therapies is the controlled clinical trial. In this kind of trial, a predetermined number of patients are assigned at random to receive either the control (standard) medication or the test (new) treatment; in other words, patients are only administered one treatment at a time, not both. This kind of design is typically used in situations when the length of the observation or treatment period is considerable, such as in chronic illnesses. But in circumstances where the goal is restricted, the patient can be treated as their own control, with each receiving the test and standard treatments for equal lengths of time. This way, the results can be compared to relatively short treatment durations, such as brief relief in asthmatic patients or subjective improvement in rheumatoid arthritis patients' signs and symptoms. A trial in which the order in which the treatments are administered is decided at random is referred to as a crossover trial. The following are typical crossover trial examples:

1. Comparison of an oral mouthwash (treatment A) with a placebo mouthwash(treatment B)
2. Study of the relative efficacy of two treatments *X* and *Y* on muscle blood flow and cardiac output in patients with intermittent effect on the novel treatment.
3. Comparison of two steroid inhalers for the treatment of bronchial asthma.
4. Comparison of a new anti-hypertensive drug with the standard drug.

9.35.1 Advantages

a. The main advantage is that the assessment of the difference between the two treatments will be more precise because the treatments are compared 'within subjects'. Thus, characteristics of individual patients will be common to both treatments, and subsequently, the effect of the characteristics of the patients on the contrast between the two treatments will usually be eliminated.
b. There would be a saving in the number of patients to be admitted. However, each patient has to be tested twice, but this is often easier to arrange than testing a large number of patients on one occasion.

9.35.2 Disadvantages

In principle, these trials may seem at first sight to be simple and straightforward, but in practice, one needs to consider their feasibility and reliability. For example, the disease condition may not be stable until the completion of the trial; patients'' cooperation may not be good enough to ensure the completion of the full course of treatments, i.e. inadequate compliance; and the presence of the 'carry-over effect', i.e. the effect of treatment persists after the end of the treatment period. To overcome this, a reasonable interval called the 'washout period' should be allowed between the two treatments.

9.36 Assumptions

1. There is no period effect, i.e. the average difference between two treatments A and B is independent of the sequence of administration, i.e., it is the same for A followed by B and for B followed by A.
2. There is **no treatment-period interaction**—the effects of treatment and period operate independently of each other, i.e. the difference in response to both treatments between the two periods is the same.

To test the hypothesis that there is **no period effect,** an independent t-test can be employed. If there is some evidence of a period effect, the overall treatment effect is taken as the mean of the two separate effects and tested for significant departure from zero with a *t*-test. The interpretation of the finding must be made with caution since the difference depends on the sequence of administration of the treatments.

Similarly, an independent t-test can be employed for testing the hypothesis that there is **no treatment-period interaction** by testing for equality of the means of the two treatment sequences, viz., A followed by B and B followed by A.

Illustration: Suppose ($n=8$) patients with mild to acute bronchial asthma were randomized such that eight received treatment A followed by treatment B and the others had B followed by A. The response of interest was the forced expired volume in 1 second (FEV_1) (Tables 9.4 and 9.5).

First, one can perform a crude overall within-patient comparison of treatments. That is, for each patient, calculate the difference (d=FEV_1 after A - FEV_1 after B). For example, the first patient had $FEV_1=1.28$ L (y_1) after A and 1.33 (y_2) after B, so the difference is -0.05 L. The mean of these 18 differences was −0.26 L, indicating that the FEV_1 tended to be somewhat lower on drug A. Test whether this mean difference in FEV_1 is different from 0.

$$t = \frac{\bar{d}}{\text{S.D. of }(d)/\sqrt{n}}\frac{0.26}{0.49/\sqrt{18}} = 2.19(P < 0.05)$$

There is evidence of lower FEV_1 with treatment A than with treatment B. CI 95% for the mean difference in FEV_1=0.02, 0.50). This analysis makes two main assumptions: no period effect and no treatment-period interaction, which should be checked.

9.37 Period Effect

To check the first assumption, the mean differences for AB will be compared with the mean differences of BA with a sign changer. This can be examined by comparing the mean difference for the two treatment orders (Table 9.6).

TABLE 9.4

Response of Interest was the Forced Expired Volume in One Second (FEV_1)

Subject Run	Group 1 (AB)		Group 2 (AB)		Group 1 (AB)		Group 2 (AB)	
	P_1 (X_1)	P_2 (Y_1)	P_1 (X_2)	P_2 (Y_2)	P_1 (X_1)	P_2 (Y_1)	P_1 (X_2)	P_2 (Y_2)
1	1.09	1.28	1.24	1.33	1.84	3.06	1.54	1.38
2	1.38	1.60	1.90	2.21	2.41	2.68	2.13	2.10
3	2.28	2.46	2.19	2.43	3.05	2.60	2.18	2.32
4	1.34	1.41	1.48	1.81	1.20	1.48	1.41	1.30
5	1.31	1.40	0.85	0.85	1.80	2.08	2.21	2.34
6	0.96	1.12	1.12	1.20	1.89	2.82	2.05	2.48
8	0.66	0.90	0.88	0.90	0.89	1.94	0.82	1.11
8	1.69	2.41	1.90	2.89	2.41	3.35	2.83	3.23

Subject: Run-in Period1 Wash-out Period2 Subject Run-in Period1 Wash-out Period2.

TABLE 9.5

Marginal Difference of Response of Interest was the Forced Expired Volume in One Second (FEV_1)

Subject Run	Group 1 (AB)		Group 2 (AB)		Group 1 (AB)		Group 2 (AB)	
	y_1 (X_1)	y_2	y_1+y_2	y_1-y_2	y_1 (X_1)	y_2	y_1+y_2	y_1-y_2
1	1.28	1.33	2.61	–0.05	3.06	1.38	4.44	1.68
2	1.60	2.21	3.81	–0.61	2.68	2.10	4.88	0.58
3	2.46	2.43	4.89	0.03	2.60	2.32	4.92	0.28
4	1.41	1.81	3.22	–0.40	1.48	1.30	2.88	0.18
5	1.40	0.85	2.25	0.55	2.08	2.34	4.42	–0.26
6	1.12	1.20	2.32	–0.08	2.82	2.48	5.20	0.26
8	0.90	0.90	1.80	0.00	1.91	1.11	3.05	0.83
8	2.41	2.89	5.20	–0.38	3.35	3.23	6.58	0.12

TABLE 9.6

Significance of Period Effect

Statistics	A followed by B	B followed by A
Mean of differences	–0.12	0.40
Standard deviation of differences	0.35	0.58
Nunber of patients	(n=8)	(n=8)

In this example, treatment A had lower FEV_1 in both sequences. The estimated overall mean period effect= $\frac{-0.12 + 0.40}{2}$ = 0.14L lower in the first period. One can test whether this is significantly different from 0 by a t-test for period effect as follows:

$$t = \frac{0.12 + 0.40}{\sqrt{\frac{(0..35)^2}{8} + \frac{(0.58)^2}{8}}} = 1.22\ (p > 0.05)$$

There is no evidence of a period effect. However, if there were evidence of a period effect, the crude overall mean treatment comparison may be slightly modified. The estimated mean difference then becomes, $\frac{-0.12 + 0.40}{2}$ = 0.26, the same as before.

The ***t*-test for treatment effect** is modified to

$$t = \frac{-0.1 - 0.40}{\sqrt{\frac{(0..35)^2}{8} + \frac{(0.58)^2}{8}}} = 2.28\ (p < 0.05)$$

To check assumption 2

For each patient calculate mean FEV1 = $\left(\frac{\text{FEV1 after A} + \text{FEV1after B}}{2}\right)$

Suppose we tested without treatment-period interaction. For each patient, calculate the mean.

$$t = \frac{1.63 - 2.14}{\sqrt{\frac{(0.63)^2}{8} + \frac{(0.66)^2}{8}}} = 1.63\ (p > 0.05)$$

There is no evidence of treatment-period interaction, and hence the original simple analysis remains valid. Sometimes, multiple treatments are compared using the above type of design (RBD). This is practicable when the intervals between the treatments are very short. For instance, if three treatments are to be compared, one might employ a Latin square design in blocks of three patients (Table 9.7).

TABLE 9.7

Assumptions Tested by Test Statistics

Statistics	A followed by B	B followed by A
Mean of patients	1.63	2.14
Standard deviation of differences	0.63	0.665
Number of patients	(n=8)	(n=8)

9.38 Factorial Trials

Even when there are more than two possible therapies for comparison, the investigators in many controlled clinical trials are only interested in comparing two treatments. The ability to distinguish changes across therapies is a significant issue because it depends on the number of individuals receiving each treatment. The researchers have to forgo three or more treatments because it is difficult to enroll a sufficient number of patients in the controlled clinical study and maintain their compliance. Taking into account factorial designs can help solve this issue to some extent without requiring more patients.

For example, in a controlled clinical trial in pulmonary tuberculosis, all 400 patients will receive once-weekly chemotherapy with streptomycin plus isoniazid for 1 year.

a. Half the patients, selected at random, will receive a third drug, PAS 6 g.
b. The streptomycin dosage will be 1 g for half the patients, selected at random, and 0.85 g for the remaining half.
c. The isoniazid dosage will be 18 mg/kg body weight for half the patients, selected at random, and 13 mg/kg body weight for the remaining half.

Thus, each of the three anti-tuberculosis drugs will be studied at two levels:

Streptomycin—1 g and 0.85 g, Isoniazid—18 mg/kg and 13 mg/kg body-weight, PAS—6 g and 0 g. The disease status at one year will be assessed as a score and compared for the various groups (Table 9.8).

Factorial trials are very efficient because every observation provides information for factors included in the study. Also, the trial is comprehensive, i.e., a systematic way of analyzing the effects of different factors either in isolation or in combination.

TABLE 9.8

Evaluation of Variance for the Randomized Controlled Trial (RCT) Data

Source	*df*
Streptomycin (S)	1
Isoniazid (I)	1
PAS (P)	1
Interaction (S * I)	1
Interaction (S * P)	1
Interaction (S * P x* I)	1
Between patients receiving the same dosage of S, I & P	392
Total	399

This is being used in research programs, particularly in industry. However, in clinical trials, the potentialities of this design have not been explored fully because it is usually larger, more complex than a single-factor trial, and, above all, difficult to find enough suitable patients to compare the combination of two or more treatments.

9.39 Multi-Center Trials

Multi-center clinical trials are trials conducted concurrently in several centers to assess the efficacy of the same treatment. It becomes necessary to undertake multi-center trials due to the scarcity of patients and the intention to complete the trial more quickly, as well as the desire to represent a broader geographic spectrum. The conclusions from multi-center trials have a broader and more representative base than those from single-center trials, as the former involve patients and clinicians from several centers. Before addressing the aspects requiring special considerations in multi-center trials, a brief review of the general features of any clinical trial (detailed below) is essential, as they are common to both multi-center and single-center trials.

9.40 Various Aspects of Clinical Trials

9.40.1 Planning

1. Design of the trial
2. Random allocation procedures
3. Design of proforma

9.40.2 Executive and Monitoring

1. Preparation and supply of random treatment assignments
2. Scrutiny of data collected
3. Rectification of errors

9.40.3 Data Analysis and Reporting

1. Database creation and validation.
2. Statistical tests and Inferences.
3. Preparation of the final report

In addition to the above-mentioned aspects, multi-center trials have some aspects that require special considerations.

9.40.4 Aspects Requiring Special Considerations

1. Random allocation of treatment to patients.
2. Standardized procedures for clinical and laboratory assessments.
3. Quality control of data collected.
4. Statistical analysis of data pooled from several centers.

The above is not an exhaustive list of aspects requiring special considerations in multi-center trials. The importance of these main aspects is explained below with illustrations.

9.41 Random Allocation

Random allocation within each center is preferable as the condition of patients and/or their response may vary between centers. Additionally, we require a balanced allocation for every '**n**' patients admitted to the trial, say, six patients. For example, if there are two groups of treatment, three should be allocated to one group, and the remaining three to the other group. Stratification within each center ensures equality of treatment numbers for every stratum (i.e. type of patient). However, as the number of strata increases, this becomes impractical. In such situations, certain special methods balance the marginal treatment totals for each type of patient, and the treatment assignment can be obtained from a Central Registration Office.

9.42 Need for Standardized Procedures for Clinical Assessments

Patients with rheumatoid arthritis were recruited for a clinical trial by two facilities, let's say A and B. Patients in both centers were required to walk a distance of 15 m. The average time to walk 15 m at center A was 25 seconds, in contrast to 15 seconds at center B. Our assumption was for a larger result, but the lower mean time at center B was unexpected because this center had more sick patients (ARC Classifications III & IV). It was found that, although patients at center A were allowed to walk independently, those at center B were permitted to do so with assistance from their attendant. Therefore, it is imperative that centers adhere strictly to defined data collection protocols and maintain standard criteria.

9.43 Need for Standardized Procedures for Laboratory Assessments

Three centers participated in a clinical trial with patients suffering from acute viral hepatitis, using the same brand of kit for biochemical evaluations and the same diagnostic criteria. In these circumstances, one would typically anticipate that the three centers would produce findings that are very comparable. The table below shows the average SGOT and SGPT scores at the three centers at the time of admission (Table 9.9).

TABLE 9.9

Vital Parameter Descriptive

		Mean Values on Admission	
Center	**Number of Patients**	**SGOT (Iu/L)**	**SGPT (Iu/L)**
A	146	80	159
B	129	401	488
C	162	255	301

Even though all three centers utilized the same brand of kit and the same diagnostic criteria, there was a significant difference in the mean SGOT and SGPT levels at admission. In particular, the readings at Center A were extremely low. This instance amply illustrates the necessity of technician uniformity and the methods that must be used prior to the trial's start.

9.44 Quality Control of Data Collection

Patients were required to visit the clinic once every 2 weeks for medication collection and clinical evaluation as part of a clinical trial. Upon reviewing the filled-out proforma sent by the center, it was found that not a single patient had ever missed a follow-up, and even more astonishingly, they had consistently shown up at the clinic on the designated date. During an impromptu site visit, it was discovered that patients had only visited the clinic once every 2 months or so, despite the fact that the data had been retroactively filled in for all prior fortnights. This underlines the importance of regular site visits to the participating centers as part of the quality control process.

9.45 Analysis of Data Pooled from Four Centers

Using the same technique, a trial was carried out at four separate locations to assess the effectiveness of a new medication. The following lists the cure rates attained at the four centers using both innovative and conventional medications:

Even though all four centers adhered to the same protocol—that is, using the same diagnostic standards, comparable patient populations, standard definitions for cure rates, etc.—differences between the centers cannot be avoided (maybe because of uncontrolled external factors). Instead of basing our judgment on the overall cure rate—that is, 81% vs. 63%—which can occasionally be misleading, it would be more appropriate to use specialized methods like Cochran's method that take into consideration the experiences at different centers to compare the new treatment with the standard. The particular approaches are advantageous as well as applicable in this case, as the comparison of overall cure rates (81% vs. 63%) results in a *P*-value of 0.05, but the unique way produces a *P*-value of 0.03, suggesting that the test is more sensitive. Considering everything that has been said so far, it is crucial to determine early on in the planning stages of a clinical trial whether or not a single-center study is adequate. If not, the following pros and cons of a multi-center trial should be taken into account (Table 9.10).

TABLE 9.10

Comparison of Standard One and New

	Cure	Rate
Center	**New Drug**	**Standard Drug**
A	68% (40/59)	58% (36/62)
B	68% (48/80)	59% (44/84)
C	69% (45/65)	62% (43/69)
D	89% (55/80)	82% (48/68)
Total	81% (188/264)	63% (181/282)

$\bar{d} = 8\%$, SE = 0.0408, P = 0.05

9.46 Advantages

a. Obviously, the main benefit of conducting a multi-center trial is that patients are accrued considerably more quickly, allowing the trial to grow larger and/or reach its target size sooner. Multi-center trials ultimately lead to speedier and more dependable findings, which accelerate overall progress in the treatment of a particular disease.
b. Since participants in a trial come from multiple centers, conclusions generated from it will have a wider, more representative base than those obtained from a single-center trial. This is because the trial incorporates patients and physicians from multiple centers. As a result, it is possible to confidently extrapolate findings to the entire community of comparable patients.
c. Ideally, clinical scientists' cooperation in a multi-center trial should lead to raised standards in the design, conduct and interpretation of the trial.

9.47 Problems

- Clearly, the planning and administration of any multi-center trial are considerably more complex than those of a single-center trial; hence, it is vital to have efficient, centralized coordination of all trial activities.
- Multi-center trials are very expensive to run, both in terms of staff and resources, so adequate funding for the study must be obtained beforehand.
- It is important to ensure that all centers strictly follow the study protocol. Adequate communication across centers at the planning stage is needed to obtain prior agreement regarding the nature of the study. In particular, any potential investigator who cannot agree to the eventual design should be encouraged to exclude themselves from further participation in the study before problems of non-compliance arise.
- The need for quality control regarding any measurements, clinical observations, and data recording requires prior recognition. Sufficient training and explanation should be provided to ensure consistency across centers.
- The collection and processing of data pose special problems in multi-center trials, which are often not anticipated. A well-organized data center is needed to receive all data and provide prompt and reliable feedback to each participating center regarding data requirements and problems

9.48 Initial Design Stage

1. Calculate the required sample size.
2. Outline the data collection schedule, quality control procedures, data analysis plans, and data entry and editing procedures.
3. Develop the organizational structure of the trial.

9.48.1 Protocol Development Stage

- Develop treatment allocation procedures.
- Develop computer programs and related procedures for receiving, processing, editing, and analyzing study data.
- Design and test data forms.
- Develop an interface for data transmission from clinics and other resource centers to the coordinating center.
- Train clinic personnel in the required data collection procedures.
- Distribute study data forms and related materials.
- Develop manuals needed for the trial, including the protocol, clinic manual of operations, etc.

9.49 Patient Recruitment Stage

- Administer treatment allocations, including checks for breakdowns in the assignment process.
- Assume a leadership role in outlining study needs for quality assurance.
- Implement editing procedures to detect data deficiencies.
- Develop performance monitoring procedures and prepare reports.
- Develop treatment monitoring and reporting procedures to detect evidence of adverse or beneficial treatment effects.
- Conduct site visits to participating clinics.
- Prepare study progress reports for submission to the sponsor.
- Update study manuals.

9.50 Treatment and Follow-Up Stage

- Prepare periodic reports on the performance of clinics.
- Carry out periodic training sessions to maintain a high level of proficiency at clinics in treatment and data collection procedure.
- Evaluate data processing procedures and modify them, if necessary.
- Assume a major role in writing a paper on design and methods.

9.51 Patient Close-Out Stage

- Develop plans for final data editing.
- Design and test computer programs needed for final data analysis.
- Develop plans for the final disposition of study data.

- Coordinate the logistics of patient disengagement from treatment.
- Assume a key role in writing papers summarizing the results of the trial.

9.52 Termination Stage

- Supervise the collection and disposal of unused study materials.
- Assume a leadership role in paper writing activities.
- Distribute draft manuscripts and published papers to participating centers.

9.53 Organization

The complex structure required to organize a large multi-center trial is illustrated by an example of the British trial of treatment for mild hypertension as described by Dollery and Brennan (1988). The trial is concerned with comparing two standard anti-hypertensive drugs with a placebo in patients with a diastolic blood pressure in the range of 90–109 mm H_g to determine if there is a subsequent reduction in mortality and morbid events such as non-fatal stroke. The trial, involving 18,000 subjects, is one of the largest multi-center trials ever undertaken in Britain, so its organizational structure, as explained below, is of particular importance (McDermid et al., 2021).

- Since the Medical Research Council is providing funding for the project, they have approved the study design and are updated on its status.

  The trial must be organized in every way, with the Chairman and Clinical Secretary in particular being in charge of this.

  The Trial Monitoring Committee is a team of skilled clinical researchers and statisticians tasked with supervising the overall direction of the trial; they are not involved in the day-to-day operations of the investigation.

  All ethical facets of the investigation are overseen by the Ethics Committee.

  The Coordinating Center comprises twelve to fifteen technical and administrative support employees, a statistician, and two clinical epidemiologists. Daily operations are conducted from this center, with a focus on data collection, subject follow-up, and result evaluation.
- There are about (n=200) trial clinics across the nation designed to screen middle-aged men and women to determine if they are willing and able to participate in the study. All biochemical testing is performed using Central Laboratory Services.

  This trial may be typical in scope given its size, but it offers a useful model for multi-center studies in terms of organizational structure. The committees might be combined in a smaller trial.

9.54 Future Line of Research Work on Medical Research

Currently, numerous medical colleges and research institutions are conducting drug trial experiments within their own facilities. The application of appropriate statistical methods minimizes random errors in the experimental data. Interim analysis plays a crucial role in monitoring confounders and outliers present in the collected research data. By leveraging the insights gained from statistical information, researchers can develop new research projects and enhance their understanding of statistical science in relation to medical research.

REFERENCES

Ashby, D. Bayesian statistics in medicine: A 25 year review. *Statistics in Medicine* **25**, 3589–3631 (2006).

Au-Yong, I., Howarth, S., Tang, T., Sayers, R. & Sivaprakasam, R. in *Cracking the MRCS Viva*, 348–351 (CRC Press, 2020). https://doi.org/10.1201/b13358-90

Bartlett, R. H. & Cornell, R. G. [Ethics and statistics in randomized clinical trials]: Comment. *Statistical Science* **6,** (2007).

Bernadt, M., Pignon, J. P., Poynard, T. & Senn, S. Statistics in clinical trials. *The Lancet* **335**, 614 (1990).

Cleophas, T. J., Zwinderman, A. H., Cleophas, T. F. & Cleophas, E. P. *Statistics Applied to Clinical Trials*, 1–559 (Springer Netherlands, 2009). https://doi.org/10.1007/978-1-4020-9523-8

Day, S. J. Statistics Notes: Blinding in clinical trials and other studies. *BMJ* **321**, 504–504 (2000).

Dollery, C., & Brennan, P. J. The Medical Research Council Hypertension Trial: The smoking patient. *American Heart Journal*, **115**(1), Part 2, 276–281 (1988).

Evans, S. R. Common statistical concerns in clinical trials. *Journal of Experimental Stroke and Translational Medicine* **3**, 1–7 (2010).

Gilmore, S. J. Evaluating statistics in clinical trials: Making the unintelligible intelligible. *Australasian Journal of Dermatology* **49,** 177–184 (2008).

Jaykaran, G., Kantharia, N. D., Preeti, Y., Bharddwaj, P. & Goyal, J. Reporting statistics in clinical trials published in Indian journals: a survey. *African Health Sciences* **10**, 204–207 (2010).

Jokura, Y., Yano, K., Watanabe, N. & Yamato, M. Bayesian statistics and clinical trial designs for human cells and tissue products for regulatory approval. *Regenerative Therapy* **5**, 86–95 (2016).

Kestin, I. Statistics in clinical trials and audit. *Anaesthesia and Intensive Care Medicine* **19**, 144–148 (2018).

Kestin, I. Statistics. *Anaesthesia and Intensive Care Medicine* **7**, 135–142 (2006).

McDermid, R. Statistics in clinical trials. *Anaesthesia and Intensive Care Medicine* **22**, 385–389 (2021).

Nauta, J. *Statistics in Clinical Vaccine Trials*, 1–153 (Springer, 2011). https://doi.org/10.1007/978-3-642-14691-6

Pocock, S. J., Clayton, T. C. & Stone, G. W. Design of major randomized trials part 3 of a 4-part series on statistics for clinical trials. *Journal of the American College of Cardiology* **66,** 2757–2766 (2015a).

Pocock, S. J., McMurray, J. J. V. & Collier, T. J. Making sense of statistics in clinical trial reports. *Journal of the American College of Cardiology* **66,** 2536–2549 (2015b).

Pocock, S. J., McMurray, J. J. V. & Collier, T. J. Statistical controversies in reporting of clinical trials part 2 of a 4-part series on statistics for clinical trials. *Journal of the American College of Cardiology* **66**, 2648–2662 (2015c).

Porkess, S. Statistics and clinical trials: It's all about the design. *Teaching Statistics* **45**, 27–35 (2023).

Rodda, B. E., Brooks, C. & Reynolds, G. Statistics and clinical trials. *Clinical Pharmacology & Therapeutics* **52,** 104–105 (1992).

Seto, W. Introduction to statistics in pharmaceutical clinical trials. *The Canadian Journal of Hospital Pharmacy* **63,** (2010).

Singh, V., Rana, R. K. & Singhal, R. Analysis of repeated measurement data in the clinical trials. *Journal of Ayurveda and Integrative Medicine* **4,** 77–81 (2013).

Ulm, J. W. & Nelson, S. F. COVID-19 drug repurposing: Summary statistics on current clinical trials and promising untested candidates. *Transboundary and Emerging Diseases* **68,** 313–317 (2021).

Wang, R., Lagakos, S. W., Ware, J. H., Hunter, D. J. & Drazen, J. M. Statistics in medicine — reporting of subgroup analyses in clinical trials. *New England Journal of Medicine* **357,** 2189–2194 (2007).

Wong, C. H., Siah, K. W. & Lo, A. W. Estimation of clinical trial success rates and related parameters. *Biostatistics* **20,** 273–286 (2019).

Wu, W. & Sargent, D. J. in *Clinical Radiation Oncology*, Third Edition, 223–237 (Elsevier, 2011). https://doi.org/10.1016/B978-1-4377-1637-5.00012-2

10

An Introduction to Sampling Techniques and Sample Size Determination in Biological Science Research

10.1 Introduction

Sampling is a common phenomenon, both in scientific studies and in everyday life. Our interest here is in the former (Qiu et al., 2024). The first question to be answered is, 'Why will sampling be necessary?' Sampling results in a reduction of cost, saving time, requires less manpower, has greater scope and flexibility, and if done well, could give more accurate results than attempts to study the entire population (Akbar et al., 2024). Besides, in some situations, it may well be impracticable to deal with the entire population. For instance, if we wish to determine the total number of mosquitoes in a town, we could not possibly think of catching every one of them, even if our resources were unlimited. Sample surveys are of two types: descriptive and objective surveys, which aim to estimate the proportion with a characteristic or the mean value of the characteristic, and analytical surveys whose purpose is to investigate some hypothesis (Mondal et al., 2024). In the context of sample surveys, the following terminology needs to be properly defined.

10.2 Population

Population is the universe; it is a group of individuals (totality of persons, objects, or specified events). These events may be real occurrences (Bujang et al., 2024), such as births, or a hypothetical visualization, such as the outcome of tossing a coin a hundred thousand times (Figure 10.1).

10.2.1 Theoretical Population

The population is defined by certain characteristics and expressions—for example, infected and uninfected populations of SARS-CoV-2. It is denoted by '*N*'.

10.2.2 Target Sample

The sample is the true representative of the population, denoted by '*n*'. An infected individual is drawn from the parental population with a specific age group (e.g. adults ages 40–50 years). Again, the sample has been divided into two types, namely realistic samples and unrealistic samples (Yoo, 2024) (Table 10.1).

TABLE 10.1
Difference between Realistic and Unrealistic Population

Realistic Sample (*n*)	Unrealistic sample (*n*)″
An individual case can express their symptoms during the window period	An individual case cannot express their symptoms during the window period
Certain attributes are definable Comorbidity	Certain attributes are not definable There is no comorbidity
Exact sample size is known	Exact sample size is not known because of carriers who do not express any symptoms
Tracking is very easy during the symptomatic and asymptomatic stages	Tracking is very complex during the symptomatic and asymptomatic stages

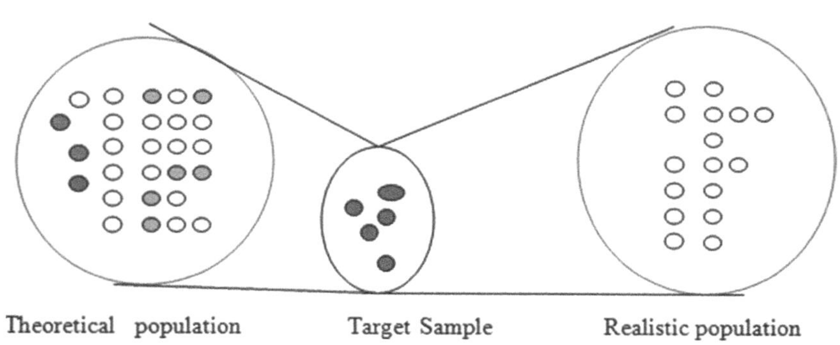

FIGURE 10.1 Structural relationship between population and sample.

DOI: 10.1201/9781032713816-10

10.3 Selection Procedure

10.3.1 Sampling Frame

This will be a list of all information (census blocks or villages), a list of households, or a list of persons (e.g. voters' list) from which the random sample for the study is drawn.

10.3.2 Sampling Unit

This is the unit of sampling and could be a town/census block, village, household or an individual person.

10.3.3 Sampling Fraction

The proportion of the population that is included in the sample—for example, 10%.

10.3.4 Sample Design

The choice of the sample design depends on numerous factors such as the degree of precision required for the estimates, the cost in terms of money and manpower, the amount of time at our disposal, and field logistics. Some of the commonly employed designs are very briefly mentioned below (Bourazas et al., 2024). The final choice of the sample design must be made in consultation with an experienced statistician.

10.4 Simple Random Sampling

Each member of the population must have the same probability of being included in the sample, from which it follows that each sample of size must have the same probability of being selected. Two factors that affect the quantity of information contained are: (i) the extent of variability of the observations and (ii) the sample size. In this context, it must be noted that the word 'random' is not used in the colloquial sense of "'haphazard' On the contrary, it means that the likelihood of any member being selected is predefined, and the choice of a particular member depends purely on chance (De Santis & Gubbiotti, 2024). A simple random sample is a sample chosen to ensure that every possible sample of a given size has an equal chance of being selected. There are two methods for drawing samples: Simple random sampling without replacement (SRSWOR) and Simple random sampling with replacement (SRSWR).

10.4.1 Simple Random Sample without Replacement (SRSWOR)

The sample's 'n units' are drawn from the population one at a time. It is drawn as $\left(\frac{1}{N}\right)$ for the first draw, $1/(N-1)$ for the second, $1/(N-r+1)$ for the third, and so on. Therefore, the probability of drawing 'n' units from a sample and its selection in the rth draw is $\left(\frac{n}{N}\right)$.

$$P(n)=\frac{n!}{N(N-1)\ldots(N-n+1)}=\frac{1}{\binom{N}{n}} \qquad (10.1)$$

The probability of selecting a specified unit at the rth draw is

$$\frac{N-1}{N}=\frac{N-2}{N-1}\cdots\frac{N-r+1}{N-r+2},\frac{1}{N-r+1}=\frac{1}{N} \qquad (10.2)$$

10.4.2 Simple Random Sample with Replacement (SRSWR)

In this method, the 'n' units of the sample are drawn one at a time from the population. The units collected at each draw are replaced in the population so that the probability of obtaining any unit in any draw is $1/N$ probability of being chosen at any draw and an n/N probability of being in the sample.

Illustration

There is a group of 6 people, and two will be chosen as leaders. Let us look at the probability through SRS, assuming that sampling is done here without replacement.

$$\binom{N=6}{n=2}=\frac{N-1}{N}=\frac{N-2}{N-1}\cdots\frac{N-r+1}{N-r+2},\frac{1}{N-r+1}=\frac{1}{N} \qquad (10.3)$$

$$\binom{N=6}{n=2}=\frac{6!}{2!}=\frac{(6)(5)}{(2)}=\frac{30}{2}=15=\frac{1}{15}=0.06$$

Therefore, the chances of one of the pairs being selected out of 15 possibilities are 0.06.

(ii) Consider that a hospital has 1,000 staff members and must allocate a night shift to 100 members. All their names will be put in a bucket to be randomly selected, as each person has an equal chance of being selected. Since we know the population size (N) and sample size (n), the calculation can be as follows (Figure 10.2).

$$\frac{N-1}{N}=\frac{N-2}{N-1}\cdots\cdot\frac{N-n}{N-(n-1)} \qquad (10.4)$$

$$\text{Cancelling } \frac{N-n}{N-(n-1)} \qquad (10.5)$$

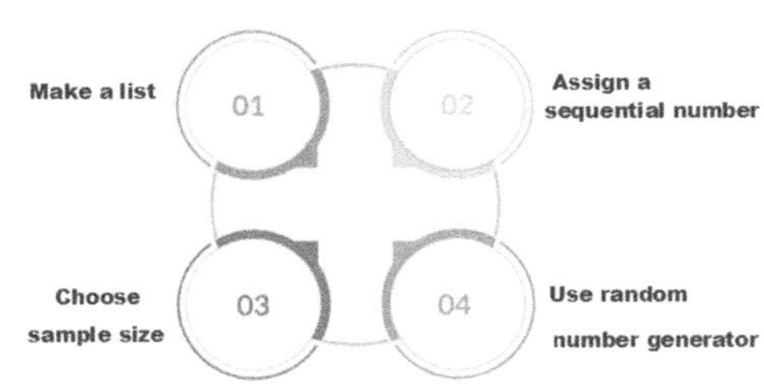

FIGURE 10.2 Outlines the steps involved in simple random sampling.

$$\frac{n}{N} = \frac{100}{1000} = 10\%$$

Simple Random Sampling Steps

Follow these steps to extract a simple random sample of 100 employees out of 500:

Step 1: Make a list of all the employees working in the organization (as mentioned above, there are 500 employees in the organization, so the record must contain 500 names).

Step 2: Assign a sequential number to each employee (1, 2, 3, … *n*). This is your sampling frame (the list from which you draw your sample).

Step 3: Determine your sample size. (In this case, the sample size is 100.)

Step 4: Use a random number generator to select the sample, using your frame (population size) from Step 2 and your sample size from Step 3. For example, if your sample size is 100 and your population is 500, generate 100 random numbers between 1 and 500 (Table 10.2).

10.4.3 Merits

Since the collection process is unbiased and free from influence, there is a fair and equal chance of being selected for all elements. This method produces random outcomes from a larger pool. Because the smaller groups in the sample represent the entire population, no further segmentation is required to filter them.

10.4.4 Demerits

The method is not completely exempt from sampling or calculation errors, which may negatively influence the final results.

10.5 Stratified Random Sampling

If the population can be separated into a number of subgroups (strata) such that the mean differences between strata are appreciably more than the differences between members within the same stratum, it is useful to employ stratified random sampling. Basically, this means that a simple random sample is selected

TABLE 10.2

Sample Error with Respect to Size of the Population

	Error Willing to Tolerate								
	Error (5%)			Error 10%			Error 15%		
	Desired Sample Size (*n*)								
	Confidence Interval			Confidence Interval			Confidence Interval		
Size (*N*)	90%	95%	99%	90%	95%	99%	90%	95%	99%
15	14	14	15	12	13	14	10	11	12
20	14	16	19	15	17	18	12	14	16
25	23	23	24	18	20	22	14	16	19
30	27	28	29	21	23	25	15	18	21
40	35	36	38	25	28	32	17	21	26
50	42	44	46	29	33	38	19	23	30
75	59	63	67	36	42	52	22	27	37
100	73	79	87	40	49	62	23	30	42
150	97	108	122	47	59	79	25	33	49
200	115	132	154	51	65	91	26	35	54
250	130	151	182	53	69	100	27	36	57
300	143	168	207	55	73	107	27	37	59
350	153	183	229	57	75	113	28	38	61
400	162	196	250	58	77	117	28	39	62
450	170	207	268	59	79	121	28	39	63
500	176	217	285	60	81	125	29	39	64
550	182	226	301	61	82	127	29	40	65
1,000	214	278	399	64	88	142	29	41	69
2,000	240	322	498	66	92	153	30	42	71
3,000	350	341	543	67	93	157	30	42	72
4,000	255	350	569	67	94	159	30	42	72
5,000	258	367	589	67	94	161	30	42	73
6,000	260	361	597	67	95	161	30	42	73
7,000	262	364	606	67	95	162	30	72	73
8,000	263	367	613	67	95	163	30	42	73
9,000	264	368	618	68	95	163	30	42	73
10,000	265	370	622	68	95	163	30	43	73

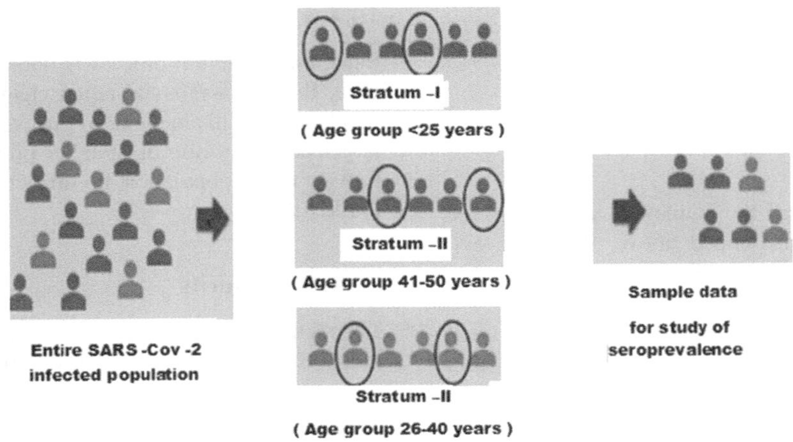

FIGURE 10.3 Selection process for sampled data using stratified random sampling.

from each of the various strata (Abdallah, 2024). This sampling design scores over simple random sampling in that: (i) It reduces the variance of the estimates. (ii) It reduces the cost of collection and analysis of data. (iii) It provides separate estimates for each stratum. Stratified sampling refers to random sampling techniques that group items of the whole population into different strata, based on their similar characteristics (Besekar et al., 2024). Then, samples from each stratum are taken, whether proportionately or disproportionately (Figure 10.3).

The stratified random sampling

$$= \left(\frac{\text{Total sample size}}{\text{Entire population}} \right) * \text{Population of sub groups}$$

The formula researchers can use to determine sample size using proportionate stratified random sampling is:

$$n_h = \frac{N_h}{N} * n \tag{10.6}$$

where n_h = sample size using proportionate stratified random sampling

N_h = Total stratum

N = Total population, n = sample size

The total population observed by researchers consisted of 500 farmers. Based on the 500 farmers, they are divided into the following groups: Small scale: 350 farmers, Medium scale: 100 farmers, and Large scale: 50 farmers (Table 10.3).

Researchers can calculate the sample using the Slovin formula based on this information. Based on the characteristics of the population, the researchers set a margin of error of 10%.

$$n = \frac{N}{1 + Ne^2} \tag{10.7}$$

N = Total population, n=sample size, e=margin of error (5% or 10%)

$$n = \frac{500}{1 + 500 * 0.10^2} = 83 \text{ farmers}$$

TABLE 10.3

Determination of Sample Size Based on the Proportion

Type of Farmers	N	$n_h = \frac{N_h}{N} * n$
Small scale	350	350/500 * 83=58 Farmers
Medium scale	100	100/500 * 83=17 Farmers
Large scale	50	50/500 * 83=8 Farmers
N	500	83 Farmers

Stratification is done when it is possible to divide the population into groups (strata) so that the within-group variance is small, ideally as small as possible. From each stratum, a sample of suitable size is drawn, usually using simple random sampling (Nakagawa et al., 2024). The greatest challenge lies in defining a suitable stratification variable. Stratified sampling is particularly useful when information is required for each stratum (e.g. each region in a country) as well as for the whole population (Ko, 2024). The allocation of sample size in stratified sampling can be equal, proportional, optimum, or follow the Neyman allocation.

10.5.1 Principles of Stratification

The principles to be kept in mind while stratifying a population are given below:

- The strata should not overlap and should together comprise the whole population.
- The strata should be homogeneous within themselves and heterogeneous between themselves with respect to the characteristics under study.
- If an investigator is facing difficulties in stratifying a population with respect to the characteristics under study, then he/she has to consider administrative convenience as the basis for stratification.
- If the limit of precision is given for a certain subpopulation, then it should be treated as a stratum.

Some notations and terminology used for stratified sampling:

N = Population size, n = sample size,

k = Numer of starta, N_i = Size of ith strata

10.6 Equal Allocations

This is a situation of considerable practical interest for reasons of administrative convenience. In this allocation method, the total sample size 'n' is divided equally among all the strata, i.e. if the population is divided into 'k' strata, then the size of the sample for each stratum will be

$$n_i = \frac{n}{k} \text{ for all } i = 1,2\ldots k \quad (10.8)$$

10.7 Applications of Sampling Allocation

Sampling allocation is primarily used in research studies to ensure that different subgroups within a population are adequately represented in a sample, allowing for more precise conclusions by dividing the population into strata and assigning an appropriate sample size to each stratum, which is particularly valuable when dealing with populations that have diverse characteristics.

Key applications of sampling allocation include

Market research:

When studying consumer behavior, researchers can use stratified sampling to ensure that different demographics like age groups, income levels, and geographic regions are proportionally represented in the sample.

Medical research:

In clinical trials, stratified sampling can be used to allocate participants across different treatment groups while ensuring that key factors like disease severity or medical history are balanced across the groups.

Political polling:

When conducting public opinion polls, researchers can use stratified sampling to ensure that different geographic regions and demographic groups are appropriately represented in the sample.

Quality control:

In manufacturing, stratified sampling can be used to select samples from different production batches or machine lines to monitor quality across the entire production process.

TABLE 10.4

Mean Score and Size of the Strata (N_i)

Hospitals	Strata Sizes (N_i)	Means Score	SD
A	400	60	20
B	200	120	80

10.8 Proportional Allocations

This allocation was originally proposed by Bowley in 1926. This allocation procedure is very common in practice because of its simplicity. As its name indicates, proportional allocation means we select a small sample from a small stratum and a large sample from a large stratum. The sample size in each stratum is fixed in such a way that for all the strata, the ratio n_i/N_i is equal to n/N, that is

$$\frac{n_i}{N_i} = \frac{n}{N} \text{ or } \frac{n}{n} * N_i = n_i \quad (10.9)$$

A sample of 60 infected cases is to be drawn from a population consisting of 600 individuals belonging to two hospitals, A and B. The means and SD of their scores are presented below. Draw a sample using proportional allocation techniques. (Table 10.4)

If we regard hospitals A and B as representing two different strata, then the problem is to draw a stratified random sample of size 30 using the techniques of proportional allocation. In proportional allocation, we have

$$\frac{n_i}{N_i} = \frac{n}{N} \text{ or } \frac{n}{n} * N_i = n_i$$

$$n_1 = \frac{60}{600} * 400 = 40$$

$$n_2 = \frac{60}{600} * 200 = 20$$

Thus, the required sample size for the two hospitals A and B, are 40 and 20, respectively.

10.9 Neyman's Allocations

This allocation of samples among different strata is based on a joint consideration of the stratum size and the stratum variance. In this allocation, it is assumed that the sampling cost per unit among different strata is the same and the size of the sample is fixed. Sample sizes are allocated by

$$n_i = n * \frac{W_i S_i}{\sum_{i=1}^{k} W_i S_i} = n \frac{N_i S_i}{\sum_{i=1}^{k} N_i S_i} \quad (10.10)$$

A sample of 100 infected cases is to be drawn from a population of hospitals A and B. The population means and the population mean squares of their monthly wages are presented below (Table 10.5).

TABLE 10.5

Mean Score and Size of the Strata (N_i) in college A & B

College	Strata Sizes (N_i)	Means Score $\bar{x}_i$	SD S_i^2
A	300	25	25
B	200	50	100

TABLE 10.6

Proportional and Neyman's Allocation Sample Size

Hospital	Proportional	Neyman's
A	60	43
B	40	57

Draw the samples using proportional and Neyman allocation and compare.

If we regard hospitals A and B as representing two different strata, then the problem is to draw a stratified sample of 100 infected cases using the techniques of Neyman's allocation.

$$n_i = n^* \frac{N_i S_i}{\sum_{i=1}^{k} N_i S_i} \quad (10.11)$$

$$\sum_{i=1}^{2} N_i S_i = 300 * 5 + 200 * 10 = 1{,}500 + 2{,}000 = 3{,}500$$

$$n_1 = 100 * \frac{300 * 5}{3{,}500} = 42.85 = 43$$

$$n_2 = 100 * \frac{200 * 5}{3{,}500} = 57.14 = 57$$

Therefore, the samples regarding hospitals A and B under both allocations are obtained (Table 10.6).

10.10 Optimum Allocation

The variance of the estimated mean depends on n_i which can be arbitrarily fixed. One more factor, which is no less important, also influences the variance of the estimated mean. The allocation is twofold. We should attain precision for the fixed cost of the survey and attain the desired degree of precision for the minimum cost. Thus, the allocation of the sample size in various strata, in accordance with these two objectives, is known as optimum allocation (Chen et al., 2024). If in any stratum the cost of the survey per sampling unit cannot be the same, then in one stratum the cost of transportation may be different from the other. Hence, it would not be wrong to allocate the cost of the survey in each stratum differently.

$$n_i = \frac{n.\ N_i \sigma_i}{N_1\sigma_1 + N_2\sigma_2 + \ldots . N_k\sigma_k}, \text{ For } i = 1,\ 2, \ldots \text{and } k \quad (10.12)$$

Illustration: A population is divided into three strata so that $N_1=6{,}000$ $N_2=3{,}000$ and $N_3=1{,}000$. Respective $SD_1=16$ $SD_2=19$ $SD_3=5$. How should a sample size $n=74$ be allocated to the three strata if we want optimum allocation using disproportionate sampling designs?

Solution

Sample size for strata with $N_1=6{,}000$

$$n_1 = \frac{74}{6{,}000(16) + 3{,}000(19) + 1{,}000\ (5)}$$

$$n_1 = \frac{7{,}1\ 04{,}000}{158{,}000} = \ 44.96 \text{ or } 45$$

Sample size for strata with $N_2=3{,}000$

$$n_2 = \frac{74(3{,}000)(19)}{6{,}000(16) + 3{,}000(19) + 1{,}000\ (5)}$$

$$n_2 = \frac{4{,}218{,}000}{158{,}000} = 26.69 \text{ or } 27$$

Sample size for strata with $N_3=1{,}000$

$$n_3 = \frac{74(1{,}000)\ (5)}{6{,}000(16) + 3{,}000(19) + 1{,}000\ (5)}$$

$$n_3 = \frac{370{,}000}{158{,}000} = 2.34 \text{ or } 2$$

Total sample size $N=n_1+n_2+n_3=45+27+2=74$

10.10.1 Merits

Sampling from each stratum guarantees that the overall sample is more representative of the whole population compared to a simple random sample. If each stratum is more homogeneous, i.e. less variable than the population as a whole with respect to key responses of interest, then estimates will be more precise. It is likely to be administratively convenient, e.g. when different sampling procedures need to be applied to different strata (see the ELUS example in Practical 2 for large-sized estates of >500 ha).

10.10.2 Demerits

It increases the costs.

There is a tendency to risk stratifying too finely.

A lack of auxiliary information can be a limitation (however, it is possible to create a stratum for units with no available information).

10.11 Cluster Sampling

A simple random sample in which the sampling unit is a collection, or cluster, of elements is called a cluster sample (Wu et al., 2023). It is useful to employ cluster sampling when (i) a detailed sample frame is not available and is too expensive to obtain and (ii) the cost, in terms of time, money, and manpower, increases with increasing distance between the sample elements—in other words, field logistics come into play. The clusters themselves could be census blocks (in a town) or a set of hamlets. Since the correlation between members of the same cluster is likely to be high, it is not profitable to select large clusters and all the elements within the cluster, even if this is administratively very convenient. Numerous clusters with relatively few members in each cluster would constitute a more efficient design. For determining the number of clusters and the number per cluster, an experienced statistician should

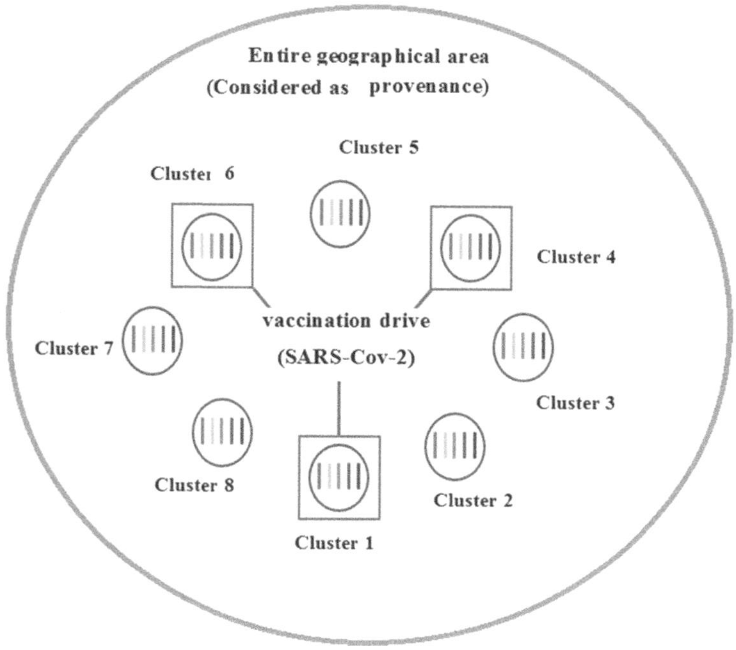

FIGURE 10.4 Schematic diagram of cluster sampling.

be consulted. In a scientific way, we describe cluster sampling as a method of obtaining a representative sample from a population that researchers have divided into groups (Louangrath, 2017). An individual cluster is a subgroup that mirrors the diversity of the whole population, while the set of clusters are similar to each other. Typically, researchers use this approach when studying large, geographically dispersed populations because it is a cost-controlling measure (Figure 10.4).

10.12 One-Stage Cluster Sampling with Equal-Sized Clusters

If all the population members in each selected cluster are used in the sample, the method is known as one-stage cluster sampling. If not, but further selection is exercised within the chosen clusters, the technique is called sub-sampling or two-stage cluster sampling. For instance, in examining the performance of school children in the educational survey referred to above, local education authorities may constitute the primary units, their schools the secondary units, classes within the schools the tertiary units, and children within the schools the members of the study population.

We start with one-stage sampling. Suppose that the population consists of a set of clusters of individual population members. There are M clusters of sizes $N_1, N_2, N_{3...}N_m \left(\sum_{i=1}^{M} N_i = N\right)$. The members of the clusters are $Y_{ij}\ (i = 1,2..M,\ j = 1,2..N_i)$, and the cluster means and variances are $\overline{Y_I}$ and S_i^2 (I = 1, 2…m), defined in the usual way. The population mean and variance are $\overline{Y_I}$ and S_i^2, respectively.

For one-stage cluster sampling with equal-sized clusters is $N_1 = N_2 = N_3 \ldots\ldots N_M = L$ then

$$N = ML \tag{10.13}$$

The one-stage cluster sample arising as SRS of m clusters has size $n = ml$, the sampling fraction

$$f = n/N = m/M \tag{10.14}$$

Suppose the observations in the sample are $Y_{ij}\ (i = 1,2..m,$ j = 1, 2..L). As an estimator $\hat{y}$, the cluster sample mean might be considered.

$$\overline{y}cl = \frac{1}{mL}\sum_{i=1}^{m}\sum_{j=1}^{L} y_{ij} \tag{10.15}$$

Since all clusters are of the same size, $\overline{y}cl$ is clearly unbiased $\overline{Y}$, i.e.

$$E(\overline{y}cl) = \overline{Y} \tag{10.16}$$

$(\overline{y}cl)$ can be expressed as

$$\overline{y}cl = \frac{1}{m}\sum_{i=1}^{m}\overline{Y_i} \tag{10.17}$$

where $\overline{Y_i}$ are the cluster means for an SR sample of m of the M clusters.

Considering the set of cluster means $\{\overline{Y_1}, \overline{Y_2}, \ldots\ldots \overline{Y_M}\}$ as the basic population from which we are sampling, this population has a mean and variance.

$$\frac{1}{M-1}\sum_{i=1}^{M}\left(\bar{Y}_i-\bar{Y}\right)^2 \quad (10.18)$$

Furthermore, its variance is easily shown to be

$$Var\left(\bar{y}_{cl}\right)=\frac{1}{m}(1-f)\sum_{i=1}^{M}\frac{\left(\bar{Y}_i-\bar{Y}\right)^2}{M-1} \quad (10.19)$$

Consider the alternative estimation of $\bar{Y}$ provided by the mean $\bar{y}$ of an SR sample of ($n=mL$) observations drawn without restriction from the total population. This is also unbiased and has a variance $(1-f)S^2/(mL)$

$$(ML-1)S^2=\sum_{i=1}^{M}\sum_{j=1}^{L}\left(Y_{ij}-\bar{Y}\right)^2 \quad (10.20)$$

$$=\sum_{i=1}^{M}\sum_{j=1}^{L}\left(Y_{ij}-\bar{Y}\right)^2+L\sum_{i=1}^{M}\left(\bar{Y}_i-\bar{Y}\right)^2 \quad (10.21)$$

$$=(ML-1)\bar{S}^2+L\sum_{i=1}^{M}\left(\bar{Y}_i-\bar{Y}\right)^2 \quad (10.22)$$

where $\bar{S}^2=\frac{1}{M}\sum_{i=1}^{M}S_i^2$ is the average within-cluster variance.

Thus,

$$\text{Var}\left(\bar{y}\right)-\text{Var}\left(\bar{y}_{cl}\right)=\frac{1-f}{mL\ (M-1)}\left\{(M-1)s^2-L\sum_{i=1}^{M}\left(\bar{Y}_i-\bar{Y}\right)^2\right\} \quad (10.23)$$

$$=\frac{(1-F)M(L-1)}{mL\ (M-1)}\left(\bar{S}^2-S^2\right) \quad (10.24)$$

The cluster sample will better than the SRS mean $\bar{y}$ if the average within-cluster variance $\bar{S}^2$ is larger than the overall population variance S^2, and vice versa.

There is another way of looking at these results. We can define a quantity called the intra-cluster correlation coeffcient.

$$\rho=2\sum_{i=1}^{M}\sum_{j<k}\left(Y_{ij}-\bar{Y}\right)\left(Y_{ik}-\bar{Y}\right)/\left[(L-1)(ML-1)S^2\right] \quad (10.25)$$

This provides an aggregate measure of the correlation between population members in the same cluster.

$$\rho=1-\left(\frac{ML}{ML-1}\right)\left(\frac{\bar{S}^2}{S^2}\right) \quad (10.26)$$

$$\text{Var}\left(\bar{y}_{cl}\right)=\frac{(1-f)}{m}*\frac{ML-1}{L^2(M-1)}S^2\left[1+(L-1)\rho\right] \quad (10.27)$$

By using cluster sampling, we determine the sample size based on the following formula. For example, a researcher wish to study the prevalence of malnutrition in children under 5 years in India across different states, 2023.

$$n=DF*\frac{Z_{1-\alpha/2}{}^2P(1-P)}{E^2}\ orn=1+\rho(m-1)*\frac{Z_{1-\alpha/2}{}^2P(1-P)}{E^2} \quad (10.28)$$

where n = required minimum sample size

df = design effect = $1+\rho(m-1)$
ρ = Intracluster correlation
m = Number of individual in each cluster
P = the estimated prevalence of an indicator
α = Level of significance
$Z_{\alpha/2}$ = the Z score corresponding to the degree of confidence
E = desired precision

10.13 Practical Considerations

The clusters should be designed so that they collectively include all members of the target population; each member must appear in one and only one cluster.

The definitions or boundaries of the clusters should be precise and unambiguous; in the case of geographical clusters, natural and man-made boundaries such as rivers and roads are often used to delimit the cluster boundaries.

10.13.1 Merits

Cluster sampling involves selecting population units that are 'close' together and does not require all the population units to be listed. Cluster sampling has two advantages:

1. It eliminates the need for a complete list of all units in the population.
2. It ensures that selected population units will be closer together, thus enumeration costs for personal interviews will be reduced, and fieldwork will be simplified.

10.13.2 Demerits

In general, cluster sampling is less accurate than SRS (for samples of the same size) because the sample obtained does not cover the population as evenly as in the case of SRS. However, it is often preferred because it is more economical.

For example, if we take a simple random sample of 10,000 households across the whole of Australia, we are more likely to cover the population more evenly, but it is more expensive than sampling 50 clusters of 200 households (Table 10.7).

Steps for Cluster Sampling

Step 1: Selection of sample

Decide on the target population and also the sample size 'n'.

TABLE 10.7

Cluster Sampling vs Stratified Sampling

Cluster Sampling	Stratified Sampling
Cost reduction	Enhanced precision
Randomly selected clusters	Randomly selected member from strata
Division naturally formed	Depends on the researchers
More errors	Reduced errors
Homogeneity externally	Homogeneity internally

Step 2: Create and evaluate sampling frames

Create a sampling frame by using either an existing framework or a new one for the target population.

Step 3: Fix the frameworks

Evaluate the frameworks based on coverage and clustering and make adjustments accordingly. These groups will vary, considering the population, and should be exclusive and comprehensive. Members of a sample are selected individually.

Step 4: Determine groups

Determine the number of groups by including the same objectives in each group. Ensure that each of these groups is distinct from one another.

Step 5: Select clusters

Choose the clusters by applying a random selection method.

Step 6: Create sub-types

Bifurcate into two-stage and multi-stage subtypes based on the number of steps researchers follow to form clusters.

10.14 Sample Size Determination for Clinical Trial (Approximates the *t*-Statistic with the *z*-Statistic)

Step 1: Calculate sample sizes without adjustment for clustering

$\alpha = 0.05$ (two-tailed): Fix the threshold probability for rejecting the null hypothesis H_0 (Type I error rate).

$\beta = 0.200$ Power of the test (Probability of failing to reject the null hypothesis under the alternative hypothesis—Type II error rate).

$q_1 = 0.500$ (Proportion of subjects that are in Group 1) (exposed)

$q_0 = 0.500$ (Proportion of subjects that are in Group 0 (unexposed); $(1 - q_1)$

$E = 0.500$ Effective size, $S = 1.00$ (Standard deviation of the outcome in the population)

The standard normal deviate for $\alpha = Z_\alpha = 1.960$

The standard normal deviate for $\beta = Z_\beta = 0.842$

$$A = (1/q_1 + 1/q_0) = 4.00$$

$$B = (Z_\alpha + Z_\beta)^2 = 7.849$$

$$B = (Z_\alpha + Z_\beta)^2 = 7.849$$

Standardized Effect Size = $(E/S) = 0.500$

10.14.1 Without Correction for Clustering

Total group size = $N_{total} = AB/(E/S)^2 = 125.58$

N_1: 63, N_0: 63, Total: 126

10.14.2 Calculate Sample Sizes with Adjustment for Clustering

The study has either (i) fixed number of clusters per group (C_1 and C_0) or (ii) a fixed cluster size m.

10.14.3 Fixed Number of Clusters per Group

By fixing the number of clusters (C_1), you limit the value of ρ. The larger C_1, the larger ρ can be. $q_1 = 0.50$, $N_1 = 62.79$

$C_1 = 6$: Number of clusters in Group 1 (must be at least 1 and less than N_1)

$\rho = 0.050$: Within-cluster correlation coefficient (must be greater than 0 and less than $C_1/N_1 = 0.095$)

Cluster size = $m = (1 - \rho)/((C_1/N_1) - \rho) = 20.85$, rounded to 21

Design Effect = $1 + (\rho(m - 1)) = 1.99$

m (rounded): 21

N_1': 126, N_0': 126, N_{total}: 252

Fixed cluster size

10.14.4 With a Fixed Cluster Size, ρ Can Take Any Value between 0 and 1

$N_{total} = 125.80$ Total group size (without adjustment)

$q_1 = 0.500$ Proportion of subjects that are in Group 1 (exposed)

$m = 20$, $\rho = 0.05$ Within-cluster correlation coefficient (must be greater than 0 and no greater than 1) Design Effect = $1 + (\rho\ (m - 1)) = 1.95$

Clusters in Group 1 = $C_1 = N_{total}$ * Design Effect * q_1/m = 6.12, rounded to 6

Clusters in Group 0 = $C_0 = N_{total}$ * Design Effect * q_0/m = 6.12, rounded to 6

N_1': 120, N_0': 120, N_{total}': 240

This study has 80.0% power to detect a $P_1 = 0.066$

Risk ratio = $P_1/P_0 = 0.220$

Odds ratio = $(P_1/(1 - P_1))/(P_0/(1 - P_0)) = 0.165$

10.15 Confidence Interval for Area under Curve ROC Given *N*

The sample sizes $N(D+)$ with disease and $N(D-)$ without disease, and the measured AUROC. This calculator returns the lower bound (LB) and upper bound (UB) of the confidence interval (Table 10.8).

Illustration: Suppose we had AUROC: 85% (expected area under the ROC curve), $N(D+)$: 50 (sample size with disease), $N(D-)$: 550 (sample size without disease) with CL: 95%

TABLE 10.8

Different Sample Size Associated with Real Clinical and Biological Research

(i) Sample Size for Proportions—Effect Size	
α (two-tailed)=0.05	Threshold probability for rejecting the null hypothesis. Type I error rate.
β=0.200	Probability of failing to reject the null hypothesis under the alternative hypothesis. Type II error rate.
P_0=0.300	Proportion in Group 0 (baseline proportion)
N_1=50	Number of subjects in Group 1
N_0=50	Number of subjects in Group 0

Solution

The standard normal deviate for $\alpha = Z_\alpha = 1.9600$, Q_1 = AUROC/(2-AUROC) = 0.7391

$Q_2 = 2(\text{AUROC})^2/(1 + \text{AUROC}) = 0.7811$

Standard Error = SE = $\sqrt{}$((AUROC * (1 − AUROC)) + ((N(D+) − 1) * (Q_1-AUROC²)) + ((N(D−) − 1) * (Q_2-AUROC²))/(N(D+) * N(D−))) = 0.0347

Lower bound = AUROC − (Z_α * SE) = 0.7820

Upper bound = AUROC + (Z_α * SE) = 0.9180

N = 600; Lower bound = 0.7820, Upper bound = 0.9180

10.15.1 Sample Size—Confidence Interval for AUROC

θ = 0.85 (AUROC)

P = 0.25 (Proportion of sample having disease, estimated)

W = 0.125 (Width of confidence interval)

CL = 0.95 (Confidence level)

Solution

$Q_1 = \theta/(2 - \theta) = 0.7391$

$Q_2 = 2\theta^2/(1 + \theta) = 0.7811$

The standard normal deviate for $\alpha = Z_\alpha = 1.9600$

Standard Error Goal = $W/2Z_\alpha = 0.031888$

$a = P(1 - P)(\text{SE goal})^2 = 0.000191$

$b = -(P(Q_1 - Q_2) + Q_2 - \theta^2) = -0.0481$

$c = Q_1 + Q_2 - \theta - \theta^2 = -0.0523$, $n = 254$

10.15.2 Correlation Sample Size

The total sample size required to determine whether a correlation coefficient differs from zero:

- α (two-tailed) = 0.05: Threshold probability for rejecting the null hypothesis (Type I error rate).
- β=0.20: Probability of failing to reject the null hypothesis when the alternative hypothesis (Type II error rate).
- r=0.20

Solution

The standard normal deviate for $\alpha = Z_\alpha = 1.9600$

The standard normal deviate for $\beta = Z_\beta = 0.8416$

TABLE 10.9

Probability of the Test Result in Non-Disease, $P(R|D^-)$.

Criteria	D^+	D^-
Total number	100	100
Total with result R	75	25
$P(R|D+)$ or $P(R|D^-)$	0.75	0.250

$C = 0.5 * \ln[(1 + r)/(1 - r)] = 0.2027$, sample size = $n = [(Z_\alpha + Z_\beta)/C]^2 + 3 = 194$

10.15.3 Conf Interval—Likelihood Ratio

This uses the general definition for the likelihood ratio of test result R, LR(R), as the probability of the test result in disease, $P(R|D^+)$, divided by the probability of the test result in non-disease, $P(R|D-)$ (Table 10.9).

LR (R) Confidence Interval: 3.000 (2.098–4.291)

Z: 1.96

ln(LR): 1.099

Variance of ln(LR): 0.033

SD of ln(LR): 0.183

ln(Lower Bound of CI): 0.741

ln(Upper Bound of CI): 1.456

Lower Bound of CI: 2.098

Upper Bound of CI: 4.291

10.15.4 Confidence Interval for a Mean

n = 80 (Sample size), m = 74 (Sample mean), S = 3 (Sample standard deviation), CL = 95% (Confidence level), Standard error of the mean (SEM) = $S/\sqrt{N}$ = 0.335

$t(\alpha, N - 1)$ = 1.99; Confidence interval = m +/− ($t(\alpha, N - 1)$*SEM)

Mean = 74; CI95%: 73.33 – 74.67

10.16 Effect Size for Before-After Study (Paired *T*-Test)

Suppose we wish to measure a continuous outcome 'y' in each subject at the start and end of the study period by various study designs. For each subject, we calculate the change $\Delta = Y_{end} - Y_{start}$. Compare the mean value of 'Δ' to 0. This requires the standard deviation $S(\Delta)$. The estimate of $S(\Delta)$ should be based on data from other subjects who were followed for similar time periods. However, it can be estimated as $S(\Delta) = S_y\sqrt{2(1 - \rho)}$, where S_y is the standard deviation of the underlying outcome measurement, assumed to be the same at both time points, and ρ is the within-subject correlation of the outcome. This formula derives the effect size in two different ways: first using the 'T' statistic (with a non-centrality parameter), then using the 'Z' statistic. The 'Z' statistic approximates the T statistic but provides an effect size that is slightly too small.

α (two-tailed) = 0.05: Threshold probability for rejecting the null hypothesis (Type I error rate).

$\beta = 0.20$: Probability of failing to reject the null hypothesis under the alternative hypothesis (Type II error rate). $n = 50$ (sample size).

$S(\Delta) = 1$ (Standard deviation of the change in the outcome). (If you don't know $S(\Delta)$)

10.16.1 Calculation Using the *T* Statistic and Non-Centrality Parameter

Degrees of freedom (DoF) = $N_{total} - 1 = 49$

The standard T value (with DoF degrees of freedom) corresponding to $\alpha = T_\alpha = 2.010$

Non-centrality parameter = $\delta = 2.85791$

$E/S(\Delta) = \delta/\sqrt{N} = 0.40417$, Effect size $E = E/S(\Delta) * S(\Delta) = 0.404$

10.16.2 Normal Approximation Using the *Z* Statistic Instead of the *T* Statistic

The standard normal deviate for $\alpha = Z_\alpha = 1.95996$

The standard normal deviate for $\beta = Z_\beta = 0.84162$

$A = 1.00000$, $B = (Z_\alpha + Z_\beta)^2 = 7.84887$, $AB/N = 0.15698$

$e/S(\Delta) = \sqrt{AB/N} = 0.39620$, Effect size (e) = $e/S(\Delta) * S(\Delta) = 0.396$

10.16.3 Post-Test Probability

The patient had some pre-test probability of the disease in question. We did a test and obtained the result '*r*'. This calculator gives the patient's new post-test probability of disease, given that result.

P-value Pre-Test = 0.012, LR(*r*) = 105; Pre-Test Odds = $P/(1 - P) = 0.012$

Post-Test Odds = (Pre-Test Odds) * LR(*r*) = 1.275

Post-test Probability = Post-Test Odds/(1+ Post-Test Odds) = 0.560

10.16.4 Sample Size—Conf Interval for a Proportion

The sample size for a descriptive study of a dichotomous variable. This formula gives both binomial and normal approximations to the sample size. $P = 0.65$ (Expected proportion), 0.10 (Total width of confidence interval), CL = 95% (Confidence interval).

10.17 Binomial Exact Calculation

Sample size = $n = 367$; Expected positive results in the sample = $x = 239$

2. Normal Approximation to the Binomial Calculation

$\alpha = (1\text{-CL})/2 = 0.025$, Standard normal deviate for $\alpha = Z_\alpha = 1.960$

Sample size = $n = 4Z_\alpha^2 P(1-P)/(W^2) = 350$; Expected positive results in the sample = $x = 228$

10.18 Sample Size—Proportions (Alternative: Specify *P*)

We will compare the proportion with a dichotomous outcome between two samples using the Chi-squared statistic (or *z*-test). The study includes Group 1 (exposed) and Group 0 (unexposed), which furnish P (the prevalence of the outcome in the population) and Q_1 (the proportion of exposed individuals in the population), and estimate the risk ratio (RR). The following mathematical formula is used for determining sample size. Finally, we derived P_1 (outcome proportion in exposed) and P_0 (outcome proportion in unexposed) and, from there, the sample sizes. Note the distinction between Q_1 (proportion of exposed individuals in the population) and q_1 (proportion of exposed subjects in your study), which should be substituted as per the study protocol.

Solution

α (two-tailed) = 0.05: Threshold probability for rejecting the null hypothesis. (Type I error rate).

$\beta = 0.20$: Probability of failing to reject the null hypothesis under the alternative hypothesis.

10.18.1 Type II Error Rate

$q_1 = 0.50$: Proportion of subjects in Group 1 (exposed)

$q_0 = 0.50$: Proportion of subjects in Group 0 (unexposed); $1 - q_1$

$Q_1 = 0.10$: Proportion of individuals in the population who are exposed.

$P = 0.05$: Prevalence of the outcome in the population.

RR = 5.00: Estimated risk ratio (P_1/P_0). For the parameters currently entered, this value must be greater than 0. See calculations below.

Q_0 = Proportion of individuals in the population who are unexposed = $1 - Q_1 = 0.9000$.

Limits on the Risk Ratio (RR):

Lower limit: Use $(P - Q_0)/(100 - Q_0) = -0.90$ if it is positive, otherwise, use 0.

Lower limit: 0.00

Upper limit: If $P > Q_1$, use $(100 - Q_1)/(P - Q_1) = -1.11$, otherwise, no upper limit.

Upper limit: None

The standard normal deviate for $\alpha = Z_\alpha = 1.9600$;

The standard normal deviate for $\beta = Z_\beta = 0.8416$

P_1 = The probability of the outcome in exposed people = $P * \text{RR}/(1 + Q_1 * (\text{RR} - 1)) = 0.1786$

P_0 = The probability of the outcome in unexposed people = $P/(\text{RR} - Q_0 * (\text{RR} - 1)) = 0.0357$

Pooled proportion = PP = $(q_1 * P_1) + (q_0 * P_0) = 0.1071$

$A = Z_\alpha \sqrt{\text{PP}(1 - \text{PP})(1/q_1 + 1/q_0)} = 1.2124$

$B = Z_\beta \sqrt{P_1(1 - P_1)(1/q_1) + P_0(1 - P_0)(1/q_0)} = 0.5065$

$C = P_1 - P_0 = 0.1429$

Total group size = $n = (A + B)^2/C^2 = 145$

Continuity correction (added to N for Group 0) = CC = $1/(q_1 * |P_1 - P_0|) = 14$

10.19 Sample Size for Survival Analysis

10.19.1 Case 1: Number of Events, Given Relative Hazard

α (two-tailed) = 0.05 Threshold probability for rejecting the null hypothesis (Type I error rate, Tables 10.10 and 10.11).

$\beta = 0.20$ Probability of failing to reject the null hypothesis under the alternative hypothesis (**Type II error rate).**

$q_1 = 0.30$: Proportion of subjects in Group 1 (exposed)

$q_0 = 0.7$: Proportion of subjects in Group 0 (unexposed); $1 - q_1$

RH = 0.667: Relative hazard (Group 1/Group 0)

The standard normal deviate for $\alpha = Z_\alpha = 1.9600$

The standard normal deviate for $\beta = Z_\beta = 0.8416$

$A = (Z_\alpha + Z_\beta)^2 = 7.8489$

$B = (\log(\mathrm{RH}))^2 q_0 q_1 = 0.0344$

Total events needed = A/B = 228

10.19.2 Case 2: Sample Size, Given Number of Events

$q_1 = 0.3$, $q_0 = 0.7$, RH = 0.667, Total events = 228

BER_0 = 0.44: Baseline event rate (events/unit time) for Group 0

ST_0 = 1.575: Median survival time in Group 0

CR = 0.40: Censoring rate – censored/unit time (assumed equal for both groups)

FU = 3: Planned average length of follow-up, in time units.

10.20 Logistic Regression—Effect Size

Compare the odds ratio from logistic regression to 1. The model is of a continuous explanatory variable and a binary outcome variable. This calculator takes the group sizes as nputs and calculates the effect size that the study has $(1 - \beta)$ power to detect. The effect size is expressed as the minimum detectable odds ratio per 1 SD increase in the continuous explanatory variable. Closer to 1 is better because it means your study is powered to detect a smaller effect size. This calculator also allows an adjustment by the variance inflation factor (Table 10.12).

$N_{1=33}$ Sample size, Group 1; $N_{0=67}$ Sample size, Group 0, $\rho^{2=0.0}$ Multiple correlation coefficient

VIF = 1: Variance inflation factor (see below), α (two-tailed) = 0.05: Threshold probability for rejecting the null hypothesis. Type I error rate, $\beta = 0.20$: Probability of failing to reject the null hypothesis under the alternative hypothesis.

TABLE 10.10

Sample Size (With Continuity Correction)

Population	N	Outcome+	Outcome−
Exposed (G_1)	87	16	71
Un exposed (G_0):	87	3	84
Total:	174	19	155

TABLE 10.11

Sample Size (Without Continuity Correction)

Population	N	Outcome+	Outcome−
Exposed (G_1)	73	13	60
Un exposed (G_0):	73	3	70
Total:	146	16	130

TABLE 10.12

Cumulative Event Rate Naïve and Study Population

	N	Events	Cumulative Event Rate
Group 1	153	57	0.370
Group 0	356	171	0.482
Total	509	228	0.448

10.20.1 Type II Error Rate

VIF = Variance Inflation Factor = $1/(1 - \rho^2) = 1$

$N_{1\mathrm{Eff}}$ = Effective sample size for Group 1 = N_1/VIF = 33.0000

$N_{0\mathrm{Eff}}$ = Effective sample size for Group 0 = N_0/VIF = 67.0000

DF = Degrees of freedom = $N_{1\mathrm{Eff}} + N_{0\mathrm{Eff}} - 2$ = 98.0000

T_α = 1.9845, NCP = Non-centrality parameter = ncp (β, DF, T_α) = 2.8293

E = Effect = NCP * $\sqrt{1/N_{1\mathrm{Eff}} + 1/N_{0\mathrm{Eff}}}$ = 0.6017

OR = exp(E) = 1.8252, (Odds Ratio: 1.8252)

10.21 Sample Size—Conf Interval for a Mean (Sample Size for a Descriptive Study of a Continuous Variable)

Solution

CL = 0.95 Confidence level (e.g. 95%), W = 5 Desired total width of confidence interval

S = 2.5 Standard deviation of the variable

W/S = 2.00, Standard normal deviate for $\alpha = Z_\alpha = 1.96$, Sample size = $n = 4Z_\alpha^2 S^2/W^2 = 4$

10.21.1 Sample Size for Before-After Study (Paired *T*-Test)

Measure a continuous outcome '*y*' in each subject at the start and end of the study period. For each subject, calculate the change $\Delta = y_{\mathrm{end}} - y_{\mathrm{start}}$. Compare the mean value of Δ to 0. This requires the standard deviation S_Δ. The estimate of S_Δ should be based on data from other subjects who were followed for similar time periods.

α (two-tailed) = 5.0%: Threshold probability for rejecting the null hypothesis. (Type I error rate).

β = 20.0: Probability of failing to reject the null hypothesis under the alternative hypothesis (Type II error rate).

E = 0.50: Effect size

S_Δ = 1.0: Standard deviation of the change in the outcome.

10.21.2 Case 1: Calculation Using the *T* Statistic and Non-Centrality Parameter

A value of $N = 33.3670$ gives the following calculations

NCP = Non-centrality parameter = $\sqrt{N} * E/S_\Delta = 2.8882$.

DF = Degrees of freedom = $N - 1 = 32.3670$.

t_α = Inverse of the two-tailed *T* distribution given a probability of $1 - (\alpha/2)$ and DF of 32.3670 = 2.0360.

Beta(t_α, DF, NCP) = 0.200003. If *N* was calculated correctly, this should closely approximate your selected value of β, (above).

The *N* thus calculated is rounded up to the next highest integer to give the group size.

Group size (*n*-34)

10.21.3 Case 2: Approximation Using the *Z* Statistic Instead of the *T* Statistic

Z_α = Standard normal deviate for $\alpha = 1.9600$.

Z_β = Standard normal deviate for $\beta = 0.8416$.

$B = (Z_\alpha + Z_\beta)^2 = 7.84$, $C = (E/S_\Delta)^2 = 0.25.N = B/C = 31.39$.

The *N* thus calculated is rounded up to the next highest integer to give the group size.

Group size n: 32

10.22 Compare Tests (McNemar's Test)

Compare the sensitivities of Test 1 (T_1) and Test 0 (T_0) in a sample of patients, all of whom are D^+ by the reference standard. This assumes a paired design in which all subjects receive both T_0 and T_1. Use Mc.Nemar's test to compare sensitivities. For example, you believe T_0 has 70% sensitivity and T_1 has 80% sensitivity. Specify those sensitivities as S_0 and S_1 below. Your null hypothesis will be $S_0 = S_1 = 75\%$, and your alternative hypothesis is $S_0 \neq S_1$. The sample size will be higher if the two tests are independent (uncorrelated), so for a first analysis, treat the two tests as independent by setting $P(T_1+ | T_0^+)$ to $S_1 = 80\%$. Then, to get the minimum sample size, treat the two tests as perfectly correlated by setting $P(T_1+ | T_0^+)$ to 100%.

10.22.1 Case 1: Compare Sensitivity

α (two-tailed) = 0.05: Threshold probability for rejecting the null hypothesis (Type I error rate).

$\beta = 0.20$: Probability of failing to reject the null hypothesis under the alternative hypothesis (Type II error rate).

$S_{0=70}$: Assumed sensitivity of test T_0, the lower sensitivity test.

$S_{1=85}$: Assumed sensitivity of test T_1, the higher sensitivity test.

$P(T_1+ | T_0^+) = 90$ Probability that test T_1 is positive given that test T_0 is positive.

The standard normal deviate for $\alpha = Z_\alpha = 1.96$

The standard normal deviate for $\beta = Z_\beta = 0.84$

$\Psi = S_0 + S_1 - (2 * S_0 * P(T_{1+}|T_{0+}) = 0.2900$. This is the probability that one test will be positive but not the other.

$V_0 = \Psi = 0.2900$. This is the variance (under the null hypothesis) of the difference between sample sensitivity in group 1 and group 0.

$V_A = \Psi - (S_1 - S_0)^2 = 0.2675$. This is the variance (under the alternative hypothesis) of the difference between sample sensitivity in group 1 and group 0.

$A = Z_\alpha * \sqrt{V_0} = 1.0555$, $B = Z_\beta * \sqrt{V_A} = 0.4353$

$C = (S_1 - S_0)^2 = 0.0225$ $n = (A + B)^2/C = 98.7721$

n (D^+ individuals in sample size): 99

10.22.2 Case 2: Compare Specificity

The standard normal deviate for $\alpha = Z_\alpha = 1.96$

The standard normal deviate for $\beta = Z_\beta = 0.84$

$\Psi = S_0 + S_1 - (2 * S_0 * P(T_1 - | T_0^-)) = 0.1700$. This is the probability that one test will be negative but not the other.

$V_0 = \Psi = 0.1700$. This is the variance (under the null hypothesis) of the difference between sample specificity in group 1 and in group 0.

$V_A = \Psi - (S_1 - S_0)^2 = 0.1600$. This is the variance (under the alternative hypothesis) of the difference between sample specificity in group 1 and in group 0.

$A = Z_\alpha * \sqrt{V_0} = 0.80$, $B = Z_\beta * \sqrt{V_A} = 0.33$

$C = (S_1 - S_0)^2 = 0.010$ $N = (A + B)^2/C = 131.04$, *N* (D^- individuals in sample size): 132.

10.23 Exponential Failure Rate—Sample Size

Estimating the sample size needed for an exponential failure rate. This formula assumes continuous time. Specify your desired confidence interval width (*W*), confidence level, and *T* = average time until failure. The failure rate (*f*) is 1/*T*. The Censoring Factor specifies the censoring rate relative to the failure rate. If the researcher can follow all subjects until they fail, set it to 0. If 20% of your subjects will be lost to follow-up, set it to 0.2. This means if you start with 5 subjects, you will see 4 failures.

$T = 3.00$, $W = 0.100$, CF = 0.00, CL = 95

f = Failure rate = Hazard = $1/T = 0.33$

Events = $N(1 - \text{CF}) = 170$

Chi²ELB = Inverse-chi-squared ((1 − CL)/2, 2*Events) = 291.49

Chi²EUB = Inverse-chi-squared (1 − ((1 − CL)/2, 2*Events) = 393.77

f = Failure rate = Hazard = 0.3333

LB = Hazard * Chi²ELB/(2*Events) = 0.28

UB = Hazard * Chi²EUB/(2*Events) = 0.38

Actual width = 0.10, Sample size (*n*) = 170

10.24 Means—Effect Size

This fundamental formula can include the group sizes as inputs and calculates the effect size that the study has $(1 - \beta)$ power to detect. The effect size is calculated in two ways: first using the *T* statistic (with a non-centrality parameter) and then the *Z* statistic. The *Z* statistic approximates the *T* statistic but provides an effect size that is slightly too small. (We provide the Z statistic calculation to allow comparison with other calculators that use the Z approximation.) α (two-tailed) = 0.05, $\beta = 0.20$, $S = 1$, $N_1 = N_2 = 60$

Total group size = $N_{total} = N_1 + N_0 = 120$

Proportion of subjects in Group 1 = $q_1 = N_1/N_{total} = 0.50$

Proportion of subjects in Group 0 = $q_0 = 1 - q_1 = 0.50$

10.24.1 Case 1: Calculation Using the *T* Statistic and Non-Centrality Parameter

Degrees of freedom (DoF) = $N_{total} - 2 = 118$

The standard T value (with DoF degrees of freedom) corresponding to $\alpha = T_\alpha = 1.980$

$k = \sqrt{1/N_1 + 1/N_0} = 0.1826$

Non-centrality parameter = $\delta = 2.82$

$E/S = k * \delta = 0.51$

This study has 80.0% power to detect an effect size of $E = S * E/S = 0.51$

10.24.2 Case 2: Normal Approximation Using the *Z* Statistic Instead of the *T* Statistic

Standard normal deviate for $\alpha = Z_\alpha = 1.9600$

Standard normal deviate for $\beta = Z_\beta = 0.8416$

$A = (Z_\alpha + Z_\beta)^2 = 7.84$, $B = 1/q_1 + 1/q_0 = 4.00$, $C = AB/N_{total} = 0.26$

$E/S = \sqrt{C} = 0.5115$, this study has 80.0% power to detect an effect size of $E = S * E/S = 0.511$. Because the formula used here is based on approximating the T statistic with a Z statistic, it will slightly underestimate the effect size for smaller values of N_{total}.

10.25 Systematic Sampling

This consists of selecting every Kth member of the sample after making a random start with a member between the first and Kth. For instance, let us suppose that there are 6,000 persons in a community and we wish to select a sample of 400 persons. As 6000/400 = 15, we select a random number between 1 and 15, e.g., 4. Then the systematic sample will consist of 4, 4 + 15, 4 + 15 + 15, etc., i.e., 4, 19, 34, … 5,974, and 5,989. The relative merits of systematic sampling and simple random sampling will now be discussed. If the elements in the population are in random order (e.g., the telephone directory), a systematic sample and a simple random sample are equally efficient. However, if the population is ordered (for example, various village panchayats are arranged in increasing order of performance statistics), a systematic sample would be better than a simple random sample, in that the estimates from it would have a lower variance. If there is some type of periodicity in the population (for instance, weekend cinema collections are heavier than those during weekdays), a systematic sample would be inferior to a simple random sample—it could be non-representative and also result in larger variance for the estimates. When conducting surveillance activities, a simple random sample may, by accident, be heavily weighted for some months, resulting in a misleading picture. This could happen in a survey of diarrhea, which is highly seasonal. On the other hand, a systematic sample with a random start will be spread uniformly throughout the year and would therefore be superior. Another alternative might be to stratify by seasons and take a simple random sample from within each season—this would constitute a stratified random sample. Finally, systematic sampling is easier to undertake in the field and often provides greater information per unit cost than simple random sampling.

$$k = \left[\frac{\text{Number of elements in the population}}{\text{Number of elements in the sample}}\right] = \left(\frac{N}{n}\right) \quad (10.29)$$

Steps of Systematic Sampling

Step 1: Divide the population size by the sample size and round the result down to the nearest whole number 'm'

Step 2: Use a random number table to obtain a number 'k' between 1 and m

Step 3: Select for the sample those numbers of the population that are numbered k, k + m, $k + 2m, \ldots, k$ (sample size − 1) m

Suppose the population N = 300 with a sample size (n = 5), random start $i = 6$, $k = \frac{N}{n} = \frac{300}{5} = 60$

The subsequent sampling units are the units in the following positions (i), $(k + i)$, $(2k + i)$, $(3k + i), \ldots, n_k)$

Five sampling units selected are

6	66	126	186	....	300

Illustration

In a group of 250 students, how will you elect a sample containing 71 students by using the systematic sampling technique?

Solution

Step 1: Prepare a sampling frame by randomly arranging the 250 students

Step 2: Assign each student a number from 1 to 250

Step 3: Find the sampling interval k, and divide the population size by the sample size 71

$$k = \frac{N}{n} = \frac{250}{71} = 3.52 \cong 4$$

Step 4: Select a number from the whole number between 0 and k + 1 using a simple random technique. The numbers between 0 and k + 1 are 1, 2, 3, and 4. This chosen value is called the random start

Step 5: Assume that the randomly selected number is 2. Use 2 as the starting number

Step 6: Select every fourth student from the sampling frame, starting from the second student

1, 2, 3, 4, 5, 6, 7, 8, 9, 10, 11, 12, 13,

↓ (2) 1st ↓ (6) 2nd ↓ (12) 3rd

10.26 Multistage Sampling

In large surveys, it is often convenient from the administrative point of view to undertake the sampling in several stages. For instance, in a national survey, a logical chain would be: (i) to select, at random, a few States in the country, (ii) within each selected State, to select a random sample of Districts, (iii) within each selected District, to select a random sample of Census blocks/Villages, and (iv) within each selected Census block/village, to select a random sample of households or individuals. Apart from field convenience, an important advantage of this type of design is that the sampling frame is required to be constituted for only a small fraction of the total population.

Step 1: Define the target population
Step 2: Form the cluster population
Step 3: Randomly select the clusters
Step 4: Randomly sample units from within the selected clusters

N : Number of primary units in the population

M_i: Number of secondary units in the ith primary unit

$$y_i = \sum_{j=1}^{M_i} y_{ij} \tag{10.30}$$

$$\text{Population total } T = \sum_{i=1}^{N}\sum_{j=1}^{M_i} y_{ij} \tag{10.31}$$

$$\mu = \frac{T}{M} \text{ Where } M = \sum_{i=1}^{N} M_i$$

n : Number of primary units selected in the first stage

M_i = Number of secondary units selected in the second stage

The sample was selected in three stages to study the refractive error of school-going children (6–15 years age group)

1. Stage 1: The state may be divided into a number of districts, and a few districts may be selected at random
2. Stage 2: Each district may be divided into a number of schools, and a sample of schools may be taken at random
3. Stage 3: Males and females were selected in the second stage

Illustration

Calculate the least-cost sample sizes for two-stage surveys for demonstrating disease freedom, where cluster sizes are unknown. This analysis calculates the number of clusters and the number of units within each cluster to be tested to provide a specified system sensitivity (probability of detecting disease) for the given unit and cluster-level design prevalences and test sensitivity, where actual cluster sizes are unknown. Test specificity is assumed to be 100% (or follow-up testing of any positive result will be undertaken to confirm or exclude disease).

Sample sizes are optimized to minimize overall cost for given cluster and unit-level testing costs. A maximum sample size per cluster must be specified, and either the number of clusters in the population or a maximum number of clusters to be tested must be specified. Numbers of units to test in each cluster are calculated using assumed binomial sampling (sample size is small relative to cluster sizes), while numbers of clusters to test are calculated using the hypergeometric distribution approximation (sampling without replacement) if the number of clusters in the population is specified or assuming binomial sampling if not (Figures 10.5 and 10.6).

Inputs for sample size

Unit-level design prevalence (as a proportion only) = 0.10

Cluster-level design prevalence as either a proportion or an integer number of clusters = 0.01

Estimated test sensitivity = 0.90

The relative (or actual) cost of testing at both cluster and unit levels = 1,000

Target system sensitivity (SSe) which is the probability of detecting disease if it is present at the specified design prevalence = 20

Target system sensitivity (confidence) = 0.95

Maximum sample size to be tested per cluster = 30

Number of clusters in the population (optional) = 5

The SeH variability was studied in the following graphical representation (Figure 10.7) (Tables 10.13–10.16).

10.27 Non-Response

No matter how efficient the sample design is, the estimate derived from the survey will not be reliable if there is non-response from a large proportion of the selected members. Sometimes, it is mistakenly thought that the ill effects of non-response can be overcome by selecting a larger sample than is required from statistical considerations. This view overlooks the fact that the members who do not respond may be very different from those who do. Therefore, intensive efforts must be made to reduce the extent of non-response. Another approach could be to take a random sample of the non-responders and make all-out efforts to collect information from them. These efforts could include offering incentives in cash or in kind (e.g., a free check-up for all the family members). From the results in the random sample, those in the total set of non-responders can be estimated. Lastly, all information available regarding those who did not respond should be compared with the corresponding information from those who did, to test whether there are any differences between the two groups.

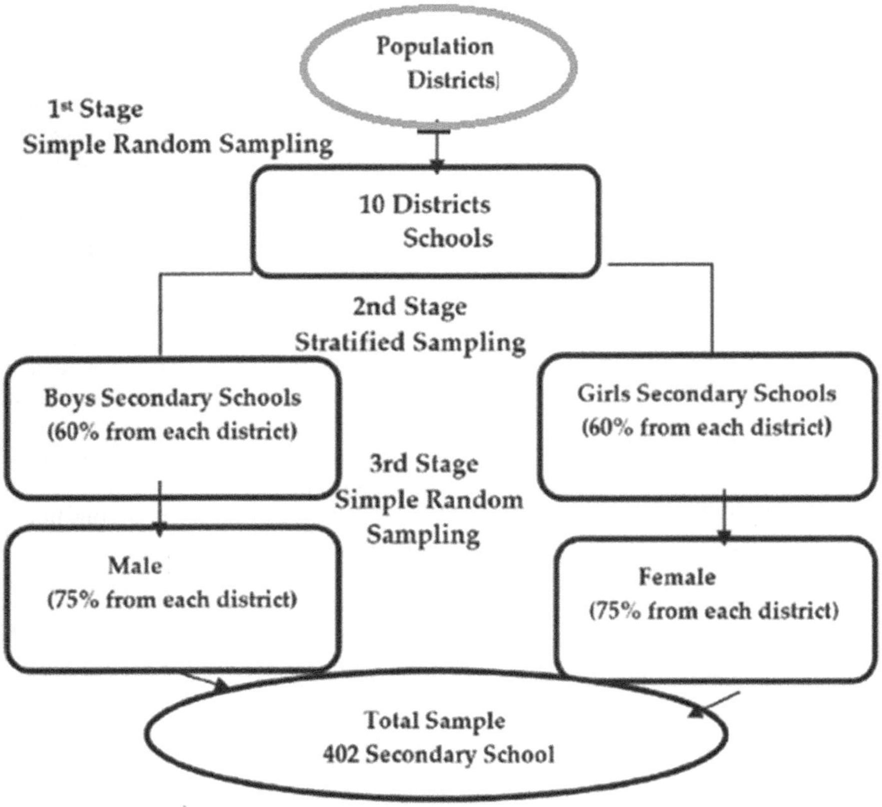

FIGURE 10.5 Selection of students through multistage sampling.

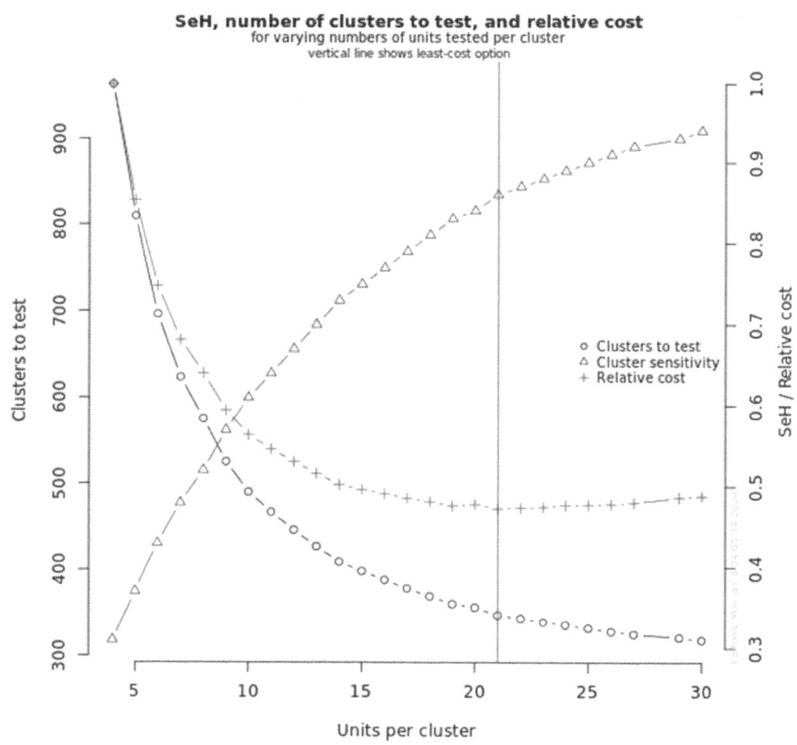

FIGURE 10.6 Sampling options alongside their relative costs.

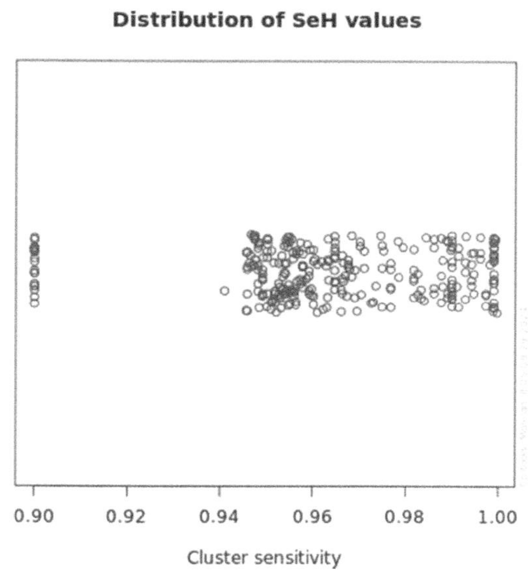

FIGURE 10.7 Scatter plot showing the distribution of the standard error of the herd.

TABLE 10.13

The Maximum Number of Clusters to be Tested=500 Based on the Two-Stage Multistage Sampling

Inputs	Values
Unit-level design prevalence	10%
Cluster-level design prevalence	1%
Number clusters	**Unknown**
Test sensitivity	0.9
Relative testing cost per cluster	1,000
Relative testing cost per unit	20
Target system sensitivity	0.95
Maximum sample size per cluster	30
Maximum number of clusters to test	1,000
Results	
Summary of least-cost option	
	Value
Clusters to test	347
Units to test per cluster	21
Total sample size (units)	7,287
Estimated cluster sensitivity (SeH)	0.86
Achieved system sensitivity (SSe)	0.9501

10.28 Other Non-Sampling Errors

Non-response, as referred to above, is only one type of non-sampling error, and often it may not be the most important one. Errors can be introduced into estimates derived from surveys due to lack of standardization of techniques, variations in the performance of the interviewers, ambiguity in questionnaire design, other biases, and last but not least, failure to monitor data as it is collected.

TABLE 10.14

Summary of Sample Size Determination (Two-Stage Sampling)

Index Cluster	Number of Clusters	Cluster Sensitivity (SeH)	Units Sampled per Cluster	Relative Cost
1	318	0.94	30	0.488
2	321	0.93	29	0.487
3	325	0.92	27	0.480
4	328	0.91	26	0.478
5	332	0.9	25	0.478
6	336	0.89	24	0.477
7	339	0.88	23	0.475
8	343	0.87	22	0.474
9	347	0.86	21	0.473
10	356	0.84	20	0.478
11	360	0.83	19	0.477
12	369	0.81	18	0.482
13	378	0.79	17	0.486
14	388	0.77	16	0.491
15	398	0.75	15	0.496
16	409	0.73	14	0.502
17	427	0.7	13	0.516
18	446	0.67	12	0.531
19	467	0.64	11	0.547
20	490	0.61	10	0.564
21	525	0.57	9	0.594
22	575	0.52	8	0.640
23	623	0.48	7	0.681
24	696	0.43	6	0.748
25	809	0.37	5	0.854
26	965	0.31	4	1

10.28.1 Optimal Size of Sample

This becomes a more complex problem. We have to consider not just the overall sample size, but how it should be distributed between clusters, and between stages of clustering. How many stages of clustering? This is usually decided on an administrative basis. Should we have many small clusters or a few large clusters? In general, the size of the lowest level of cluster (e.g. the number of households in a village, the number of plots in a field) will be determined by the amount of work that one enumerator (or one team) can do in a day. Many small clusters will usually give a better estimate than a few large clusters. Should we take the same number of units from each cluster, or more from the larger ones? We cannot consider any stage of a sample design in isolation, but must consider the overall effect. This will be discussed further in the section on 'self-weighting' designs below.

10.28.2 Selection with Probability Proportional to Size (PPS)

Most often, clusters will be of very different sizes, and *we* may wish to give the bigger clusters a higher probability of selection. This can be achieved most easily by making a list of cumulative totals of cluster sizes, and then taking a systematic sample from the list. (This may lead to a very large cluster being selected twice, in which case we take two independent

TABLE 10.15

Matrix of Selection of Lumpy Skin Disease Infected Animals in Different Clusters After Intervention of Disease Outbreak for the Vaccination

Cluster ID	Tested	Cluster Size	Cluster ID	Tested	Cluster Size	Cluster ID	Tested	Cluster Size	Cluster ID	Tested	Cluster Size
1	30	175	35	30	151	69	25	25	103	30	91
2	30	54	36	30	52	70	30	44	104	30	84
3	30	188	37	30	113	71	30	127	105	1	1
4	30	77	38	30	49	72	30	179	106	30	119
5	30	152	39	30	36	73	30	35	107	30	98
6	20	20	40	30	58	74	5	5	108	30	146
7	24	24	41	14	14	75	30	74	109	30	161
8	30	51	42	30	43	76	30	112	110	30	86
9	30	108	43	30	80	77	30	58	111	30	111
10	30		44	30	151	78	30	163	112	30	140
11	30	123	45	5	5	79	30	49	113	30	93
12	30	193	46	30	127	80	5	5	114	20	20
13	30	45	47	30	78	81	30	69	115	5	5
14	30	53	48	30	125	82	30	67	116	30	179
15	30	169	49	30	138	83	30	42	117	30	34
16	30	35	50	15	15	84	18	18	118	8	8
17	30	51	51	30	148	85	30	137	119	30	71
18	13	13	52	30	172	86	30	167	120	30	100
19	30	144	53	30	93	87	30	65	121	30	107
20	30	111	54	30	94	88	30	196	122	30	70
21	30	177	55	30	128	89	30	82	123	4	4
22	30	111	56	30	169	90	30	169	124	8	8
23	19	19	57	30	154	91	30	168	125	30	37
24	30	180	58	30	199	92	30	107	126	17	17
25	30	52	59	30	79	93	30	129	127	24	24
26	30	44	60	30	188	94	30	135	128	30	76
27	30	35	61	30	109	95	30	94	129	30	47
28	30	86	62	30	190	96	30	142	130	30	123
29	30	199	63	30	142	97	30	137	131	30	152
30	30	163	64	3	3	98	30	76	132	8	8
31	30	181	65	30	130	99	30	43	133	30	168
32	30	111	66	30	198	100	30	44	134	7	7
33	24	24	67	30	86	101	0	0	135	30	73
34	30	42	68	30	72	102	30	59	136	30	37

Cluster ID	Tested	ClusterSize	Cluster ID	Tested	Cluster Size	Cluster ID	Tested	Cluster Size
137	30	46	171	30	176	205	30	144
138	30	162	172	30	199	206	6	6
139	30	101	173	30	168	207	30	123
140	30	40	174	30	138	208	30	192
141	30	31	175	30	150	209	7	7
142	9	9	176	30	87	210	30	51
143	30	136	177	30	39	211	30	115
144	30	155	178	30	38	212	30	188
145	30	79	179	30	105	213	30	120
146	30	78	180	30	112	214	30	153
147	30	195	181	23	23	215	23	23
148	30	83	182	16	16	216	25	25
149	30	119	183	30	88	217	7	7
150	21	21	184	30	42	218	30	173
151	30	174	185	30	107	219	30	64
152	30	149	186	14	14	220	30	190
153	30	140	187	30	66	221	30	104

(Continued)

TABLE 10.15 (*Continued*)

Matrix of Selection of Lumpy Skin Disease Infected Animals in Different Clusters After Intervention of Disease Outbreak for the Vaccination

Cluster ID	Tested	ClusterSize	Cluster ID	Tested	Cluster Size	Cluster ID	Tested	Cluster Size
154	30	196	188	30	55	222	30	159
155	25	25	189	30	36	223	30	132
156	10	10	190	30	175	224	30	52
157	30	191	191	30	62	225	30	48
158	30	118	192	30	136	226	30	134
159	30	171	193	30	30	227	30	199
160	30	66	194	6	6	228	4	4
161	30	95	195	30	33	229	30	168
162	30	192	196	30	128	230	30	69
163	30	53	197	30	85	231	30	33
164	30	199	198	30	55	232	30	61
165	30	165	199	30	109	233	30	93
166	16	16	200	30	32	234	0	0
167	25	25	201	30	53	235	30	194
168	30	141	202	30	177	236	30	65
169	30	79	203	30	187	237	30	197
170	30	114	204	30	169	238	30	160

TABLE 10.16

Multistage Sampling Output Based on Level of Prevalence

Inputs	Value
Unit-level design prevalence	10%
Cluster-level design prevalence	1%
Clusters in population	Unknown
Test sensitivity	0.9
Prior probability of freedom	0.5
Summary Results	
System sensitivity	0.9368
Probability of freedom	0.9406
SeH minimum	0
SeH 2.5% percentile	0.9
SeH 5% percentile	0.9
SeH 50% percentile	0.961
SeH 95% percentile	0.999
SeH 97.5% percentile	0.999
SeH maximum	0.9997
SeH mean	0.9543
SeH standard deviation	0.101
Number of clusters tested	**288**

SeH, Standard error of Herd.

subsamples from that cluster). In some cases, it may be simpler (and just as effective) to stratify by size of cluster (e.g. small, medium, large) and just take a bigger sampling fraction from the stratum of large clusters.

10.28.3 Self-Weighting Designs

A self-weighting design is one in which each unit at the final level (e.g. plot, person, household) has the same probability of being selected. This probability is the product of the probabilities of selection (over all draws) at each stage of the design. For example, suppose we are planning a two-stage survey from a population of villages of different sizes. We will take a sample of villages, and within each selected village, take a subsample of households. We can make the design self-weighting by either.

a. Choosing villages with probability proportional to size, and then choosing the same number of households in each selected village, or
b. Choosing a simple random sample of villages, and then selecting a number of households in each selected village that is proportional to the size of that village.

The advantage of a self-weighting design is that the straightforward sample mean $\bar{x} = \frac{1}{n}\sum x_1$ is an unbiased estimator of the true population mean. This is not the case for a non-self-weighting design, where the correct formula must take into account the different probabilities of selection.

10.29 Estimation

Assume PSUs are sampled with replacement, and that if a PSU is re-visited, a totally independent subsample is chosen. Then each PSU can be considered as an independent replicate sample, and the ith sampled primary unit (PSU) provides an independent unbiased estimate of the population total X_T given by

$$X_i^x = m\sum_j \frac{x_{ij}}{P_{ij}}$$

where 'm' is the number of primary units sampled

X_{ij} is the measurement made on the jth unit in the ith PSU;

P_{ij} is the overall probability of selection of the jth unit in the ith PSU;

and the summation takes place over all selected x_{ij}'s in the PSUs, incorporating all draws and all stages of selection. Then we can estimate the overall total X_T by

$$X^* = \frac{1}{m}\sum_{i=1}^{m} X_i^* \tag{10.32}$$

$$\text{Var}\left(X^*\right) = \frac{1}{m(m-1)}\sum_{i=1}^{m}\left(X_i^* - X_i^*\right)^2 \tag{10.33}$$

10.30 Sample Size

The question that is most frequently posed to a statistician by any investigator is, "How large should the sample size be"?" There can be no universal answer to this question. The required sample size depends on the underlying assumptions and the investigator's requirements regarding Type I error and Power. There are also other important considerations such as cost, time-frame, feasibility, and losses due to non-response that must be taken into account when deciding on the final size of the study. In other words, sample size estimation is not an exclusively mathematical or statistical exercise. It involves assumptions that could be subjective, and therefore close interaction between the investigator and the statistician is necessary for the exercise to be meaningful and pragmatic. Sample size determination is considered for the following situations:

1. To demonstrate that a difference between two proportions or mean values (e.g., in a clinical trial or an observational study) is significantly different from zero.
2. To demonstrate that relative risk (e.g. in a case-control study) is significantly different from one.
3. To estimate a proportion or mean value (e.g. in a sample survey) with a pre-specified degree of precision.
4. to estimate the magnitude of an effect (e.g., a difference between two proportions or two mean values in a prospective study or intervention trial) with a pre-specified degree of precision. Demonstrating a Statistically Significant Difference Between Two Proportions

Suppose a standard regimen has an efficacy of 80%, and a new regimen that has been evolved would be of interest if it is at least 10% more effective. The former figure may be known from previous experience, but the latter must be decided by the investigator, i.e., whether a 10%, 15%, or 20% difference is of practical interest. The magnitude of Type I error (usually 0.05) and whether the test considers the possibility of the new treatment being different (better or worse) from the standard treatment or only being better must be decided, i.e., whether to employ a two-tailed probability (factor of 1.96 for a 5% significance level) or a one-tailed probability (factor of 1.645 for a 5% significance level). The power of the test, i.e., the probability of obtaining a statistically significant result if the difference exceeds a pre-specified amount, must also be specified by the investigator. It is often taken to be 90%, but could be less (80%, 75%, etc.). The trial size is given by the formula

$$n = \frac{2\left(Z_\alpha + Z_\beta\right)^2\left(P_1Q_1 + P_2Q_2\right)}{\left(P_1 - P_1\right)^2} \tag{10.34}$$

where

P_1: Proportion of successes in Group 1

Q_1: Proportion of failures in Group 1 ($= 1-P_1$)

P_2: Proportion of successes in Group 2

Q_2: Proportion of failures in Group 2 ($= 1-P_2$)

Z_α: Factor corresponding to Type I error of α^2 (Usually, α is taken as 5%, and Z_α is then 1.645 for a one-tailed test and 1.960 for a two-tailed test).

Z_β: Factor corresponding to Type II error of β, i.e. Power of $1-\beta$ (usually $1-\beta$ is taken to be 90%, and Z_β is then 1.282)

In the example above, P_1=0.8 or 80%, P_2=0.9 or 90%. Z_α(1-tail)=1.645, Z_β=1.282, and consequently

$$n = \frac{2(1.64+1.28)^2(80\times 20+90\times 10)}{(80-90)^2}$$

In Table 10.15, the sample size is set out for various combinations of P_1, P_2, Z_α, and Z_β. It may be seen that: (i) As the difference in efficacy increases, the required sample size decreases; (ii) As the power of the test decreases, the required sample size decreases; and (iii) As the significance level decreases (e.g., from 0.05 to 0.01), the required sample size increases. Finally, it should be noted that the sample size is also influenced by the absolute values of the two proportions. For instance, for α=0.05, power 90%, and a difference in efficacy=10%, the required sample size is 300 when P_1=0.95; 428 when P_1=0.90; and 540 when P_1=0.85 (Tables 10.17 and 10.18).

TABLE 10.17

Total Trial Size α (1 − Tail) = 0.05

Efficiency		Difference	Power of the Design		
New (%)	Standard (%)	in Efficacy	90%	75%	50%
95	85	10	300	188	94
	80	15	158	100	50
	75	20	64	64	52
90	80	10	428	268	136
	75	15	212	132	66
	70	20	128	80	40
85	75	10	540	338	170
	70	15	258	162	82
	65	20	152	96	48

TABLE 10.18

Total Trial Size α (1 − Tail) = 0.01

Efficiency		Difference in Efficacy	Power of the Design		
New (%)	Standard (%)		90%	75%	50%
95	85	10	456	316	190
	80	15	240	166	100
	75	20	151	106	64
90	80	10	652	450	270
	75	15	322	222	134
	70	20	196	136	82
85	75	10	820	568	342
	70	15	390	270	162
	65	20	232	160	96

10.31 Pragmatic Approaches to Determining Sample Size

Sometimes the number of patients available for study may be limited, either due to the non-availability of a large number of eligible patients or because of resource constraints (cost, time, and staff). In such situations, the problem of sample size may be formulated in other ways (Gentile & Sambucini, 2024). For instance, suppose only 300 patients are available in the situation described above. One might rephrase the question as: 'What is the smallest superiority that can be detected with a power of 90%?' or 'What is the power with which a 10% superiority can be detected?' It can be computed that the answer to the former question is 11.6% (i.e. the new regimen must be at least 91.6% effective), while the answer to the latter is 80% power.

10.32 Demonstrating That a Relative Risk Is Significantly Different from Unity

In a case-control study, the proportion of individuals who are exposed to the risk factor and the relative risk that is of practical interest must be specified, together with Type I error and power, for determining the study size. For instance, consider a situation where the proportion of subjects who ate contaminated food is 0.3 (30%), and our interest is in detecting a relative risk of diarrhea of at least 3 in these subjects compared to those who did not eat the contaminated food. If P_2 is the proportion of diarrhea cases who consumed contaminated food,

Relative Risk (RR) $=\dfrac{0.7\,P_2}{0.3(1-P_2)}=3$, from which P_2 can be derived to be 0.5625, and the problem reduces to comparing two proportions, 0.30 and 0.5625, for which the study size can be found to be 140, i.e., 70 cases and 70 controls, for a two-tailed test. If only 50 cases are available, the sample size question could be reformulated as: 'What is the power of detecting an RR of 3?' or 'What is the smallest RR that can be detected with a power of 90%?' The answer to the former is 75%, and for the latter, it is an RR of 3.6 (Table 10.19).

TABLE 10.19

Consumed Contaminated Food

Consumed Contaminated Food	Case (Diarrhea)	Control (No Diarrhea)
Yes	P_2	0.3
No	$1-P_2$	0.70
Total	1.0	1.0

Demonstrating a Statistically Significant Difference Between Two Means

The study size, n, is given by the formula

$$n=\frac{2\left(Z_\alpha+Z_\beta\right)^2\left(S^2{}_1+S_2{}^2\right)}{\left(m_1-m_1\right)^2} \qquad (10.35)$$

where m_1 and m_2 are the mean values in the two groups, and S_1^2 and S_2^2 are the variances in the two groups. Suppose the mean birth weight is 3,000 g in one district with an SD of 500 g, and our interest is in detecting a 200 g difference either way in another district where the SD is also 500 g. This situation would correspond to a two-tail test, so that $Z=1.96$. For a power of 90%, as before, $Z_\beta=1.28$.

$$n=\frac{2(1.96+1.28)^2\left(500^2+500^2\right)}{(3{,}200-3{,}000)^2}$$

From the formula above, it is observed that

1. The larger the stipulated difference between the two districts (m_1-m_2), the larger the denominator and consequently, the smaller the study size;
2. The larger the sd in the two groups, the larger the numerator and consequently, the larger the study size;
3. The greater the power, the larger the value of Z_β and the numerator, and consequently, the greater the study size.
4. The lower the significance level (Type I error), the larger the value of Z_α and the numerator, and consequently, the greater the study size.

Estimating a Proportion with a Pre-Specified Degree of Precision. If 'P' is the proportion to be estimated and Q is its complement ($1-P$), and the limit of accuracy (L) required is defined in terms of a percentage of the estimate (e.g., 10% of P), and Z is the confidence limit factor (usually taken to be 1.96, corresponding to 95% confidence limits), the number of subjects required to be included in the sample is given by

$$n=\frac{Z_\alpha{}^2PQ}{L^2} \qquad (10.36)$$

For instance, if the proportion of deaths due to diarrhea is believed to be of the order of 2%, and the limit of accuracy required is 25% (of 2%, i.e. $L=0.5\%0$ and the degree of confidence is 95% (i.e. $Z_\alpha=1.96$)

$$n=\frac{1.96^2 2\times 98}{(0.50)^2}=3{,}012$$

TABLE 10.20

Sample Size Deployed to Mortality Rate

Mortality (%)	P	L	N	L	N
10	0.10	±0.0250	553	±0.010	3,457
5	0.05	±0.0125	1,168	±0.005	7,299
2	0.05	±0.0050	3,012	±0.002	18,824
1	0.01	±0.0025	6,085	±0.001	38,032

From the formula above and the table below, it is clear that the sample size increases as P decreases (e.g. 3012 for P-2% and 6,085 for $P=1\%$); it also increases as L (the limit of accuracy) decreases (e.g. 3,012 for $L=25\%$ and 18,824 for $L=10\%$) (Table 10.20).

The computation above pertains to a situation where simple random sampling is employed to select the study subjects. In practice, this technique is not very convenient, and investigators take recourse to operationally feasible methods such as cluster sampling or multi-stage sampling. While the latter methods are easier to implement, the estimates they provide have larger sampling errors. Consequently, a multiplication factor (usually called the 'Design Effect') has to be used to escalate the study sizes. For instance, if cluster sampling is to be employed and has a design effect of 2, the required sample size would be $3{,}012\times 2=6{,}024$. Estimating a Difference Between Two Proportions with a Pre-Specified Degree of Precision. If the limit of accuracy required in estimating a difference between two proportions P_1 and P_2 is L, and X_α is the confidence factor (1.96 for 95% confidence), the required sample size (N) in each group is given by the formula:

$$n=\frac{Z_\alpha^{\,2}\left(P_1Q_1+P_2Q_2\right)}{L^2} \tag{10.37}$$

If $P_1=30\%$, $P_2=15\%$, $L=10\%$ and $Z_\alpha=1.96$

$$n=\frac{1.96^2\left(30\times 70+15\times 85\right)}{10^2}=135$$

The study size is therefore $135\times 2=270$

Estimating a mean With a Required Degree of Precision

If the standard deviation is s, and the mean (m) is to be estimated with an accuracy of $L\%$, and the confidence factor is Z_α (1.96 for 95% confidence).

$$n=\frac{Z_\alpha^2S^2)}{\left(L\%\ \text{of}\ m\right)^2},$$

If $m=5{,}000$ g, $s=500$ g, and $L=1\%$,

$$n=\frac{1.96\ ^2500^2)}{(50)^2}=384$$

Estimating a Difference Between Two Means With a Specified Degree of Precision. If m_1 and m_2 denote two means whose difference is to be estimated with a limit of accuracy of $L\%$, and a confidence factor of Z_α, the sample size in each group.

$$n=\frac{2\ Z_\alpha^{\,2}S^2)}{\left(L\%\ \text{of}\ m_1-m_2\right)^2} \tag{10.38}$$

If $m_1=2{,}000$ g, $m_2=3{,}200$ g and $L=50\%$, and $s=500$ g and $Z_\alpha=1.96$

$$n=\frac{2\ 1.96_\alpha^2500^2)}{\left(L\%\ \text{of}\ 2{,}000-3{,}200\right)^2}=192$$

Study size $=2\ N=2\times 192=384$

10.33 Concluding Remarks

The following steps constitute a pragmatic approach to decision-making on sample size:

- Remember that there is no stock answer.
- Initiate an early dialogue with an applied statistician
- Discuss your assumptions with the statistician—consider various possibilities, such as a range of values if you are uncertain about your assumed estimates.
- Examine feedback from the statistician.
- Consider other factors as well, such as the availability of cases, cost, and time.

10.34 Make a Balanced Choice

Ask if this number gives you a reasonable prospect of arriving at useful conclusions.

If yes, proceed; if no, reformulate your problem for study.

If a trial design is employed wherein each patient receives both treatments (e.g., on two separate occasions to the same eye, with a wash-out period between them, or on the same occasion but with one treatment to each eye), the trial design will be more efficient, and the required trial size will be smaller. The gain will depend on the magnitude of the correlation between the results in the same patient.

10.35 Comparison of Proportions

If the outcome measure is dichotomous (e.g. success/failure), the difference between the two treatments will be tested by a paired chi-square test (McNemar's test), which is essentially based on the pairs in which one treatment succeeds while the other fails (discordant pairs). One, therefore, needs to have an estimate of f, the proportion of such pairs. If $P_2-P_1=d$, the number of patients (N) required is given by the formula

$$n=\frac{\left[Z_\alpha\sqrt{f}+Z_\beta\sqrt{\left(f-d^2\right)^2}\right]}{d^2} \tag{10.39}$$

If $P_1=0.8$ and $P_2=0.9$, $d=0.1$; f is unknown, but it can be shown that its highest value will be $[(P_1+P_2-2\ P_1 \times P_2)-d^2]$ i.e. 0.26−0.01 =0.25. This is obtained when the findings for the two eyes are uncorrelated. Setting $d=0.1$, $f=0.25$, $Z_\alpha=1.645$, and $Z_\beta=1.282$, $N=210$, i.e. 420 eyes are required. This is similar to the 428 patients (eyes) that would have required had only one treatment been administered to each patient. However, if f takes a value lower than 0.25 because there is a correlation between the results of the two treatments, there will be a decrease in sample size. For example, if $f=0.2$, $N=168$, i.e. 336 eyes.

10.36 Comparison of Mean Values

If the outcome measure is quantitative (e.g. IOP), the difference between the two treatments will be tested by a paired t-test, i.e. the mean of the differences between the two treatments in individual patients is tested to be equal to zero. If SD^2 denotes the variance of these differences, the number of patients required is given by

$$N = \frac{SD^2\left[Z_\alpha + Z_\beta\right]^2}{d^2} \tag{10.40}$$

If SD^2 is unknown, it may be estimated by $2s^2\ (1-r)$, where 'r' is the estimated interclass correlation coefficient, and s is the estimated standard deviation for one of the treatments (e.g. control).

10.37 Demonstrating That Two Treatments Are Equally Effective

Until now, sample sizes are estimated on the premise that a new treatment is better than the standard treatment. However, there are situations where the interest is to demonstrate that two treatments are equally effective. For example, in breast cancer, one may want to know whether simple mastectomy is as effective as radical mastectomy. Suppose the response is p with the standard treatment. Then one chooses a value d such that the upper 95% confidence limit of the difference in the responses of the new and standard treatments should not exceed d with Power $(1-\beta)$. Suppose $p=90\%$, $d=5\%$, α (Type 1 error)=5%, and β (type II error)=10%, i.e. Power=90%.

$$n = \frac{SD^2\left[Z_\alpha + Z_\beta\right]^2 PQ}{d^2} \tag{10.41}$$

$$n = \frac{2[1.645+1.28]^2(90)(10)}{5^2}, n = 620$$

The sample size required in each group $n=620$. This illustrates that many patients are required to establish the equivalence of two treatments.

10.38 Pragmatic Approaches with Modified Power Estimates

Sometimes, the number of patients available for study may be limited, either due to the non-availability of large numbers of eligible patients or because of resource constraints (cost, time, staff). In such situations, the problem of sample size may be formulated in other ways. For instance, suppose only $n=300$ patients are available in the situation described earlier. One might rephrase the question as, "What is the Power with which a 10% superiority can be detected?" It can be computed that the answer to this question is 80% Power. Suppose only $n=270$ patients are available; then the Power is 75%. If only $n=100$ patients are available, the Power is 50%. Very large values of Type II error, say 50% (i.e. Power of 50%), are not really acceptable because the chances of missing the true difference become too high. A realistic assessment of the sample size with corresponding Power has to be made.

10.39 Concluding Remarks

The following steps constitute a pragmatic approach to decision-making on sample size:

- Remember that there is no stock answer.
- Initiate an early dialogue with an applied statistician.
- Discuss your assumptions with the statistician—consider various possibilities, for instance, a range of values if you are not sure of your assumed estimates.
- Examine feedback from the statistician.
- Consider other factors as well, such as the availability of cases, cost, and time.
- Make a balanced choice.
- Ask if this number gives you a reasonable prospect of arriving at useful conclusions. If yes, proceed; if no, reformulate your problem for study.

REFERENCES

Abdallah, J. M. A review of sample size determination for common experimental designs: Further simplified equations. *An-Najah University Journal for Research - A (Natural Sciences)* **38**, 8–16 (2024).

Akbar, M. H., Ali, S., Shah, I. & Alqifari, H. N. Sample size determination for time-to-event endpoints in randomized selection trials with generalized exponential distribution. *Heliyon* **10**, e27013 (2024).

Besekar, S., Jogdand, S. & Naqvi, W. Navigating sample size determination in educational research: A rapid review unveiling strategies, challenges, and recommendations. *F1000Research* **12,** 1291 (2024).

Bourazas, K., Consonni, G. & Deldossi, L. Bayesian sample size determination for detecting heterogeneity in multi-site replication studies. *Test* (2024). doi:10.1007/s11749-023-00916-4

Bujang, M. A., Omar, E. D., Foo, D. H. P. & Hon, Y. K. Sample size determination for conducting a pilot study to assess reliability of a questionnaire. *Restorative Dentistry and Endodontics* **49**, e3 (2024).

Chen, Y., Lam, K. F. & Xu, J. Sample size calculation for multi-arm parallel design with restricted mean survival time. *Statistical Methods in Medical Research* **33,** 130–147 (2024)

De Santis, F. & Gubbiotti, S. On the distribution of the power function for the scale parameter of exponential families. *Statistics in Medicine* **43,** 1973–1992 (2024).

Gentile, S. & Sambucini, V. Frequentist and Bayesian sample size determination for single-arm clinical trials based on a binary response variable: A shiny app to implement exact methods. *Open Journal of Statistics* **14,** 90–105 (2024).

Ko, F. S. The Simon's two-stage design accounting for genetic heterogeneity. *Communications in Statistics - Theory and Methods* **53,** 2661–2669 (2024).

Louangrath, P. I. Sample size determination for non-finite population. *SSRN Electronic Journal* (2017). https://doi.org/10.2139/ssrn.2710455

Mondal, D., Vanbelle, S., Cassese, A. & Candel, M. J. J. M. Review of sample size determination methods for the intraclass correlation coefficient in the one-way analysis of variance model. *Statistical Methods in Medical Research* **33,** 532–553 (2024).

Nakagawa, S., Lagisz, M., Yang, Y. & Drobniak, S. M. Finding the right power balance: Better study design and collaboration can reduce dependence on statistical power. *PLoS Biology* **22,** e3002423 (2024).

Qiu, S. F. et al. Sample size determination for interval estimation of the prevalence of a sensitive attribute under non-randomized response models. *British Journal of Mathematical and Statistical Psychology* (2024). https://doi.org/10.1111/bmsp.12338

Wu, P. Y., Ou, J. H. & Liao, C. T. Sample size determination for training set optimization in genomic prediction. *Theoretical and Applied Genetics* **136,** 57 (2023).

Yoo, H. Sample size calculations for clustered count data based on zero-inflated discrete Weibull regression models. *Communications for Statistical Applications and Methods* **31,** 55–64 (2024).

11

A Comprehensive Overview of Machine Learning and Artificial Neural Network Modeling

11.1 Introduction

By using sample data or historical information, machine learning teaches a computer how to maximize a performance criterion. Our model is only partially defined, and learning is the process of using training data or past knowledge to run a computer program and optimize the model's parameters (Cock et al., 2010). The model can be predictive to predict future events or descriptive to learn from data, or it can be both at the same time. Arthur Samuel, an IBM employee and pioneer in artificial intelligence and computer gaming, coined the phrase 'machine learning' for the first time in 1959 (Grada & Weinbrecht, 2013). 'The field of study that gives computers the ability to learn without being explicitly programmed' is how he defined machine learning. However, there is not a single, agreed-upon definition of machine learning. The term is defined differently by different authors.

11.2 Definition of Learning

A computer program is said to learn from experience (E) with respect to some class of tasks (T) and performance measure (P) if, with experience (E), its performance at tasks (T), as determined by (P), improves (Lee et al., 2011).

Examples I: Learning a handwriting recognition problem

- Task T: Identifying and categorizing handwritten words in pictures
- Performance P: Words correctly classified as a percentage
- Experience in training E: A collection of handwritten words with predetermined categories

ii. A learning problem for robot driving
  - Task T: Using vision sensors to drive on highways
  - Performance P: Average distance traveled prior to an error
  - Experience E: A series of pictures and steering instructions captured while watching a human driver

iii. A chess education exercise
  - Task T: Chess play
  - Performance P: The percentage of victories over adversaries
  - Experience E: Competing against oneself in practice games

The learning process, whether by a human or a machine, can be divided into four components, namely data storage, abstraction, generalization, and evaluation. The steps and different parts of the learning process are shown in (Figure 11.1).

11.3 Data Repository

Large-scale data storage and retrieval capabilities are essential to the learning process. Both computers and humans use data storage as the basis for sophisticated reasoning. Information is stored in the brain and retrieved by electrochemical impulses in humans. Computers store data on hard drives, flash memory, random access memory, and similar devices, retrieving it using cables and other technologies (Liu et al., 2012).

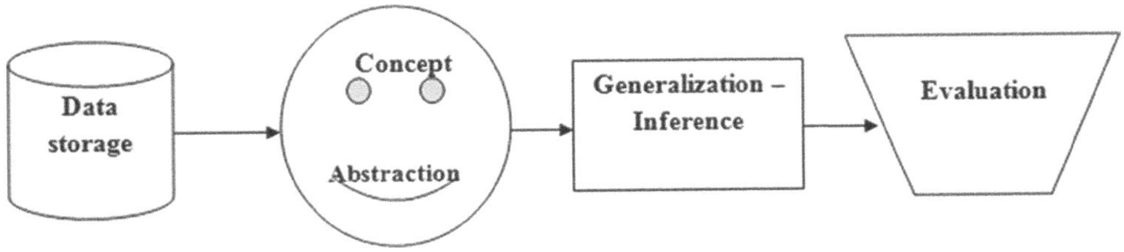

FIGURE 11.1 Computer learning process in machine learning.

DOI: 10.1201/9781032713816-11

11.4 Distraction

The second component of the learning process is known as abstraction. The process of extracting knowledge about stored data is called abstraction. Developing broad notions about the data as a whole is required for this. The application of existing models and the development of new models are both necessary for knowledge creation. Training is the process of adapting a model to a dataset. The data is converted into an abstract form that condenses the original information once the model has been trained.

11.5 Broad Generalization

Generalization is the third step in the learning process. The process of transforming information about stored data into a format that can be applied to future actions is known as generalization. Work that is comparable to previous tasks—but not exactly the same—must be done to complete these actions. Finding the aspects of the data that will be most useful for upcoming tasks is the aim of generalization.

11.6 Evaluation

Evaluation is the last component of the learning process. It is the process of giving feedback to the user to measure the utility of the learned knowledge. This feedback is then utilized to effect improvements in the whole learning process.

11.7 Models for Machine Learning

The goal of machine learning is to create models that are appropriate for the tasks at hand by utilizing the appropriate features. Based on the core idea of learning models, three categories have been developed. The set of all possible solutions for a particular problem is known as the sample space or instance space.

Using a logical expression (Logical models)

Using the geometry of the instance space (Geometric models)

Using probability to classify the instance space (Probabilistic models)

11.7.1 Grouping and Grading

Logical models use a logical expression to divide the instance space into segments and hence construct grouping models. An expression that yields a Boolean value, or a True or False result, is considered logical. After employing a logical expression to organize the data, it is separated into homogeneous groupings according to the problem we are attempting to solve. In a classification problem, for instance, every instance in the group is a member of the same class. Logical models can be broadly classified into two categories: rule models and tree models. An assortment of implications, or IF-THEN rules, makes up rule models. In tree-based models, a segment is defined by the 'if-part', and the model's behavior within that segment is defined by the 'then-part'. Rule models make use of the same logic (Mardis, 2013).

To get a deeper understanding of logical models, we must grasp the concept of concept learning.

Concept Learning involves learning logical expressions or concepts from examples. The idea of concept learning fits in well with the idea of machine learning, i.e. inferring a general function from specific training examples. Concept learning forms the basis of both tree-based and rule-based models. More formally, concept learning involves acquiring the definition of a general category from a given set of positive and negative training examples of the category. The process of inferring a Boolean-valued function from training examples of its input and output is known as the formal definition of concept learning. In concept learning, we only learn a description for the positive class and label everything that doesn't satisfy that description as negative (Martin & Wang, 2011) (Table 11.1).

11.7.2 Geometric Models

Features in geometric models can be thought of as points in a three-dimensional space (x, y, and z) or as points in two dimensions (the x and y axes). It is possible to model features in a geometric way even if they are not inherently geometric (temperature as a function of time, for instance, can be modeled on two axes). There are two ways we could enforce similarity in geometric models (Metzker, 2010).

We could use geometric concepts like lines or planes to segment (classify) the instance space. These are called linear models.

Alternatively, we can use the geometric notion of distance to represent similarity. In this instance, two points can be

TABLE 11.1

Algorithm of ML for Diseases Status

#	Disease status	Climatic Condition for Virus/Bacteria	Humidity	Contact Rate	Precaution	Prediction	Treatment
1	Symptomatic	Warm	Normal	Strong	Yes	Infective	Yes
2	Symptomatic	Warm	High	Strong	Yes	Infective	Yes
3	a- Symptomatic	Rainy	High	Strong	Yes	Un infection	No
4	Symptomatic	Warm	High	Strong	Yes	infective	Yes
	Attributes						Concept

categorized as similar if they are near each other and have similar feature values. We call these models "distance-based models."

11.7.3 Linear Models

Simpler models are those that are linear. In this case, the function is represented as a linear combination of its inputs. As a result, a linear combination of X_1 and X_2 is represented by a X_1+bX_2, if X_1 and X_2 are two scalars or vectors of the same dimension, and '*a*' and '*b*' are arbitrary scalars.

In the simplest case, where $f(x)$ represents a straight line, we have an equation of the form

$$f(x)=m\,x+c \quad (11.1)$$

where c represents the intercept and m represents the slope.

Because linear models are parametric, they have a fixed form and a limited number of numerical parameters that must be determined from the available data. For instance, the parameters that we are attempting to learn from the data are m and c in $f(x)=m\,x+c$. This methodology differs from tree or rule models in that the model's structure (i.e., the characteristics to use in the tree and where to use them) is not static. This usually indicates that the differences extend to the remainder of the tree. Linear models are characterized by low variance and high bias due to their relatively small number of parameters. This suggests that, compared to certain other models, linear models have a lower propensity to overfit the training set (Raza & Ahmed, 2017). However, they are more prone to underfitting. For example, linear models are unlikely to provide a satisfactory approximation if we need to learn the borders between nations based on labeled data.

11.7.4 Distance-Based Models

The second type of geometric models is those that are based on distance. Distance-based models are based on the geometry of the data, just like linear models. Distance-based models, as their name suggests, focus on the idea of distance. Within machine learning, the definition of distance extends beyond the sheer physical separation of two points. Alternatively, we might incorporate the means of transportation between two points when calculating the distance (e.g. Euclidean, Minkowski, Manhattan, and Mahalanobis). Distance is used in relation to neighbors and role models. Referred to as exemplars, neighbors are points that are close together in terms of the distance metric. Examples include medoids or centroids, which locate the center of mass based on a selected distance measure that identifies the data point most centrally positioned. The arithmetic mean, which minimizes the squared Euclidean distance to every other point, is the most widely used centroid. The centroid denotes the geometric center of a flat figure, which is the spot where all of the form's points are arithmetically averaged to be located. Any object in n-dimensional space can be defined in this way: its centroid is the average position of all its points.

11.7.5 Probabilistic Models

The third family of machine learning algorithms is probabilistic models. We have seen before that the k-nearest neighbor algorithm uses the idea of distance (e.g. Euclidean distance) to classify entities, and logical models use a logical expression to partition the instance space. In this section, we see how probabilistic models use the idea of probability to classify new entities. Probabilistic models view features and target variables as random variables (Sanger & Coulson, 1975; Sanger et al. 1977). The process of modeling represents and manipulates the level of uncertainty with respect to these variables. There are two types of probabilistic models: predictive and generative. Predictive probability models use the idea of a conditional probability distribution $P\,(Y|X)$ from which Y can be predicted from X. Generative models estimate the joint distribution $P\,(Y, X)$. Once we know the joint distribution for generative models, we can derive any conditional or marginal distribution involving the same variables. Thus, the generative model is capable of creating new data points and their labels, knowing the joint probability distribution. The joint distribution looks for a relationship between two variables. Once this relationship is inferred, it is possible to infer new data points. Naïve Bayes is an example of a probabilistic classifier

$$P\left(A_i/B_i\right)=(P\left(B_i/A_i\right)*P\left(A_i\right)/(P\left(B_i\right) \quad (11.2)$$

The Naïve Bayes algorithm is based on the idea of conditional probability. Conditional probability involves finding the probability that something will happen, *given that something else* has already happened. The task of the algorithm, then, is to look at the evidence and determine the likelihood of a specific class, assigning a label accordingly to each entity (Table 11.2).

11.8 Grouping and Grading

Grading versus grouping is an orthogonal categorization to geometric-probabilistic-logical-compositional. Grouping models break the instance space into groups or segments and, in each segment, apply a very simple method (such as the majority class).

For example, decision tree, KNN. Grading models form one global model over the instance space.

For example, linear classifiers–neural networks

TABLE 11.2

Difference between Geometric, Probabilistic and Logical Models

Geometric Model	Probabilistic Models	Logical Models
E.g. K-nearest neighbors, linear regression, support vector machine, logistic regression, …	Naïve Bayes, Gaussian process regression, conditional random field, …	Decision tree, random forest

Formulation of Designing Learning Systems

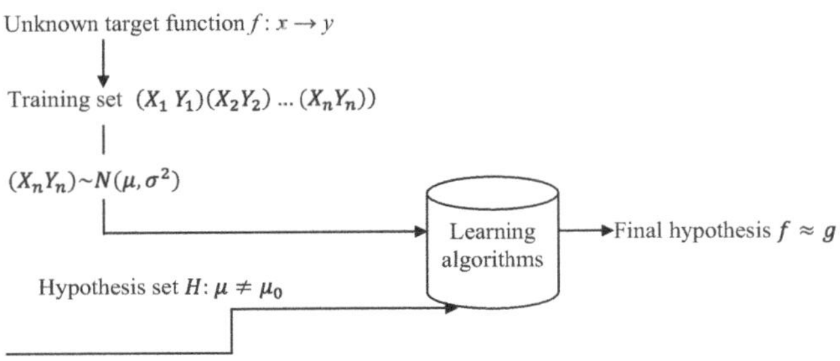

Task T, training experience E, performance measure P, and the goal of locating an unknown target function are the first steps in the learning process. The precise knowledge that must be acquired during training is known as the target function. For instance, in the case of credit approval, the learning system's task would be to categorize client application data as experience and determine if the particular customer application is qualified for a loan. Thus, in this instance, the training examples may be expressed as (X_1, Y_1) (X_2, Y_2). (X_n, Y_n), where 'X' stands for the contents of the customer's application and y for the credit acceptance status.

11.9 Design of a Learning System

We recently examined the learning process and understood the learning objective. We must take into account a few designs in order to create a learning system that mirrors the learning process.

Decisions: The design decisions will determine the following essential elements:

1. Type of training experience
2. Selecting the goal function
3. Selecting a target function representation
4. Selecting an approach for the target function approximation
5. The complete design

We will examine the learning problem of the game of checkers and apply the aforementioned design choices. The three components of a checkers learning problem are as follows:

1. Task T: Checkers play
2. Performance metric P: The overall percentage of games won in the competition
3. Training experience E: A set of games played against itself

11.10 Type of Training Experience

During the design of the checker's learning system, the type of training experience available for a learning system will have a significant effect on its success or failure.

11.10.1 Direct or Indirect Training Experience

In the case of direct training experience, individual board states and the correct moves for each board state are given. In the case of indirect training experience, the move sequences for a game and the final result (win, loss, or draw) are given for a number of games. Assigning credit or blame to individual moves is referred to as the credit assignment problem.

11.10.2 Supervised

The training experience will be labeled, which means all the board states will be labeled with the correct moves. Thus, the learning takes place in the presence of a supervisor or teacher.

11.10.3 Unsupervised

The training experience will be unlabeled, which means all the board states will not have associated moves. The learner generates random games and plays against itself with no supervision or teacher involvement.

11.10.4 Semi-Supervised

The learner generates game states and asks the teacher for help in finding the correct move if the board state is confusing.

11.10.5 Training Experience

Are the training examples representative of the distribution of instances used to calculate the system's final performance? When training and test samples come from the same or a similar distribution, performance is at its best. By competing with themselves, checkers players gain knowledge. They have an indirect experience. They might not come across movements that are typical of expert human play. Selecting the Target Function will be the following design stage after the appropriate training experience is provided.

11.10.6 Choosing the Target Function

Whenever you play the game of checkers, you must decide which move out of the available options is the best one. You reflect and apply the knowledge you have learned from the encounter. The goal is to move a checker such that your board state moves toward a winning scenario for a given board. It is now necessary to define this learning in terms of the target function. There are two factors to consider here: direct and indirect experience.

11.10.7 During the Direct Experience

All that the checkers learning system needs to learn is how to select the optimal move from a wide range of possible moves. To select the optimal course of action among options, we must identify a target function. This function can be named Choose Move and denoted with the notation Choose Move: $B \to M$ to show that it takes any board from the set of legal board states B as input and returns a move from the set of legal moves M as output.

11.10.8 Indirect Experience

It becomes more challenging to pick up such a function. Consider assigning a true score to the board state. Thus, the function would be $V: B \rightarrow R$, meaning that it takes any board from the set of valid board states B as input and returns a real score as the result. Better board states receive higher scores according to this function.

The system can choose the optimal move from any board position with ease if it can effectively learn such a goal function V.

As a result, let us define the target value $V(b)$ as follows for any board state b in B:

1. $V(b)=100$ if b is the ultimate board state that is won.
2. $V(b)=-100$ if b is the last lost board state.
3. If b is the last drawn state of the board, then $V(b)=0$.
4. If b is not a game-ending position, then $V(b)=V(b')$, where b' is the greatest possible final board state that can be attained by starting at b and continuing to play as optimally as possible until the game is over.

Since (4) is a recursive definition, it searches ahead for the best line of play until the end of the game in order to find the value of $V(b)$ for a given board state. We refer to this concept as being operational because our checkers-playing algorithm cannot efficiently compute it. Finding an operational description of V, or a description that the checkers-playing program can use to assess states and choose moves within reasonable time constraints, is the learning objective in this scenario. In general, mastering such an operational form of V might be exceedingly challenging. We anticipate that learning algorithms will only obtain a rough representation of the target function $\hat{V}$.

11.10.9 Choosing a Representation for the Target Function

The learning algorithm will employ a representation to describe the function V that it will learn after we have specified the ideal target function V. We have a wide range of possibilities, just like in previous design decisions. One such approach would be to provide the software with the ability to display a large table with individual entries for each unique board condition, each indicating a different value. Alternatively, we may allow it to represent itself using an artificial neural network, a quadratic polynomial function of predetermined board features, or a set of rules that compare against board state features. Generally speaking, there is a significant trade-off in this representational decision. On the one hand, we want to use a highly expressive representation so that we may depict something close to the best estimate of the optimal target function V. However, the more expressive the representation is, the more training data the computer will need to select among the various hypotheses it might represent. Let us use a straightforward representation to keep the discussion concise: for any given board state, the function $\hat{V}$ will be computed as a linear combination of the board properties listed below:

$X_1(b)$: number of black pieces on board b
$X_2(b)$: number of red pieces on b
$X_3(b)$: number of black kings on b
$X_4(b)$: number of red kings on b
$X_5(b)$: number of red pieces threatened by black (i.e. peices that can be taken on black's next turn)
$X_6(b)$: number of black pieces threatened by red

$$\hat{V} = w_0 + w_1 * x_1(b) + w_2 * x_2(b) + w_3 * x_3(b) + w_4 * x_4(b) + w_5 * x_5(b) + w_6 * x_6(b) \quad (11.5)$$

where w_0 through w_6 are numerical coefficients or weights to be obtained by a learning algorithm. Weights w_1 to w_6 will determine the relative importance of different board features.

11.10.10 Specification of the Machine Learning Problem at This Time

So far we have worked on choosing the type of training experience, selecting the target function, and determining its representation. The checkers learning task can be summarized as follows:

Task T: Play Checkers
Performance Measure: Percentage of games won in the world tournament
Training Experience E: Opportunity to play against itself
Target Function: V: Board ->R

Target Function Representation

$$\hat{V} = w_0 + w_1 * x_1(b) + w_2 * x_2(b) + w_3 * x_3(b) + w_4 * x_4(b) + w_5 * x_5(b) + w_6 * x_6(b) \quad (11.6)$$

The first three items above correspond to the specification of the learning task, whereas the final two items constitute design choices for the implementation of the learning program.

11.11 Choosing an Approximation Algorithm for the Target Function

11.11.1 Generating Training Data

To train our learning program, we need a set of training data, each describing a specific board state b and the training value Vtrain (b) for b. Each training example is an ordered pair <b, $\bar{V}$ train (b). For example, a training example may be ($X_1=3$, $X_2=0$,

$X_3=1$, $X_4=0$, $X_5=0$, $X_6=0$) +100. This is an example where black has won the game since $x_2=0$ (or red has no remaining pieces). However, such clean values of *V* train (*b*) can be obtained only for board states *b* that are a clear win, loss, or draw. In the above case, assigning a training value *V*train(*b*) for the specific boards b that are a clear win, loss, or draw is straightforward, as they are direct training experiences. But in the case of indirect training experience, assigning a training value *V* train(*b*) for intermediate boards is difficult. In such cases, the training values are updated using temporal difference learning.

Temporal Difference (TD) learning is a concept central to reinforcement learning, in which learning happens through the iterative correction of estimated returns towards a more accurate target return.

Let Successor (*b*) denote the next board state following *b*, for which it is again the program's turn to move. $\hat{V}$ is the learner's current approximation to *V*. Using this information, the training value of *V*train (*b*) for any intermediate board state *b* is assigned as follows:

$$\overline{v} \text{ train } (b) \leftarrow \overline{v} \text{ Sucessor } (b)$$

11.11.2 Adjusting the Weights

Now it's time to define the learning algorithm for choosing the weights that best fit the set of training examples. One common approach is to define the best hypothesis as the one that minimizes the squared error *E* between the training values and the values predicted by the hypothesis $\hat{V}$

$$E = \sum_{(B,V(B))\int \text{Training}} \left(V_{\text{train}}(b) - \hat{V}(b) \right)^2 \qquad (11.7)$$

The learning algorithm should incrementally refine weights as more training examples become available and it needs to be robust to errors in the training data. The Least Mean Square (LMS) training rule is a training algorithm that adjusts weights by a small amount in the direction that reduces the error.

For each training example '*b*'

1. Compute the error (b) $= V_{\text{train}}$

$$V_{\text{train}}(b) - \hat{V}(b) \qquad (11.8)$$

2. For each board feature x_i update W_i,

$$w_i \leftarrow w_i + \mu.\text{error}(b).x_i \qquad (11.9)$$

Here 'μ' is a small constant (e.g. 0,1) that moderates the size of the weight update.

11.11.3 Final Design for Checkers Learning System

The final design of our checkers learning system can be naturally described by four distinct

program modules that represent the central components in many learning systems.

11.12 The Performance System

Takes a new board as input and outputs a trace of the game it played against itself.

11.13 The Critic

Takes the trace of a game as input and outputs a set of training examples for the target function.

11.14 The Generalizer

Takes training examples as input and outputs a hypothesis that estimates the target function. Good generalization to new cases is crucial.

11.15 The Experiment Generator

Takes the current hypothesis (currently learned function) as input and outputs a new problem (an initial board state) for the performance system to explore.

11.16 Types of Learning

In general, machine learning algorithms can be classified into three types:

(i) Supervised learning, (ii) Unsupervised learning, (iii) Reinforcement learning

11.16.1 Supervised Learning

A training set of examples with the correct responses (targets) is provided and, based on this training set, the algorithm generalizes to respond correctly to all possible inputs. This is also called learning from exemplars. Supervised learning is the machine learning task of learning a function that maps an input to an output based on example input-output pairs. In supervised learning, each example in the training set is a pair consisting of an input object (typically a vector) and an output value. A supervised learning algorithm analyzes the training data and produces a function, which can be used for mapping new examples. In the optimal case, the function will correctly determine the class labels for unseen instances. Both classification and regression.

Example

i. Consider the following data regarding patients entering a clinic. The data consists of the gender and age of the patients, and each patient is labeled as 'healthy' or 'sick'.

Gender	M	M	F	M	F	M	F	M	M
Age	45	48	55	56	54	58	41	40	38
Label	Sick	Healthy	Sick	Healthy	Sick	Healthy	Sick	Sick	Sick

11.16.2 Unsupervised Learning

Correct responses are not provided; instead, the algorithm tries to identify similarities between the inputs so that inputs with something in common are categorized together. The statistical approach to unsupervised learning is known as density estimation. Unsupervised learning is a type of machine learning algorithm used to draw inferences from datasets consisting of input data without labeled responses. In unsupervised learning algorithms, a classification or categorization is not included in the observations. There are no output values, so there is no estimation of functions. Since the examples given to the learner are unlabeled, the accuracy of the structure output by the algorithm cannot be evaluated. The most common unsupervised learning method is cluster analysis, which is used for exploratory data analysis to find hidden patterns.

Example: Consider the following data regarding patients entering a clinic. The data consists of the gender and age of the patients.

Gender	M	M	M	M	F
Age	56	48	65	58	54

Based on this data, can we infer anything regarding the patients entering the clinic?

11.16.3 Reinforcement Learning

This is somewhere between supervised and unsupervised learning. The algorithm is told when the answer is wrong, but does not get told how to correct it. It has to explore and try out different possibilities until it works out how to get the answer right. Reinforcement learning is sometimes called learning with a critic because of this monitor that scores the answer but does not suggest improvements. Reinforcement learning is the problem of getting an agent to act in the world so as to maximize its rewards. A learner (the program) is not told what actions to take, as in most forms of machine learning, but instead must discover which actions yield the most reward by trying them. In the most interesting and challenging cases, actions may affect not only the immediate reward but also the next situation and, through that, all subsequent rewards. Example – Consider teaching a dog a new trick: we cannot tell it what to do, but we can reward or punish it if it does the right or wrong thing. It has to find out what it did that made it get the reward or punishment. We can use a similar method to train computers to do many tasks, such as playing backgammon or chess, scheduling jobs, and controlling robot limbs. Reinforcement learning is different from supervised learning. Supervised learning is learning from examples provided by a knowledgeable expert.

11.17 Issues in Machine Learning

Our checkers example raises a number of generic questions about machine learning. The field of machine learning, and much of this book, is concerned with answering questions such as the following:

- What algorithms exist for learning general target functions from specific training examples? In what settings will particular algorithms converge to the desired function, given sufficient training data?
- Which algorithms perform best for which types of problems and representations?
- How much training data is sufficient? What general bounds can be found to relate the confidence in learned hypotheses to the amount of training experience and the character of the learner's hypothesis space?
- When and how can prior knowledge held by the learner guide the process of generalizing from examples? Can prior knowledge be helpful even when it is only approximately correct?
- What is the best strategy for choosing a useful next training experience, and how does the choice of this strategy alter the complexity of the learning problem?
- What is the best way to reduce the learning task to one or more function approximation problems? Put another way, what specific functions should the system attempt to learn? Can't his process itself be automated?
- How can the learner automatically alter its representation to improve its ability to represent and learn the target function? Figure 11.2, (Table 11.3)

11.18 *k*-Means Clustering Algorithm

One of the most commonly used clustering algorithms is *k-means*. It allows grouping the data into k clusters based on the existing similarities among them, with *k* being given as input to the algorithm. I'll start with a simple example. Let's imagine we have 5 objects (say 5 people), and for each of them, we know two features (height and weight). We want to group them into $k=2$ clusters.

Person No	Height (*H*)	Weight (*W*)
#1	167	55
#2	120	32
#3	113	33
#4	175	76
#5	108	25

First of all, we have to initialize the value of the centroid for our clusters. For instance, let's choose Person 2 and Person 3 as the two centroids c_1 and c_2, so that $c_1=(120, 32)$ and $c_2=(113, 33)$.

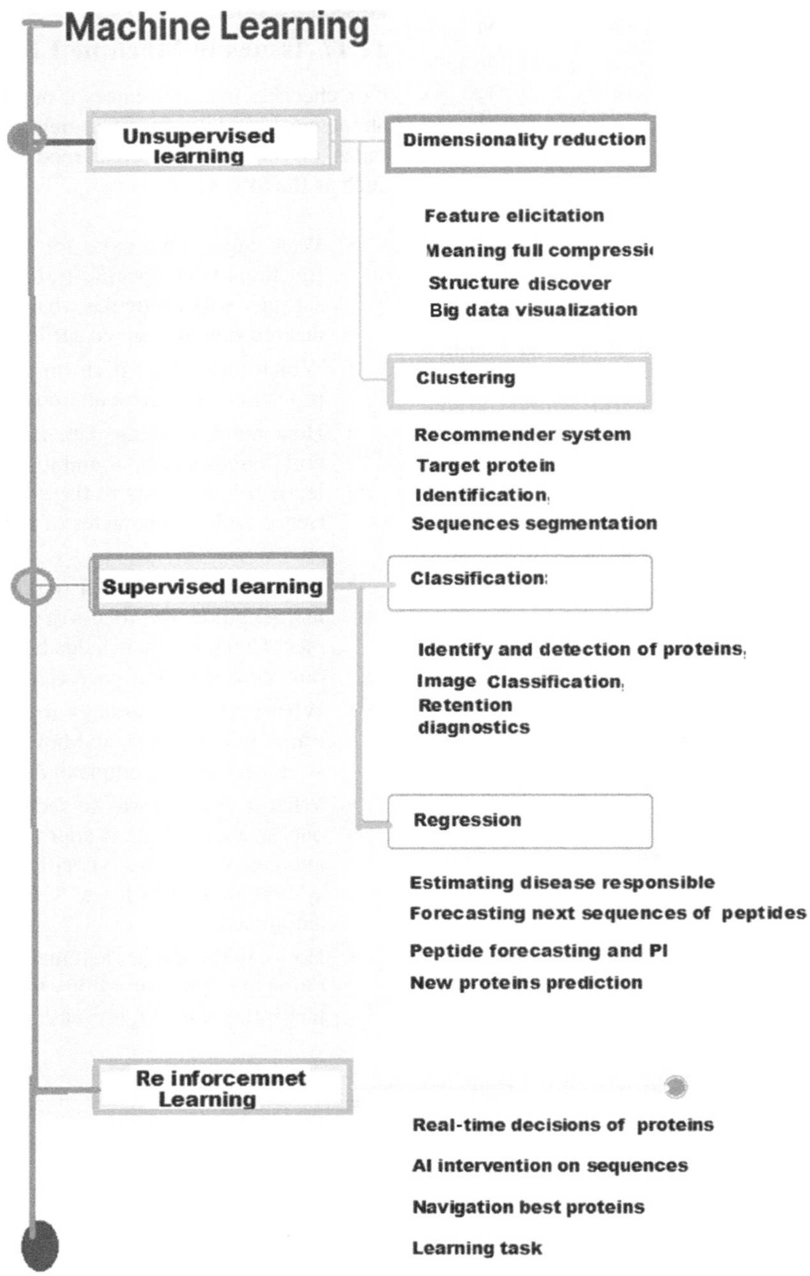

FIGURE 11.2 Classification within machine learning.

TABLE 11.3

Salient Difference between Supervised and Unsupervised Learning

Supervised Learning	Unsupervised Learning
Algorithms are trained using labeled data.	It is a learning algorithms that are trained using unlabeled data.
It takes direct feedback to check if it is predicting the correct output or not,	It does not take any feedback.
It predict the output.	It finds the hidden patterns in data.
In supervised learning, input data is provided to the model along with the output.	In unsupervised learning, only input data is provided to the model.
The goal is to train the model so that it can predict the output when given new data.	The goal is to find hidden patterns and useful insights from the unknown dataset.

(Continued)

TABLE 11.3 (*Continued*)

Salient Difference between Supervised and Unsupervised Learning

Supervised Learning	Unsupervised Learning
It needs supervision to train the model.	It does not need any supervision to train the model.
It can be categorized into **Classification** and **Regression** problems.	It can be classified in Clustering and Association problems
It can be used for cases where we know the input as well as the corresponding outputs.	It can be used for those cases where we have only input data and no corresponding output data.
It produces an accurate result.	It may give less accurate resultscompared to supervised learning.
It is not close to true artificial intelligence, as in this case, we first train the model with each data, and only then can it predict the correct output.	It is closer to the true artificial intelligence, as it learns similarly to how a child learns daily routine things through experiences.
It includes various algorithms such as Linear Regression, Logistic Regression, Support Vector Machines, Multi-class Classification, Decision Trees, Bayesian Logic, etc.	It includes various algorithms such as Clustering, KNN, and the Apriori algorithm.

TABLE 11.4

Calculation of Distance from C_1 to C_2

Person No.	Distance of Object from C_1	Distance of Object from C_2
#1	52.30	58.30
#2	0.00	7.10
#3	7.10	0.00
#4	70.40	75.40
#5	13.90	9.40

TABLE 11.5

Algorithmic Flow Table

Person No.	Cluster #
#1	1
#2	1
#3	2
#4	1
#5	2

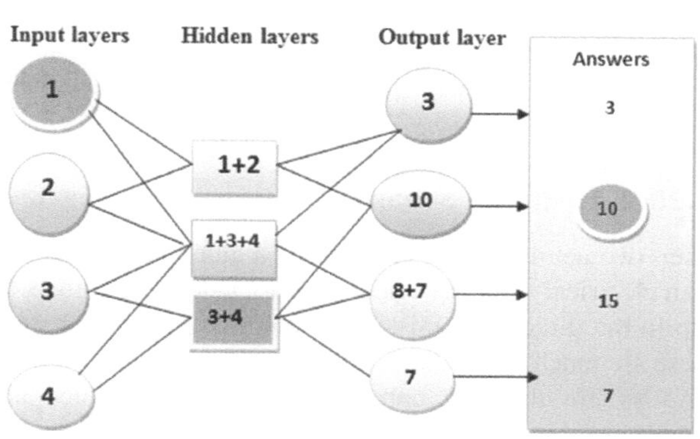

FIGURE 11.3 Numerical example illustrating the summation of four numbers.

Now we compute the Euclidean distance between each of the two centroids and each point in the data (Table 11.4). If you did all the calculations, you should have come up with the following numbers:

At this point, we will assign each object to the cluster it is closer to, based on the two computed distances for each object. We can arrange the points as follows:

Let's iterate, which means redefining the centroids by calculating the mean of the members of each of the two clusters (Table 11.5).

So, $C_1 = (167 + 120 + 175)/3$, $(55 + 32 + 76)/3$ (154,54)

C_2's $= (113 + 108)/2$, $(33 + 25)/2 = (110.50,29)$

Then, we calculate the distance again and reassign the points to the new centroids. We repeat this process until the centroids do not move anymore (or the difference between them is under a certain small threshold). In the above case, the results we get are given in the figure presented.

11.19 Artificial Neural Networks (ANNs)

An existing science and technology, we transfer newer technology for the scientific community. Out-of- the-box analytical techniques will be formulated based on fundamental, theoretical, and practical postulates to identify solutions for research problems. Nowadays, the entire world is living without walls, and we can reach fractions of a second worldwide and access scientific, business, and other related problems. In this pragmatic approach, the present section will attempt to build artificial neural network (ANN) modeling and its fundamental mechanisms. This chapter will also provide different aspects of statistical modeling on ANNs and how to formulate the model and trained data in connection with biological and life science research. In scientific terms, 'Neural networks (ANNs) are defined as the cyclic network system for interconnection between different neurons. Biologically, neurons or nerve cells are the basic building blocks of the brain, which are considered as networks. This neuron network functionally works and completes various tasks with different commands. For example, the brain takes multiple commands in fractions of a second, processing events to be completed in an appropriate manner without any overlapping. For instance, numerical aptitude tasks will be completed with different commands, yielding success and failure binary outcomes. In this manner, the human brain works on numerical solutions for completing mental tasks (Figure 11.3).

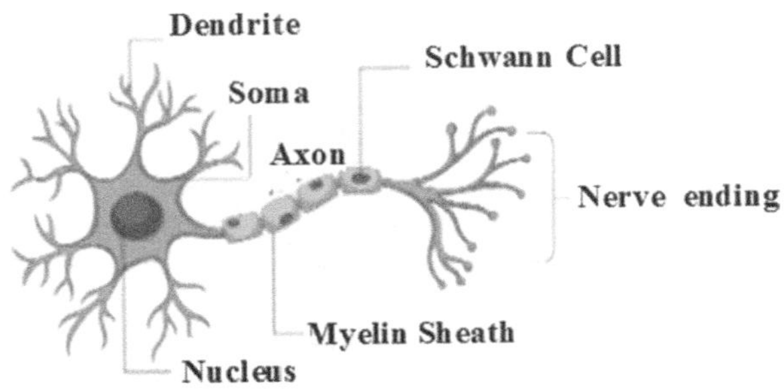

FIGURE 11.4 Outlines of the anatomical structure of neurons.

Basically, these AI techniques have been developed based on the human brain. The human brain can store millions of data points and simultaneously produce output data in the form of hidden layers Figure 11.4. For example, we have simplified numerical solutions for partial differential equations and numbers. Each solution in mathematics can be routed to work on the basis of different commands, and carry the signals; commands will retrieve information in the form of various responses (hidden layers) or output of the events. The structure of a neuron is presented in Figure 11.4.

Neurons are the building blocks of the nervous system. They receive and transmit signals to different parts of the body, carried out in both physical and electrical forms. There are several different types of neurons that facilitate the transmission of information Figure 11.4. Sensory neurons carry information from sensory receptor cells present throughout the body to the brain, whereas motor neurons transmit information from the brain to the muscles. Interneurons transmit information between different neurons in the body. In physical science, we can use these principles for constructing ANN networks each signal will be considered as random variables (X_1, X_2 … X_n) distributed normally with mean (μ) and common variance (σ^2). Each layer varies with the size of the input data depending on its error components (ϵ_{ij}). Biologically, the neuron structure (Figure 11.4) can vary in shape and size depending on its function and location. All neurons have three different parts: dendrites, cell body, and axon.

11.19.1 Dendrites

These are branch-like structures that receive messages from other neurons and allow the transmission of messages to the cell body. As per this part, physically, neurons can receive input data in the form of signals.

11.19.2 Cell Body

Each neuron has a cell body with a nucleus, Golgi body, endoplasmic reticulum, mitochondria, and other components.

11.19.3 Axon

The axon is a tube-like structure that carries electrical impulses from the cell body to the axon terminals, which pass the impulses to another neuron. In the case of AI, each signal transmits messages in the form of different layers called hidden and output layers.

11.19.4 Synapse

It is the chemical junction between the terminal of one neuron and the dendrites of another neuron. It is a widely connected network of neurons using hidden and output layers. Again, layers are divided into supervised and unsupervised learning processes. Based on the learning process, biologically, it can blend with neuron types, which are used for the construction of three types of ANNs. Different types of neurons are described in the following headings.

11.19.5 Sensory Neurons

Sensory neurons convert signals from the external environment into corresponding internal stimuli. The sensory inputs activate the sensory neurons and carry sensory information to the brain and spinal cord. They are pseudounipolar in structure.

11.19.6 Motor Neurons

These are multipolar and are located in the central nervous system, extending their axons outside the central nervous system. This is the most common type of neuron and transmits information from the brain to the muscles of the body.

11.19.7 Interneurons

They are multipolar in structure. Their axons connect only to nearby sensory and motor neurons. They help in passing signals between two neurons.

11.20 Neuron Functions

11.20.1 Chemical Synapse

In chemical synapses, the action potential affects other neurons through a gap present between two neurons, known as the synapse. The action potential is carried along the axon to a postsynaptic ending that initiates the release of chemical messengers, known as neurotransmitters. These neurotransmitters excite the postsynaptic neurons, generating an action potential of their own.

11.20.2 Electrical Synapse

When two neurons are connected by a gap junction, it results in an electrical synapse. These gaps include ion channels that help in the direct transmission of a positive electrical signal. These are much faster than chemical synapses.

ANNs are the best computational tools, modeled after brains (biological connections between the human brain and networks). Usually, they are made up of an interconnected structure of artificially produced neurons (APNs) that function as pathways for data transfer. Researchers attempt to design ANNs to solve a variety of problems: (i) pattern recognition, (ii) quantitative prediction, (iii) image patterns of proteomics, (iv) metabolomics optimization through Venn diagrams, (v) identification of various chemical pathways, and (vi) the study

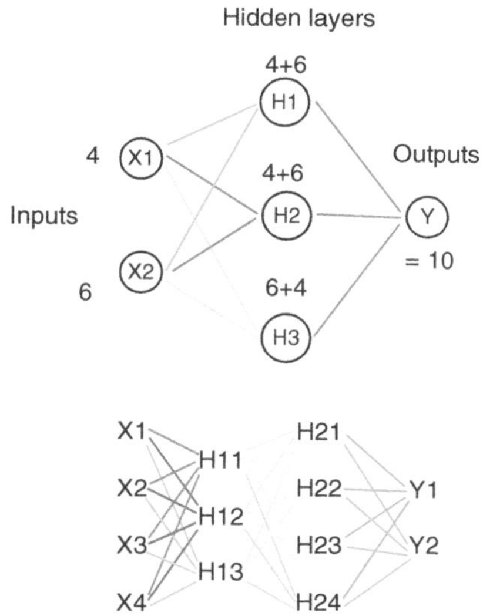

FIGURE 11.5 Flowchart depicting the input and output processes of AI.

of genetic structures of populations associated with gene pools. Eventually, it is the second-best way to form interconnected neurons. These ANN networks are treated as model brains to perform specific computational tasks (Figure 11.5). Successful ANN applications will have the capability of character recognition. The following salient applications of neural networks are presented in the following headings

11.21 Applications of Artificial Neural Networks

11.21.1 Medical Science

- ANNs are used in medical science for the recognition of image patterns (X-ray, CT, and Doppler).
- They are used for robotic surgery, electronic health records, and evidence-based medicine.
- They help in the identification of rare genes and genetic drift in individual diseases

11.21.2 Agricultural Science

- Estimation of agricultural risk is based on weather parameters.
- ANNs are widely used for agronomical trials and plant physiology experiments for the estimation of water-use efficiency and transpiration.
- They are used in plant breeding programs to formulate breeding designs and compile complex breeding data sets (heritability, repeatability, polygenic inheritance, and estimation of breeding value using additive genetic effect V_A).
- They are used for farm inventory analysis and economic theory.

11.21.3 Life Science

- Currently, ANN models are used for chemical analysis and the grouping of C and N terminal amino acids.
- They are used for the estimation of retention time in the binding of amino acids (AAs).
- They are applied for the grouping of covalent bonds and the categorization of proteins based on different peptides.
- They are used in various applications of chromatography, etc.

11.21.4 Physical Science

- Artificial neural network applications have been used in the field of solar energy for modeling and designing a solar steam-generating plant.
- They are useful in system modeling, such as in implementing complex mapping and system identification.
- ANNs are used for the estimation of heating loads of buildings, parabolic-trough collectors' intercept factor, and local concentration ratio.
- ANNs are used in diverse applications, including control, robotics, pattern recognition, forecasting, biomedicine, power systems, manufacturing, optimization, signal processing, and social/psychological sciences. They are applied in the study of nanoparticle interactions.
- They have been used for the prediction of airflows in a naturally ventilated test room and for the prediction of the energy consumption of solar buildings.
- They can handle noisy and incomplete data and deal with non-linear problems.
- Artificial neural networks are used in ventilating and air-conditioning systems, refrigeration, modeling, heating, load forecasting, control of power generation systems, and solar radiation.
- They provide an alternative way to tackle complex problems, as they are among the newest signal-processing technologies.
- This newer technique offers real solutions that are difficult to match with other technologies. Neural network-based solutions are very efficient in terms of development, time, and resources, etc.

11.22 Merits of ANNs

a. A neural network can perform tasks that a linear program cannot perform.
b. When an element of the neural network fails, it can continue without any problem due to its parallel nature.
c. A neural network does not need to be reprogrammed, as it learns by itself.

d. It can be implemented easily without any problem.
e. As adaptive, intelligent systems, neural networks are robust and excel at solving complex problems. Neural networks are efficient in their programming, and scientists agree that the advantages of using ANNs outweigh the risks.
f. It can be implemented in any application.

11.23 Demerits of ANNs

a. ANNs require training to operate.
b. They require high processing time for large neural networks.
c. The architecture of a neural network differs from the architecture and history of microprocessors, so they must be emulated.
d. An artificial neural network is developed with a systematic step-by-step procedure that optimizes a criterion commonly known as the learning rule.
e. The input/output training data is fundamental for these networks as it communicates the information necessary to discover the optimal operating point. The non-linear nature of a neural network makes its processing elements flexible within their system.
f. An artificial neural network is a system, and this system is a structure that receives an input, processes the data, and provides an output. The input in a data array can be a sound wave, data from an image file, or any kind of data that can be represented in an array. Once an input is presented to the neural network, the required target response is set at the output, and from the difference between the desired response and the output of the real system, an error is obtained. The error information is fed back to the system, and it makes many adjustments to its parameters in a systematic order, commonly known as the learning rule. This process is repeated until the desired output is achieved.

11.24 Benefits of ANNs

1. Artificial neural networks are flexible and adaptive.
2. Artificial neural networks are used in sequence and pattern recognition systems, data processing, robotics, modeling, etc.
3. ANNs acquire knowledge from their surroundings by adapting to internal and external parameter and they solve complex problems that are difficult to manage.
4. They generalize knowledge to produce adequate responses to unknown situations.
5. Flexibility: Artificial neural networks are flexible and have the ability to learn, generalize, and adapt to situations based on their findings.
6. Non-linearity: This function allows the network to efficiently acquire knowledge by learning. This is a distinct advantage over a traditionally linear network, which is inadequate when it comes to modeling non-linear data.
7. An artificial neural network can have greater fault tolerance than a traditional network. Without the loss of stored data, the network is able to regenerate a fault in any of its components.
8. An artificial neural network is based on adaptive learning.

11.25 Types of ANNs

There are different types of Artificial Neural Networks (ANNs). Depending on the functions of human brain neurons and networks, an artificial neural network or ANN performs tasks in a similar manner. Most artificial neural networks have some resemblance to their more complex biological counterparts and are very effective at their intended tasks, such as segmentation or classification.

11.25.1 Feedback ANNs

In this type of ANN, the output goes back into the network to achieve the best-evolved results internally. The feedback network feeds information back into itself and is well-suited to solving optimization problems, according to the University of Massachusetts, Lowell Center for Atmospheric Research. Feedback ANNs are used for internal system error correction.

11.25.2 Feed-Forward ANNs

A feed-forward network is a simple neural network consisting of an input layer, an output layer, and one or more layers of neurons. Through the evaluation of its output by reviewing its input, the power of the network can be observed based on the group behavior of the connected neurons, and the output is determined. The main advantage of this network is that it learns to evaluate and recognize input patterns.

11.26 Classification-Prediction ANNs

It is a subset of feed-forward ANN, and the classification-prediction ANN is applied to data-mining scenarios. The network is trained to identify particular patterns and classify them into specific groups and then further classify them into "**novel patterns,**" which are new to the network.

An artificial neural network is a computational simulation of a biological neural network. These possess the behavior of neurons and the electrical signals through which they communicate between input such as from the eyes or nerve endings in the hand, to the brain's output, such as reacting to light, touch, or heat. Scientists have been researching the design of artificial neural networks and the creation of artificial intelligence to understand how neurons semantically communicate.

11.27 Statistical Model Setting

Artificial neural networks are relatively crude electronic networks of neurons based on the neural structure of the brain. They process records one at a time and learn by comparing their prediction of the record (largely arbitrary) with the known actual record. The errors from the initial prediction of the first record are fed back to the network and used to modify the network's algorithm for the second iteration. These steps are repeated multiple times.

A neuron in an artificial neural network is:

1. A set of input values (x_i) with associated weights (w_i).
2. An input function (g) that sums the weights and maps the results to an output function (y).

11.28 Statistical Model Formulation

Artificial neural networks are relatively crude electronic networks of neurons based on the neural structure of the brain. They process records one at a time and learn by comparing their prediction of the record (largely arbitrary) with the known actual record (Figure 11.6). The errors from the initial prediction of the first record are fed back to the network and used to modify the network's algorithm for the second iteration. These steps are repeated multiple times.

A neuron in an artificial neural network is:

1. A set of input values (x_i) with associated weights (w_i)
2. An input function (g) that sums the weights and maps the results to an output function (y).

Neurons are organized into the following layers: input, hidden, and output. The input layer is composed not of full neurons but simply of the values in a record that are inputs to the next layer of neurons. The next layer is the hidden layer, of which there could be several. The final layer is the output layer, where there is one node for each class (Figure 11.7). A single sweep forward through the network results in the assignment of a value to each output node. The record is assigned to the class node with the highest value.

11.29 Artificial Neural Network Training

In the training phase, the correct class for each record is known (*i.e.* supervised training), and the output nodes can be assigned correct values: '1' for the node corresponding to the correct class, and '0' for the others. Results have been found using values of 0.9 and 0.1, respectively. As a result, it is possible to compare the network's calculated values for the output nodes to these correct values and calculate an error term for each node. These error terms are then used to adjust the weights in the hidden layers (H_i) so that the next time around, the output values will be closer to the correct values.

11.30 The Iterative Learning Process

A key feature of neural networks is an iterative learning process in which records (rows) are presented to the network one at a time, and the weights associated with the input values are adjusted each time. After all cases are presented, the process often starts over again. During this learning phase, the network trains by adjusting the weights to predict the correct class label of input samples. Advantages of neural networks include their high tolerance to noisy data, as well as their ability to classify patterns on which they have not been trained. The most popular neural network algorithm is the back-propagation algorithm proposed in the 1980s. Once a network has been structured for a particular application, it is ready to be trained. To start this process, the initial weights are chosen randomly. Next, the training begins. The network processes the records in the training data one at a time using the weights and functions in the hidden layers, then compares the resulting outputs against the desired outputs. Errors are then propagated back through the system, causing the system to adjust the weights for the next record. This process occurs again as the weights are continually tweaked. During the training of a network, the same set of data is processed many times as the connection weights are continually refined.

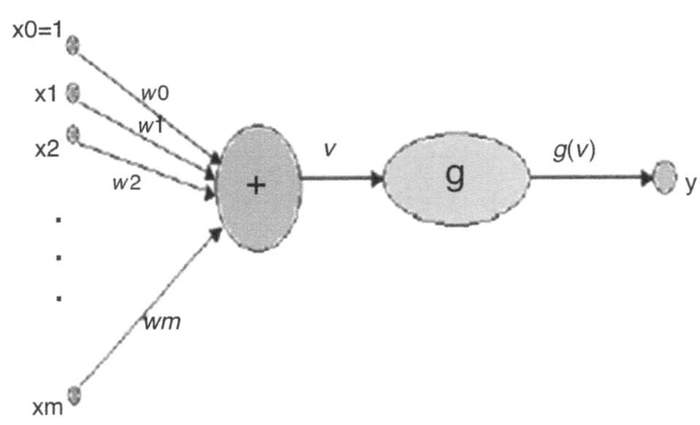

FIGURE 11.6 ANN Statistical Model Settings.

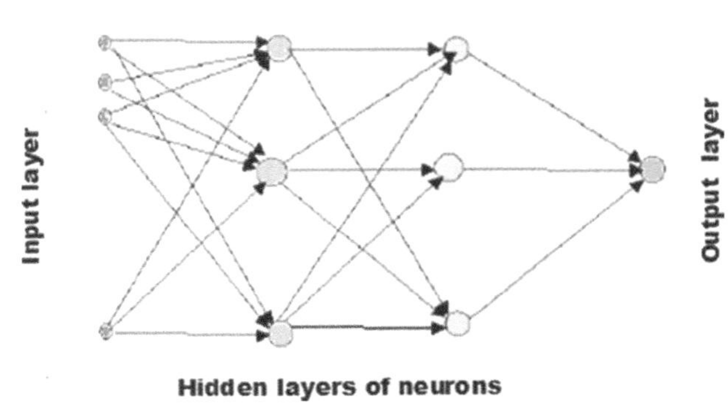

FIGURE 11.7 ANN Statistical Model Settings: Input and Output Layer.

11.31 Feed-Forward, Back-Propagation

The feed-forward, back-propagation architecture was developed in the early 1970s by several independent sources (Werbor, Parker, Rumelhart, Hinton, and Williams). This independent co-development was the result of a proliferation of articles and debates at various conferences that stimulated the entire industry. Currently, this synergistically developed back-propagation architecture is the most popular and effective model for complex, multi-layered networks. Its greatest strength is in non-linear solutions to ill-defined problems. The typical back-propagation network has an input layer, an output layer, and at least one hidden layer. Theoretically, there is no limit on the number of hidden layers, but typically there are just one or two. Some studies have shown that the total number of layers needed to solve problems of any complexity is five (one input layer, three hidden layers, and one output layer). Each layer is fully connected to the succeeding layer. The training process normally uses some variant of the Delta Rule (Δ), which starts with the calculated difference between the actual outputs and the desired outputs. Using this error, connection weights are increased in proportion to the error times a scaling factor for global accuracy. This means that the inputs, the output, and the desired output must all be present at the same processing element. The most complex part of this algorithm is determining which input contributed the most to an incorrect output and how to modify the input to correct the error. (An inactive node would not contribute to the error and would not need to change its weights.) To solve this problem, training inputs are applied to the input layer of the network, and desired outputs are compared at the output layer (Figure 11.8). During the learning process, a forward sweep is made through the network, and the output of each element is computed layer by layer. The difference between the output of the final layer and the desired output is back-propagated to the previous layer(s), usually modified by the derivative of the transfer function. The connection weights are normally adjusted using the Delta Rule. This process proceeds for the previous layer(s) until the input layer is reached.

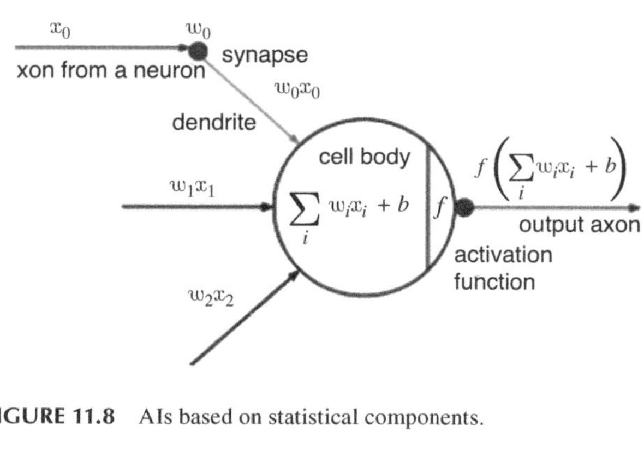

FIGURE 11.8 AIs based on statistical components.

11.32 Structuring the Network

The number of layers and the number of processing elements per layer are important decisions. In a feed-forward, back-propagation topology, these parameters are also the most ethereal—they are the art of the network designer. There is no quantifiable, best answer to the layout of the network for any particular application. There are only three general rules picked up over time and followed by most researchers and engineers applying this architecture to their problems.

11.32.1 Rule One

As the complexity in the relationship between the input data and the desired output increases, the number of processing elements in the hidden layer should also increase.

11.32.2 Rule Two

If the process being modeled is separable into multiple stages, then additional hidden layer(s) may be required. If the process is not separable into stages, then additional layers may simply enable memorization of the training set, and not a true general solution.

11.32.3 Rule Three

The amount of training data available sets an upper bound for the number of processing elements in the hidden layer (H_i). We calculate this upper bound by using the number of cases in the training set and dividing that number by the sum of the number of nodes in the input and output layers in the network.

$$\left(\text{Upper bound} = \frac{\text{Number of cases in the training sets}}{\sum \text{Nodes in the input and output layers in the network}} \right) \tag{11.10}$$

Then, divide that result again by a scaling factor between five and ten. Larger scaling factors are used for relatively less noisy data. If too many artificial neurons are used, the training set will be memorized, not generalized, and the network will be useless on new data sets.

11.33 Ensemble Methods

Analytically, we offer two powerful ensemble methods for use with Neural Networks: bagging (bootstrap aggregating) and boosting. The Neural Networks Algorithm on its own can be used to find one model that results in good predictions for new data. We can view the statistics and confusion matrices of the current predictor to see if our model is a good fit for the data, but how would we know if there is a better predictor just waiting to be found for the solution (better predictors)? However, ensemble methods allow us to combine multiple weak neural networks which, when taken together, form new, more accurate, strong neural network. These methods work by creating multiple diverse networks, by taking different samples of the original dataset, and then pooling their outputs. (Outputs may be combined by several techniques, for example, majority vote for classification and averaging for prediction.) This combination of models effectively reduces the variance in the strong model. The two different types of ensemble methods offered in Analytic Solver Data Science (bagging and boosting) differ on three items

1. The selection of training data for each predictor or weak model
2. How the weak models are generated
3. How the outputs are combined. In all three methods, each weak model is trained on the entire training set to become proficient in some portion of the data set.

11.34 Bagging

Bagging (bootstrap aggregating) was one of the first ensemble algorithms written. Bagging generates several training sets by using random sampling with replacement (bootstrap sampling), applies the neural network algorithm to each data set, and then takes the average among the models to calculate the predictions for the new data. The biggest advantage of bagging is the relative ease with which the algorithm can be parallelized, making it a better selection for very large data sets.

11.35 Boosting

Boosting builds a strong model by successively training models to concentrate on records receiving inaccurate predicted values in previous models. Once completed, all predictors are combined by a weighted majority vote. Analytic Solver Data Science offers three different variations of boosting as implemented by the AdaBoost algorithm (one of the most popular ensemble algorithms in use today): M_1 (Freund), M_1 (Breiman), and SAMME (Stage-wise Additive Modeling using a Multi-class Exponential).

AdaBoost.M_1 first assigns a weight (w_b (i) to each record or observation. This weight is initially set to a value $\left(\frac{1}{n}\right)$ and will be updated on each iteration of the algorithm. An original neural network is created using this first training set (T_b), and an error is calculated as:

$$eb - \sum_{i=1}^{n} w_b\ (i) I\ \left(C_b(x_i) \neq y_i\right) \tag{11.11}$$

where the $I()$ function returns 1 if true and 0 if not.

The error of the network in the bth iteration is used to calculate the constant α_b. This constant is used to update the weight $w_b(i)$. In AdaBoost.M_1 (Freund), the constant is calculated as:

$$a_b = \ln\left(1 - eb/eb\right) \tag{11.12}$$

In AdaBoost. M_1 (Brieman), the constant is calculated as:

$$\alpha_b = ½ \ln\left(\left(1 - eb\right)/e\ b\right) \tag{11.13}$$

In SAMME, the constant is calculated as:

$$\alpha_b = ½ \ln\left(1 - eb\right)/e\ b + \ln\left(k - 1\right) \tag{11.14}$$

where 'k' is the number of classes.

If the number of categories is equal to two, SAMME behaves the same as AdaBoost Breiman. In any of the three implementations (Freund, Breiman, or SAMME), the new weight for the (b+ 1)th iteration will be:

$$wb + t(i) - w_b\ (i) I\ \left(C_b(x_i) \neq y_i\right) \tag{11.15}$$

Afterwards, the weights are all readjusted to sum to 1. As a result, the weights assigned to the observations that were assigned inaccurate predicted values are increased, and the weights assigned to the observations that were assigned accurate predicted values are decreased. This adjustment forces the next neural network to put more emphasis on the records that were assigned inaccurate predictions. (This constant is also used in the final calculation, which will give the network with the lowest error more influence.) This process repeats until (b=Number of weak learners, controlled by the user). The algorithm then computes the weighted average among all weak learners and assigns that value to the record. Boosting generally yields better models than bagging; however, it does have a disadvantage as it is not parallelizable. As a result, if the number of weak learners is large, boosting would not be suitable.

11.36 Unsupervised Learning

One of the statistical approaches for unsupervised learning is the method of moments. In the method of moments, the unknown parameters (of interest) in the model are related to the exploration of moments of one or more random variables. These unknown parameters can be estimated given the moments 'Mdx'.

$$\mu_{DX} = \int (x - \bar{x})^p\ (y - \bar{y})^q\ dy\ dx \text{ Sample} \tag{11.16}$$

$$\mu_{Dx} = \int \left(X_i - \mu_x\right)^p\ \left(Y_i - \mu_y\right)^q\ dy\ dx \text{ Population} \tag{11.17}$$

where, $\mu_x = \frac{\sum_{x=1}^{N} X_i}{N}$, $\mu_y = \frac{\sum_{x=1}^{N} Y_i}{N}$, p and q=the power of unsupervised trainer data

11.37 Application of Unsupervised Learning

The MI-based features of each CT image are extracted using the Hu invariant moments function from the Open-CV library. *Hu* moments have specific features such as being scaling, rotation-insensitive, and capable of being translated,. They are effective at extracting contour and shape characteristics from images. *Hu* moments of an image are statistical outcomes produced by weighing the average image intensities of the pixels, and the moments are insensitive to image transformations. This signifies that the same moment values are produced irrespective of the image alterations. Hence, to enhance the 'CNN' in terms of invariance to the limitations of image transformations, the Hu moments of the image can be fed to the FC neural network. The central moments are provided as shown in the following equation (11.18) (Tables 11.6 and 11.7)

$$\mu_{pq} = \int_{-\infty}^{\infty}\int (x-\bar{x})^p (y-\bar{y})^q f(c,y)\,dx\,dy \tag{11.18}$$

$$\gamma = \left(\frac{p+q}{N}\right) \tag{11.19}$$

Log odds −2.68, odd ratio <1

- Unsupervised learning allows the model to discover patterns and relationships in unlabeled data.
- Clustering algorithms group similar data points together based on their inherent characteristics.

TABLE 11.6
CT Image Model

	Type of Image		
CT Image	***p*-Raster**	***q*-Normal**	**Total**
p–visible (%)	$p_i * p_i$	$p_i * q_i$	$\sum p_i$
q-not visible (noise) (%)	$q_i - p_i$	$q_i * q_i$	$\sum q_i$
Total	$\sum p_i$	$\sum q_i$	N

TABLE 11.7
p-Raster and *q*-Normal Statistics

CT image	*p*-Raster	*q*-Normal	Total
p–visible (%)	88(62.11)	12(37.89)	100
q-not visible (noise) (%)	30(55.89)	60(34.11)	90
Total	118	72	190
Statistics	Rate	Risk ratio	Odd ratio
Group1	0.12	0.18	0.136
Group2	0.66		2

- Feature extraction captures essential information from the data, enabling the model to make meaningful distinctions.
- Label association assigns categories to the clusters based on the extracted patterns and characteristics.

11.38 Types of Unsupervised Learning

Unsupervised learning is classified into two categories of algorithms:

11.38.1 Clustering

A clustering problem is one where you want to discover the inherent groupings in the data, such as grouping customers by purchasing behavior (Figure 11.9). Clustering is a type of unsupervised learning that is used to group similar data points together

11.38.2 Types of Clustering

Broadly speaking, there are two types of clustering that can be performed to group similar data points:

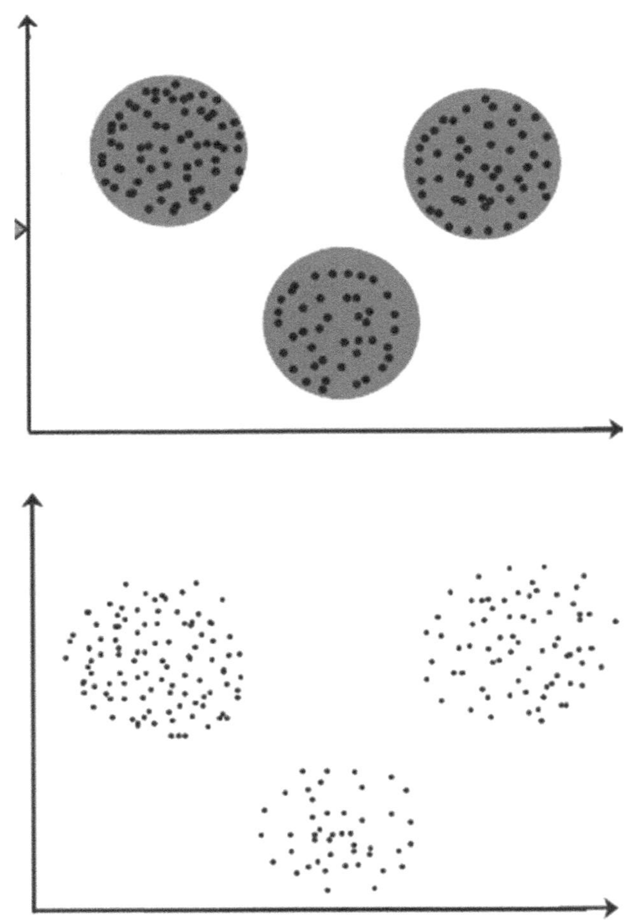

FIGURE 11.9 Scatter plot of clustering.

11.38.3 Hard Clustering

In this type of clustering, each data point belongs to a cluster completely or not. For example, let's say there are 4 data points and we have to cluster them into two clusters. So, each data point will either belong to cluster 1 or cluster 2.

A	B	C	D
C_1	C_2	C_2	C_1

11.39 Soft Clustering

In this type of clustering, instead of assigning each data point to a separate cluster, a probability or likelihood of that point belonging to each cluster is evaluated. For example, let's say there are 4 data points and we have to cluster them into 2 clusters. In this case, we will be evaluating the probability of each data point belonging to both clusters. This probability is calculated for all data points.

Probability	A	B	C	D
C_1	0.91	0.30	0.17	1.00
C_2	0.09	0.70	0.83	0.00

11.40 Uses of Clustering

Before we begin with the types of clustering algorithms, let's go through the use cases of clustering algorithms. Clustering algorithms are majorly used for:

11.40.1 Market Segmentation

Businesses use clustering to group their customers and use targeted advertisements to attract more audiences.

11.40.2 Market Basket Analysis

Shop owners analyze their sales and figure out which items are most commonly bought together by customers. For example, in the USA, according to a study, diapers and beer were often bought together by fathers.

11.40.3 Social Network Analysis

Social media sites use your data to understand your browsing behavior and provide you with targeted friend recommendations or content recommendations.

Medical imaging: Doctors use clustering to identify diseased areas in diagnostic images like X-rays.

11.40.4 Detection of Anomalies

Clustering can be used to find outliers in a stream of real-time data or to forecast fraudulent transactions. Simplify working with large datasets – Each cluster is given a cluster ID after clustering is complete. Now, you can reduce a feature set's entire feature set into its cluster ID. Clustering is effective when it can represent a complicated case with a straightforward cluster ID. Using the same principle, clustering data can make complex datasets simpler. There are many more use cases for clustering, but these are some of the major and most common ones. Moving forward, we will be discussing clustering algorithms that will help you perform the above tasks.

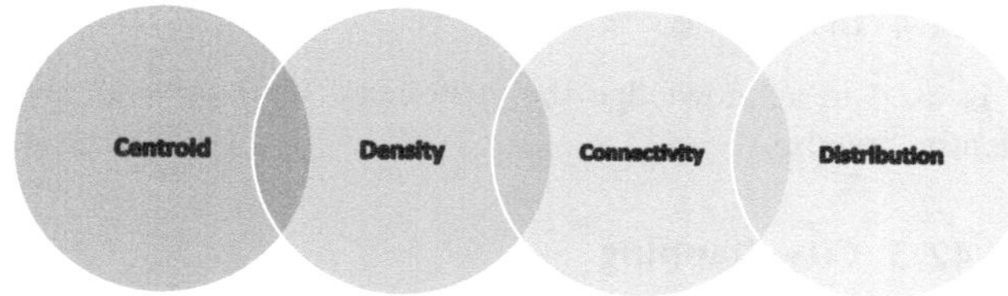

FIGURE 11.10 Types of clustering based on centroid.

11.41 Types of Clustering Algorithms

At the surface level, clustering helps in the analysis of unstructured data. Graphing, the shortest distance, and the density of the data points are a few of the elements that influence cluster formation. Clustering is the process of determining how related the objects are based on a metric called the similarity measure. Similarity metrics are easier to locate in smaller sets of features. It becomes harder to create similarity measures as the number of features increases. Depending on the type of clustering algorithm being utilized in data mining, several techniques are employed to group the data from the datasets (Figure 11.10). In this part, the clustering techniques are described. Various types of clustering algorithms are

- Centroid-based Clustering (Partitioning methods)
- Density-based Clustering (Model-based methods)
- Connectivity-based Clustering (Hierarchical clustering)
- Distribution-based Clustering

11.42 Applications of Clustering in Different Fields

11.42.1 Marketing

It can be used to characterize and discover customer segments for marketing purposes.

11.42.2 Biology

It can be used for classification among different species of plants and animals.

11.42.3 Libraries

It is used in clustering different books based on topics and information.

11.42.4 Insurance

It is used to acknowledge the customers, their policies and identify frauds.

11.42.5 City Planning

It is used to group houses and study their values based on their geographical locations and other factors.

11.42.6 Earthquake Studies

By learning from earthquake-affected areas, we can determine the dangerous zones.

11.42.7 Image Processing

Clustering can be used to group similar images together, classify images based on content, and identify patterns in image data.

11.42.8 Genetics

Clustering is used to group genes that have similar expression patterns and identify gene networks that work together in biological processes.

11.42.9 Finance

Clustering is used to identify market segments based on customer behavior, identify patterns in stock market data, and analyze risk in investment portfolios.

11.42.10 Customer Service

Clustering is used to group customer inquiries and complaints into categories, identify common issues, and develop targeted solutions.

11.42.11 Manufacturing

Clustering is used to group similar products together, optimize production processes, and identify defects in manufacturing processes.

11.42.12 Medical Diagnosis

Clustering is used to group patients with similar symptoms or diseases, which helps in making accurate diagnoses and identifying effective treatments.

11.42.13 Fraud Detection

Clustering is used to identify suspicious patterns or anomalies in financial transactions, which can help in detecting fraud or other financial crimes.

11.42.14 Traffic Analysis

Clustering is used to group similar patterns of traffic data, such as peak hours, routes, and speeds, which can help in improving transportation planning and infrastructure.

11.42.15 Social Network Analysis

Clustering is used to identify communities or groups within social networks, which can help in understanding social behavior, influence, and trends.

11.42.16 Cyber Security

Clustering is used to group similar patterns of network traffic or system behavior, which can help in detecting and preventing cyber attacks.

11.42.17 Climate Analysis

Clustering is used to group similar patterns of climate data, such as temperature, precipitation, and wind, which can help in understanding climate change and its impact on the environment.

11.42.18 Sports Analysis

Clustering is used to group similar patterns of player or team performance data, which can help in analyzing player or team strengths and weaknesses and making strategic decisions.

11.42.19 Crime Analysis

Clustering is used to group similar patterns of crime data, such as location, time, and type, which can help in identifying crime hotspots, predicting future crime trends, and improving crime prevention strategies.

11.42.20 Divisive Approach

This approach of hierarchical clustering follows a top-down approach, where we consider that all the data points belong to one large cluster and try to divide the data into smaller groups based on a termination logic or a point beyond which there will be no further division of data points. This termination logic can be based on the minimum sum of squares of error inside a cluster, or for categorical data, the metric can be the 'Gini' coefficient inside a cluster. Hence, iteratively, we are splitting the data, which was once grouped as a single large cluster, into "*n*" smaller clusters to which the data points now belong. It must be taken into account that this algorithm is highly "rigid" when splitting the clusters – meaning, once a clustering is done inside a loop, there is no way for the task to be undone (Figure 11.11).

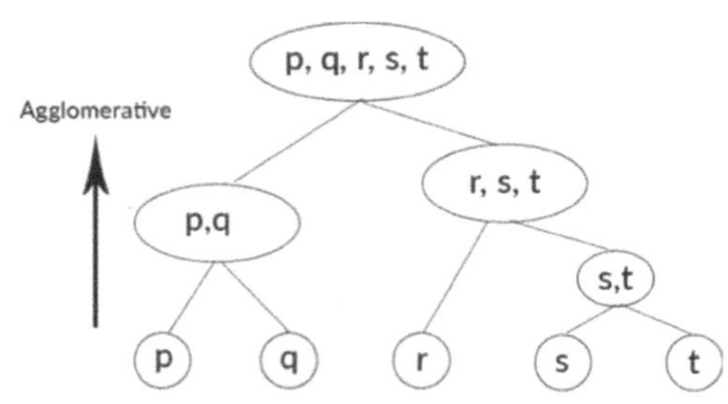

FIGURE 11.11 Decisive approach to agglomerative classification.

TABLE 11.8
Agglomerative Classification

	Trainer Data Interaction					
Agglomerative Class	p	q	r	s	t	Sum $\sum R_i$
P	p^2	pq	pr	ps	pt	R_1
Q	qp	q^2	qr	qs	qt	R_2
R	rp	rq	r^2	rs	rt	R_3
S	sp	sq	sr	s^2	st	R_4
T	tp	tq	tr	st	t^2	R_5
Sum $\sum C_i$	C_1	C_2	C_3	C_4	C_5	Grand total

$p^2 = q^2 = r^2 = t^2 = 1$ (Homogeneous interaction).

By using Bayesian law of probability, we predicted the probable chance for the interaction of trainers' data to understand the agglomerative relation (Table 11.8).

$$P\left(R_1/C_1\right) = \text{Proba } (C_1/R_1) * \text{Prob } \left(R_1\right) / \text{Prob } \left(C_1\right)$$

$$P\left(R_1/C_2\right) = \text{Proba } (C_2/R_1) * \text{Prob } \left(R_1\right) / \text{Prob } \left(C_2\right)$$

$$P\left(R_1/C_3\right) = \text{Proba } (C_3/R_1) * \text{Prob } \left(R_1\right) / \text{Prob } \left(C_3\right)$$

$$P\left(R_i/C_k\right) = \text{Proba } (C_k/R_i) * \text{Prob } \left(R_i\right) / \text{Prob } \left(C_k\right)$$

$$P\left(R_i/C_k\right) = \left(\prod_{i=1}^{k} \text{Proba } (C_k/R_i)\right) * \text{Prob } \left(R_i\right) / \text{Prob } \left(C_k\right)$$

Large language models (LLMs) are a type of artificial intelligence (AI) program that can recognize and generate text. They are trained on large amounts of data, such as internet-scale datasets with hundreds of billions of parameters, and are built on machine learning, specifically a type of neural network called a transformer model. LLMs are capable of many tasks, including identifying sequences of words that are likely to appear in the next sentences. LLM models like GPT-3 have revolutionized language processing by using neural networks to understand and generate human text, surpassing rule-based systems' limitations. A large language model is a groundbreaking artificial intelligence (AI) innovation that has revolutionized how computers understand and generate human language. This type of neural network possesses remarkable versatility to comprehend, analyze, and produce text like a human. In the past, language processing heavily relied on rule-based systems that followed pre-defined instructions. However, these systems faced limitations in capturing the intricate and nuanced aspects of human language. A significant breakthrough came with the emergence of deep learning and neural networks. One notable transformer architecture, exemplified by models like GPT-3 (Generative Pre-trained Transformer 3), brought about a transformative shift (Table 11.9).

11.43 Generative AI

It refers to AI systems that can generate new content or data that is not explicitly derived from existing examples. This can include generating text, images, music, and more. Generative

TABLE 11.9
Difference between AI and Predictive Analytics

AI	Predictive Analytics
AI is the simulation of human intelligence processes by machines, especially computer systems. Specific applications of AI include robotics process automation (RPA), natural language processing (NLP), and ML.	Predictive analytics is a form of advanced analytics that uses historical data to forecast activity, behavior and trends. It involves applying statistical analysis techniques, analytical querries, and automated ML algorithms to datasets to create predictive models that assigs a numerical value to the likelihood of an event happening.
Applications	Applications
• Robotics process automation automates repetitive and rules-based data processing tasks, especially in the case of chatbots. • NLP has found adaptation in sentiment analysis and enterprise search. • ML is especially used for the automation of predictive analytics and has a wide range of uses, including self-driving cars and fraud detection.	• In biological trials, we have adopted predictive models to identify the best cause of spurious variables and concurrent variables under the researcher's interest. These tools will help us make decisions based on real data. • Development of large language models using predictive techniques.

AI often uses techniques like Generative Adversarial Networks (GANs) and Variational Autoencoders (VAEs).

LLMs are a class of Artificial Intelligence (AI) that can understand, interpret, and generate text. Based on extensive training on vast datasets, LLMs are capable of understanding and generating text at a level indistinguishable from humans (Figure 11.12). LLMs have become increasingly popular due to their exceptional ability in a wide range of NLP tasks and natural language understanding (NLU) tasks, such as translation and text summarization.

LLM: GPT is a specific application of Generative AI, specializing in natural language

Generative AI: Leverages DL techniques to create new content

Deep learning: it is a specialized tool within the ML complex, unstructured data

Machine learning: It is a key component of AI and a fundamental building block

AI: It is the umbrella that encompasses all these subfields

Modern LLMs emerged in 2017 and use transformer neural networks, commonly referred to as transformers. With a large number of parameters and the transformer model, LLMs can rapidly understand and generate accurate responses, making the AI technology broadly applicable across many different domains.

LLMs can be viewed as a subset of foundation models that focus on language-related tasks. Foundation models are large deep learning neural networks trained on large datasets and

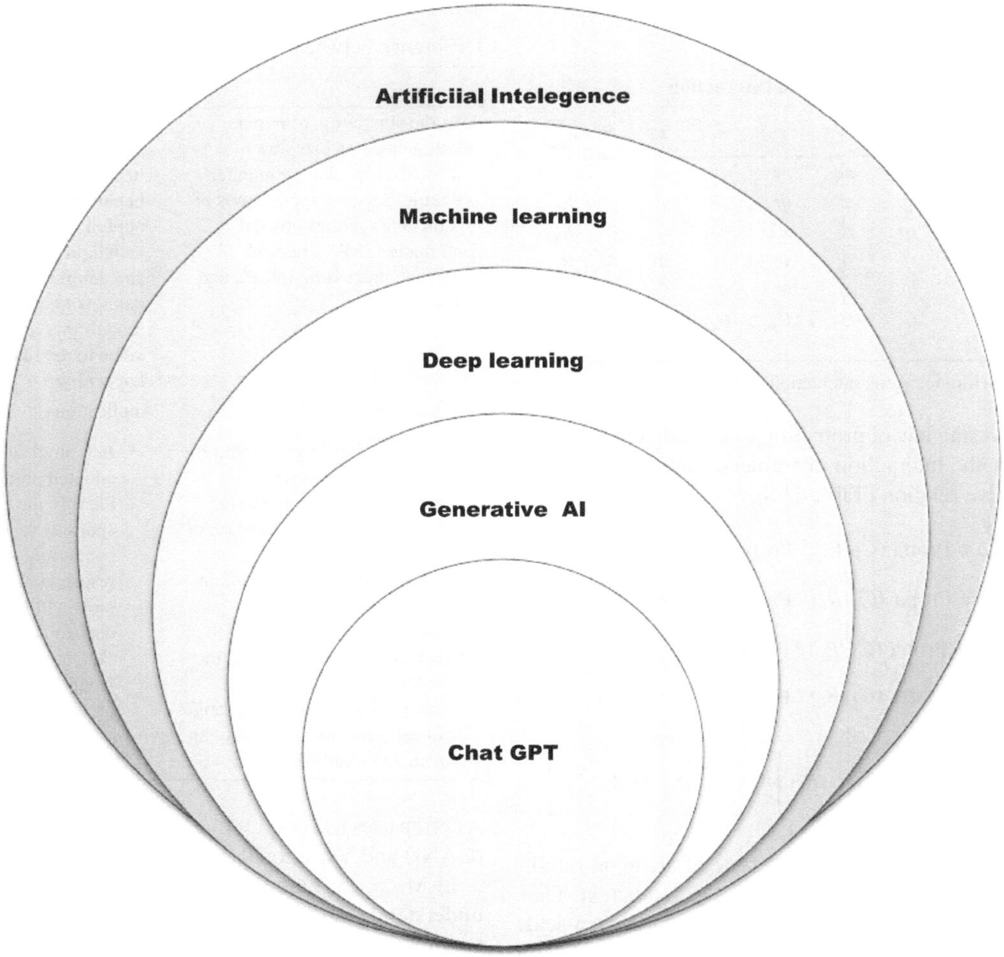

FIGURE 11.12 Features a workflow diagram for Chat GPT.

serve as a fundamental building block for various applications. They can produce wide and various outputs (e.g. text, images, and audio) or can generally apply a pretraining objective to the dataset, so that the foundation model becomes good at that objective (e.g. image creation). Foundation models can be used as a 'base' for other models, which can be built on top of the foundation model. A foundation model is so large and impactful that it serves as the foundation for further optimizations and specific use cases.

Large language models are based on deep learning architectures, with a specific focus on so-called transformer models. Transformers are neural network architectures that use self-attention mechanisms to process input data, enabling them to handle long-range dependencies in language effectively. The following sections detail the components and training process of large language models.

Components of Large Language Models

1. Parameters: The core components of a large language model are its parameters, which include weights and biases. These parameters are adjusted during training to minimize the difference between the model's predictions and the real values.
2. Layers: A large language model comprises several layers, each responsible for extracting and processing different levels of information from the input data. These layers typically include input and output layers, as well as multiple hidden layers.
3. Attention mechanism: The attention mechanism is a critical component of large language models, allowing them to selectively focus on relevant parts of the input data. This mechanism helps the model capture dependencies between words and phrases, even when they are far apart in the text (Figure 11.13).

11.44 Applications of LLM

- Content generation
- Translation and localization
- Search and recommendation
- Virtual assistant
- Code development
- Sentiment analysis
- Question answering

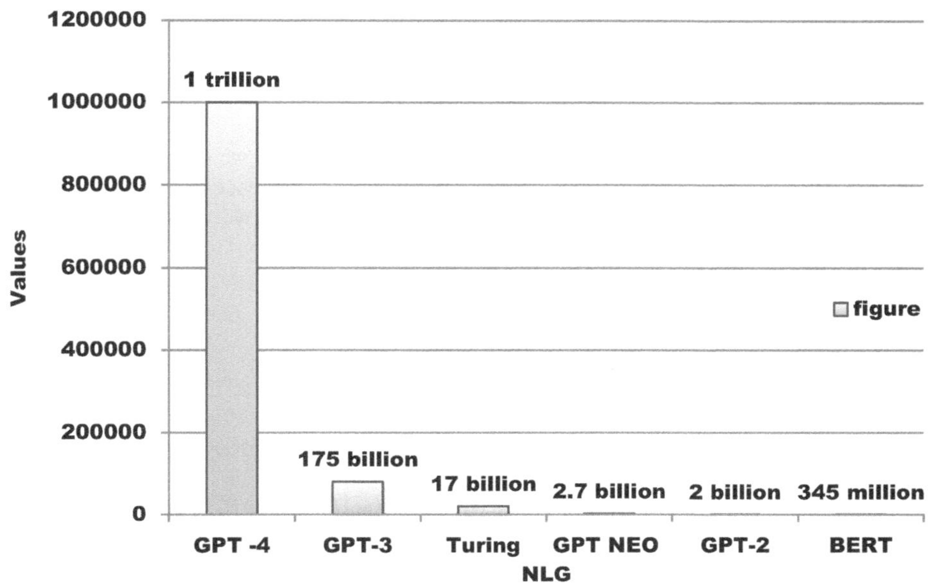

FIGURE 11.13 Outlines of the parameters of a transformer-based language model.

TABLE 11.10

Difference between Statistical and Neural Large Language Model

Statistical LM	Neural LM
Which includes probabilistic models that can predict the next word in a sequence	Which uses neural works to make predictions
Unigram: Simple and Independent	Speech recognition, ability to process speech Used by the likes of Alexa and Siri
N–gram: Also known as *Q* grams Simple, scalable–Exponential model	Machine translation, AI translations Used by Google translators and Microsoft translate- Parsing tools

- Biological research
- Education
- Classification (Table 11.10)

11.45 Unigram

Among the simplest types of LM Are unigram models. Unigram models evaluate terms independently and do not require any conditioning context to make a decision. They are used for NLP tasks such as information retrieval

The unigram model follows the chain rule.

$$P(x_1,x_2....x_n)=P(x_1)P(x_2/x_1)\ P(x_3/x_1,\ x_2).....P(x_n/x_1..\ x_{n-1}) \tag{11.20}$$

$$P(\text{"about five minutes"})=P(about)P(\text{five/about})$$
$$P(\text{minutes/about five}).....P(\text{from/about 5 min })$$

Probability of words in sentences:

$$P(w_1,w_2....w_n)=\prod_{n=i}^{1}P(w_i/w_1,w_2.w_{3...}w_{i-1}) \tag{11.21}$$

Assume the corpus word with the highest probability is 'dinner'. The LLM sentence could be 'about five minutes dinner'.

This model estimates the probability of a word in a sentence, typically based on the words that have come before it.

Given sentence *P* ("*i* have a dream [END]"

Estimates: *P* (*i*)

P (have | *i*)

P (al i have)

P (dream | i have a)

P ([**END**] | i have a dream)

Statistics is a unigram ($n=1$), machine learning is a bigram ($n=2$), and natural language processing is a trigram ($n=3$). For longer n-grams, people just use their lengths to identify them, such as 4-g, 5-g, and so on.

$$P_{\text{train}}\left(\text{dream/i have a}\right)=P_{\text{train}}\ (\text{dream})=\frac{n_{\text{train}}(\text{dream})}{N_{\text{train}}}$$

↓

Total number of words in training text

(11.22)

$$P_{eval}(\text{i have a dream [END]}) = P_{train}(\text{i})\ P_{train}(\text{have} \mid \text{i})\ P_{train}(\text{a} \mid \text{i have})\ P_{train}(\text{dream} \mid \text{i have})\ P_{train}(\text{[END]})\ \text{i have a dream}$$
$$= P_{train}(\text{i})\ P_{train}(\text{have})\ P_{train}(\text{i})\ (\text{a})\ P_{train}(\text{dream})\ P_{train}(\text{[END]})$$

We can go further than this and estimate the probability of the entire evaluation text, such as *dev1* or *dev2*. Under the naive assumption that each sentence in the text is independent of other sentences, we can decompose this probability as the product of the sentence probabilities, which, in turn, are nothing but products of word probabilities.

$$\text{Evaluation text: "I have a dream. We are free at last."}$$
$$P_{eval}(\text{text}) = P_{eval}(\text{i have a dream[END]})\,P_{eval}(\text{we are free at last[END]})$$
$$P_{eval}(\text{i have a dream[END]}) = P_{train}(\text{i})P_{train}(\text{have})P_{train}(\text{a})P_{train}(\text{dream})P_{train}(\text{[END]})$$
$$P_{eval}(\text{we are free at last[END]}) = P_{train}(\text{we})P_{train}(\text{are})P_{train}(\text{free})P_{train}(\text{at})P_{train}(\text{last})P_{train}(\text{[END]})$$
$$\Rightarrow P_{eval}(\text{text}) = P_{train}(\text{i})P_{train}(\text{have})P_{train}(\text{a})P_{train}(\text{dream})P_{train}(\text{[END]})\ P_{train}(\text{we})P_{train}(\text{are})P_{train}(\text{free})P_{train}(\text{at})P_{train}(\text{last})P_{train}(\text{[END]}) \quad (11.23)$$

11.46 Evaluation Metric: Average Log Likelihood

When we take the log of both sides of the above equation for the probability of the evaluation text, the log probability of the text (also called log likelihood) becomes the sum of the log probabilities for each word. Lastly, we divide this log likelihood by the number of words in the evaluation text to ensure that our metric does not depend on the number of words in the text.

$$P_{eval}(\text{text}) = \prod_{\text{word}} P_{train}(\text{word})$$
$$\log(P_{eval}(\text{text})) = \sum_{\text{word}} \log(P_{train}(\text{word}))$$
$$\text{Average log likelihood}_{eval} = \frac{\sum_{\text{word}} \log(P_{train}(\text{word}))}{\underset{\substack{\uparrow \\ \text{total number of words in} \\ \text{evaluation text}}}{N_{eval}}} \quad (11.24)$$

11.47 Dealing with Unknown Unigrams

There is a *big* problem with the above unigram model: for a unigram that appears in the evaluation text but not in the training text, its count in the training text, and hence its probability, will be zero. This will completely implode our unigram model: the log of this zero probability is negative infinity, leading to a negative infinity average log likelihood for the entire model.

$$n_{train}(\text{unigram}) = 0$$
$$\Rightarrow P_{train}(\text{unigram}) = \frac{0}{N_{train}} = 0 \quad (11.25)$$
$$\Rightarrow \log(P_{train}(\text{unigram})) = \log(0) = -\infty$$

11.48 Laplace Smoothing

We add an artificial unigram called [UNK] to the list of unique unigrams in the training text. This represents all the unknown tokens that the model might encounter during evaluation. Of course, the count for this unigram will be zero in the training set, and the unigram vocabulary size—the number of unique unigrams—will increase by 1 after this new unigram is added. Next, we add a **pseudo-count of *k*** to all the unigrams in our vocabulary. The most common value of *k* is 1, and this is known by the intuitive name of 'add-one smoothing'.

$$P_{train}(\text{unigram}) = \frac{n_{train}(\text{unigram}) + k}{N_{train} + k\underset{\substack{\uparrow \\ \text{unigram vocabulary size} \\ \text{(number of unique unigrams in training text)}}}{V}} \quad (11.26)$$

Similar unigram distributions between training and evaluation texts

	A	B
Train	0.1	0.9
Eval	0.3	0.7

$$\begin{aligned}&\text{Average log likelihood}_{\text{eval}}\\&= P_{\text{eval}}(A)\log\left(P_{\text{train}}(A)\right) + P_{\text{eval}}(B)\log\left(P_{\text{train}}(B)\right)\\&= 0.3\log(0.1) + 0.7\log(0.9)\\&= 0.3(-3.32) + 0.7(-0.15)\\&= -1.10\end{aligned}$$

Different unigram distributions between training and evaluation texts

	A	B
Train	0.9	0.1
Eval	0.3	0.7

$$\begin{aligned}&\text{Average log likelihood}_{\text{eval}}\\&= P_{\text{eval}}(A)\log\left(P_{\text{train}}(A)\right) + P_{\text{eval}}(B)\log\left(P_{\text{train}}(B)\right)\\&= 0.3\log(0.9) + 0.7\log(0.1)\\&= 0.3(-0.15) + 0.7(-3.32)\\&= -2.37\end{aligned} \tag{11.27}$$

interpolation weight of un-smoothed unigram model

$$\rightarrow a_{\text{unigram}} = \frac{N_{\text{train}}}{N_{\text{train}} + kV} \Leftrightarrow k = \frac{N_{\text{train}} - N_{\text{train}}a_{\text{unigram}}}{Va_{\text{unigram}}} \tag{11.28}$$

a_{unigram}	k
0	∞
0.2	106
0.4	40
0.6	18
0.8	7
1	0

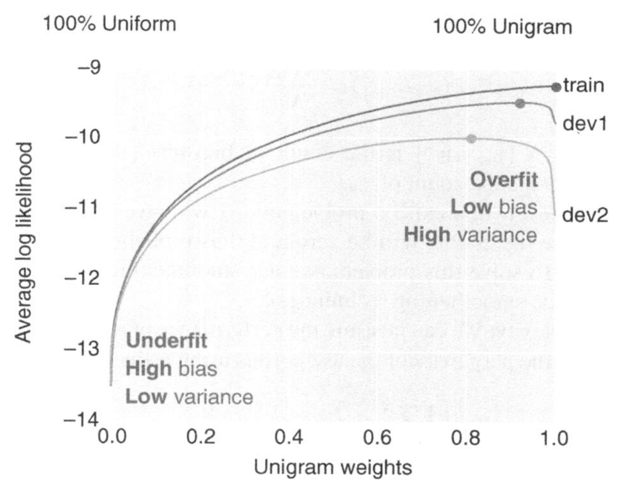

FIGURE 11.14 Smoothing of the unigram language model.

11.49 *N*-Gram

N-grams (*Q*-grams) are simple and scalable. These models create a probability for predicting the next item in a sequence. *N*-gram models are widely used in probability, communication theory, and statistical natural language processing.

N-grams of texts are extensively used in text mining and natural language processing tasks. An *n*-gram is a contiguous sequence of *n* items from a given sample of text or speech. An *n*-gram of size 1 is referred to as a 'unigram'; size 2 is a 'bigram'; size 3 is a 'trigram'. When $N>3$, this is usually referred to as four-grams or five-grams, and so on (Figure 11.14).

Formula to calculate number of *N*-grams in a sentence:

$\text{Ngramk} = X-(N-1)$

where *f* X=Number of words in a given sentence, N is the bag of words. Example: Sentence: I want to learn Machine Learning.

11.50 Unigram

Now calculate the number of unigrams in a sentence using the formula.

Here, $X=6$ and $N=1$ (for unigram)

$N\text{gramk} = X-(N-1)$

$\text{Ngramk} = 6-(1-1) = 6$ (i.e. unigram is equal to number of words in a sentence)

[I][want][to][learn][Machine] [Learning]

11.51 Bi Gram

Here, $X=6$ and $N=2$ (for biagram)

$\text{Ngramk} = X-(N-1)$

Ngramk=6−(2−1)=5

[I want] [want to] [to learn] [learn Machine] [Machine Learning]

11.52 Trigram

Here, $X=6$ and $N=3$ (for trigram)

Ngramk=$X-(N-1)$

Ngramk=6−(3−1)=4

[I want to][Want to learn][to learn Machine][Learn Machine Learning]

We can also generate for $N=4$, 5, 6, and so on.

$$p(w^N{}_1)=\prod_{n=1}^{N} p\left(w_n/w^{n-1}{}_0\right) \quad (11.29)$$

$$\text{Unigram } p\left(w_1{}^N\right)=\prod_{n=1}^{p} p\left(w_n\right) \quad (11.30)$$

$$\text{Bigram } p\left(w_1{}^N\right)=\prod_{n=1}^{p} p\left(w_n/w_{n-1}\right) \quad (11.31)$$

$$\text{Trigram } p\left(w_1{}^N\right)=\prod_{n=1}^{p} p\left(w_n/w_{n-2},w_{n-1}\right) \quad (11.32)$$

Training: Using MLE criteria, these probabilistic can be estimated using counts, e.g. the bigram model:

$$p\left(w_n/w_{n-1}\right)=\frac{N\left(w_{n-1},w_n\right)}{N\left(w_{n-1}\right)} \quad (11.33)$$

where, $N\left(w_{n-1},w_n\right)$ is the count of bigrams $\left(w_{n-1},w_n\right)$ and $N\left(w_{n-1}\right)$ is the count of w_{n-1}

However, this can be problematic if we have unseen data because the counts will be zero and thus probability is undefined. To solve this problem, we use smoothening techniques (Laplace smoothening techniques).

Perplexity: We can measure the performance of a LM by computing the perplexity of the test corpus using trained m grams.

$$P=\left[\prod_{n=1}^{N} p\left(w_n/w_{n-m+1}^{n-1}\right)\right]^{-1/N} \quad (11.34)$$

Unigrams ($n=1$): *dog, that, barks, does, not, bite*

Bigrams ($n=2$): *dog that, that barks, barks do, does not, not bite*

Trigrams ($n=3$): *dog that barks, that barks do, barks do not, does not bite* etc.

11.53 TF and IDF

TF-IDF stands for Term Frequency—Inverse Document Frequency. It is a metric for calculating the importance of a term in a corpus (a set of documents). An n-gram is 'a consecutive sequence of n words in a sentence'. For example: 'Natural language processing is an emerging field of AI'.

One-grams: NLP is considered a key branch of artificial intelligence, utilizing machine learning techniques to analyze and generate human language.

A one-gram, or unigram, in natural language processing (NLP) is a sequence of one word. For example, "cat" and "dog" are unigrams.

Two-grams:

- Natural language
- language processing
- processing is
- is an
- an emerging
- emerging field
- field of
- of AI

11.54 Skip-Gram

Skip-gram is exactly the opposite. In this model, the input is the word, and it tries to predict the surrounding words. N-gram is a simple technique usedto split sentences/documents into words or phrases (when $n>1$). For example, if your sentence is: 'The cat that was curious got itself killed'. When you apply N-gram with $n=1$, you will get

['The', 'cat', 'that', 'was', 'curious', 'got', 'itself', 'killed']

You can see how it simply gives you a list of all the words/tokens in the sentence/document.

When $n=2$, you get

['The cat', 'cat that', 'that was', 'was curious', 'curious got', 'got itself', 'itself killed']

Now you can guess what happens when $n=3$ and so on.

Not good=word2vec["not"]+word2vec["good"]

Skip-gram model computes the probability of the center word appearing with the context words by computing the similarity using the dot product. It then converts the similarity into a probability by passing it through the soft-max functions (Figure 11.15).

Goal: Predict surrounding words within a window of each word, we maximize the probability of any context word given the current word

$$w_1,w_2\ldots,w_{t-1}\ldots.w_t,\ w_{t+1},w_{t+m}\ldots,\ w_{T-1}\ldots.w_T \text{ (Context)} \quad (11.35)$$

$$p\left(w_{0,1},w_{0,2}\ldots,w_{0,C}/w_I\right)=\prod_{C=1}^{C} p\left(w_{0,C}/w_I\right) \quad (11.36)$$

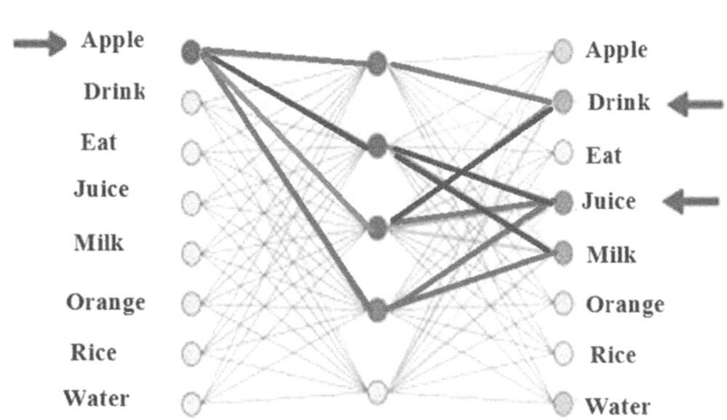

FIGURE 11.15 Pattern recognition capabilities of an AI tool.

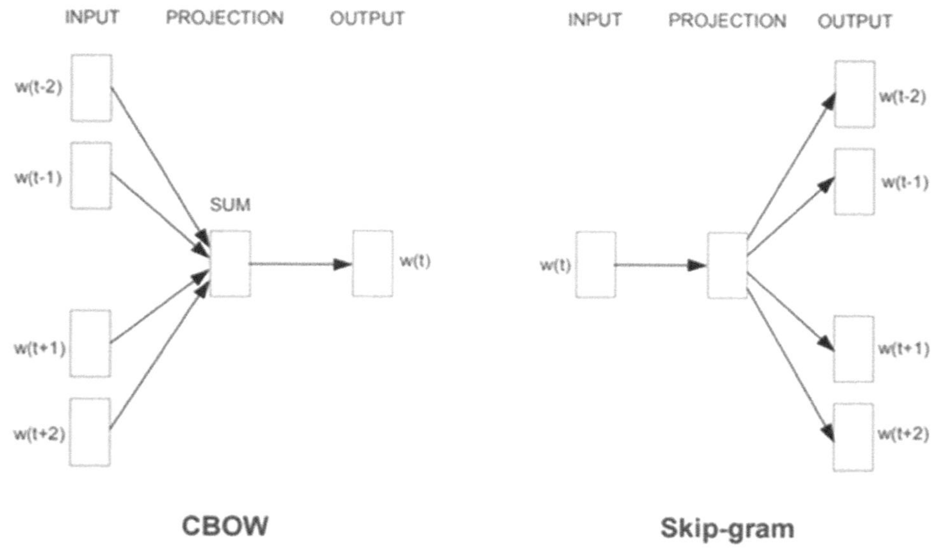

FIGURE 11.16 Skip Word Continuous Bag of Words (CBOW) diagram.

Out side

$$C(\theta) = -\sum \sum_{c=1}^{C} \log p(w_{0,c}/w_I)\ p\left(\frac{w_{0,c}}{w_I}\right) = \frac{\exp\left(v'^{T}_{w0} v_{w_I}\right)}{\sum_j Exp\left(v'^{T}_{w0} v_{w_I}\right)} \tag{11.37}$$

Target word

Bag of words: As the name implies, these are a set of words and their frequency counts in a document. Imagine a bag with unique words and frequency attached to each of the word.

Unigram: On the other hand, a unigram is a phrase that has ONE word (Figures 11.16 and 11.17). When comparing bag of words versus unigrams, here's an example: (Unigrams are space-separated in the example).

11.55 Exponential

Also referred to as maximum entropy models, these models evaluate via an equation that combines feature functions and n-grams. Essentially, it specifies features and parameters of the desired results and leaves the analysis parameter vague.

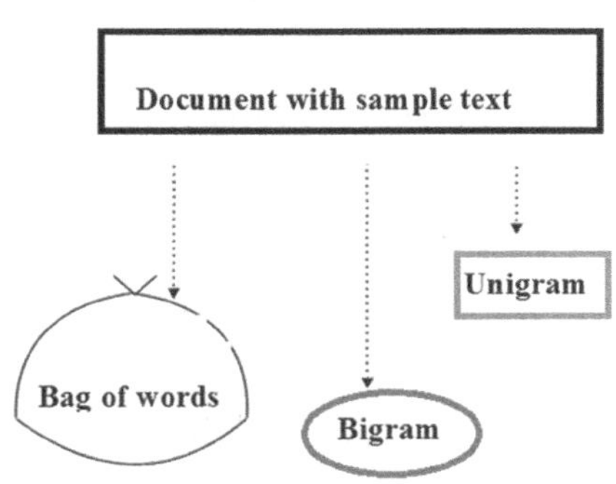

FIGURE 11.17 Generative model of the Bag of Words.

They are most commonly found in computational linguistic applications. The model is based on the principle of entropy, which states that the probability distribution with the most entropy is the best choice. Exponential models have fewer statistical assumptions, which means the chances of having accurate results are higher (Wu et al., 2012).

11.56 Continuous Space

In this type of statistical model, words are arranged as a non-linear combination of weights in a neural network. The process of assigning a weight to a word is known as word embedding. This type of model proves helpful in scenarios where the dataset of words continues to grow and includes unique words. In cases where the dataset is large and consists of rarely used or unique words, linear models such as n-grams do not work. This is because, with an increasing number of words, the possible word sequences increase, and thus the patterns predicting the next word become weaker. An association rule learning problem is one where you want to discover rules that describe large portions of your data, such as 'people who buy *X* also tend to buy *Y*'.

11.57 ANCOVA (Analysis of Covariance) Applications in ML

It is a statistical technique used to analyze the relationship between a dependent variable and one or more independent variables while adjusting for the effects of one or more covariates. It can be thought of as a combination of ANOVA and regression analysis, as it allows you to test for differences in means while controlling for the effects of other variables (Table 11.11).

TABLE 11.11

Difference between ANOVA and ANCOVA

Charectistics	ANOVA	ANCOVA
Purpose	Compare means of two or more groups	Compare means of two or more groups while controlling for the effects of one or more covariates.
Assumptions	Normality and equality of variances	Normality, equality of variances, and linearity between the dependent and independent variables are assumed.
Null hypothesis	The means of all groups are equal.	The means of all groups are equal after adjusting for the effects of the covariates.
Alternative hypothesis	Means of at least one group is different	Means of at least one group is different after adjusting for the effects of the covariates.
Post-hoc tests	Tukey's HSD, Scheffé's test, etc. can be used for post-hoc analysis	Tukey's HSD, Scheffé's test, etc., or tests specific to the covariates used can be used for post-hoc analysis.

11.58 Key Differences between ANOVA and ANCOVA

1. ANOVA (Analysis of Variance) is a statistical method used to compare the means of two or more groups, while ANCOVA (Analysis of Covariance) is a method that combines ANOVA with regression analysis.
2. ANOVA is used to test for significant differences in means among groups, while ANCOVA is used to test for significant differences in means while controlling for the effects of one or more covariates.
3. ANOVA is used for comparing the means of a categorical independent variable and a continuous dependent variable, while ANCOVA is used for comparing the means of a categorical independent variable and continuous dependent variable with one or more continuous covariates.
4. ANOVA is a parametric test, assuming that the data is normally distributed, while ANCOVA is also a parametric test, assuming that the data is normally distributed and that the relationship between the dependent variable and the covariate(s) is linear.
5. ANOVA tests the equality of population means while ANCOVA tests the equality of population means while adjusting for the effect of one or more covariates
6. ANOVA requires that the observations within each group are independent and have equal variances, while ANCOVA requires that the observations within each group are independent, have equal variances, and that the relationship between the dependent variable and the covariate(s) is linear.
7. ANOVA is typically used when the goal is to understand the main effect of a categorical independent variable on a continuous dependent variable, while ANCOVA is typically used when the goal is to understand the main effect of a categorical independent variable on a continuous dependent variable while controlling for the effects of one or more continuous covariates.
8. ANOVA is a simpler method and is used when there are only two or more groups to compare, while ANCOVA is more complex and is used when there are more than two groups to compare and you want to adjust for the effects of one or more covariates.

11.59 Advantages of ANCOVA

1. ANCOVA allows for the control of extraneous variables, which can help isolate the effects of the independent variable on the dependent variable.
2. It can be used to analyze data from a randomized experiment or a quasi-experiment.
3. ANCOVA can be used to test for differences in means between groups, while controlling for the effects of other variables.
4. The technique allows for the use of both continuous and categorical variables as covariates.
5. It can be useful in situations where there is non-constant variability in the dependent variable.
6. ANCOVA can help identify interactions between the independent and covariate variables.
7. It can be useful in situations where the assumptions of traditional parametric tests, such as homogeneity of variances, are not met.

ANOVA compares means across groups, while ANCOVA compares means while controlling for a covariate.

11.60 Disadvantages of ANCOVA

1. ANCOVA requires that certain assumptions be met, such as linearity and homogeneity of regression slopes, which may not be appropriate for all datasets.
2. It can be sensitive to outliers and violations of assumptions, leading to inaccurate conclusions.
3. The technique can be computationally complex and may require specialized software or expertise to implement.
4. ANCOVAmay not be appropriate for small sample sizes, as the power to detect effects may be low.

5. It can be difficult to interpret the results of an Ancova, especially in situations where interactions or non-linear effects are present.
6. It may not be appropriate when the covariate is highly correlated with the independent variable, as this can lead to unstable estimates of the effects.
7. In some cases, it may be more appropriate to use alternative techniques, such as multivariate regression or structural equation modeling, which can provide more flexibility in modeling the relationships between variables.

REFERENCES

Cock, P. J. A., Fields, C. J., Goto, N., Heuer, M. L., & Rice, P. M. The Sanger FASTQ file format for sequence quality score and the Solexa/Illumina FASTQ variants. *Nucleic Acids Research* **38**, 1767–1771 (2010). https://doi.org/10.1093/nar/gkp1137

Grada, A. & Weinbrecht, K. Next-generation sequencing: Methodology and application. *Journal of Investigative Dermatology* **133**, e11 (2013). https://doi.org/10.1038/jid.2013.248

Lee, H. C. et al. Bioinformatics tools and databases for analysis of NGS data. *Briefings in Functional Genomics* **11**, 12–24 (2011).

Liu, L. et al. Comparison of next generation sequencing systems. *Journal of Biomedicine and Biotechnology* 2012, 251364 (2012). https://doi.org/10.1155/**2012**/251364

Mardis, E. R. Next-generation sequencing platforms. *Annual Review of Analytical Chemistry* **6**, 287–303 (2013). https://doi.org/10.1146/annurev-anchem-062012-092628

Martin, J. A. & Wang, Z. Next-generation transcriptome assembly. *Nature Reviews Genetics* **12**, 671–682 (2011). https://doi.org/10.1038/nrg3068

Metzker, M. L. Sequencing technologies-the next generation. *Nature Reviews Genetics* **11**, 31–46 (2010). https://doi.org/10.1038/nrg2626

Raza, K. & Ahmed, H. Recent advancement in next generation sequencing techniques and its computational analysis. *International Journal of Bioinformatics Research and Applications* **13**, 1–14 (2017)

Sanger, F. & Coulson, A. R. A rapid method for determing sequences inDNA by primed synthesis with DNA polymerase. *Journal of Molecular Biology* **94**, 441–448 (1975). https://doi.org/10.1016/0022-2836(75)90213-2

Sanger, F., Nicklen, S. & Coulson, A. R. DNA sequencing with chain terminating inhibitors. *Proceedings of the National Academy of Sciences of the United States of America* **74**, 5463–5467 (1977). https://doi.org/10.1073/pnas.74.12.5463

12

Applying Meta-Analysis in the New Trend of Scientific Scenarios

12.1 Introduction

A recent field known as 'meta-analysis' aims to critically evaluate and statistically combine the findings of earlier research studies and reports. A review of the literature indicates that meta-analysis is becoming more and more important in the field of clinical trials and public health. The goals of a meta-analysis are to: (i) improve the estimate of the size of the effect; (ii) resolve uncertainty when reports differ; (iii) raise additional questions that are not addressed at the outset of the study (Zipkin & Isenburg Von, 2022). Because too few patients are enrolled in each trial and too few are being studied, these functions are especially useful in randomized controlled clinical trials (RCTs) to identify clinically significant differences. Study design, combinability, control of bias, statistical analysis, sensitivity analysis, problems applicability, and control of bias are important characteristics of any meta-analysis (Nyaga & Arbyn, 2023). Understanding a study's findings in light of all other studies is the aim of a synthesis. These problems can be addressed by meta-analysis but not by narrative reviews. Firstly, we provide an example of how a narrative review and a meta-analysis would tackle the same problem. We then use this example to illustrate the main distinctions between the two approaches (Sørensen et al., 2021). (Table 12.1)

The first thing we must ascertain is whether the effect size is constant throughout the corpus of data (Allen, 2020). We seek to determine the effect size with the highest level of accuracy possible and to document its robustness across the different study types included in the synthesis if it continues to be consistent. However, if there are significant differences between studies, we should measure the variance and take the implications into account (Freeman et al., 2019).

TABLE 12.1

Criteria of Research Review

Criteria	Narrative Review	Systematic Review	Meta Analysis
Author	An individual	A team-based work	A team-based work
Search strategy	Individual	Based on a protocol	Based on a protocol
Summary Conclusion	Authors Judgment	Can be qualitative	Summary by statistical techniques
	Review articles written		

12.2 Steps in Meta-Analysis

Step 1: Clearly state the specific hypothesis and research question.

Step 2: Establish the standards for which studies are included and are not excluded.

Step 3: Track down research papers.

Step 4: Ascertain which studies can be included.

Step 5: Categorize and code significant study characteristics (such as sample size, follow-up, definition of outcome, medication brand, and dosage).

Step 6: Use a standard metric to choose or interpret each study's findings.

Step 7: Combine results from several studies to create weighted, pooled effect size estimates.

Step 8: Assess the pooled studies' statistical homogeneity

Step 9: To ascertain the impact of reducing or removing unpublished studies, studies of lesser quality, out-of-date studies, etc., perform sensitivity analysis.

12.3 Assessing Study Quality

The evaluation of study quality should be grounded in the merit of research papers and their applicability. There is a significant need for a broader scope to ascertain the quality of these research papers (Polanin et al., 2017). Each paper included in the assessment must be meticulously reviewed by team members, and appropriate interventions should be implemented in subsequent processes (Demirel & Dağyar, 2016).

12.4 Publication Bias

Was the study described as randomized (this includes words such as randomly, random and randomization)? Rating 0/1

Was the method used to generate the sequence of randomization described and appropriate (table of random numbers, computer-generated, etc.) Score 0/1

Was the study described as double-blind? Score 0/1

Was the method of double blinding described and appropriate (identical placebo, active placebo dummy, etc.) Score 0/1

Was there a description of withdrawals and dropouts? Score 0/1

If the process for creating the randomization sequence was explained but was not appropriate, deduct one point. Rating of 0/1

DOI: 10.1201/9781032713816-12

If the study was labeled as double-blind but the blinding technique was improper, deduct one point. Points 0/1

Research question: Suppose we wish to determine if streptokinase prevents acute myocardial infarction-related death. What is the best way to organize a search strategy? We will look only through the PubMed database.

12.4.1 Standards for Evaluation

Research with noteworthy findings is more likely to

- Appear in print
- Be released in English
- Be referenced by additional parties
- Create several publications

Publication bias may be introduced by using only published studies.

The majority of reviews do not examine publications for bias.

12.5 Approaches to Bias of Publication

12.5.1 Fail Safe N

The 5% of studies with type I errors are published, while the 95% of studies with non-significant effects are kept on file.

12.5.2 Funnel Plots

Trim and fill method: small studies are taken out of the symmetric funnel plot.

To balance them, the small studies should then be replaced with studies on the opposite side of the funnel (Figures 12.1–12.3).

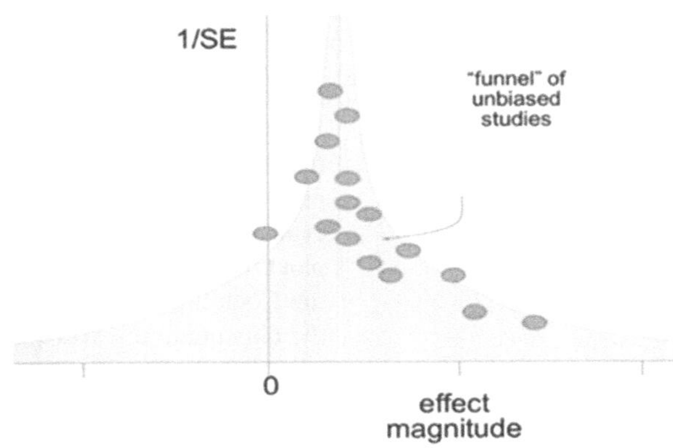

FIGURE 12.1 Funnel plot pertaining to the removal of small studies.

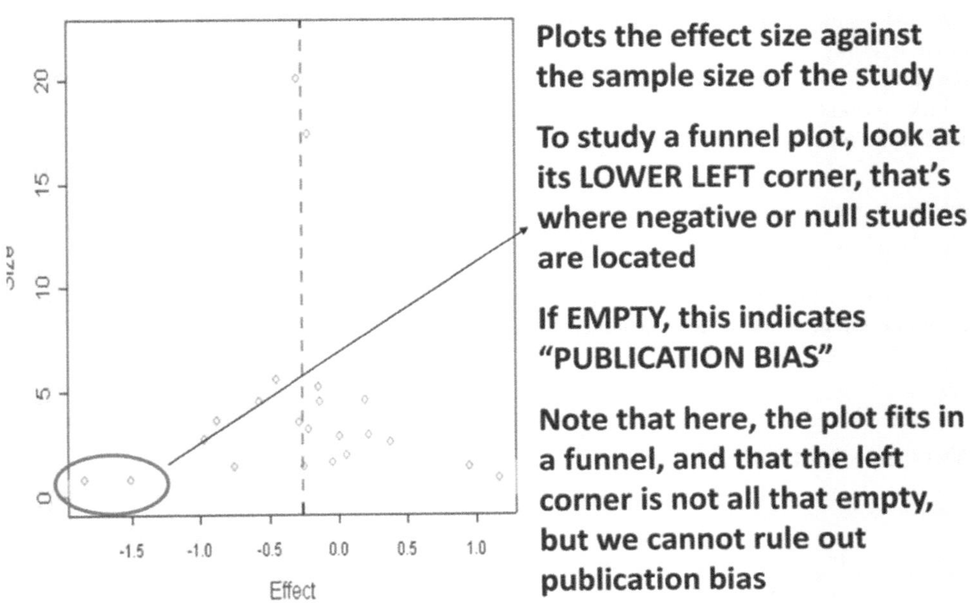

FIGURE 12.2 Funnel plot related to the removal of small studies on the effect estimates from smaller studies.

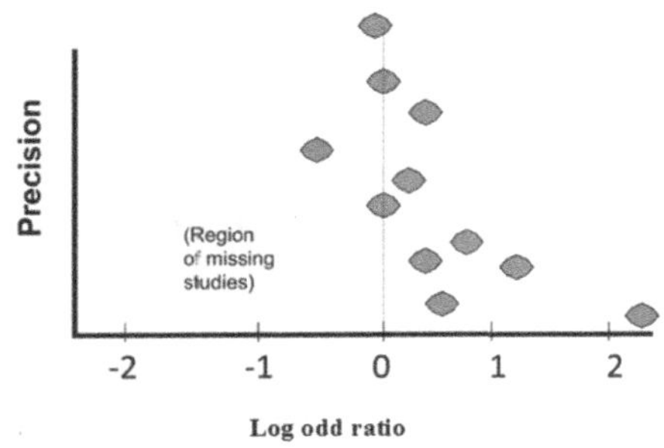

FIGURE 12.3 Asymmetrical plot indicative of publication bias.

12.6 Quality Score (Score: Yes = 1; No = 0, Coded)

1. Published in a peer-reviewed journal
2. Conducted by experts in research
3. Study supported by an unbiased organization
4. Conducted by unbiased researchers
5. Subjects chosen at random from the general population
6. Subjects chosen at random from the general population
7. Subjects divided into treatments at random
8. A high percentage of participants started and/or completed the research
9. Subjects who are blind to the treatment
10. Blind data collectors
11. Analysis carried out by binding

12.7 Averaging the Studies

Starting with summary statistics for each study, how should we combine these?

Each study is given the same weight by a simple average (Koricheva & Gurevitch, 2014).

From an intuitive point of view, this is incorrect.

There is a greater likelihood for some studies to yield results that are more accurate than others in terms of the effect.

12.8 Weighting Studies

Give more weight to the studies that provide more details, such as greater numbers of participants, events, and variance, computed utilizing the variance and effect estimates

12.9 Inverse Variance Method

See Table 12.2.

$$\left(\text{Weight} = \frac{1}{\text{Varaiance of estimate}} = \frac{1}{\text{SE}^2}\right) \tag{12.1}$$

TABLE 12.2

Procedure for Weights Assigning

Headache	Caffeine	Decaf	Weight
A et al. (2000)	2/31	10/34	6.6%
B et al. (2004)	10/40	9/40	21.90%
C et al. (1999)	12/53	9/61	22.20%
D et al. (1991)	3/15	1/17	2.90%
E et al. (1991)	19/613	9/64	26.40%
F et al. (1991)	4/35	2/37	5.10%
G et al. (2003)	13/35	6/37	14.90%

$$\text{Pooled estimate} = \frac{\sum_{i=1}^{n}(\text{estiamte} * \text{weight})}{\text{Sum of weights}} \tag{12.2}$$

12.10 Heterogeneity

Heterogeneity will be assessed by bringing together all studies conducted by various individuals, in various contexts, across various nations, on various individuals, using various methods, for varying amounts of time, allowing us to examine diverse results (Chang et al., 2013). We also reviewed the clinical, methodological, and statistical heterogeneity of published materials during the heterogeneity process. Visual evaluation is possible through the summary effect and forest plot CI of study overlap. We used the Mantel-Haenszel test or the Cochran Q test to statistically assess heterogeneity using the chi-square test (Gasparrini et al., 2012). The key hypotheses are examined to determine whether heterogeneity exists, whether there is any poor power, and whether the individual effects deviate significantly from the common effects beyond what would be predicted by chance. A p-value of less than 0.01 indicates significant heterogeneity. Careful attention should be given to the identification of heterogeneity. Avoid pooling all publications at once, take heterogeneity into account and use a fixed-effect model, let heterogeneity in and use a random-effects model, and investigate the study intervention (Deeks et al., 2011) (Table 12.3).

Conduct a sensitivity analysis. Several aspects of the meta-analysis can be changed to determine how robust the results are: (i) changing the inclusion criteria, incorporating and excluding unpublished studies, weighting the analysis based on study quality, experimenting with various methods to impute missing data, and eliminating each study individually to observe the change (Hartung et al., 2008).

12.11 Effective Size Calculation

Three approaches are utilized for effective size (ES), and the effect size has been determined using the following formula (Figure 12.4).

$$\Delta_{\text{Glass}} = \frac{\bar{Y}_{\text{Exp}} - \bar{Y}_{\text{Control}}}{\text{SD}_{\text{Control}}} \tag{12.3}$$

TABLE 12.3

Difference between Fixed and Random Effect Model

Fixed Effect Model	Random Effect Model
To conduct, assume that there is no heterogeneity and that the treatment effect size is the same (fixed) for all studies, with variation coming from chance and pooling by the Mantel-Haenszel OR	If heterogeneity is present, conduct pooling using the Der Simonian-Laird method (DSL) for odds ratios, assuming that the treatment effect size varies among studies.
$\left(\text{Weight} = \frac{1}{\text{Varaiance of estimate}} = \frac{1}{\text{SE}^2}\right)$	$\left(\text{Weight} = \frac{1}{\text{Varaiance of estimate}}\right)$ $\left(\text{Weight} = \frac{1}{\left(\text{SE}^2 + \text{inter-trialvaraiane}\right)}\right)$
We obtain pooled estimates when heterogeneity exists, which could overvalue large-scale studies. Adjust the narrow confidence interval and the p-value that are too small. As an example, the M-Haenszel method, Peto's method, MLE, and exact interval estimation techniques.	We obtain distinct pooled estimates with a different interpretation, a larger P-value, and a wider confidence interval when heterogeneity is present. For example, lease square statistical analysis methods, both weighted and unweighted MLE's restricted MLE's and the exact approach to random effects of binary data.

Note: 25% low heterogeneity, 50% moderate, and 75% high (0–100).

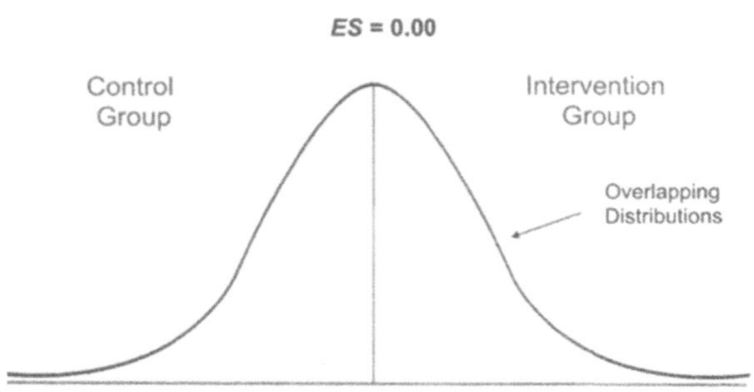

FIGURE 12.4 Calculation of zero effect size.

$$d_{\text{Cohen}} = \frac{\bar{Y}_{\text{Exp}} - \bar{Y}_{\text{Control}}}{\sqrt{\left(\text{SD}_e^2 + \text{SD}_c^2\right)/2}} \quad (12.4)$$

$$G_{\text{Hedges}} = \frac{\bar{Y}_{\text{Exp}} - \bar{Y}_{\text{Control}}}{\sqrt{\frac{\left(N_e - 1\right)\text{SD}_e^2 + \left(N_c - 1\right)\text{SD}_c^2}{\left(N_{\text{Tot}} - 2\right)}}} * \left(1 - \frac{3}{4\left(N_e + N_c - 9\right)}\right) \quad (12.5)$$

Precautions to be taken: the reviewing process should be conducted by at least two reviewers who should sift and sift again. To determine which studies to retrieve in full, the first step in the prescreening process is to sort. The second step in the selection process is to go over these studies once more and determine which ones should be included in your review. Avoid gathering outcome data concurrently with eligible information (Hedges et al., 1999) (since the latter may be excluded from the study later on, and the results may influence the decision). Lastly, keep an eye out for duplicate publications (Petrricrew & Robert, 2006).

12.11.1 Quality Assessment of MA

The PRISMA statement (formerly QUOROM) is the preferred reporting item for systematic reviews and meta analysis (MA).

TABLE 12.4

Binary Data Table

Study 'i'	Event	No Event	Total
Experimental	a_i	b_i	n_{1i}
Control	c_i	d_i	n_{2i}

12.12 The MOOSE Assertion

- Moose is a framework for software analysis that can also be used for any data.
- It includes mechanisms and frameworks for importing, meta-modeling, and exchanging data with external tools.
- It includes a framework for parsing that uses technologies like parsing expression grammar.
- It can supports visualization through engines for expressing graph visualizations and charts.
- The philosophy of Moose is to enable analysts to create new analysis tools and customize the flow of analysis.

12.13 Statistical Techniques for Meta-Analysis to Combine the Papers That Have Been Published

Take into consideration a meta-analysis of 'K' studies. When the study results are dichotomous, the percentage of participants in each of the two groups (naive and study) who do not experience the event can be shown in a 2×2 table for each study (Egger et al., 2001) (Table 12.4).

The number of participants in each of the two groups, their mean response, and the standard deviation of their responses are needed to conduct a meta-analysis if the outcome is a continuous measure (Table 12.5).

TABLE 12.5

Mean Response of Dose

Study 'i'	Size of the Group	Mean Response	SD
Experimental	n_{1i}	m_{1i}	SD_{1i}
Control	n_{2i}	m_{2i}	SD_{2i}

TABLE 12.6

OE Method Calculation Group Size

Study 'i'	O-E	Var (O-E)	Size of the Group (Experimental)	Size of the Group (Control)
	Z_i	V_i	n_{1i}	n_{2i}

TABLE 12.7

Linear Estimation of Group Size (MLEs)

Study i	Estimate of Effect	SE	Size of the Group (Experimental)	Size of the Group (Control)
	$\hat{\theta}_i$	$SE\{\hat{\theta}_i\}$	n_{1i}	n_{2i}

In the event that the results are analyzed using the log-rank approach, which compares observed with expected values, O-E statistics and their variances are necessary to carry out the meta-analysis. Group size may also be entered by the review author, but it is not involved in the analysis (Table 12.6).

An all-encompassing method can be employed for alternative results, whereby the user provides the values of the intervention effect estimate and its standard error for every study (the standard error could be computed using a confidence interval). 'Difference' measures of effect, such as risk difference and differences in means, are typically expressed on their natural scale, whereas 'ratio' measures of effect, such as odds ratio, risk ratio, hazard ratio, and ratio of means, are typically expressed on a logarithmic basis. Group sizes are optional for the review author and are not utilized in the analysis.

12.14 Generic Data

See Table 12.7.

Formulae for individual studies: Estimates for individual studies concerning dichotomous outcomes.

12.15 Peto Odds Ratio

For study I, denote the cell counts as in Table 12.1, with $n_{1i} = a_i + b_i, n_2 = c_i + d_i$ and let $N_i = n_{1i} + n_{2i}$. For the Peto method, the individual odds ratios are given by

$$\text{OR}_{\text{peto}} = \exp\left\{\frac{Z_i}{V_i}\right\} \tag{12.6}$$

The logarithm of odd ratio has SE

$$\text{SE}\left\{\ln\left(\text{OR}_{\text{peto}}\right)\right\} = \exp\left\{\frac{Z_i}{V_i}\right\} \tag{12.7}$$

where Z_i is the $O - E'$ statistics

$$Z_i = a_i - E(a_i) \tag{12.8}$$

$$E(a_i) = \frac{n_{1i}(a_i + c_i)}{N_i} \tag{12.9}$$

(the expected number of events in the experimental intervention group) and

$$V_i = \frac{n_1 n_2 (a_i + c_i)(b_i + d_i)}{N_i^2 (N_i - 1)} \tag{12.10}$$

The hypergeometric variance of a_i.

For study i, denote the cell counts as in Table, with $n_{1i} = a_i + b_i, n_{2i} = c_i + d_i, N_i = n_{1i} + n_{2i}$

For methods other than the 'Peto' method, the odd ratio for each study is given by

$$\text{OR}_i = \frac{a_i d_i}{b_i c_i} \tag{12.11}$$

The SE of the log odd ratio being

$$\text{SE}\left\{\ln\left(\text{OR}_i\right)\right\} = \sqrt{\frac{1}{a_i} + \frac{1}{b_i} + \frac{1}{c_i} + \frac{1}{d_i}} \tag{12.12}$$

The risk ratio for each study is given by

$$\text{RR}_i = \frac{a_i / n_{1i}}{c_i / n_{2i}} \tag{12.13}$$

$$\text{SE}\{\ln(\text{RR}_i)\} = \sqrt{\frac{1}{a_i} + \frac{1}{b_i} + \frac{1}{c_i} + \frac{1}{d_i}} \tag{12.14}$$

The risk difference for each study

$$\text{RD}_i = \frac{a_i}{n_{1i}} - \frac{c_i}{n_{2i}} \tag{12.15}$$

With standard error

$$\text{SE}\{(\text{RD}_i)\} = \sqrt{\frac{a_i b_i}{n_{1i}^3} + \frac{c_i d_i}{n_{2i}^3}} \tag{12.16}$$

Empty cells: Where zeros cause problems with the computation of effects or standard errors, 0.5 is added to all cells (a_i, b_i, c_i, d_i) for that study, except when $a_i = c_i = 0$ or $b_i = d_i = 0$, when the relative effect measures OR_i and RR_i are undefined.

12.16 Individual Study Estimates

Continuous outcomes $N_i = n_{1i} + n_{2i}$ and

$$s_i = \sqrt{\frac{(n_{1i} - 1)\,\text{SD}_{1i}^2 + (n_{2i} - 1)\text{SD}_{2i}^2}{N_i - 2}} \tag{12.17}$$

Be pooled standard deviation across the two groups

12.17 Difference in Means

The difference in means is given by

$$\text{MD}_i = m_{1i} - m_{2i} \tag{12.18}$$

$$\text{SE}\{\text{MD}_i\} = \sqrt{\frac{\text{SD}_{1i}{}^2}{n_{1i}} + \frac{\text{SD}_{2i}{}^2}{n_{2i}}} \tag{12.19}$$

12.18 Standardized Difference in Means (SMD)

There are several popular formulations of the standardized mean difference. The one implemented by Hedges is adjusted, which is very similar to Cohens 'd' but includes an adjustment for small sample bias.

$$\text{SMD}_i = \frac{m_{1i} - m_{2i}}{s_i}\left(1 - \frac{3}{4\,N_i - 9}\right) \tag{12.20}$$

$$\text{SE}\{\text{SMD}_i\} = \sqrt{\frac{N_i}{n_1, n_2} + \frac{\text{SMD}_i^{\,2}}{2\,(N_i - 3.94)}} \tag{12.21}$$

Individual study estimates (O-E) and variance

For study i, the effect estimates is given by

$$\hat{\theta}_i = \frac{Z_i}{V_i} \tag{12.22}$$

$$\text{SE}\{\hat{\theta}_i\} = \sqrt{\frac{1}{V_i}} \tag{12.23}$$

The effect estimate is either a log odds ratio or a log hazard ratio, depending on how the observed and expected values are derived.

12.19 Individual Study Estimates

Generic method: Since the user enters the intervention effect estimates and their standard errors directly, no additional processing is needed. The intervention effects can be of any type, but this method may be particularly useful when the effects have been calculated with special consideration for design (e.g. cluster randomized and crossover trials), when the effects have been adjusted for other effects (e.g. adjusted effects from non-randomized studies), or when the effects have not been addressed by other methods (e.g. ratios of means, relative event rates).

12.20 Meta-Analysis Methods

All summations are over i, from 1 to the number of studies, unless otherwise specified.

Mantel-Haenszel methods for combining results across studies:

The Mantel Haenszel summary log odds ratio is given by

$$\ln(\text{OR}_{\text{MH}}) = \ln\left(\frac{\sum w_{\text{MH}i,}\text{OR}_i}{\sum w_{\text{MH},i}}\right) \tag{12.24}$$

and the MH summary odds ratio by

$$(\text{OR}_{\text{MH}}) = \ln\left(\frac{\sum w_{\text{MH}i,}\text{OR}_i}{\sum w_{\text{MH},i}}\right) \tag{12.25}$$

where each study's odds ratio is given weight

$$w_{\text{MH},i} = \frac{b_i c_i}{N_i} \tag{12.26}$$

The summary log odds ratio has a standard error given by

$$SE\{ln(OR_{MH})\} = \sqrt{\frac{1}{2}\left(\frac{E}{R^2} + \frac{F+G}{RS} + \frac{H}{S^2}\right)} \tag{12.27}$$

$$R = \sum \frac{a_i d_i}{N_i}, S = \sum \frac{b_i c_i}{N_i}$$

$$E = \sum \frac{(a_i + d_i)a_i d_i}{N_i^2}, F = \sum \frac{(a_i + d_i)b_i c_i}{N_i^2},$$

$$G = \sum \frac{(b_i + c_i)a_i d_i}{N_i^2}, H = \sum \frac{(b_i + c_i)b_i c_i}{N_i^2}$$

12.21 Risk Ratio

The MH summary log risk ratio is given by

$$\ln(\text{RR}_{\text{MH}}) = \ln\left(\frac{\sum w_{\text{MH},i}\text{RR}_i}{\sum w_{\text{MH},i}}\right) \tag{12.213}$$

and the MH summary risk ratio by

$$(\text{RR}_{\text{MH}}) = \left(\frac{\sum w_{\text{MH},i}\text{RR}_i}{\sum w_{\text{MH},i}}\right) \tag{12.29}$$

where each study's MH risk ratio is given weight

$$w_{\text{MH},i} = \frac{c_i(a_i + b_i)}{N_i} \tag{12.30}$$

The summary log risk ratio has a standard error given by

$$\text{SE}\{\ln(\text{RR}_{\text{MH}})\} = \sqrt{\frac{P}{\text{RS}}} \tag{12.31}$$

where

$$P = \sum \frac{n_1, n_2(a_i + c_i) - a_i c_i N_i}{N_i^2}; R = \sum \frac{a_i n_{2i}}{N_i}; S = \sum \frac{C_i n_{1i}}{N_i}$$

12.22 Risk Difference

The MH summary risk difference is given by

$$\text{RD}_{\text{MH}} = \frac{\sum w_{\text{MH},i}, \text{RD}_i}{\sum w_{\text{MH},i}} \tag{12.32}$$

where each study's risk difference is given weight.

The summary risk difference has a standard error that is given weight.

$$w_{\text{MH},i} = \frac{n_{1i} n_{2i}}{N_i} \tag{12.33}$$

The summary risk difference has a standard error given by

$$SE\{RD_{MH}\} = \sqrt{\frac{J}{K^2}} \quad (12.34)$$

$$J = \sum \frac{a_i b_i n_{2i}^2 + c_i d_i n_{1i}^3}{n_1 n_2, N_i^2}, K = \sum \frac{n_{1i} n_{2i}}{N_i} \quad (12.35)$$

12.23 Test for Heterogeneity

The heterogeneity test statistics is given by

$$Q_{MH} = \sum w_i \left(\hat{\theta}_i - \hat{\theta}_i\right)^2 \quad (12.36)$$

where $\hat{\theta}$ represents the log odds ratio, log risk ratio, or risk difference, and they are the weights calculated as $\frac{1}{SE\left\{\hat{\theta}_i\right\}^2}$ rather than the weights used for the MH meta-analysis. Under the H0 that there are no differences in intervention effect among studies, this follows a chi-squared distribution $(k-1)df$ (where 'k' is the number of studies contributing to the meta-analysis).

The statistics I^2 is determined as

$$I^2 = \max\left\{100\% * \frac{Q_{MH} - (k-1)}{Q_{MH}}, 0\right\} \quad (12.37)$$

This measures the extent of inconsistency among the study results and is interpreted as approximately the proportion of total variation in study estimates that is due to heterogeneity rather than sampling error.

12.24 Inverse-Variance Methods

Inverse-variance methods for combining results across studies: For binary data, inverse-variance methods are used to combine intervention effect estimates in the generic method and to pool all mean differences and standardized mean differences. As an additional analysis option, they are also used to pool log odds ratios, log risk ratios, and risk differences. In the general formula, the intervention effect estimate is denoted by $\hat{\theta}_i$, which is the study's log odds θ_i ratio, log risk ratio, risk difference, mean difference or standardized mean difference, or the estimate of intervention effect in the generic method. Each effect size is weighted based on the reciprocal of its variance, which is determined by squaring the standard error provided in the section above for each individual study.

$$w_i = \frac{1}{SE\left\{\hat{\theta}_i\right\}^2} \quad (12.38)$$

These are combined to give a summary estimates:

$$\hat{\theta}_{IV} = \frac{\sum w_i \hat{\theta}_i}{\sum w_i} \quad (12.39)$$

$$SE\left\{\hat{\theta}_{IV}\right\} = \frac{1}{\sqrt{\sum w_i}} \quad (12.40)$$

The heterogeneity statistics is given by a similar formula as for MH test.

$$Q_{IV} = \sum w_i \left(\hat{\theta} - \hat{\theta}_{IV}\right)^2 \quad (12.41)$$

Under the H_0 that there is no difference in intervention effect among studies, this follows a chi-squared distribution with $(k-1)\,df$ (where 'k' is the number of studies contributing to the meta-analysis I^2 statistics

$$I^2 = \max\left\{100\% * \frac{Q_{IV} - (k-1)}{Q_{IV}}\right\} \quad (12.42)$$

12.25 (O – E) and Variance Method for Combining Studies

This is an implementation of the Peto method, which allows its application to time-to-event data as well as binary data. The summary effect estimate is given by

$$\hat{\theta} = \frac{\sum V_i \hat{\theta}_i}{\sum V_i} \quad (12.43)$$

where the estimate $\hat{\theta}$, from study i, is calculated from Z_i and V_i as for individual studies. The summary effect is either a log odds ratio or a log hazard ratio (the user should specify which). The effect estimate (on a non-log scale) is given by

$$\text{Effect estimates} = \exp\left\{\frac{\sum V_i \hat{\theta}_i}{\sum V_i}\right\} \quad (12.44)$$

And is either an odds ratio or a hazard ratio. The effect estimates (on the log scale) have SE

$$SE\,\hat{\theta} = \frac{1}{\sqrt{V_i}} \quad (12.45)$$

The heterogeneity statistics is given by

$$Q_{peto} = \sum V_i \left(\hat{\theta}_i^2 - \hat{\theta}^2\right) \quad (12.46)$$

Under the null hypothesis that there are no differences in intervention effect among studies this follows a chi-squared distribution with $k-1$ degrees of freedom (where k is the number of studies contributing to the meta-analysis). I^2 is calculated as

$$I^2 = \max\left\{100\% * \frac{Q_{peto} - (k-1)}{Q_{peto}}, 0\right\} \quad (12.47)$$

12.27 Der Simonian and Laird Random-Effects Models

Under the random-effects model, the assumption of a common intervention effect is relaxed, and the effect sizes are assumed to have a distribution $\theta_i \sim N(\theta, \tau^2)$

The estimate of τ^2 is given by

$$\tau^2 = \max\left\{\frac{Q-(k-1)}{\sum w_i - \left(\sum w_i^2\right)/\sum w_i}, 0\right\} \quad (12.48)$$

where the w_i are the inverse variance weights, calculated as $w_i = \frac{1}{\text{SE}\,(\theta_i)^2}$

For continuous data and for the generic method, Q is Q_{IV}. For binary data, either Q_{IV} or Q_{MH} may be taken. Both are implemented in RevMan (and this is the only difference between random-effects methods under the 'Mantel-Haenszel' and 'inverse-variance' options). Again, for odds ratios, risk ratios, and other ratio effects, the effect size is taken on the natural logarithmic scale.

12.28 Each Study's Effect Size Is Given Weight

$$w_i' = \frac{1}{SE\,\{\theta_i\}^2 + \hat{\tau}^2} \quad (12.49)$$

The summary effect size is given by

$$\widehat{\theta_{\text{DL}}} = \frac{\sum w_i'\widehat{\theta_i}}{\sum w_i'} \quad (12.50)$$

$$\text{SE}\left\{\widehat{\theta_{\text{DL}}}\right\} = \frac{1}{\sqrt{w_i'}} \quad (12.51)$$

Note that in the case where the heterogeneity statistic Q is less than or equal to its degrees of freedom $(k-1)$, the estimate of the between-study variation, $2\hat{\tau}$, is zero, and the weights coincide with those given by the inverse-variance method.

12.29 Confidence Interval

The 100 $(1-\alpha)$ % confidence interval for θ_i is given by

$$\hat{\theta} - \text{SE}\left\{\hat{\theta}\right\}\Phi\left(1-\alpha/2\right) \text{ to } \hat{\theta} + \text{SE}\left\{\hat{\theta}\right\}\Phi\left(1-\alpha/2\right) \quad (12.52)$$

where $\widehat{\theta_i}$ is the log odds ratio, log risk ratio, risk difference, mean difference, standardized mean difference, or generic intervention effect estimate, and Φ is the standard normal deviate. For log odds ratios, log risk ratios, and generic intervention effects entered on the log scale (and identified as such by the review author), the point estimate and confidence interval limits are exponentiated for presentation.

Test for presence of an overall intervention effect: In all cases, the test statistic is given by

$$Z = \frac{\hat{\theta}}{\text{SE}\left(\hat{\theta}\right)} \quad (12.53)$$

where the odds ratio, risk ratio, and other ratio measures are again considered on the log scale. Under the null hypothesis that there is no overall effect of intervention effect, this follows a standard normal distribution (Tables 12.8–12.10).

TABLE 12.8

Quality of Reporting Meta-Analysis – QUOROM for RCT's

Heading	Subheading	Descriptor
Title		Identify the report as meta-analysis of RCTs
Abstract		Use a structured format
	Objectives	The clinical question explicitly
	Data source	The databases (i.e. list) and other information sources
	Review methods	The selection criteria (population, intervention, outcomes, and study design), methods for validity assessment, data abstraction and characteristics, and quantitative data analysis (in sufficient detail to permit replications)
	Results	Characteristics of RCT included and excluded (qualitative and quantitative findings—point estimation, etc.)
	Conclusion	The main results
Text introduction		The explicit clinical problem, biological rationale for the intervention and rationale for review
Methods	Searching	The information sources in detail (e.g. databases, registers, personal files, expert informants, agencies, hand searching) and any restrictions (years considered, publication status, language of publication)
Selection		The inclusion and exclusion criteria employed
Methods		Elaborate the methods

12.30 Weakness of MA

A significant amount of work-quality difference between studies is not captured.

Even with a strong meta-analysis, poorly designed studies will still produce subpar statistics.

Selection bias: Predisposition to focus on general inquiries that might not be immediately relevant to specific patients.

TABLE 12.9

Method for Data Analysis of MA

Data Type	Outcome Measures	Other Method
Continuous	Descriptive statistics (Mean and SD)	Meta regression, regression smoothening, and Boot strap techniques with boolneal form
Dichotomous (binary)	Odds ratio, hazard risk, risk ratio, risk difference	
Non-parametric test	Mann-Whiteny U test, Chi-square test, Fisher F test	

TABLE 12.10

Meta Analysis Software

Free Ware	Commercial
RevMan (reviewer manager)	Comprehensive meta-analysis Met Win, We ASY ma
Meta analyst	General statistical packages
Epimeta	Stat
Easy MA	SAS
Meta Test	S-Plus
Meta stat	R and Python

12.31 Strength of MA

Detailed search approach: We found information from several sources.

Using an explicit methodology will guarantee transparency and reproducibility.

Emphasis on all clinically significant outcomes, including those pertaining to the effectiveness, safety, and tolerability of the interventions being evaluated.

Reducing error: Two reviewers at every significant stage reduce bias and boosts accuracy. There are currently more than 200 published meta-analyses of RCTs in a variety of clinical and statistical journals, and many more are released annually. The extensive use of this analysis has led to a number of challenges, with regard to establishing its validity, for statisticians and those responsible for health care policy. As the analysis deals with data that are already collected, caution should be exercised while interpreting the results.

12.32 Standards for Papers and Reports to Be Included

In order to combine and pool the results, data from published papers and reports on RCTs will be incorporated into the analysis. Using ranks or a scoring sheet, each paper will be assessed by a minimum of two investigators following a listing of the meta-analysis's priority elements. The identity of the journal and the authors of the papers will be hidden from the investigators. An item will receive a rating of 'adequate' if it has been fully addressed, 'partial' if it is missing, and 'none or unknown' if it was not dealt with. A conference with the other investigators will settle any disagreements, if any, regarding ratings. The most recent publication will be analyzed if the results are the same or updated in two publications.

12.33 Method of Analysis

12.33.1 Study Design

This includes mainly the following: (i) protocol, (ii) literature search, (iii) list of trials analyzed, (iv) treatment assignment, and (v) characteristics of patients, diagnoses, and treatments.

The quantity (and percentage) of trials that were carried out in accordance with a preset protocol should be known. To include as many pertinent trials as possible in the analysis, a thorough search strategy must be used, such as consulting current contents, reviews, textbooks, and experts in the specific field of study. Together with the list of trials excluded and the justifications for exclusion, the number (and percentage) of trials analyzed should be given. Any information that is not used must be explained to the reader, especially if it conflicts with the hypothesis. Details about the variety of patient characteristics, such as age, sex, pertinent socioeconomic information, or other illnesses, should be included. It is also necessary to meet the requirements for the ranges, dosage, mode of administration, frequency, and length of treatment. In a similar vein, specifics regarding the categories of diagnoses, diagnostic standards, and illness stages or grades covered are also crucial.

12.34 Combinability

The ability to meaningfully combine the outcomes of independent trials is a significant issue when pooling data. It is necessary to address the pooling criteria. It is also necessary to discuss how the differences, if any, affect the conclusions. Heterogeneity is another issue. If each study uses a sample from the same population, then experimental error or study variability can be blamed for the differences that are seen. Alternatively, if we assume that each study comes from a distinct population, then the variations result from variations in the populations (variability between studies). There are homogeneity testing techniques that require care in how the results are interpreted. Therefore, it is critical to understand how often the analysis uses the test for homogeneity. Potential sources of unconscious bias should be taken into consideration whenever feasible when conducting a meta-analysis. Looking at the procedures rather than the end product will help to mitigate some of the bias in the selection and rejection of papers. To prevent measurement bias, it is also possible to prevent bias in data extraction by having multiple investigators extract data and measuring their agreement. To give the conclusions appropriate weight, the source of funding or financial support must also be mentioned.

12.34.1 Statistical Analysis

We consider here any accepted statistical technique for pooling, not just the sum of the trials' results. Regression analysis, the Mantel-Haenszel test, standardized average effect size, and significance tests are a few statistical techniques used in pooling. We will also discuss potential Type I and Type II error issues. We will also take into account the major outcomes' confidence intervals for the differences. Consideration will also be given to the degree of subgroup analysis. Sensitivity analysis, which examines whether the same set of data under various tests, criteria, and assumptions yields different results, is also employed (Plana et al., 2022). To formulate appropriate recommendations, the methodological rigor and scientific quality of the papers should be evaluated. Additionally, the quality features that need to be addressed are randomization, patient compliance, patient and investigator blinding, statistical analysis techniques, and withdrawal management. Studies that show a negative correlation with published results are less likely to be published than those that show a positive correlation. As a helpful indicator of the strength of the published evidence in this case, a straightforward rule like the number of unpublished negative studies needed to contradict the published evidence is put forth. Once the meta-analysis is finished, the researcher needs to decide whether the novel treatment is superior to the standard of care for all patients or just a subset of them. Additionally, they need to decide if the results are conclusive or point to the need for more research. The economic impact of adapting new methods of treatment of diagnosis is also important.

The benefits and drawbacks of different pooling techniques, as well as statistical concerns, must all be taken into consideration. The *P*-values or test statistics can be combined using various techniques. Nevertheless, these approaches do not differentiate between small and large studies, nor do they offer a measure of the effect size. A variation of the Mantel-Haenszel method that is used for pooling individual 2×2 tables is gaining traction. Two advantages of these methods are that each treatment can be compared to its own control, and a weighted estimate representing the study's sample size can be produced. Although opinions on the suitability of these approaches are divided, they also allow for the estimation of the effect size and homogeneity testing. The literature on psychiatry describes another way to calculate effect size as the standard deviation of the control group divided by the mean difference between the experimental and control groups. These effect sizes can be combined because they are converted into common units. However, clinical interpretation is challenging. A few reports have tried using log-linear models or multivariate techniques to account for study differences. Readers will benefit from uniformity in the presentation of meta-analysis results in the form of scientific papers, complete with sections on methods and results (Figures 12.5 and 12.6). The reader will then be better equipped to determine for themselves whether the results are legitimate as a result of this. To summarise, meta-analysis ought to be conducted similarly to a scientific experiment, beginning with a comprehensive protocol. It is necessary to investigate inter-observer and intra-observer variation and make an effort to locate and predicted value expressed as 13.7 and 13.8. The quantitative synthesis of data in RCTs can be more helpful to practicing physicians and clinicians than a traditional descriptive article, even though the validity of pooling is questioned. Naturally, for the synthesis studies to merit serious consideration, they must be conducted corrected value was seen 13.9 (Tables 12.11 and 12.12).

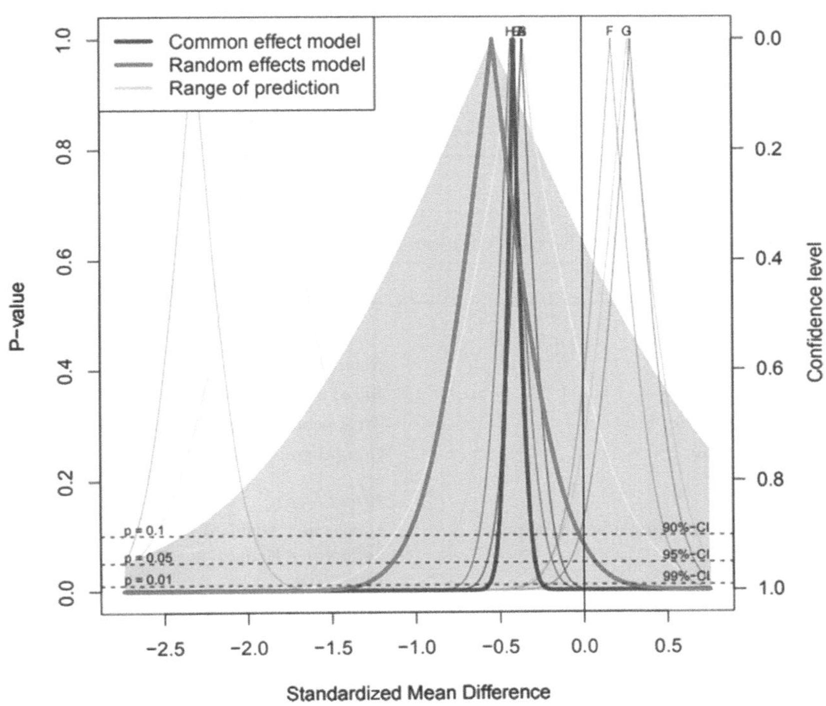

FIGURE 12.5 Comparisos of the standardized mean difference (SMD)—fixed and random effects models.

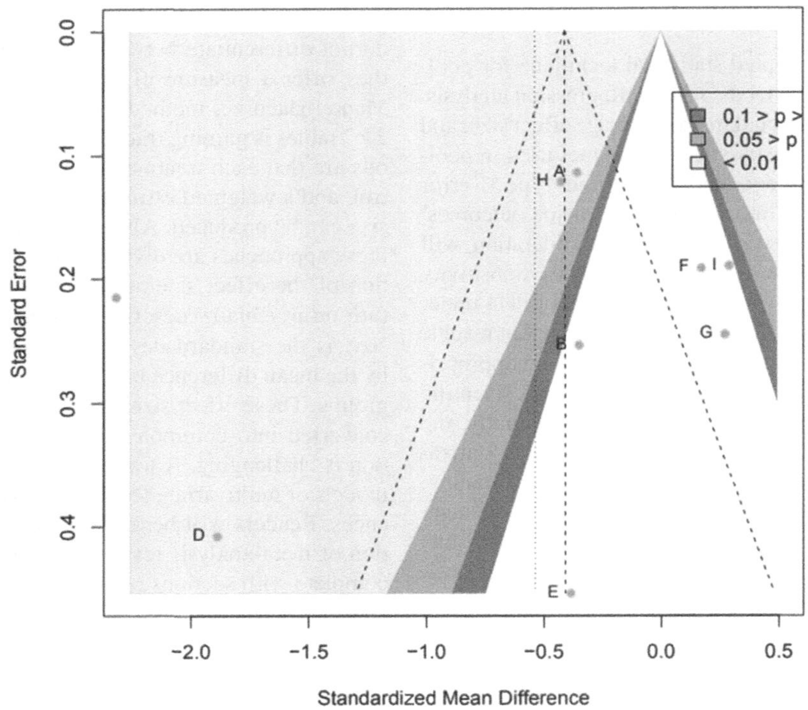

FIGURE 12.6 Funnel plot representing the standardized mean difference (SMD) and (SE).

TABLE 12.11

Mean and SD of G1 and G2

Study	N_1	$Mean_1$	SD_1	N_2	$Mean_2$	SD_2	Subgroup
A	155	55	47	156	75	64	1
B	31	27	7	32	29	4	1
C	75	64	17	71	119	29	1
D	113	66	20	113	137	413	2
E	13	14	13	13	113	11	2
F	57	19	7	52	113	4	2
G	34	52	45	33	41	34	3
H	110	21	16	1133	31	27	3
I	60	30	27	52	23	20	3

TABLE 12.12

Different Methods Used for Effect Size Calculation

Hedges '*d*' Conversion	Effect Size (Ne)
Regression Method	
Unstandarized Co eff of B: 7.4	Standardized Coeff. of B: 3.3
SD of dependent variable: 7	SD of dependent variable: 7
Treatment group sample size: 110	Treatment group sample size: 100
Control group sample size: 250	Control group sample size: 150
Effect Size	*Effect Size*
Conversion: Unstandarized regression coefficient to effect size Hedges' g	Conversion: standardized regression coefficient to effect size Hedges' g
Effect Size: 1.206	Effect Size: 2.126
Standard Error: 0.123	Standard Error: 0.160
Variance: 0.015	Variance: 0.025
Lower CI: 0.965	Lower CI: 1.1312
Upper CI: 1.44	Upper CI: 2.441
Weight: 66.130	Weight: 313.1306

(Continued)

TABLE 12.12 (*Continued*)

Different Methods Used for Effect Size Calculation

Hedges '*d*' Conversion	**Effect Size (Ne)**
Point Biserial Correlation	
r: 0.35	P-value: 0.045
n_1: 44	**n_1: 120**
n_2: 54	**n_2: 210**
Point-biserial 'r' to effect size Hedges' g	Conversion: point-biserial 'r' to effect size Hedges' g
Hedges' g	Effect Size: 0.229
Effect Size: 0.745	Standard Error: 0.114
Standard Error: 0.210	Variance: 0.013
Variance: 0.044	Lower CI: 0.004
Lower CI: 0.333	Upper CI: 0.454
Upper CI: 1.157	Weight: 75.1397
Weight: 22.663	
Mean and SE	
Sample Size Group 1:350	Sample Size Group 2:350
Mean: 44.37	Mean : 44.37
SE: 2.13	SE:2.13
Sample Size Group 2 :390	Sample Size Group 2 :390
Mean: 27.139	Mean: 27.139
SE: 4.6	SE: 4.6
Conversion: Hedges' g	Conversion: Hedges' g
Effect Size: 4.270	Effect Size: 0.219
Standard Error: 0.133	Standard Error: 0.073
Variance: 0.017	Variance: 0.005
Lower CI: 4.009	Lower CI: 0.074
Upper CI: 4.531	Upper CI: 0.364
Weight: 56.276	Weight: 1133.357
Contingency Table	
Group 1	Group 2
TP: 350	FP: 400
FN:69	TN:139
2×2 table (OR) to effect size Hedges' g	
Effect Size: 0.96133	
Standard Error: 0.01331	
Variance: 0.0069	
Lower CI: 0.13054	
Upper CI: 1.1311	
Weight: 144.131332	
Chi-Square	
Chi-square: 9.9	Total sample size (n=150)
P-value: 0.024	
Total sample size (n=150)	
Conversion: chi-squared-value to effect size Hedges' g	Conversion: chi-squared-value to effect size Hedges' g
Effect Size: 0.529	Effect Size: 0.373
Standard Error: 0.169	Standard Error: 0.166
Variance: 0.0213	Variance: 0.027
Lower CI: 0.197	Lower CI: 0.047
Upper CI: 0.1360	Upper CI: 0.6913
Weight: 35.025	Weight: 36.226
Student t-Test	
t: 3.3	p: 0.01
Total sample size: 150	Total sample size: 200
Conversion: t-value to effect size Hedges' g	Conversion: t-value to effect size Hedges' g
Effect Size: 0.536	Effect Size: 0.366
Standard Error: 0.166	Standard Error: 0.142
Variance: 0.027	Variance: 0.020
Lower CI: 0.210	Lower CI: 0.0136
Upper CI: 0.1362	Upper CI: 0.645
Weight: 36.1136	Weight: 49.1613

(*Continued*)

TABLE 12.12 (*Continued*)

Different Methods Used for Effect Size Calculation

Hedges 'd' Conversion	Effect Size (Ne)
One-Way ANOVA	
F-value : 16.46, n_1=110, n_2: 100	
Conversion: F-value (one-way-ANOVA) to effect size Hedges' g	
Effect Size: 0.5513	
Standard Error: 0.140	
Variance: 0.019	
Lower CI: 0.2132	
Upper CI: 0.1334	
Weight: 50.405	

Review:	Model: SMD - Fixed & Random Effect			
SMD	95%-CI	%W (Common)	%W (Random)	Subgroup
A	−0.3552 [−0.5792; −0.1311]	29.1	11.9	Subgroup 1
B	−0.3479 [−0.134513; 0.1499]	5.9	11.1	Subgroup 1
C	−2.3176 [−2.7393; −1.139513]	13.2	11.4	Subgroup 1
D	−1.1313130 [−2.61374; −1.013136]	2.3	10.0	Subgroup 2
E	−0.31340 [−1.2740; 0.5060]	1.13	9.6	Subgroup 2
F	0.1721 [−0.2044; 0.54137]	10.3	11.5	Subgroup 2
G	0.2721 [−0.2092; 0.7533]	6.3	11.2	Subgroup 3
H	−0.4246 [−0.6636; −0.11356]	25.6	11.13	Subgroup 3
I	0.21396 [−0.013313; 0.6629]	10.5	11.5	Subgroup 3
Number of studies combined: k=9				
Number of observations: o=11,513				
SMD	95%-CI	z|t|p-value		
Common effect model	−0.40713 [−0.52137; −0.21369] −6.61	<0.0001		
Random effects model	−0.5362 [−1.2490; 0.1767]−1.73	0.1211		
Quantifying Heterogeneity				
tau²=0.713137 [0.3241; 3.10613]; tau=0.1313131 [0.5693; 1.7626]				
I^2=93.5% [139.7%; 95.9%]; H=3.91 [3.12; 4.91]				
Test of Heterogeneity				
Q df	p-value			
122.42	13<0.0001			
Results for Subgroups (Common Effect Model)				
k	SMD	95%-CI	*Q*	I^2
Subgroup=subgroup1	3	−0.7272 [−0.9111; −0.5434]	67.45	97.0%
Subgroup=subgroup2	3	−0.2252 [−0.5434; 0.0929]	21.02	90.5%
Subgroup=subgroup3	3	−0.1442 [−0.3299; 0.0415]	13.35	135.0%
Test for Subgroup Differences (Common Effect Model)				
	Q df	*P*-value		
Between groups	20.60	2<0.0001		
Within groups	101.132	6<0.0001		
Results for Subgroups (Random Effects Model)				
k	SMD	95%-CI	tau²	tau
Subgroup=subgroup1	3	−1.0042 [−3.13209; 1.13124]	1.2423	1.1146
Subgroup=subgroup2	3	−0.66139 [−3.3270; 1.91392]	1.0267	1.0133
Subgroup=subgroup3	3	0.0170 [−1.02213; 1.0567]	0.1507	0.313132
Test for Subgroup Differences (Random Effects Model)				
	Q df	*P*-value		
Between groups	2.137	2 0.2376		

Details on meta-analytical method:

- Inverse variance method
- Restricted maximum-likelihood estimator for tau^2
- *Q*-Profile method for confidence interval of tau^2 and tau
- Hartung-Knapp (HK) adjustment for random effects model (*df*=13)
- Hedges'g (bias corrected standardized mean difference; using exact formulae) (Figures 12.7 and 12.8)

Study or Subgroup	Experimental Mean	Experimental SD	Experimental Total	Control Mean	Control SD	Control Total	Weight	Std. Mean Difference IV, Random, 95%CI
subgroup= subgroup 1								
A	55.00	47.0000	155	75.00	64.0000	156	11.9%	−0.36 [−0.58;−0.13]
B	27.00	7.0000	31	29.00	4.0000	32	11.1%	−0.35 [−0.85;0.15]
C	64.00	17.0000	75	119.00	29.0000	71	11.4%	−2.32 [−2.74;−1.90]
Total (95%CI)			261			259	34.4%	−1.00 [−3.82;1.81]
Heterogeneity: Tau^2 =1.2423; Chi^2 = 67.45, df = 2 (P < 0.01); I^2= 97%								
Subgroup = Subgroup 2								
D	66.00	20.0000	18	137.00	48.0000	18	10.0%	−1.89 [−2.69;−1.09]
E	14.00	8.0000	8	18.00	11.0000	13	9.6%	−0.38 [−1.27;0.51]
F	19.00	7.0000	57	18.00	4.0000	52	11.5%	0.17 [−0.20;0.55]
Total (95%CI)			83			83	31.0%	−0.67 [−3.33;1.99]
Heterogeneity: Tau^2 = 1.0267; Chi^2 = 21.02, df = 2 (P <0.01); I^2= 90%								
Subgroup=Subgroup3								
G	52.00	45.0000	34	41.00	34.0000	33	11.2%	0.27 [−0.21;0.75]
H	21.00	16.0000	110	31.00	27.0000	183	11.8%	−0.42 [−0.66;−0.19]
I	30.00	27.0000	60	23.00	20.0000	52	11.5%	0.29 [−0.08;0.66]
Total(95%CI)			204			268	34.6%	0.02 [−1.02;1.06]
Heterogeneity: Tau^2 = 0.1507; Chi^2 = 13.35, df = 2 (P < 0.01); I^2 = 85%								
Total (95%CI)			548			610	100.0%	−0.54 [−1.25;0.18]
Heterogeneity: Tau^2 = 0.7887; Chi^2 = 122.42, df = 8 (P < 0.01); I^2 = 93% Test for subgroup differences: Chi^2 = 2.87, df = 2 (P = 0.24)								

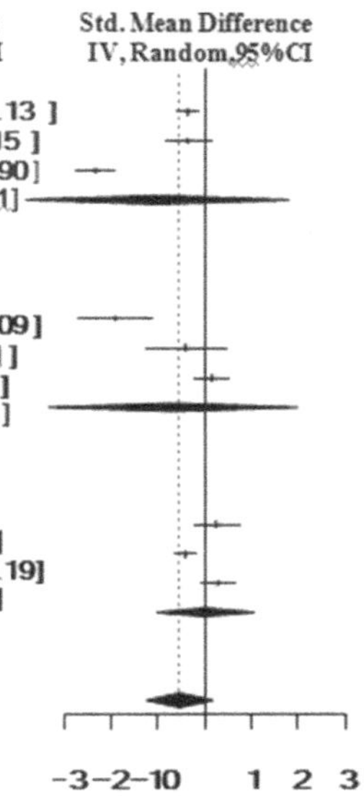

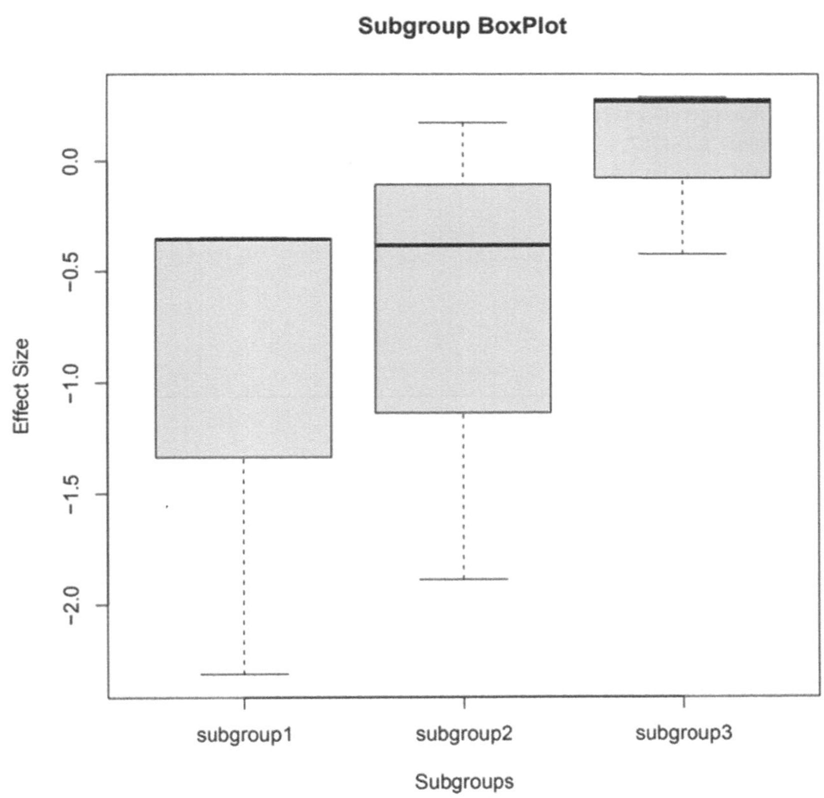

FIGURE 12.7 Box plot for subgroup analysis.

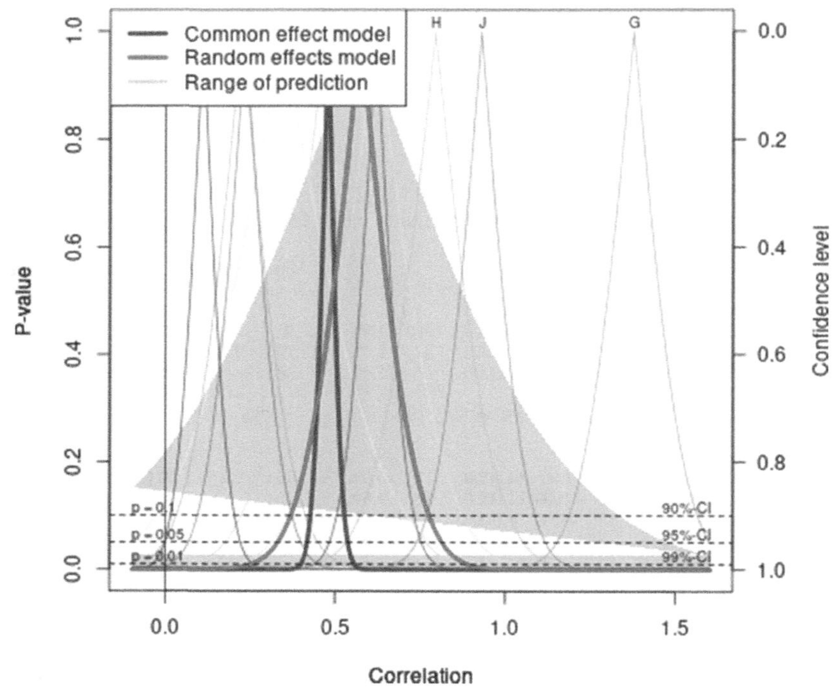

FIGURE 12.8 Funnel plot comparing various models.

Review:		**Model: SMD - Fixed and Random Effect**	
Linear regression test of funnel plot asymmetry			
Test result: $t=-0.50$, $df=7$, p-value$=0.6300$			
Sample Estimates			
Bias	SE.Bias	Intercept	SE.Intercept
−1.7502	3.4754	−0.1102	0.6429
Mixed-Effects Model ($k=9$; tau^2 estimators: REML)			
tau^2 (estimated amount of residual heterogeneity): 0.71345 (SE=0.4915)			
tau (square root of estimated tau^2 value): 0.131357			
I^2 (residual heterogeneity/unaccounted variability): 94.62%			
H^2 (unaccounted variability/sampling variability): 113.513			
R^2 (amount of heterogeneity accounted for): 0.53%			
Test for Residual Heterogeneity			
QE($df=6$)=101.131131, p-value< 0.0001			
Test of Moderators (Coefficients 2:3)			
$F(df1=2, df2=6)=1.02013$, p-value=0.4153			

Model Results:

	Estimate	**SE**	***t* value**	***df***	***p* value**	**ci.lb**	**ci. ub**
Intercept	−1.0026	0.5257	−1.907		0.105	−2.21313	0.2133
Subgroup=subgroup 2	0.3427	0.7634	0.4413	6	0.669	−1.525	2.210
Subgroup=subgroup 3	1.0419	0.7425	1.403	6	0.210	−0.774	2.13513

Correlation

Study	***N***	***r***	**Subgroup**
A	23	0.33	Subgroup 1
B	47	0.44	Subgroup 1
C	44	0.55	Subgroup 1
D	713	0.22	Subgroup 2

(*Continued*)

(Continued)

Study	N	r	Subgroup
E	311	0.11	Subgroup 2
F	144	0.23	Subgroup 2
G	79	0.1313	Subgroup 3
H	59	0.66	Subgroup 3
I	214	0.55	Subgroup 3
J	110	0.73	Subgroup 3

Review:		Model: COR - Fixed and Random Effect		
COR	**95%-CI**	**%W (Common)**	**%W (Random)**	**Subgroup**
A	0.3300 [−0.0951; 0.6533]	1.9	13.2	Subgroup 1
B	0.4400 [0.1749; 0.6456]	4.1	9.6	Subgroup 1
C	0.5500 [0.3025; 0.72130]	3.13	9.5	Subgroup 1
D	0.2200 [−0.0027; 0.4219]	7.0	10.1	Subgroup 2
E	0.1100 [−0.0012; 0.21135]	213.5	10.9	Subgroup 2
F	0.2300 [0.0690; 0.3793]	13.1	10.6	Subgroup 2
G	0.131300 [0.131131; 0.92113]	7.0	10.2	Subgroup 3
H	0.6600 [0.41361; 0.71336]	5.2	9.9	Subgroup 3
I	0.5500 [0.4490; 0.6371]	19.6	10.7	Subgroup 3
J	0.7300 [0.62137; 0.13069]	9.9	10.4	Subgroup 3

Number of studies combined: $k=10$

Number of observations: $o=1{,}109$

COR	95%-CI	z|t p-value
Common effect model	0.4445 [0.3954; 0.4911]	15.70<0.0001
Random effects model	0.5171 [0.21353; 0.69113]	4.64 0.0012

Quantifying Heterogeneity

$tau^2=0.1406$ [0.05137; 0.41330]; tau=0.3750 [0.2422; 0.6950]

$I^2=93.9\%$ [90.13%; 96.0%]; $H=4.07$ [3.30; 5.01]

Test of Heterogeneity

Q df	*p*-value
1,413.72	9<0.0001

Results for Subgroups (Common Effect Model)

k	COR	95%-CI	*Q*	I^2
Subgroup=subgroup 1	3	0.46513 [0.3035; 0.60113]	1.10	0.0%
Subgroup=subgroup 2	3	0.15136 [0.0742; 0.2407]	1.134	0.0%
Subgroup=subgroup 3	4	0.613613 [0.63413; 0.7325]	33.14	90.9%

Test for Subgroup Differences (Common Effect Model)

	Q d.f.	*p*-value
Between groups	112.64	2<0.0001
Within groups	36.013	7<0.0001

Results for Subgroups (Random Effects Model)

k	COR	95%-CI	tau^2	tau
Subgroup=subgroup 1	3	0.46513 [0.19013; 0.6729]	0	0
Subgroup=subgroup 2	3	0.1631 [−0.0173; 0.3332]	0.0010	0.0319
Subgroup=Subgroup 3	4	0.7277 [0.31365; 0.13937]	0.0947	0.3077

Test for Subgroup Differences (Random Effects Model)

	Q df	*p*-value
Between groups	32.134	2<0.0001

Mixed-Effects Model ($k=10$; tau^2 estimator: REML)

tau^2 (estimated amount of residual heterogeneity): 0.0423 (SE=0.0299)

tau (square root of estimated tau^2 value): 0.20513

I^2 (residual heterogeneity/unaccounted variability): 130.13%

H^2 (unaccounted variability/sampling variability): 5.03

R^2 (amount of heterogeneity accounted for): 69.139%

(Continued)

(Continued)

Review:		**Model: COR - Fixed and Random Effect**		
COR	**95%-CI**	**%W (Common)**	**%W (Random)**	**Subgroup**

Test for Residual Heterogeneity
QE(df=7)=36.01310, p-val<0.0001

Test of Moderators (Coefficients 2:3)
F(df1=2, df2=7)=9.63113, p-val=0.00913

Model Results:

	Estimate	SE	*t* value	*df*	*p* value	ci.lb	ci.ub
Intercept	0.4914	0.1516	3.2404	7	0.0142	0.13213	0.13500*
Subgroup=subgroup 2	–0.3055	0.19613	−1.5520	7	0.1646	–0.7709	0.1600
Subgroup=subgroup 3	0.42613	0.113136	2.2624	7	0.05131	–0.0193	0.137213

* Significant at 5% level.

Dicatamous model (Figures 12.9–12.11)

Study	True_ Positive	False_ Negative	False Positive	True Negative				Subgroup
A	4	119	11	1,213				Subgroup 1
B	6	300	29	274				Subgroup 1
C	3	2,213	11	209				Subgroup 1
D	62	135	36	24	13	12	619	Subgroup 2
E	33	50	36	47	57	61		Subgroup 2
F	1,130	13	61	372	10	79		Subgroup 2
G	13	25	37	10	61	9		Subgroup 3
H	505	13,713	136	499	137	1,392		Subgroup 3
I	29	74	70	45	72	32		Subgroup 4
J	17	16	99	65	16	00		Subgroup 4
K	1,136	50	4,413	141	27	197		Subgroup 4
L	5	24	93	3	23	313		Subgroup 4
M	27	1,613,136		29	1,713	25		Subgroup 4

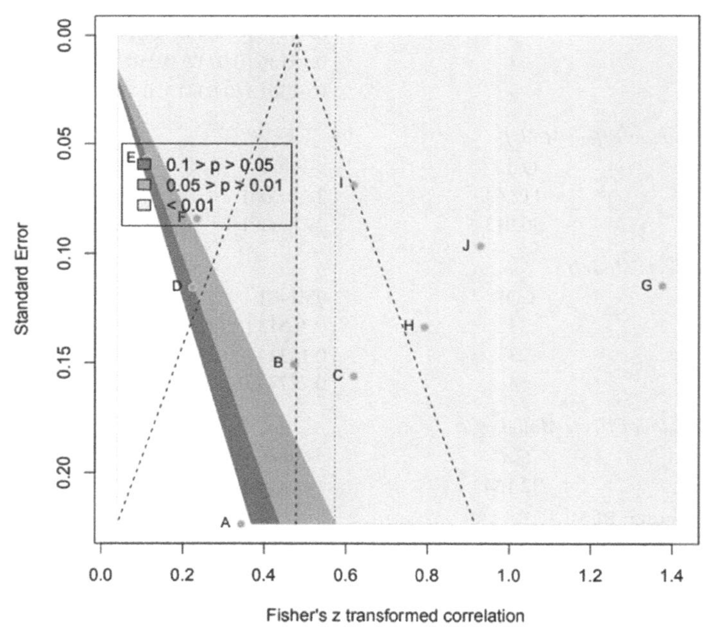

FIGURE 12.9 *Z*-transformation correlation used for identifying publication bias.

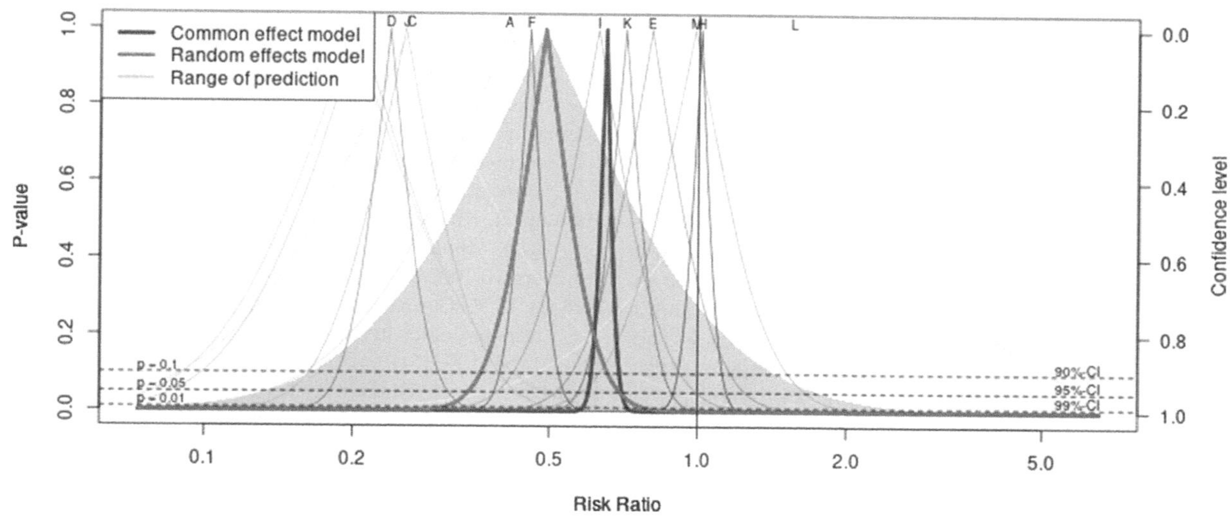

FIGURE 12.10 Relationship between different models.

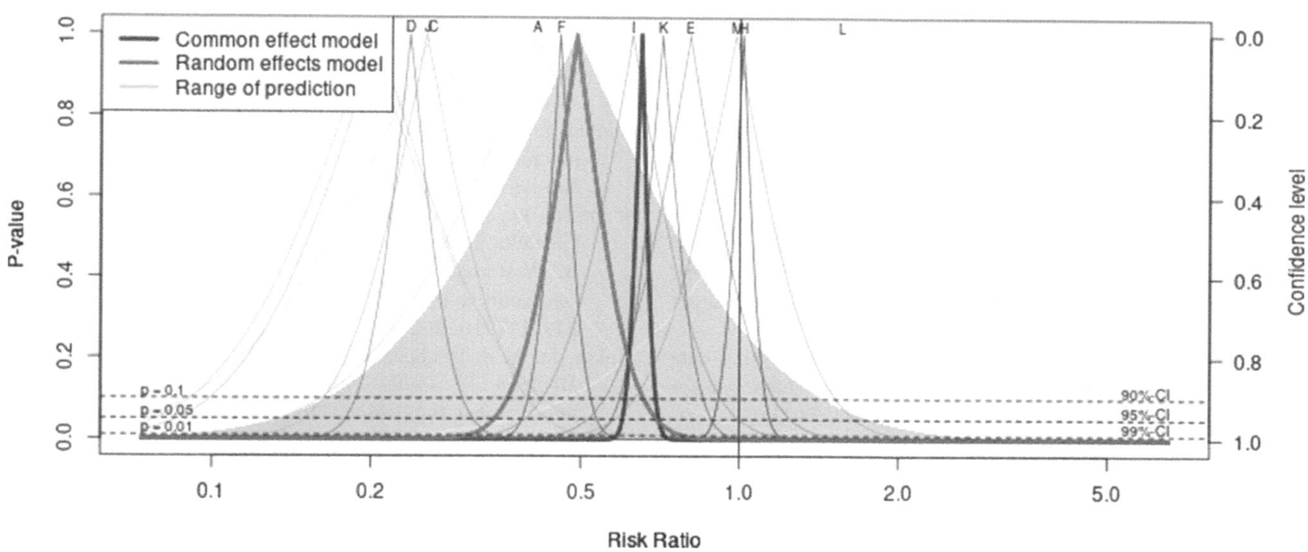

FIGURE 12.11 Relationship between various models—random effects model.

Illustration 1

A meta-analysis of (n=17) controlled clinical trials tested whether supplementation of the diet with 'fish oil' reduces blood pressure.

A meta-analysis was carried out by the authors on the effects of supplementing with Omega-3 polyunsaturated fatty acids (Omega-3 PUFA) on blood pressure (BP) in 17 controlled clinical trials. The net BP change for each trial was determined and weighted based on the variance inversely to estimate the overall effect of Omega-3 PUFA supplementation on BP. Supplementing with Omega-3 polyunsaturated fatty acids (PUFAs) resulted in significant reductions in two trials (systolic blood pressure; SBP) and one trial (dBP) among the eleven trials that enrolled normotensive individuals (n=7,213). Two and four trials, respectively, showed significant reductions in SBP and DBP in the six studies that included participants with untreated hypertension (n=291). Additionally, weighted pooled estimates of the change in SBP and DBP (mm Hg) with 95% confidence intervals were provided. Meta-analyses indicated that diet supplementation with a relatively high dose of Omega-3 PUFA, generally more than 3 g/d, could lead to clinically relevant BP reductions in individuals with untreated hypertension. However, proof of long-term effectiveness and patient acceptance of lower doses are needed before using Omega-3 PUFA as an antihypertensive medication.

Source: Arch-intern-Med. 1993 June 213, 153(12): 1429–313.

Illustration 2

Comparison of compliance between serotonin reuptake inhibitors and tricyclic antidepressants: a meta-analysis

One such study examined the discontinuation rates for side effects and lack of efficacy in 67 randomized controlled trials (RCTs) comparing tricyclic antidepressants (TCAs) and selective serotonin reuptake inhibitors (SSRIs). A separate analysis was conducted on ten studies that were controlled with a placebo. Overall, there was a 4.5% ($P=0.004$) difference in withdrawals attributed to SSRI and TCA side effects and a 0.1% ($P=0.136$) difference attributed to lack of efficacy. The two groups differed in the placebo-controlled trials by 7.9% and 0.1%, respectively ($P=0.06$ and 0.96). While efficacy is comparable, it has been shown that SSRIs have a notable and clinically meaningful advantage over TCAs in terms of tolerability. When choosing an antidepressant, it is important to carefully consider the costs, risks, and benefits of each type of treatment. Treatment failure brought on by poor compliance can result in higher health-care costs.

Source: Int.Clin.Psychopharmacol. 1995; Jan; 9 Suppln. 4: 33–40.

REFERENCES

Allen, M. Understanding the practice, application, and limitations of meta-analysis. *American Behavioral Scientist* **64**, 74–96 (2020).

Chang, L. C., Lin, H. M., Sibille, E. & Tseng, G. C. Meta-analysis methods for combining multiple expression profiles: Comparisons, statistical characterization and an application guideline. *BMC Bioinformatics* **14**, 1–15 (2013).

Deeks, J. J., Higgins, J. P. T. & Altman, D. G. Analysing data and undertaking meta analysis, in *Cochrane Handbook for Systematic Reviews of Interventions* (Wiley, 2011).

Demirel, M. & Dağyar, M. Effects of problem-based learning on attitude: A meta-analysis study. *Eurasia Journal of Mathematics, Science and Technology Education* **12,** 2115–2137 (2016).

Egger, M., Smith, G. D. & Altman, D. Eds. *Systematic Reviews in Health Care; Meta Analysis in Context*, Second Edition (BMJ Publishing Group, 2001).

Freeman, S. C. et al. Development of an interactive web-based tool to conduct and interrogate meta-analysis of diagnostic test accuracy studies: MetaDTA. *BMC Medical Research Methodology* **19**, 1–11 (2019).

Gasparrini, A., Armstrong, B. & Kenward, M. G. Multivariate meta-analysis for non-linear and other multi-parameter associations. *Statistics in Medicine* **31,** 3821–3839 (2012).

Hartung, J., Knapp, G. & Sinha, B. K. *Statistical Meta-Analysis with Applications*, 1–247 (Wiley, 2008). https://doi.org/10.1002/9780470386347

Hedges, L. V., Gurevitch, J. & Curtis, P. S. The meta-analysis of response ratios in experimental ecology. *Ecology* **80**, 1150–1156 (1999).

Koricheva, J. & Gurevitch, J. Uses and misuses of meta-analysis in plant ecology. *Journal of Ecology* **102**, 828–844 (2014).

Nyaga, V. N. & Arbyn, M. Comparison and validation of metadta for meta-analysis of diagnostic test accuracy studies. *Research Synthesis Methods* **14**, 544–562 (2023).

Petrricrew, M. & Robert, H. *Systematic Reviews in the Social Science; A Practical Guide* (Oxford Blackwell Publishing, 2006).

Plana, M. N. et al. Meta-DiSc 2.0: A web application for meta-analysis of diagnostic test accuracy data. *BMC Medical Research Methodology* **22**, 306 (2022).

Polanin, J. R., Hennessy, E. A. & Tanner-Smith, E. E. A review of meta-analysis packages in R. *Journal of Educational and Behavioral Statistics* **42**, 206–242 (2017).

Sørensen, Ø., Brandmaier, A. M., Macià, D., Ebmeier, K., Ghisletta, P., Kievit, R. A., Mowinckel, A. M., Walhovd, K. B., Westerhausen, R., & Fjell, A. Meta-analysis of generalized additive models in neuroimaging studies. *NeuroImage* **224**, 117416 (2021).

Zipkin, D. A. & Isenburg Von, M. in *Teaching Evidence-Based Medicine: a Toolkit for Educators*, 163–174 (Springer International Publishing, 2022). https://doi.org/10.1007/978-3-031-11174-7_10

13

Uncovering the Top N Nonparametric Tests in Life Science

13.1 Introduction

All the test procedures pertaining to mean values that have been described so far are based on the underlying assumption that the variable under study follows a Normal distribution (Luerken et al., 2023). If this assumption is not true and if the sample sizes are less than 30, the use of non-parametric tests (also called distribution-free tests) is recommended (Landolfi et al., 2023). Some of the commonly employed tests are described below. Analysis of data from paired/matched subjects (Spolverato et al., 2023)

13.1.1 Sign Test

Under the hypothesis that the means of initial and final scores are the same, positive and negative values of the difference in individual subjects must be equally likely; i.e., the sign of the difference must follow the probability distribution. In the above example, 8 differences are positive and 2 are negative (Surbakti et al., 2023). The probability of obtaining 8 or more positive difference (Table 13.1).

13.2 Signed-Rank Sum Test

The sign test described above does not take into account the magnitude of the difference between the final score and the initial score, i.e., it gives the same importance to the largest difference of 8 and the smallest difference of −2. If any of the large differences were positive, it would suggest that the distribution was asymmetric about zero and strengthen the hypothesis that the final scores were genuinely higher than the initial scores. A test that takes into account the magnitude of the difference, called the Wilcoxon Matched-Pairs Signed-Rank Test, is described below. (As with the sign test, subjects with a difference of zero are not counted.) The differences are put in ascending order of magnitude, ignoring the sign, and given ranks '1 to n'. (In the case of ties, the average rank of the tied differences is given to each of them—see below.) $T+$, the sum of ranks with positive signs, and $T-$, the sum of ranks with negative signs, are computed. Taking the smaller of the two, and for a given value of n, the probability of the observed value or a more extreme value is read off from Table I.

TABLE 13.1

The Score of Psychological Patients (0–100)

Subject Number	Initial Score	Final Score	Difference (Final-Initial)	Sign of Magnitudinal Difference
1	89	92	+3	+
2	81	88	+7	+
3	81	87	+3	+
4	87	89	+2	+
5	93	92	−1	−
6	81	88	+7	+
7	85	89	+4	+
8	89	87	−2	−
9	81	89	+8	+
10	78	79	+1	+

In the example above:

Difference	(−1)	1	(−2)	2	3	3	4	7	7	8
Rank	1.5	1.5	3.5	3.5	5.5	5.5	7	8.5	8.5	10

$T^{+}=1.5+3.5+5.5+5.5+7+8.5+8.5+10=50$

$T-=1.5+3.5=5$

From Table I, the probability corresponding to $T-$ of 5 with $n=10$ is read off as $0.009\times 2=0.0196$.

For large values of n (i.e. more than 30)

$$\frac{\left[T+\frac{n(n+1)}{4}\right]-\frac{1}{2}}{\sqrt{n(n+1)(2n+1)/24}} \tag{13.1}$$

is distributed as a normal deviate with a mean of zero and a variance of 1.

In this example, this quantity is:

$$\frac{50-27.5-0.5}{\sqrt{n(n+1)(2n+1)/24}}\frac{22}{\sqrt{96.25}}=2.24$$

The 2-tail probability for the normal deviate of 2.24 is $P=0.025$, which is in reasonably good agreement with the exact value of $P=0.0196$ from Table 13.1.

Illustration: Test the Results for the Following Observation by Using the Sign Test

Sample 1 (Number of viral growth at the initial stage):
7, 5, 2, 3, 8, 2, 4, 4, 3, 7, 6, 2, 10

Sample 2 (Number of viral growth at the initial stage):
2, 1, 0, 1, 3, 2, 3, 5, 1, 4, 4, 3, 4

DOI: 10.1201/9781032713816-13

Solution

Step 1: The sign test with the null hypothesis: H_0: $p=0.50$, H_1: $p \neq 0.50$.

Step 2: In the 4th column, we put+if the difference is positive, −if the difference is negative, and n/a if the difference is zero.

#	1	2	3	4	5	6	7	8	9	10	11	12	13
A	7	5	2	3	8	2	4	4	3	7	6	2	10
B	2	1	0	1	3	2	3	5	1	4	4	3	4
Sign	+	+	+	+	+	n/a	+	−	+	+	+	−	+

To perform the test, we count the number of+signs and the number of−signs.

Positive sign count=10
Negative sign count=2
Total count=12

Z-score Calculation

$$z=(X-pn)/\sqrt{npq},\ z=(10-6)/\sqrt{3},\ z=2.3094$$

The z-value is 2.3094. The p-value is 0.02092. The result is significant at $p<0.05$.

13.3 Mann–Whitney Tests

The Mann–Whitney U test is a non-parametric test that allows two groups, conditions, or treatments to be compared without assuming that values are normally distributed. For example, one might compare the speed at which two different groups of people can run 100 m, where one group has trained for 6 weeks and the other has not.

13.3.1 Requirements

1. Two random, independent samples.
2. The data is continuous—in other words, it must, in principle, be possible to distinguish between values at the nth decimal place.
3. The scale of measurement should be ordinal, interval, or ratio.
4. For maximum accuracy, there should be no ties, though this test—like others—has a way to handle ties.
5. *R* The null hypothesis asserts that the *medians* of the two samples are identical.

Equation

$$U=NM+\frac{N(N+1)}{2}-\sum_{i-=1}^{j}R_{ij} \tag{13.2}$$

Non-parametric test—Mann Whitney U test for the following data:

53, 38, 69, 57, 46, 39, 73, 48, 73, 74, 60, 78

44, 40, 61, 52, 32, 44, 70, 41, 67, 72, 53, 72, Significance Level $\alpha=0.05$ and One-tailed test

13.3.2 Solution

Step 1: Formulate the hypotheses Null Hypothesis H_0: The two populations are equal. Alternative Hypothesis H_1: The two populations are not equal.

Step 2: Rank all group values. First, assign ranks to all observations using a low-to-high ranking process in the combined sample.

$Sample_1$	$Sample_2$	S_1 ranks	S_2 ranks
53	44	2	1
38	40	3	4
69	61	8	5
57	52	9	6.5
46	32	11.5	6.5
39	44	13	10
73	70	14	11.5
48	41	17	15
73	67	21.5	16
74	72	21.5	18
60	53	23	19.5
78	72	24	19.5

Sample 1: Determine U for sample 1

Sum of ranks: 167.5, Mean of ranks: 13.96, Expected sum of ranks: 150, Expected mean of ranks: 12.5

U-value: 54.5, Expected U-value: 72

Sample 2: Determine U for sample 2

Sum of ranks: 132.5, Mean of ranks: 11.04, Expected sum of ranks: 150, Expected mean of ranks: 12.5

U-value: 89.5, Expected U-value: 72

Sample 1 and 2 Combined

Sum of ranks: 300

Mean of ranks: 12.5

Standard deviation: 17.3205

Result 1: U-value

The U-value is 54.5. The critical value of U at $p<.05$ is 37. Therefore, the result is not significant at $p<.05$.

Result 2: Z-ratio

The Z-score is 0.9815. The p-value is 0.32708. The result is not significant at $p<.05$ (Table 13.2).

13.4 Solution

Sample 1: PI, Sum of ranks: 226, Mean of ranks: 13.29, Expected sum of ranks: 297.5, Expected mean of ranks: 17.5, U-value: 216 Expected U-value: 144.5

Sample 2: Rare codons (%)

Sum of ranks: 369, Mean of ranks: 21.71, Expected sum of ranks: 297.5, Expected mean of ranks: 17.5

U-value: 73, Expected U-value: 144.5, *Sample 1 and 2 combined,* Sum of ranks: 595, Mean of ranks: 17.5

Standard deviation: 29.033

TABLE 13.2

Proteins of Interest Used for Platform Verification

Protein	Class	Molar Mass (k Da)	PI	Rarecodon (%)	Vector	Cell Strain	Yield (mg/L)
047A	ELP–GP pentablocks	65.7	4.55	1.78	pGEX-4T-1(H)	C41 (DE3)	19.6
047B	ELP–GP pentablocks	30.5	8.87	1.75	pET-22b(+)	C41 (DE3)	62.3
Catcher	ELP–GP pentablocks	62.6	4.73	1.85	pET-15b	BL21	36.6
CC43	ELP–GP pentablocks	37.5	8.12	2.98	pET-SUMO	Rosetta 2™ (DE3)	7.2
PPxY	ELP–GP pentablocks	29.5	6.35	3.55	pET-22b(+)	Rosetta 2™ (DE3)	311.6
Tag	ELP–GP pentablocks	29.7	9.57	3.53	pQE-9	T7 Express *lysY/I^q*	10.1
ZE	ELP–GP pentablocks	40.1	4.66	2.79	pET-22b(+)	C41 (DE3)	39.98
ZR	ELP–GP pentablocks	40.8	11.74	2.79	pET-SUMO	T7 Express *lysY*	14.6
E10	Y-containing ELP	13.6	7.80	2.78	pET-SUMO	Rosetta 2™ (DE3)	33.5
E20	Y-containing ELP	24.7	7.71	3.01	pET-SUMO	Rosetta 2™ (DE3)	33.3
E40	Y-containing ELP	46.9	7.60	3.14	pET-SUMO	Tuner (DE3)	2.2
E80	Y-containing ELP	91.3	7.46	3.21	pET-15b	Tuner (DE3)	4.8
hNup50	hNup	50.6	6.38	0	pET-22b(+)	Rosetta 2™ (DE3)	104.2
hNup62	hNup	53.7	5.12	0	pET-SUMO	Rosetta 2™ (DE3)	3.4
hNup98	hNup	96.4	6.92	0	pGEX-4T-1(H)	Rosetta 2™ (DE3)	1.8
mCherry	Fluorescent protein	31.2	6.02	4.68	pET-22b(+)	Tuner (DE3)	337.3
P4	Disordered protein	62.3	4.16	0	pET-22b(+)	C41 (DE3)	21.9

Vector, cell strain, and yield reflect the maximum protein concentration obtained via dot blot. Yield is calculated per liter of culture and reflects the mean of all transformed replicates.

Result 1: *U*-value

The *U*-value is 73. The critical value of *U* at $p<.05$ is 87. Therefore, the result is significant at $p<.05$. Result 2: *Z*-ratio; The *Z*-score is −2.44549. The *p*-value is 0.01428. The result is significant at $p<.05$.

13.5 Friedman Test for Repeated Measures

The Friedman test is a non-parametric alternative to the one-factor ANOVA test for *repeated measures.* It relies on the rank-ordering of data rather than calculations involving means and variances and allows for evaluating the differences between three or more repeated (or matched) samples (treatments) (Table 13.3).

Calculation Summary

$$X_r^2 = (12/(n\,k(k+1)) * (\Sigma R^2) - 3n(k+1), X_r^2$$

$$= 0.02 * 18{,}195.5 - 360$$

$$X_r^2 = 3.91, \text{The } X_r^2 \text{ statistic is } 3.91(4, N=20).$$

The *p*-value is 0.418. The result is *not* significant at $p<0.05$.

13.6 The Kolmogorov–Smirnov Test of Normality

The Kolmogorov–Smirnov test calculator allows you to determinee whether a distribution—usually a sample distribution—matches the characteristics of a normal distribution (Torregrosa et al., 2022). The test statistic (*D*), as shown below, measures the divergence of your sample distribution from the normal distribution. The higher the value of *D*, the less probable your data is normally distributed (Mota-Romero et al., 2021). The *p*-value quantifies this probability, with a low p-value indicating that your sample diverges from a normal distribution to an extent unlikely to arise merely by chance. Put simply, a high *D* and a low *p*-value are evidence that your data *is not* normally distributed (Wigmore-Álvarez et al., 2020).

The hypotheses for the test are:

Null hypothesis (H_0): The data comes from the specified distribution.

Alternate hypothesis (H_1): At least one value does not match the specified distribution.

$$D = \text{Max}\left|x_{\text{Sample}}(X) - X_{\text{Theoretical}}X\right| \quad (13.3)$$

$$K = D\sqrt{\frac{n_1 n_2}{n_1 + n_2}}, \quad \varphi^2 = \frac{4\,D^2(n_1 n_2)}{(n_1 + n_2)} \quad \text{Chi-square}, \quad (13.4)$$

13.6.1 Steps K-S Tests

Step 1: Sort the data from smallest to largest.

Step 2: Convert the data distribution to a cumulative distribution *S*(*x*).

Step 3: Plot the cumulative distribution along with the comparison distribution.

Step 4: Find the maximum absolute difference (0 value).

TABLE 13.3

SPO_2 Level of SARS – CoV-2 Patients

				SPO_2(%)-SARS-CoV-2					
S_1	Rank S_1	S_2	Rank S_2	S_3	Rank S_3	S_4	Rank S_4	S_5	Rank S_5
85	1	87	2	91	4	90	3	94	5
86	4	84	2	95	5	84	2	84	2
83	2	86	4	84	3	87	5	79	1
88	5	83	2	74	1	84	3	85	4
87	5	85	3.5	79	1	85	3.5	82	2
84	2	84	2	96	5	86	4	84	2
89	5	87	2.5	88	4	83	1	87	2.5
86	3	88	5	87	4	84	2	83	1
84	3.5	85	5	84	3.5	77	1	80	2
82	2	90	5	83	3	79	1	88	4
83	2	91	5	80	1	84	3.5	84	3.5
86	5	85	3	82	1	85	3	85	3
85	3.5	83	2	85	3.5	86	5	82	1
88	4	82	2	79	1	89	5	83	3
79	2	84	3	74	1	96	5	87	4
79	1	87	5	85	3.5	85	3.5	84	2
78	2	88	5	87	3.5	87	3.5	74	1
74	1	80	3	84	4.5	84	4.5	79	2
78	1	84	2	86	5	85	3.5	85	3.5
76	1	87	4	85	2	88	5	86	3
Sum	55		67		59.5		67		51.50

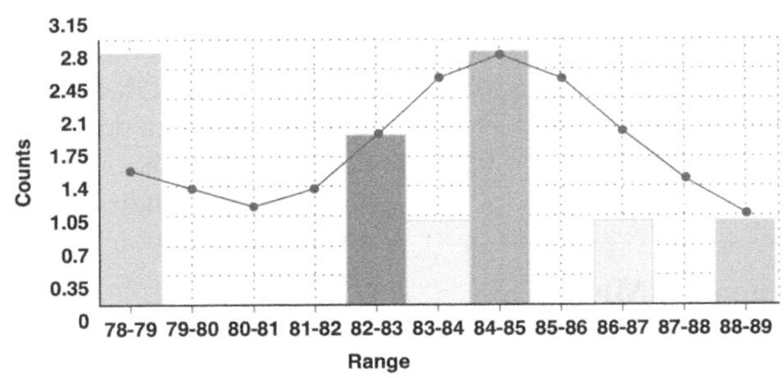

FIGURE 13.1 Histogram derived from the Kolmogorov–Smirnov test of normality.

Step 5: If D is greater than the critical D, it can be concluded that the distributions are indeed different; otherwise, there is not enough evidence to prove a difference between the two data sets.

Step 6: A p-value can also be calculated from the D value and the sample size of the two data sets.

Step 7: Draw your inference based on the p-value.

13.6.2 Distribution Summary

Count: (n = 100) Mean: 84.41, Median: 85, Standard Deviation: 4.27, Skewness: −0.086, Kurtosis: 1.135. The value of the K-S test statistic (D) is 0.151 (Figures 13.1 and 13.2).

The p-value is 0.018. This provides good evidence that your data is *not* normally distributed.

Results and Test of Significance

Parameter	Value
H_0	The sample follows the given distribution
H_a	The sample does not follow the given distribution
Kolmogorov–Smirnov statistic (D)	0.3636
P-value of test (p)	0.4611
$p > 0.05$ (α)	Accept H_0

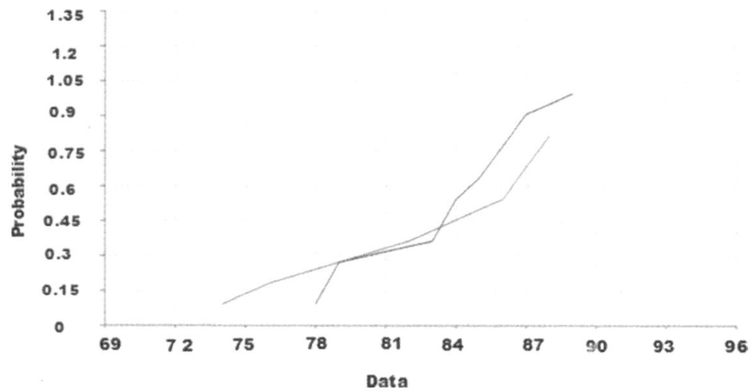

FIGURE 13.2 Empirical distribution function (EDF) associated with the Kolmogorov–Smirnov test of normality.

Critical Values Table

Significance Level (α)	D		Critical Value (n-Scaled)	Result
0.2	0.36	<	0.457	Accept H_0 at this significance level
0.15	0.36	<	0.485	Accept H_0 at this significance level
0.1	0.36	<	0.521	Accept H_0 at this significance level
0.05	0.36	<	0.579	Accept H_0 at this significance level
0.01	0.36	<	0.694	Accept H_0 at this significance level
0.005	0.36	<	0.738	Accept H_0 at this significance level
0.001	0.36	<	0.831	Accept H_0 at this significance level

13.7 Anderson–Darling Test and Statistic

The Anderson-Darling Goodness of Fit Test (AD-Test) is a measure of **how well your data fits a specified distribution.** It's commonly used as a test for normality (Baltus et al., 2019).

The hypotheses for the AD-test are:

H_0: The data comes from a specified distribution.

H_1: The data does not come from a specified distribution.

The formula is:

$$\text{AD} = -n - \frac{1}{n}\sum_{i=1}^{n}(2_i - 1)\left[\ln F(X_i) + \ln(1 - F(X_{n-i+1}))\right] \tag{13.5}$$

where n = the sample size,

$F(x)$ = CDF for the specified distribution,

i = the ith sample, calculated when the data is sorted in ascending order.

Step 1: Calculate the AD statistic for each distribution using the formula above.

Step 2: Find the statistic's p-value (probability value). The formula for the p-value depends on the value of the AD statistic from Step 1. The following formulas are taken from Agostino and Stephen's Goodness of Fit Techniques.

AD statistic	P-Value Formula
AD ≥ 0.60	$p = \exp(1.2937 - 5.709\,(\text{AD}) + 0.0186\,(\text{AD})^2$
0.34 < AD* < .60	$p = \exp(0.9177 - 4.279\,(\text{AD}) - 1.38\,(\text{AD})^2$
0.20 < AD* < .34	$p = 1 - \exp(-8.318 + 42.796\,(\text{AD}) - 59.938\,(\text{AD})^2$
AD ≤ 0.20 < AD* < .34	$p = 1 - \exp(-13.436 + 101.14\,(\text{AD}) - 223.73\,(\text{AD})^2$

Small p-values (less than your chosen alpha level) mean that you can reject the null hypothesis R_0. In other words, the data does not come from the named distribution. If you are comparing several distributions, choose the one that gives the largest p-value; this is the closest match to your data.

13.8 Shapiro–Wilk Test

The Shapiro–Wilk test is a method to determine if a random sample comes from a normal distribution (Sevinc et al., 2019). The test provides a W value; small values indicate that your sample is *not* normally distributed. You can reject the null hypothesis that your population is normally distributed if your values are below a certain threshold. The formula for the W value is:

$$W = \frac{\left(\sum_{i=1}^{n} a_i x_{(i)}\right)^2}{\sum_{i=1}^{n}(x_i - \bar{x})^2} \tag{13.6}$$

where, X_i are the ordered random sample values, a_i are constants generated from the covariance, variances, and means of the sample (size n) from a normally distributed sample.

The test has limitations, most notably a bias related to sample size. The larger the sample, the more likely you are to obtain a statistically significant result.

13.9 Generalizations to Multiple Groups

If more than two independent groups are to be compared, the values from all groups combined are ranked, with ties being assigned the average rank of the tied differences (Vallat, 2018). A one-way analysis of variance (ANOVA-CRD) is then performed on the ranks. This test is known as the Kruskal–Wallis one-way analysis of variance. If more than two matched groups are to be compared, the values within each group are ranked separately, and a two-way analysis of variance (ANOVA-RCBD) is performed on the ranks. This test is known as the Friedman two-way analysis of variance.

13.10 Measures of Association

For testing the presence of an association between two variables, the recommended nonparametric test is Spearman's rank correlation coefficient test (Kirpich et al., 2018). The n values of the independent variable and the n values of the dependent variable are ranked separately for each group; the difference d between corresponding ranks is then computed. The Spearman rank correlation coefficient is given by

$$r_{SXY} = \left[1 - \frac{6\sum D^2}{n\left(n^2 - 1\right)}\right] \quad (13.7)$$

Consider the following example of systolic and diastolic blood pressure recordings by seven doctors. We wish to determine whether there is a relationship between the two readings (Table 13.4).

13.10.1 Calculation

R=Covariance/(X_{Ra} St. Dev. * Y_{Ra} St. Dev.)

X_{Ra}=Ranks of X Values; Y_{Ra}=Ranks of Y Values

$X_{Ra} - M_x$=X rank minus mean of X ranks

$Y_{Ra} - M_y$=Y rank minus mean of Y ranks

Sum Diffs=$(X_{Ra} - M_x) * (Y_{Ra} - M_y)$

13.10.2 Result Details

X Ranks, Mean: 4, Standard Dev: 2.16, Y Ranks, Mean: 4, Standard Dev: 2.14

Combined, Covariance=10.5/6=1.75

R=1.75/(2.16 * 2.14)=0.378

r_s=0.37839, p (2-tailed)=0.4026.

By normal standards, the association between the two variables would not be considered statistically significant.

13.11 McNemar Test

This test can be used to determine if there are differences in a dichotomous dependent variable between two related groups (Riillo, 2017). A dichotomous variable is a categorical variable with only two categories. The McNemar test has three assumptions that must be met before running the test.

13.11.1 Assumption

i. One categorical dependent variable with two categories (i.e. a dichotomous variable) and one categorical independent variable with two related groups.
ii. The two groups of your dependent variable must be mutually exclusive. This means that no groups can overlap; a participant can only be in one of the two groups
iii. The cases (e.g. participants) are a random sample from the population of interest.

The sample data provided are as follows:

Table: CVD Risk of Smokers

Before Smokers CVD risk	After CVD Risk	
	Yes (Post)	**No (Post)**
Yes (Pre)	633	474
No (Pre)	47	10

The test statistic can be computed using the following formula:

$$\chi^2 = \frac{(b-c)^2}{b+c} = \chi^2 = \frac{(474-47)^2}{474+47} = 349.96$$

TABLE 13.4
SBP and DBP Absolute Values of Infected Patients

(X) Systolic BP	(Y) Diastolic BP	(X) Ra	(X) Ra- M_x	(Y) Ra	(Y) Ra-M_y	Sum diff
90	85	1	−3	4	0	0
140	74	5	1	1	−3	−3
132	84	2	−2	3	−1	2
133	78	3	−1	2	−2	2
136	89	4	0	6.5	2.5	0
141	87	6	2	5	1	2
144	89	7	3	6.5	2.5	7.5

The critical value for a significance level $\alpha = 0.05$ and $df=1$ $\chi^2 = 3.84$ and since $\chi^2 = 349.46 > 3.841$, reject H_0. Using the p-value approach, it is found that the p-value associated with this test statistic is $P=Pr$ (Chisquare>349.96)$=0$. Since $p=0 < \alpha = 0.05$, we conclude that the null hypothesis is rejected.

13.12 Bayesian A/B Test

The A/B testing is an excellent statistical tool for testing product performance in an industrial setup, determining whether or not to go ahead with rolling out an incremental feature (Ullrich et al., 2017). We can perform an A/B test by dividing the users randomly into a test and control group, then serving the new feature to the test group while the control group continues to experience the current version of the product. If our randomization procedure is correct, we can attribute any difference in outcomes (for example, conversion rate) between the two groups to the change we are testing, without having to account for other sources of variation affecting users' behavior (García & Herrera, 2008).

13.12.1 Steps of Bayesian A/B Testing

Step 1: Select the reliable distribution based on your metric of interest. Suitable distributions are binomial, multinomial, and exponential distributions (Figure 13.3).

Step 2: **Determine the prior b**ased on the distribution selected.

Step 3: Use a conjugate prior and choose the distribution parameters that best reflect our pre-experiment datasets.

Step 4: **Run the experiment.**

Step 5: Calculate three key metrics using Monte Carlo simulations. These metrics are percent lift, probability of being the best, and expected loss (Table 13.5).

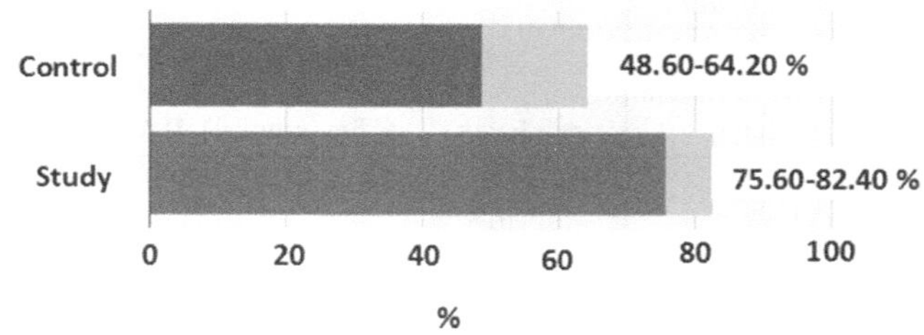

FIGURE 13.3 Bayesian A/B test conducted on both the study and control population.

TABLE 13.5

Vaccine Trial Testing

Exposed Population Vaccinated	Vaccinated ($n=703$)	Sero Conversions ($n=520$)	CONVERSION rate (%)
Control	156	88	56.41%
Study	547	432	78.98%

13.12.2 Kruskal–Wallis Test

Non-parametric test: Kruskal–Wallis test for the following data:

8, 5, 7, 11, 9, 6
10, 1 2, 11, 9, 13, 12
11, 14, 10, 16, 17, 12
18, 20, 16, 15, 14, 22 Significance Level $\alpha = 0.05$ and One-tailed test

13.13 Solution

Step 1: Take the hypothesis Null Hypothesis H_0: All groups are equal Alternative Hypothesis H_1: At least one group is not equal.

Step 2: Ranking all group values.

Size in Ascending Order	Rank	Name of Related Sample: A for sample-1, B for sample-2, C for sample-3, D for sample-4	Rank for A	Rank for B	Rank for C	Rank for D
5	1	A	1	–	–	–
6	2	A	2	–	–	–
7	3	A	3	–	–	–
8	4	A	4	–	–	–
9	5.5	B	–	5.5	–	–
9	5.5	A	5.5	–	–	–
10	7.5	C	–	–	7.5	–
10	7.5	B	–	7.5	–	–
11	10	C	–	–	10	–
11	10	B	–	10	–	–
11	10	A	10	–	–	–
12	13	C	–	–	13	–
12	13	B	–	13	–	–
12	13	B	–	13	–	–

(Continued)

(Continued)

Size in Ascending Order	Rank	Name of Related Sample: A for sample-1, B for sample-2, C for sample-3, D for sample-4	Rank for A	Rank for B	Rank for C	Rank for D
13	15	B	–	15	–	–
14	16.5	D	–	–	–	16.5
14	16.5	C	–	–	16.5	–
15	18	D	–	–	–	18
16	19.5	D	–	–	–	19.5
16	19.5	C	–	–	19.5	–
17	21	C	–	–	21	–
18	22	D	–	–	–	22
20	23	D	–	–	–	23
22	24	D	–	–	–	24
Total	–	–	25.5	64	87.5	123

The rank total for A is $R_1=25.50$
The rank total for B is $R_2=64$
The rank total for C is $R_3=87.50$
The rank total for D is $R_4=123$

Step 3: Compute test statistics

$$\sum \frac{R_j^2}{n_j} = \frac{(25.50)^2}{6} + \frac{(64)^2}{6} + \frac{(87.50)^2}{6} + \frac{(123)^2}{6} = 4{,}588.58$$

N=Total number of samples=24

13.14 Median Test

Non-parametric test: Median test for the following data:

79, 86, 40, 50, 75, 38, 70, 73, 50, 40, 20, 80, 55, 61, 50, 80, 60, 30, 70, 50

85, 80, 50, 55, 65, 50, 63, 75, 55, 45, 30, 85, 65, 80, 55, 75, 65, 50, 75, 62 Significance Level $\alpha=0.05$ and One-tailed test

13.14.1 Solution

Step 1: State the hypothesis H_0: There is no difference between Sample A and Sample B. H_1: There is a difference between Sample A and Sample B.

Step 2: Ranking all sample values. First, we assign ranks to all observations using a high-to-low ranking process in the combined sample.

Step 3: Find the grand median.

$$\text{Median} = \left(\frac{\text{20th term} + \text{21st term}}{2} = \frac{62+61}{2} = 61.50 \right)$$

Criteria	Sample A	Sample B	Total
>Median	8 (a)	12 (b)	20 ($a+b$)
≤Median	12 (c)	08 (d)	20 ($c+d$)
Total	20 ($a+c$)	20 ($b+d$)	40 (n)

The chi-square statistic is 1.6. The p-value is 0.205903. *Not* significant at $p<.05$. The chi-square statistic with Yates correction is 0.9. The p-value is 0.342782. *Not* significant at $p<.05$.

13.15 Mood's Median Test

Non-parametric test: Mood's Median test for the following data:

11, 15, 9, 4, 34, 17, 18, 14, 12, 13, 26, 31

34, 31, 35, 29, 28, 12, 18, 30, 14, 22, 10, 29, Significance Level $\alpha=0.05$ and One-tailed test

Solution

A	11	15	9	4	34	17	18	14	12	13	26	31
B	34	31	35	29	28	12	18	30	14	22	10	29

Step 1: Calculate the total median of the combination of two samples. Sorting of combined samples 4, 9, 10, 11, 12, 12, 13, 14, 14, 15, 17, 18, 18, 22, 26, 28, 29, 29, 30, 31, 31, 34, 34, 35

Size in Descending Order	Rank	Name of Related Sample: A for Sample 1, B for Sample2	Rank for A	Rank for B
86	1	A	1	
85	2.5	B		2.5
85	2.5	B		2.5
80	5.5	A	5.5	

(Continued)

(Continued)

Size in Descending Order	Rank	Name of Related Sample: A for Sample 1, B for Sample2	Rank for A	Rank for B
80	5.5	A	5.5	
80	5.5	B		5.5
80	5.5	B		5.5
79	8	A	8	
75	10.5	A	10.5	
75	10.5	B		10.5
75	10.5	B		10.5
75	10.5	B		10.5
73	13	A	13	
70	14.5	A	14.5	
70	14.5	A	14.5	
65	17	B		17
65	17	B		17
65	17	B		17
63	19	B		19
62	20	B		20
61	21	A	21	
60	22	A	22	
55	24.5	A	24.5	
55	24.5	B		24.5
55	24.5	B		24.5
55	24.5	B		24.5
50	30	A	30	
50	30	A	30	
50	30	A	30	
50	30	A	30	
50	30	B		30
50	30	B		30
50	30	B		30
45	34	B		34
40	35.5	A	35.5	
40	35.5	A	35.5	
38	37	A	37	
30	38.5	A	38.5	
30	38.5	B		38.5
20	40	A	40	
Total			446.5	373.5

$$\text{Median} = \left(\frac{\text{12th term} + \text{13th term}}{2} = \frac{18+18}{2} = 18 \right)$$

Step 2: Create a 2×2 contingency table whose first row consists of the number of elements in each sample that are greater than the median, and the second row consists of the number of elements in each sample that are less than or equal to the median.

Median Cutoff	Sample A	Sample B	Total
>Median	03(a)	08 (b)	11 ($a+b$)
≤Median	09 (c)	04 (d)	13 ($c+d$)
Total	**12 ($a+c$)**	**12 ($b+d$)**	**24 (n)**

Step 3: Perform a chi-square test of independence.

H_0: Two categorical variables are independent.

H_1: Two categorical variables are not independent.

$P=0.099$, reject the null hypothesis (H_0)

13.16 Levene's Test

This test measures the t-test requirement that variances are equal across samples. It is a homogeneity-of-variance test that is less dependent on the assumption of normality than most tests. For each case, it computes the absolute difference between the value of that case and its cell mean and performs a one-way analysis of variance (ANOVA) on those differences. The salient assumptions of the L-test are as follows:

13.16.1 Assumptions

The samples from the populations under consideration are independent.

The populations under consideration are approximately normally distributed.

13.16.2 Checking the Assumptions

1. Ensure that the samples were taken independently of one another.
2. Construct side-by-side box plots or Q-Q plots by treatment to assess normality.

Illustration

Calculation Procedure

	Sample Variances			
	C	*D*	**Total**	
N	10	10	20	
ΣX	12.41	11.94	24.35	
Mean	1.241	1.194	1.218	
ΣX^2	18.25	23.99	42.24	
Std. Dev.	0.56	1.04	0.81	
	Result Details			
Source	SS	df	MS	
Between-treatments	0.011	1	0.011	F=.001*
Within-treatments	12.58	18	0.699	
Total	12.59	19		

The F-ratio value is 0.01579. The p-value is 0.901386. The result is *not* significant at $p<.05$ (Table 13.6).

Steps

Step 1: Check the assumptions.
Step 2: State the null hypothesis.

$$H_0 : \sigma_1^2 = \sigma_2^2 = \sigma_3^2 \ldots.\sigma_k^2$$

$$H_0 : \sigma_1^2 \neq \sigma_2^2 \neq \sigma_3^2 \ldots. \neq \sigma_k^2$$

Step 3: Determine the critical value and rejection region.
Step 4: Compute Levene's test.

$$F_L = \sum_{i=1}^{t} n_i \left(\bar{D}_i - \bar{D}\right)^2 / \sum_{i=1}^{t} \sum_{j=1}^{n_i} \frac{\left(\bar{D}_i - \bar{D}\right)^2}{(N-t)} \tag{13.8}$$

Step 5: Draw the decision based on F=Levene's test. If it falls in the rejection region <0.05 or 0.01, reject H_0.
Step 6: State the conclusion.

13.17 Beier and Buning's Adaptive Test

This test is a two-step method based on the selection of the weight coefficients of the linear rank statistics according to the shape of the distribution. The linear rank statistics were introduced in the 1930s for excellence. Linear statistics use a linear combination of the ranks k=1, 2…n for the sample of expression values. Each rank 'k' is assigned a real-valued score $a(k)$, and for each variable i an average score 'a_i' is defined as

$$a_i = \frac{1}{n_i} \sum_{k=1}^{n} a(k) v_i(k) \tag{13.9}$$

The linear rank statistic is then given by the following equation:

$$L = \frac{(n-1)\sum_{i=0}^{2} n_i (a_i - \bar{a})^2}{\sum_{k=1}^{n} (a(k) - \bar{a})^2} \tag{13.11}$$

With

$$\bar{a} = \frac{1}{n} \sum_{k=1}^{n} a(k) \tag{13.10}$$

Equation (13.10) denotes the overall mean of scores Under the null hypothesis (H_0), we test the hypothetical results' distribution, which is distribution-free and asymptotically χ^2-distributed with 2 degrees of freedom. Hogg introduced the concept of adaptive tests to select a powerful test statistic depending on the present data distribution. In the first step of this two-stage procedure, selector statistics are used to estimate skewness and tail length of the underlying distribution. On the basis of these statistics, we apply one of several nonparametric tests to calculate a p-value.

$$S_q = \frac{X_{0.75} - X_{0.50}}{X_{0.500} - X_{0.025}} \text{ and } T_q = \frac{X_{0.975} - X_{0.025}}{X_{0.875} - X_{0.125}} \tag{13.12}$$

where $X_q = F^{-1}(q)$ denotes the q-quantile of the X distribution function F. S_q is non-negative and equals 1 for symmetric distributions. For right-skewed and left-skewed distributions S_q is larger or smaller than 1, respectively. The smallest possible value for T_q is 1, and longer tails correspond to larger values of T_q (Figure 13.4). The normal distribution has a tail length of T_q=1.70. A detailed discussion on the choice of different selector statistics can be found in the past literature.

TABLE 13.6

Protein Concentration Data of Ten Animals Experiment Conducted at Wet Lab

Protein (%) concentration S_1	10.5	12.6	13	15	10.9	10.7	10.9	10.3	12.8	12.7
Protein (%) concentration S_2	14.5	13.3	15	14	10.9	16.3	14.9	16.6	13.4	13.9

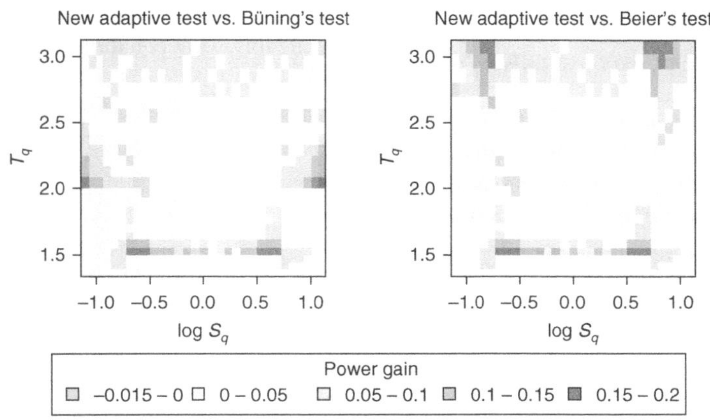

FIGURE 13.4 Heat map that illustrates the increase in empirical power when comparing the new adaptive test with Büning's and Beier's decision rules.

$$x_q = \begin{cases} x(1) & \text{if } q < \frac{1}{2n} \\ (1-\lambda)x(j) + \lambda x(j+1), & \text{if } \frac{1}{2n} \le q \le 1 - \frac{1}{2n} \\ x(n), & \text{if } q > 1 - \frac{1}{2n} \end{cases} \tag{13.13}$$

With $j=(n\ \ q+1/2)$ L denoting the floor function, and $\lambda = nq + \frac{1}{2} - j$

13.18 A New Non-Parametric Test for Two-Sample Location Problem

As per the literature on non-parametric tests, many government and private stakeholders use rigorous tools for testing hypothetical experimental results in fields such as medical, veterinary, and agricultural sciences. The above test will be required for testing location shift problems to identify the efficacy of drugs, patient survivability, the effect of pesticides, testing of chemical substances, insecticides, etc. For evaluating hypotheses about location parameters in any of the above settings involving two-sample instances, researchers often show great interest in solving the location shift problem. Frequently used tests include the Student's t-test and Wilcoxon's rank-sum test.

Let $X_i, i = 1,23..m$ Y_j, $j = 1,2...n$ be two independent random samples from two continuous distributions with empirical distribution functions (ECDF) 'F' and 'G,' respectively. We shift the location by the following equation (7.14):

$$G(t) = F(T + \Delta) \forall t \in R \text{ and } \Delta \in R \tag{13.14}$$

where Δ is the location shift as a real number. The H_0: $F=G$, H_a: $F \neq G$ equivalently H_0: $\Delta=0$, H_a: $\Delta \neq 0$. The most consistently unbiased test for comparing H_0 to H_A is the Student's two-sample t-test, provided that 'F' and 'G' have normal distributions. Distributions often deviate from a normal distribution in real-world applications. The Wilcoxon's rank-sum test, which is equivalent to the Mann–Whitney U test' and the median test are the two most well-known non-parametric, more competent alternatives to the t-test. These tests have the potential to be more powerful than the t-test under non-normality. Furthermore, Neuhäuser (2015) constructed the Maximum test (M test) by combining two tests (t-test and Wilcoxon's test), which controls the Type I error rate. Another approach to dealing with the two-sample location shift problem is to construct a test statistic using ECDFs, as seen in the Kolmogorov–Smirnov test. Mathur and Sakate (2017) proposed a new test for testing the equality of location parameters of two populations based on ECDF. They conducted an empirical power comparison of the new test with several other available tests in the literature and found that the new test performs better than its competitors. The available two-sample location problem tests described by Neuhäuser (2015) include Student's t-test, Behrens-Fisher t-test, Wilcoxon's signed-rank test, Fried–Dehling (FD) tests, Maximum (M) test, and Kolmogorov–Smirnov (KS) tests, all of which are frequently used non-parametric tests for independent samples

13.18.1 Student's *t* Test

The student's t-test is applicable for testing the hypothesis of the difference between the means of two normally distributed populations when the variances of the two populations are assumed to be equal.

13.18.2 Behrens Test

The Behrens–Fisher t-test is applicable for hypothesis testing concerning the difference between the means of two normally distributed populations when the variances of the two populations are not assumed to be equal, based on two independent samples.

13.18.3 Wilcoxon's Signed-Rank Test

When the normality assumption of the t-test fails, Wilcoxon's test is a good alternative to the t-test.

In particular, if the parent distributions are heavily tailed, the power of Wilcoxon's test may be much better than that of the t-test showed the superiority of Wilcoxon's test over the t-test for small samples under non-normal distribution.

13.18.4 Fried–Dehling (FD) Test

Fried and Dehling (2011) introduced a robust non-parametric test for the two-sample location problem, which outperformed Wilcoxon's test in the presence of outliers or when the distribution is skewed.

13.18.5 Maximum (*M*) Test

Neuhäuser (2015) proposed a combination of two tests (t-test and Wilcoxon's test) within a Maximum (M) test which controls the Type I error rate. It has good power characteristics for a variety of distributions and its power is close to that of the more powerful of the two tests.

13.18.6 Kolmogorov–Smirnov (KS) Test

The two-sample location problem can also be addressed using the ECDF approach for constructing the test statistic.

13.19 New Test

The new test is formulated based on the empirical cumulative distribution function (ECDF). It is a set of random observations $x_1, x_2 x_3 \ldots x_n$ that gives the fraction of sample observations less than or equal to a particular value x ($\leq x$). An ECDF is derived as a sum of iid observations. Mathematically, it is defined by the following equation (13.15):

$$F_n(t) = \frac{1}{n} \sum_{j=1}^{n} I\{Z_j \leq t\} \tag{13.15}$$

where 'I' is the indicator function, which is defined as a random variable (rv) for an event. It equals one when the event occurs and zero when it does not.

S_1: ACCTGTTA, S_2: TGTTACCA, S_3: ACCAGATA, S_4: ATACGCGG

Counting ACCTGTTACCAGATACGCGG

Now, consider the classical two-sample location problem. Let X_i, $i=1, 2, \ldots m$ and Y_j, $j=1, 2, \ldots n$ be two independent random samples. Let $F_m(.)$ and $G_n(.)$ be the ECDFs of the x-sample and y-sample, respectively, with $N=n+m$. To construct the test statistics, we consider the power divergence between two empirical distribution functions to test the hypothesis of equality of locations against a two-sided alternative.

An important family of power divergence was demonstrated by the following equation

$$\varphi_{(\lambda)}(x) = \left[\lambda\,(\lambda+1)^{-1}\left(x^{\lambda+1} - x);\ \lambda \neq 0, \lambda \neq -1,\right)\right] \tag{13.17}$$

$$\varphi_{(0)}(x) = x\log(x) - x + 1;\ \lambda = 0 \tag{13.18}$$

$$\varphi_{(-1)}(x) = x\log(1/x) - x - 1;\ \lambda = -1 \tag{13.19}$$

The power divergence measure between two 'n' component probability vectors p_1 and p_2 is given by Pardo (2005):

$$\varphi_{(\lambda)}(p_1, p_2) = \sum_{i=1}^{n} p_{1i}\ \varphi\left[\frac{p_{2i}}{p_{1i}}\right]$$

Let $t_i, i = 1, 2, 3 \ldots N$ be the ith observation in the pooled sample. Substituting $p_{1i} = F_m(t_i)$and $p_{2i} = G_n(t_i)$ in equation (13.19), we will get

$$D_{\varphi(\lambda)}(p_1, p_2) = \sum_{i=1}^{N} F_m(t_i)\ \varphi\ \frac{G_n(t_i)}{F_m(t_i)} \tag{13.20}$$

Hence, from equation (13.20) with $\lambda = 0$, we have

$$\varphi_{(0)}(x) = (p_1, p_2)$$

$$= \sum_{i=1}^{N} F_m(t_i)\left[\frac{G_n(t_i)}{F_m(t_i)} \log\left(\frac{G_n(t_i)}{F_m(t_i)}\right) - \frac{G_n(t_i)}{F_m(t_i)} + 1\right] \tag{13.21}$$

$$\varphi_{(0)}(x) = (p_1, p_2) = \sum_{i=1}^{N} G_n(t_i)\ log\left[\frac{G_n(t_i)}{F_m(t_i)} - G_n(t_i) - F_m(t_i)\right] \tag{13.22}$$

The new statistic (denoted as D), based on power divergence between the two ECDFs, tests the H_0. The equation (13.22) becomes

$$D = \sum_{i=1}^{N} G_n(t_i)\ log\left[\frac{G_n(t_i)}{F_m(t_i)} - G_n(t_i) - F_m(t_i)\right] \tag{13.23}$$

The above equation (13.23) does not depend on the parent distribution in its construction and does not use non-parametric density estimates to carry out the inference procedure.

13.19.1 Practical Applications

In the new test, divergence has been used effectively in hypothesis testing, and it does not suffer from sample size distortion. Although the power divergence is defined using probabilities, it can also be used to measure the divergence between ECDFs. If G differs from F only in location, then the power divergence between $F_m(t_i)\ G_n(t_i)$ is very large. Thus, large values of D support H_1, while small values of D favor of H_0.

13.19.2 Test Statistics

If the probability associated with the occurrence under H_0 is equal to or less than the significance level, then the null hypothesis is rejected. Finally, draw your appropriate inference.

13.19.3 Hypothesis State

Null hypothesis (H_0): There is no significant location difference between the two samples $H_0 = \Delta = 0$.

Alternative hypothesis (H_1): There is a significant location difference between the two samples $H_1 = \Delta \neq 0$. The level of significance is fixed at 1% and 5% levels based on your experiments.

13.19.4 Test Statistics

$$D = \sum_{i=1}^{N} G_n(t_i)\ log\left[\frac{G_n(t_i)}{F_m(t_i)} - G_n(t_i) - F_m(t_i)\right] \quad (13.24)$$

Theorem 1

The sampling distribution D is free from the distribution from which the samples are drawn.

Proof: To show that the sampling distribution of 'D' is free from F and G, it is enough to show that the distribution of D is free from F and G.

$$D_i = G_n(t_i) log\left[\frac{G_n(t_i)}{F_m(t_i)}\right] - G_n(t_i) - F_m(t_i) \quad (13.25)$$

For simplicity, we omit the suffix i in this proof. Under H_0 $F=G$, so denote the common distribution function as H. Hence, $t = H^{-1}(z), z\epsilon(0,1)$

Consider the CDF of D_i

$$P(D_i \le d) P\left[\begin{array}{c} G_n\ H^{-1}(Z) log\left(\frac{G_n H^{-1}(z)}{G_n H^{-1}(z}\right) - \\ G_n\ H^{-1}(Z) - F_m\left(H^{-1}(z) \le d\right) \end{array}\right] \quad (13.26)$$

Since

$$G_n\ H^{-1}(Z) = \frac{1}{n}\sum_Y I(Y \le H^{-1}(Z) = \frac{1}{n}\sum_Y I(Y) \le z) \quad (13.27)$$

$$F_m\ H^{-1}(Z) = \frac{1}{m}\sum_X I(X \le H^{-1}(Z) = \frac{1}{m}\sum_X I(X) \le z) \quad (13.28)$$

where 'I' is the indicator function. By the probability integral transform, $H(X)$ and $H(Y)$ are uniform variates.

Theorem 2

Under H_0, D is approximately distributed as a sum of two linear combinations of independent and identical chi-square 'rv'.

Proof: As in Theorem 1, under H_0, $F=G$ here denotes the common distribution function as H. The Taylor series approximation of D_i up to the second order around $F(t_i)$ and $G(t_i)$ is

$$D_i \cong \left[\begin{array}{c} D_i\left\{F(t_i), G(t_i) + (1 - \frac{G(t_i)}{F(t_i)}(F_m(t_i) - F(t_i) + \right\} \\ log\left(\frac{G(t_i)}{F^2(t_i)}\{G_n(t_i) - G(t_i)\}\right) + \\ \frac{1}{2\ G(t_i)}\left(G_n(t_i) - G(t_i)^2\right) \end{array}\right] \quad (13.29)$$

Substituting the common distribution under H_0, $H(t_i)$ for $F(t_i)$ and $G(t_i)$, and adjusting the terms, we get

$$D_i \cong \frac{(1 - H(t_i))}{2}\left[\frac{1}{m}U_i + \frac{1}{n}V_i\right] \quad (13.30)$$

where,

$$U_i = \left[\frac{\sqrt{m}\ (F_m\ (t_i) - H(t_i))}{\sqrt{H(t_i)(1 - H(t_i)}}\right]^2, V_i = \left[\frac{\sqrt{n}\ (G_n\ (t_i) - H(t_i))}{\sqrt{H(t_i)(1 - H(t_i)}}\right] \quad (13.31)$$

For $i=1, 2,\ldots, N$ U_i and V_i are 'iid' distributed as chi-square.

Hence, the statistics 'D' can be expressed

$$D_i \cong \sum_{i=1}^{N} C_1 i U_i + \sum_{i=1}^{N} C_2 i V_i, \quad \text{Where,} \quad C_1 i = \frac{1 - H(t_i)}{2m}, C_2 i = \frac{1 - H(t_i)}{2n} \quad (13.32)$$

This indicates that, under H_0, the test statistics 'D' is approximately distributed as a sum of two linear combinations of independent and identical random variables.

Also, in a particular case, for $m=n$

$$D_i = \frac{(1 - H(t_i))}{2n} W_i,$$

Where $W_i = U_i + V_i$ and $W_i \sim \chi^2$. Hence, the statistics 'D' can be expressed as $D_i \cong \sum_{i=1}^{N} C_i W_i$ where

$$C_i = \frac{(1 - H(t_i))}{2n} \quad \forall = 1,2..N \quad (13.33)$$

indicating that 'D' is approximately a linear combination of iid's chi-square variables.

13.20 Merits

13.20.1 Non-Parametric

The test is constructed without relying on assumptions about the parent distribution, thereby enhancing its robustness across various data types.

13.20.2 Flexibility

It applies to a wide range of scenarios and datasets.

13.20.3 Empirical Nature

Utilizing ECDFs in the test statistic makes it more data-driven and reflective of observed patterns instead of relying on complicated theories.

13.20.4 Sensitivity to Differences

The logarithmic term in the test statistic emphasizes the relative differences between the ECDFs, making the test sensitive to shifts in location parameters.

13.20.5 Applicability to Small Samples

In some cases, traditional parametric tests might require larger sample sizes for reliable results, whereas this non-parametric test may be more suitable for smaller sample sizes.

13.21 Demerits

13.21.1 Complexity of Interpretation

The formula involves logarithmic terms and summation, making it potentially challenging to interpret the results.

13.21.2 Assumption of Independence

The test might assume the independence of observations, and violations of this assumption could affect its validity.

13.21.3 Limited Availability of Software Implementation

If there's a lack of readily available software implementations for the test, users might face challenges in applying it, particularly if programming skills are required for implementation.

13.22 Applications

13.22.1 Environmental Studies

Assessing the impact of environmental factors on different locations or time periods.

13.22.2 Manufacturing and Industry

Assessing the impact of different manufacturing processes or materials on product quality and performance.

13.22.3 Biomedical Research

Comparing clinical outcomes, responses to treatments, or physiological measurements between two groups of patients.

13.22.4 Epidemiological Studies

Investigating disease patterns and risk factors by comparing health-related data across diverse populations (Table 13.7).

In most cases, the *D* test has larger power and performs well

TABLE 13.7

Monte Carlo Power at 5% Level of Significance Based on 10,000 Run

Test	Normal	Exponential (Scale=1)	Pearson III (Shape=2, scale=1)	Cauchy (Scale=1)	Pearson II (Shape=0.5, Scale=4)	Uniform	Laplace (Scale=1)	Logistic (Scale=1)
$\Delta=0$								
t	0.0495	0.0451	0.0562	0.0452	0.0463	0.0482	0.0504	0.052
W	0.0471	0.0423	0.0525	0.0461	0.042	0.0422	0.0458	0.0464
M	0.0484	0.053	0.054	0.0473	0.0486	0.0516	0.0461	0.0532
FD	0.0511	0.0458	0.0539	0.046	0.0453	0.0494	0.0477	0.0522
KS	0.055	0.0492	0.0506	0.0535	0.0535	0.1713	0.0554	0.1713
D	0.0497	0.0419	0.0562	0.0493	0.0444	0.0516	0.0468	0.0511
$\Delta=0.5$								
t	0.1917	0.2388	0.1321	0.0621	0.1029	0.9659	0.1307	0.1035
W	0.1883	0.3121	0.1325	0.0825	0.1179	0.9123	0.1478	0.1028
M	0.1767	0.3127	0.1465	0.0823	0.1243	0.9632	0.1307	0.1051
FD	0.1804	0.2656	0.1225	0.082	0.0742	0.9023	0.1414	0.1023
KS	0.1665	0.3502	0.1358	0.0993	0.0944	0.9601	0.1583	0.2458
D	0.1671	0.4584	0.1677	0.082	0.1254	0.912	0.1351	0.1018

(Continued)

TABLE 13.7 (*Continued*)

Monte Carlo Power at 5% Level of Significance Based on 10,000 Run

Test	Normal	Exponential (Scale=1)	Pearson III (Shape=2, scale=1)	Cauchy (Scale=1)	Pearson II (Shape=0.5, Scale=4)	Uniform	Laplace (Scale=1)	Logistic (Scale=1)
$\Delta=1$								
t	0.5744	0.6377	0.3634	0.1081	0.2889	1	0.3803	0.2409
W	0.5507	0.7153	0.3869	0.1878	0.2903	1	0.4223	0.2389
M	0.5529	0.7299	0.4189	0.1937	0.3018	1	0.3811	0.2445
FD	0.5286	0.7576	0.3573	0.1967	0.1561	1	0.4317	0.2374
KS	0.4735	0.8109	0.3759	0.2456	0.2422	1	0.4337	0.4227
D	0.5026	0.8724	0.499	0.1843	0.303	1	0.392	0.2219

13.23 Terpstra–Magel Test

The test statistic does not focus on pairwise information. Instead, it uses the information present in the $N^* = n_1, n_2, n_3 \ldots n_k$, k-tuples, where a k-tuple includes one observation from each treatment group. More specifically, the TM test is based on the following equation (7.34). The new nonparametric test for decreasing alternatives is

$$T = \sum_{j=1}^{n1}\sum_{j=1}^{nk} k\left(x_{1\ j_1}, x_{2\ j_2} \ldots .. x_{k\ jk}\right) \tag{13.34}$$

where $k\left(x_{1\ j_1}, x_{2\ j_2} \ldots .. x_{k\ jk}\right) = \sum_{j=1}^{k} I\ R(x_i) = i, R(x_i)$ denotes the rank of X_i with respect to X_1, X_2, $X_3 \ldots . X_k$ and I (.) denotes the indicator function (the indicator function '0' indicates failure and '1' indicates success). The remainder of this section presents and derives results pertaining to the null distribution of the proposed test statistic. We assume throughout this section that the observed data, $\{X_{ij}\}$, is essentially a random sample from some continuous probability distribution function F. Hence, the possibility of ties has a probability of zero. In principle, the test statistic uses the k-tuple method of Terpstra and Magel. Additionally, under the null hypothesis, each $k\left(x_{1\ j_1}, x_{2\ j_2} \ldots .. x_{kjk}\right)$ follows the Binomial (k, $1/k$) distribution. For these reasons, we will refer to this test as the KTMB test.

13.24 The Exact Null Distribution

Let N denote the sum of the sample sizes for each treatment. Namely, let $N = n_1 + \cdots + n_k$. Here, we have $N!/(n_1! \cdots n_k!)$ partitions of the numbers 1… N. The null distribution of T means each one of these partitions is equally likely, so the mean and variance can be calculated directly by multiplying each possible value of T with its probability. When the number of partitions is small, we can easily calculate the exact distribution by hand or with a computer. Table 13.1 shows the probabilities, means, and variances of the test statistic T for sample size arrangements (2, 1, 1), (2, 1, 2), and (1, 1, 3) respectively.

Real-world cases are not always as simple as the above illustration. For example, when $k=4$, $n_1 = n_2 = n_3 = n_4 = 5$, we have $20!/(5!5!5!5!) = 1.1733 \times 10^{10}$ partitions. Such a distribution function cannot be calculated even with the most efficient personal computers. We will, therefore, introduce a Monte Carlo approximation to the null distribution. On the other hand, if the distribution of T can be approximated or shown to converge to a well-known distribution, we can avoid computational complexity altogether. This test proposes a new nonparametric test for the ordered alternative problem. The new test statistic is based on calculating all $k\left(x_{1\ j_1}, x_{2\ j_2} \ldots .. x_{kjk}\right)$ in proper (ascending) order. Due to the small number of groups and sample sizes, we tabulated and listed their distribution as well as the exact mean and variance of the null distribution. From the equation for the exact mean and variance of the null distribution, the asymptotic null distribution is normal.

13.24.1 The Mean and Variance

If the asymptotic null distribution of a test statistic is normal and the exact mean and variance of T under H_0 in standard form can be established, we can then standardize T by using the exact mean and variance to obtain Z_{KTMB}, where

$$Z_{\text{KTMB}} = \{T - E_0(T)\}/\sqrt{V_0(T)} \tag{13.35}$$

In this case, we can find critical values from the standard normal table. We will start by finding the mean value of T, $E_0(T)$, noting that (T) is nothing but a sum of the $k * \prod_{i=1}^{k} n_i$ Bernoulli $(1/k)$ distribution. It is straightforward to get

$$E_0(T) = \prod_{i=1}^{k} n_i \tag{13.36}$$

Here, and in the following we let $n^* = \prod_{i=1}^{k} n_i$,

$$V_0(T) = v_0{}^2 + \sum_{i=1}^{k-1} v_i{}^2 + v_k{}^2 \tag{13.37}$$

where

$$v_0^2 = \begin{Bmatrix} 1-(1/k)\prod_{i=1}^{k} n_i \text{ for no tie}, v_k^2 \\ = n^* k(k-1)\left\{k - 2/k! - \frac{1}{k^2}\right\} \end{Bmatrix} \quad (13.38)$$

for 'k' ties for $i \neq j$. For the case of i ties, we present an algorithm for the computation of $\sum_{i=1}^{k-1} v_i^2$. Terpstra et al. (2011) proposed the k-tuplet Terpstra page (KTP) test based on the statistics

$$KTP = \sum_{i_1=1}^{n_1} \ldots \sum_{i_2=1}^{n_2} r_s\left(X_{1i_1}, X_{2i_2} \ldots X_{kik}\right) \quad (13.39)$$

where $r_s\left(X_{1i_1}, X_{2i_2} \ldots X_{kik}\right)$ denotes the Spearman rank correlation coefficient based on $(1, X_{1i_1}),(2, X_{2i_2} \ldots (k, X_{kik})$. Under H_0, the statistics KTP is approximately distributed, and its mean and variance are $E(KTP) = 0$ and

$$V(KTP) = \frac{144\left(N^*\right)^2}{k^2\left(k^2-1\right)^2} S \quad (13.40)$$

where,

$$S = \sum_{i_1=\frac{1}{2}=}^{k-1} \sum_{i_1+1}^{k} \left[\frac{(i_2-i_1)^2(n_{i1}+n_{i2}+1)}{12 n_{i1} n_{i2}}\right] + \sum_{i1=1}^{k-2} \sum_{i2=i1+1}^{k-1} \sum_{i3=i2+1}^{k} \left[\frac{(i_2-i_1)(i_3-i_1)}{6n_{i1}} + \frac{(i_3-i_2)(i_1-i_2)}{6n_{i2}} + \frac{(i_1-i_3)(i_2-i_3)}{6n_{is}}\right] \quad (13.41)$$

In the KTP test, the Spearman rank correlation coefficient 'r' is given by the following formula:

$$r_s = 1 - \left(\frac{6\sum_{i=1}^{k} d_i^2}{k\left(k^2-1\right)}\right) \quad (13.42)$$

where d_i represents the difference between the ranks assigned to the values of the variable for each item of the particular data y_i. This formula is applied in cases where there are no tied observations. The formula to use when there are tied observations is

$$r_s = \frac{\sum_{i=1}^{k}(y_i-\bar{y})(x_i-\bar{x})}{\sqrt{\sum_{i=1}^{k}(y_i-\bar{y})^2 \sum_{i=1}^{k}(x_i-\bar{x})^2}} \quad (13.43)$$

where $(y, x)(1, X_{1i_1}),(2, X_{2i_2} \ldots (k, X_{kik})$ and x_i is the rank of X_i. Note that if all of x_i values are equal, the $\sum(x_i-\bar{x})^2$ result is zero. This result also applies to the Kendall correlation coefficient (Table 13.8).

13.25 Practical Exercise

Illustration 1

Consider a Phase II clinical trial designed to investigate the effectiveness of a new drug in reducing symptoms of asthma in children. A total of $n=10$ participants are randomized to receive either the new drug or a placebo. Participants are asked to record the number of episodes of shortness of breath over a 1-week period following receipt of the assigned treatment. The data are shown below.

Placebo	7	5	6	4	12
New Drug	3	6	4	2	1

Is there a difference in the number of episodes of shortness of breath over a 1-week period between participants receiving the new drug and those receiving the placebo? By inspection, it appears that participants receiving the placebo have more episodes of shortness of breath, but is this statistically significant?

Illustration 2

A new approach to prenatal care is proposed for pregnant women living in a rural community. The new program involves in-home visits during the course of pregnancy in addition to the usual or regularly scheduled visits. A pilot randomized trial with 15 pregnant women is designed to evaluate whether women participating in the program deliver healthier babies than women receiving usual care. The outcome is the APGAR score measured 5 minutes after birth. Recall that APGAR scores range from 0 to 10 with scores of 7 or higher considered normal (healthy), 4–6 low, and 0–3 critically low. The data are shown below.

Usual Care	8	7	6	2	5	8	7	3
New Program	9	9	7	8	10	9	6	

Is there statistical evidence of a difference in APGAR scores between women receiving the new and enhanced prenatal care versus usual prenatal care? We run the test using the five-step approach.

TABLE 13.8

Empirical Cumulative Distribution Function (ECDF) of KTMB

Sample Sizes	Test Statistic Value (KTMB)											Mean and Variance	
(n_1, n_2, n_3)	0	1	2	3	4	5	6	7	8	9	12	E[T]	V[T]
(2, 1, 1)	2/12	3/12	4/12	1/12	1/12	–	1/12	–	–	–	–	2	2.667
(2, 1, 2)	2/30	2/30	6/30	4/30	6/30	2/30	3/30	2/30	2/30	–	1/30	4	6.867
(1, 1, 3)	2/20	3/20	4/20	6/20	1/20	1/20	1/20	1/20	–	1/20	–	3	5.000

Illustration 3

A clinical trial is run to assess the effectiveness of a new anti-retroviral therapy for patients with HIV. Patients are randomized to receive either a standard anti-retroviral therapy (usual care) or the new anti-retroviral therapy and are monitored for 3 months. The primary outcome is viral load, which represents the number of HIV copies per milliliter of blood. A total of 30 participants are randomized, and the data are shown below.

Standard therapy	7,500	8,000	2,000	550	1250	1,000	2,250	6,800	3,400	400
New therapy	400	250	800	1,400	8,000	7,400	1,020	6,000	920	undetectable

Illustration 4

A new chemotherapy treatment is proposed for patients with breast cancer. Investigators are concerned with patient's ability to tolerate the treatment and assess their quality of life both before and after receiving the new chemotherapy treatment. Quality of life (QOL) is measured on an ordinal scale, and for analysis purposes, numbers are assigned to each response category as follows: 1=Poor, 2=Fair, 3=Good, 4=Very Good, 5=Excellent. The data are shown below.

Patient	QOL Before Chemotherapy Treatment	QOL After Chemotherapy Treatment
1	3	2
2	2	3
3	3	4
4	2	4
5	1	1
6	3	4
7	2	4
8	3	3
9	2	1
10	1	3
11	3	4
12	2	3

The question of interest is whether there is a difference in QOL after chemotherapy treatment compared to before.

Illustration 5

A clinical study is designed to assess differences in albumin levels in adults following diets with different amounts of protein. Low-protein diets are often prescribed for patients with kidney failure. Albumin is the most abundant protein in blood, and its concentration in the serum is measured in grams per deciliter (g/dL). Clinically, serum albumin concentrations are also used to assess whether patients get sufficient protein in their diets. Three diets are compared, ranging from 5% to 15% protein, with the 15% protein diet representing a typical American diet. The albumin levels of participants following each diet are shown below.

5% Protein	10% Protein	15% Protein
3.1	3.8	4.0
2.6	4.1	5.5
2.9	2.9	5.0
	3.4	4.8
	4.2	

Is there a difference in serum albumin levels among subjects on the three different diets? For reference, normal albumin levels are generally between 3.4 and 5.4 g/dL. By inspection, it appears that participants following the 15% protein diet have higher albumin levels than those following the 5% protein diet. The issue is whether this observed difference is statistically significant. In this example, the outcome is continuous, but the sample sizes are small and not equal across comparison groups ($n_1=3$, $n_2=5$, $n_3=4$). Thus, a nonparametric test is appropriate. The hypotheses to be tested are given below, and we will use a 5% level of significance.

Illustration 6

A personal trainer is interested in comparing the anaerobic thresholds of elite athletes. Anaerobic threshold is defined as the point at which the muscles cannot get more oxygen to sustain activity or the upper limit of aerobic exercise. It is also a measure related to maximum heart rate. The following data are anaerobic thresholds for distance runners, distance cyclists, distance swimmers and cross-country skiers.

Distance Runners	Distance Cyclists	Distance Swimmers	Cross-Country Skiers
185	190	166	201
179	209	159	195
192	182	170	180
165	178	183	187
174	181	160	215

Is there a difference in anaerobic thresholds among the different groups of elite athletes?

REFERENCES

Baltus, V., Jankauskaite-Jureviciene, L. & Zebrauskas, T. Parametric architecture today and tomorrow. *Journal of Sustainable Architecture and Civil Engineering* **25**, 85–89 (2019).

Fried, R., & Dehling, H. Robust nonparametric tests for the two-sample location problem. *Stat Methods Appl* **20**, 409–422 (2011).

García, S. & Herrera, F. An extension on "statistical comparisons of classifiers over multiple data sets" for all pairwise comparisons. *Journal of Machine Learning Research* **9**, 2677–2694 (2008).

Kirpich, A. S. et al. SECIMTools: A suite of metabolomics data analysis tools. *BMC Bioinformatics* **19**, 151 (2018).

Landolfi, E., Salvi, A., Troiano, A. & Natale, C. Hardware-in-the-loop validation of an adaptive model predictive control on a connected and automated vehicle. *International Journal of Adaptive Control and Signal Processing* **37**, 1459–1491 (2023).

Luerken, L. et al. Diagnostic accuracy of indocyanine green clearance test for different stages of liver fibrosis and cirrhosis. *Diagnostics* **13**, (2023).

Mathur, S. K., & Sakate, D. M. A new test for two-sample location problem based on empirical distribution function. *Communications in Statistics – Theory and Methods* **46**(24), 12345–12355 (2017).

Mota-Romero, E. et al. Frailty, complexity, and priorities in the use of advanced palliative care resources in nursing homes. *Medicina* **57**, 1–15 (2021).

Neuhauser, Karyn L. The Global Financial Crisis: What have we learned so far. *International Journal of Managerial Finance* **11**(2), 134–161 (2015).

Riillo, C. A. F. Beyond the question "Does it pay to be green?": How much green? and when? *Journal of Cleaner Production* **141**, 626–640 (2017).

Sevinc, B., Cetintav, B., Esemen, M. & Gurler, S. RSSampling: A pioneering package for ranked set sampling. *R Journal* **11**, 401–415 (2019).

Spolverato, Y. C. et al. Biohumoral predictors of advanced pancreatic carcinoma not shown at pre-operative imaging. *Indian Journal of Surgery* **85**, 545–551 (2023).

Surbakti, A., Nyorong, M. & Aini, N. Pengaruh penyuluhan Keselamatan Dan Kesehatan Kerja terhadap Perilaku Penggunaan alat Pelindung diri. *Jurnal Kesehatan dan Kedokteran* **2**, 29–38 (2023).

Torregrosa, S., Champaney, V., Ammar, A., Herbert, V. & Chinesta, F. Surrogate parametric metamodel based on optimal transport. *Mathematics and Computers in Simulation* **194**, 36–63 (2022).

Ullrich, A. et al. Quality of life, psychological burden, needs, and satisfaction during specialized inpatient palliative care in family caregivers of advanced cancer patients. *BMC Palliative Care* **16**, 31 (2017).

Vallat, R. Pingouin: Statistics in python. *Journal of Open Source Software* **3**, 1026 (2018).

Wigmore-Álvarez, A., Ruiz-Lozano, M. & Fernández-Fernández, J. L. Management of University Social Responsibility in business schools. An exploratory study. *International Journal of Management Education* **18**, 100382 (2020).

14

Statistical Ethics and Future Line of Research Work

14.1 Introduction

Throughout history, statistics and its related fields have maintained a significant connection. As early as 300 BC, the Indian monarch Ashoka Chakravarthi, ruler of Kalinga, pioneered the use of statistical methods for population enumeration, thereby establishing a distinct category known as vital statistics (Lesser & Nordenhaug, 2004). The fundamental nature of statistics revolves around the collection of data concerning both living and non-living entities (Williamson et al., 2000). The absence of statistical information would severely impede the operations of the real world. Daily, we engage with and apply statistical concepts that assist us in measuring essential and theoretical principles. The versatility of statistical knowledge and its applications is widely recognized and appreciated by researchers across various disciplines, both academic and investigative (Miranda-Novales & Villasís-Keever, 2019). A layperson may interpret the essence of statistics differently, viewing it primarily as a discipline focused on equations and the assessment of probabilities through analytical frameworks. Navigating complex data analysis, such as predicting the likelihood of specific traits, can prove to be a daunting task (Saltelli, 2020). From the viewpoint of a statistician, statistics not only imparts theoretical knowledge but also offers valuable insights into diverse applications and the behavioral patterns of individuals. By employing statistical techniques in our everyday lives, we can enhance our understanding of how to harmoniously interact with the ecosystem for the collective benefit (Brenner et al., 2020). Historically, ancient civilizations such as those in the Indus Valley and Mesopotamia employed statistical methods in their daily activities, operating under the belief that human existence is initiated by statistics and culminates in facts or evidence. Statistics uniquely enables us to glean insights from human society (Andrade, 2020). This chapter aims to explore the role and importance of statistics, addressing the practical challenges and ethical considerations present in various research domains (Utts, 2021). The primary goal of this chapter is to comprehend the effective utilization of statistics and the appropriate conduct in their application.

14.2 Salient Steps for Applying Statistical Methods

1. The term 'procedure' is derived from two components: 'proce', which signifies methods, and 'dure', which refers to the manner of conducting research studies utilizing experimental materials. In statistical contexts, procedures pertain to the estimation of probabilities through appropriate statistical measures (Razak et al., 2021). It is evident that numerous researchers globally may struggle to select and implement suitable measures during the analysis of experimental data. The following steps are recommended for researchers to consider prior to data analysis:
   - Seek appropriate statistical guidance from subject matter experts.
   - Carefully examine and establish study objectives.
   - Collaborate with statisticians to formulate the research topic, facilitating an easier study design.
   - Calculate the sample size using various statistical formulas, employing both univariate and multivariate methods.
   - Define the study population based on diverse characteristics that can enhance the rationale of the study.
   - Carefully evaluate and select relevant parameters pertinent to the study.
   - Initiate the research study in accordance with the Standard Operating Procedures (SOP).
   - Systematically gather research data.
   - Assess the effects of multicollinearity within and between data series.
   - Estimate probabilities using various relevant statistical measures.
   - Draw effective conclusions and inferences regarding the study objectives and hypothesis testing.

14.3 Statistics in Research Integrity (Culture)

The field of research integrity encompasses all scientific disciplines where research is conducted. It includes all stages of research, from the initial idea and design of experimental materials to the actual conduct of the research and its dissemination. It is essential to ensure that research environments encourage good research practices rather than hindering them. These components collectively make up research integrity. Before commencing the research, it is crucial to ensure that all components of the experimental materials are homogeneous and selected without any bias using probability sampling techniques (Georgious, 2022). Each experimental material, whether it is a replication or a treatment, should be chosen impartially. Study designs should be implemented systematically in the

DOI: 10.1201/9781032713816-14

designated research sites or units. If any bias is detected in the experimental materials, appropriate actions should be taken at the earliest stage possible (Heston, 2023). Standard operating procedures must be followed when selecting research objectives to ensure that they are well-defined and address any research gaps (Hernández-Zambrano et al., 2023). Eliminate sampling and human biases from the experimental units (Kambhampati et al., 2023).

14.3.1 Integrity

Integrity in all facets of research, such as the way goals, intentions, and conclusions are presented; how research methods and procedures are reported; how data is gathered; how other researchers' work is used and acknowledged; how legitimate interpretations are presented; and how claims based on research findings are justified.

14.3.2 Scope

Rigor, in keeping with accepted disciplinary norms and standards, in conducting research and employing suitable methodologies, in following a protocol when necessary, in interpreting and extrapolating conclusions from the study, and in disseminating the findings.

14.3.3 Openness

Openness and transparency are important when it comes to disclosing potential conflicts of interest, reporting research data collection procedures, analyzing and interpreting data, disseminating research findings (including publishing or otherwise sharing negative or null results to acknowledge their value as part of the research process), and presenting the work to the public and other researchers.

14.3.4 Confidentiality

Researchers must treat the integrity of the research record with care and respect, as well as that of all research participants, subjects, users, and beneficiaries, including people, animals, the environment, and cultural artifacts.

14.3.5 Accountability

In order to create a research environment where individuals and organizations are empowered to take charge of the research process, funders, employers, and researchers must collaborate. Researchers also need to ensure that institutions and individuals are held responsible when actions depart from the standards set forth by this concordat (Figure 14.1).

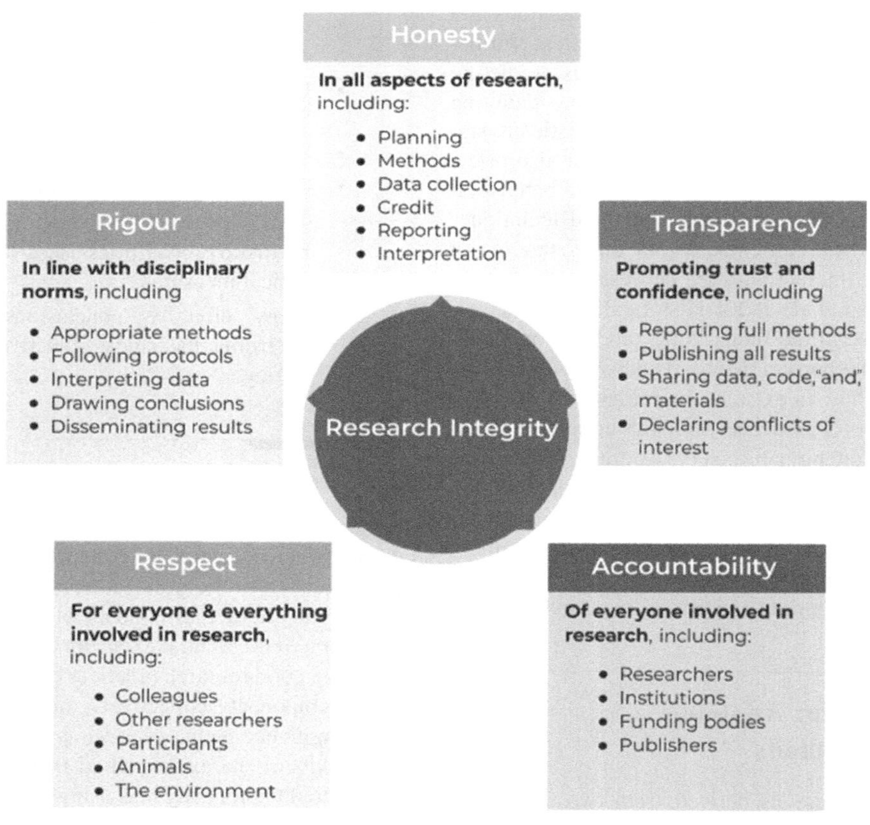

FIGURE 14.1 Flowchart depicting research integrity.

14.4 Ethics

A set of moral guidelines or standards that specify what is and is not appropriate make up ethics. Individuals utilize it as a set of rules or a code of conduct to determine what conduct is proper and improper (Wester et al., 2023). Ethics, in general, refers to the rules or standards of conduct that individuals use to determine whether or not a certain behavior is appropriate. A person's community's laws and religious teachings often influence their ethical beliefs. For instance, a society's ethics may dictate that it is improper to steal from another person. It is possible that the ethics you personally adhere to differ from those of your community or society (Drolet et al., 2023). For example, you may move from a country where owning a gun is legal to one where it is not. In this case, the morality of the country you have moved to may conflict with your personal ethics, i.e., your belief that owning a gun is acceptable. In philosophy, the study of what affects a person's judgment of what is right and wrong is referred to as ethics. Philosophers also use the term 'morality' to refer to the same idea. However, morality and ethics are viewed as two different categories of values outside of philosophy. The general distinction between ethics and morals is that ethics is what a community believes to be right and wrong, while morals are what an individual believes to be good, bad, or evil. The word ethics comes from the Greek word 'ethos', which means 'character or conduct'.

14.4.1 Real Nature of Ethics

Ethics is the study of what morality requires of people, usually in terms of duties, rights, justice, and particular virtues. It has to do with matters of propriety, or what is proper in an ethical context and improper in an unethical one (Correia et al., 2023). Occasionally, in place of right and ethical, other terms that are frequently used are 'proper', 'fair', and 'just'. As far as the practical aspects of the study of ethics are concerned, it may be urged that if ethics is a theoretical study of moral problems. The average person on the street might find no use for such a declaration. It is said that theoretical knowledge cannot possibly elevate a person's ethical standard; experience alone can enrich the ethical aspect of the human mind. Virtue, according to Socrates, is knowledge. He was attempting to make the point that, without a true understanding of what a moral act entails, no one could be moral by accident. According to Jaina ethics, which is recognized as the highest virtue and the practical application of non-violence, 'knowledge must precede compassion'. These claims alone demonstrate that virtue awareness serves as the cornerstone for the practice of virtue.

14.4.2 Merits of Studying Ethics

Ethics is a very important subject for both the individual and the society. We are able to achieve life's ideals and virtues because of this subject. Good behavior and character development in both individuals and society follow specific paths. Ethics is significant because it aids in our ability to discriminate between good and bad as well as right and wrong. In contrast, ethics is especially important in countries like India because it helps establish higher values and virtues in life and dispels the darkness of evil.

Ethics is important for two reasons. The first is its constructive value, which results in the development of a moral and critical attitude in humans. One of the most valuable aspects of the human personality is moral sense. Our overall personality development must be aided by ethics. The significance of ethics' second facet is its destructive value. It eliminates and destroys humanity's evils and ills because of this feature. The applications of studying ethics are as follows.

- Character Development
- Understanding of principles, values, and virtues
- Responsibility and duty
- Social Concerns
- Law, justice, and order
- Usefulness in the domains of religion, politics, and economy
- Final result
- Usefulness in the study of psychology

14.5 Moral Judgment

Moral judgment is the main cognitive factor in moral consciousness. It involves the intuition of the moral standard by reason and the comparison of a voluntary action with this standard. It evaluates the rightness or wrongness of actions. It consists in the apprehension of the moral quality of an action with reference to the moral standard. However, the consciousness of rights and duties, virtues and vices, merit and demerit, responsibility, or accountability is also involved in moral consciousness. Moral judgment is a judgment of value, as distinguished from a judgment of fact. A judgment of fact is a judgment of what is. A judgment of value is a judgment of what ought to be. The former is a descriptive judgment, while the latter is an appreciative or critical judgment. Moral judgment is the mental act of discerning and pronouncing a particular action to be right or wrong.

14.5.1 Logical Judgment

Logical judgment is the judgment of an argument. Since logic is the science of reasoning, it is concerned with the ideal of truth. It judges the validity or invalidity, truth or falsehood of an argument. In other words, logical judgment deals with finding the rules of correct thinking and the pursuit of truth. Thus, logical judgement refers to the ideal of truth. Like moral judgment, logical judgment is also normative because normative science deals with life's norms, ideals, and values. It is distinguished from those of natural or positive science whose judgments are descriptive in nature. The judgement of positive science is factual while that of normative science is axiological. Logical judgment is also called evaluative judgment as it evaluates the validity or invalidity of an argument.

TABLE 14.1

Difference between Moral and Social Judgment

Moral	Social
Moral judgment refers to the pursuit of good.	Logical judgment refers to the pursuit of truth.
Moral judgment is concerned with the rightness or wrongness and goodness or badness of the conduct.	Logical judgment is concerned with the validity or invalidity and the truth or falsehood of intellectual pursuits.
The moral judgment refers to the moral worth of an action.	Logical judgment refers to the logical worth of an action.
Moral judgment is concerned with the ideal of conduct.	Logical judgment is concerned with the standard of correct thinking and reasoning.
Moral judgment is accompanied by the consciousness of moral obligation and responsibility.	Logical judgment lacks in moral obligation and moral feeling.
Moral judgment is a judgment upon an action. It judges the rightness or wrongness of our actions.	Logical judgment is a judgment of reasoning. It judges the validity or the invalidity, soundness, or fallacy in an argument.
Moral judgment is obligatory in character.	Logical judgment does not imply any such obligation.

Logical judgement is also appreciative in nature because it appraises the logical worth of an argument. Thus, logical judgment is appreciative-normative and evaluative. It is the judgment of worth and value rather than facts. It is concerned with what ought to be rather than what is. Hence, logical judgment involves the intellectual aspect because in order to know the truth and falsity, we need to touch the intellectual aspect (Table 14.1).

14.6 Ethics in Research

A set of guidelines known as research ethics guarantees that studies are carried out with honesty, integrity, and respect for the subjects and participants. Among the crucial facets of research ethics are:

14.6.1 Informed Permission

Participants should give their informed consent before research begins.

14.6.2 Confidentiality

Participants' privacy and identities should be safeguarded by researchers. This can involve making people and places anonymous. From the moment they are approached about potentially participating in a study, even if they decline, until the end of their involvement, people should be treated with respect. This includes the following points

i. Preserving their confidentiality and honoring their right to privacy, honoring their freedom to change their mind, determine that the research is not in line with their interests, and withdraw without incurring any penalties.
ii. Letting them know about any new information that might surface during the study (which could alter their evaluation of the advantages and disadvantages of taking part).
iii. Monitoring their well-being and, in the event that they have negative reactions, unanticipated effects, or changes in their participation, ensuring they receive the right care and, if required, are removed from the study.
iv. Telling them what the study's findings revealed.

14.6.3 Data Security

Researchers ought to give thorough explanations of the procedures used for gathering, storing, analyzing, and dispersing data.

14.6.4 Integrity of Research

In addition to avoiding conflicts of interest and manipulating research materials or findings, researchers should employ sound research methodologies. They should also ensure they select samples impartially and remain on topic.

14.6.5 Honesty

It is important for researchers to be truthful, open, responsible, and reliable. In addition, they ought to write their thesis in their native tongue and abstain from plagiarism.

14.6.6 Acceptance from an Ethical Standpoint

Before beginning any research, whether or not it uses human subjects, it should be approved by an ethical council.

14.6.7 Control Statistical Bias

Many scientists believe that the opinions of statistical experts are not reliable, as this is a flaw that can lead to an increase in subject random errors and publication bias. Prior to beginning any research, the investigator should gather the relevant data as prescribed by foundational theories. It is important for all researchers to remember that research involves global interaction and new approaches and objectives. Many research studies in the real world fail to produce satisfactory results because of the negligence of the researchers; instead, they have to spend more money to satisfy the desire for research.

A discrepancy between the true and expected values of a population parameter under estimation is known as statistical bias. An alternative way to conceptualize it is as the underestimation or overestimation of a population value. Analysis or estimation techniques may lead to statistical bias. For instance, estimated treatment effects could be skewed if significant prognostic factors are not taken into account in the statistical analysis.

TABLE 14.2

The Statistical Bias in Research

Selection Bias	Survivorship Bias	Non-response Bias	Observer Bias	Confirmation Bias
• A statistical bias in the selection of sample units. To obtain meaningful data and results, it is necessary to avoid this bias. For example, in marketing, selection bias jeopardizes the objectivity of customer surveys and other market research methods.	• A type of bias in statistics where the researcher only looks at the parts of the data set that have already been pre-selected.	• A potential source of bias in census counts and in long-form estimates. When the traits of respondents are different from those of non-respondent bias arises.	• A kind of bias in statistics caused by the subjectivity of the observer. There will always be observer bias because no human can be totally objective.	• Among the most prevalent types of bias in research. It occurs when data is interpreted based on preconceived notions or theories rather than allowing conclusions to be shared by research findings.

14.7 Statistical Ethics

The American Statistical Association's Ethical Guidelines for Statistical Practice are meant to assist statisticians in making moral decisions. They also seek to encourage responsibility by educating individuals who use statistics. The study of statistics lays the groundwork by bridging the gap between observational skills and decision-making abilities (Table 14.2).

14.7.1 To Establish a More Informed Society

Regardless of their training, profession, or title, statisticians have an obligation to conduct themselves in a professional, competent, courteous, and ethical manner because society depends on well-informed decisions supported by statistical methods. Good statistical practice is based on transparent assumptions, repeatable results, and valid interpretations. Although certain situations might necessitate prioritizing principles based on context, stakeholders are always obligated to act with integrity, behave in a manner that aligns with these guidelines, and encourage others to do the same.

14.7.2 To Avoid Unethical to Use Statistics for Immoral Goals

Any form of professional or scientific misconduct, including but not limited to bullying, sexual harassment, other forms of harassment, discrimination based on personal characteristics, or intimidation, is not acceptable in the context of ethical statistical practice. The principles expressed here should guide both those whose primary occupation is statistics and those in all other disciplines who use statistical methods in their professional work. Therefore, throughout these guidelines, the term 'statistician' includes all practitioners of statistics and quantitative sciences regardless of job title or field of degree. This includes members of various professions who use and report statistical analyses and their implications, as well as statisticians at all levels of the profession.

14.7.3 Professional Integrity and Accountability

An ethical statistician employs methods and data that are appropriate and relevant, free from bias or partiality, and designed to yield reliable, understandable, and repeatable outcomes. An ethical statistician will never willingly take on work for which they are insufficiently qualified, will be upfront with the client about any areas in which they are not comfortable, and will seek the advice of other statisticians when needed or uncertain. It is imperative that statisticians show consideration for others.

1. Recognizes and reduces any preferences that the investigators or data providers may have that could affect or predetermine the analyses or findings.
2. Uses analytical techniques and methods of selection or sampling that are appropriate and valid for the particular question being addressed, ensuring that results, under reasonable assumptions, extend beyond the sample to a population relevant to the objectives with little error.
3. Honors and recognizes other people's contributions and intellectual property.
4. Attempts to be transparent about the rationale behind authorship placement, should it be decided upon factors other than intellectual contribution, when assigning credits for posters, papers, and other scholarly works.
5. Reports financial and non-financial conflicts of interest and handles them in accordance with established (local, regional, and institutional) laws and regulations.
6. Takes complete accountability for their work performance. Only offers written work, oral presentations, and expert testimony that they are willing to have peer-reviewed.
7. Shows respect for others by not participating in or endorsing bullying, harassment, or intimidation based on personal traits, or by engaging in any form of unwanted physical or sexual contact. When others engage in these unethical behaviors, they should be held accountable.

14.8 Integrity of Data and Methods

Any known or suspected restrictions, flaws, or biases in the data that might compromise the accuracy or dependability of the statistical analysis are openly disclosed by the ethical statistician. The degree of reliability and integrity of the data must be acknowledged by the underlying analysis in order for the results to be interpreted objectively and validly.

- Acknowledges statistical and substantive assumptions made in the execution and interpretation of any analysis. Recognizes data editing processes, including any imputation and missing data mechanisms, when reporting on the accuracy of the data used.
- Outlines the constraints of statistical inference and potential sources of inaccuracy.
- In publications, reports, or testimony, identifies who is responsible for the statistical work if it would not otherwise be apparent.
- Describes the sample(s) that were actually used, reports the sources and the data's assessed adequacy, and accounts for all the data taken into consideration in a study.
- Clearly and fully reports the steps taken to preserve data integrity and valid results.
- Addresses potential confounding variables not included in the study when appropriate.
- Communicates the results in publications and reports in a way that is honest and relevant to the user/reader. This includes tables, models, and graphics.
- Identifies the study's ultimate financial sponsor, purpose, and intended use of the study results in publications or testimony.
- When reporting analyses of volunteer data or other data that might not be representative of a defined population, includes the proper disclaimers and, if applicable, weighting.
- To aid in replication and peer review, shares the data used in the analyses when appropriate and practical. Takes reasonable precautions to safeguard proprietary and confidential data, including any information that might unintentionally reveal respondent identities.
- Makes an effort to fix any mistakes found during the final report's production or after it has been published as soon as possible. Distributes the correction to the public or to those who are depending on the results, as appropriate.

14.9 Responsibilities to Science

Keeping in mind the interests of the public, funder, client, or customer (as well as professional colleagues, patients, the public, and the scientific community), the ethical statistician supports valid inferences, transparency, and good science in general. The statistician with an ethical code:

- Provides a client or employer, to the greatest extent feasible, with options from legitimate alternative statistical approaches that differ in terms of scope, cost, or precision.
- Strives to explain any expected adverse consequences of failure to follow through on an agreed-upon sampling or analytic plan.
- Uses scientific statistical sampling and analysis techniques without biasing the results
- Aims to broaden the dissemination of new statistical knowledge to benefit society as a whole as well as beyond their own area of expertise.
- Recognizes and complies with any use limitations set forth by the data provider, as well as the confidentiality requirements of data collection, release, and dissemination
- Protects confidential information belonging to a client, funder, or employer, safeguarding data use and disclosure in accordance with the degree mandated by law.

14.9.1 Responsibilities to Research Subjects

At every stage of their participation in a project, human and animal subjects' rights and interests are safeguarded and respected by ethical statisticians. This includes respondents to physically or psychologically invasive research, those whose data are found in administrative records, census or survey subjects, etc.

The ethical statistician:

1. Remains knowledgeable about and complies with relevant laws, permits, and regulations pertaining to the safety and well-being of both human and animal participants.
2. Makes well-informed recommendations for study size in an effort to prevent the use of an excessive or insufficient number of research subjects as well as an excessive risk to research subjects (in terms of health, welfare, privacy, and ownership of their own data).
3. Preserves the privacy and confidentiality of research participants and the information about them, whether collected directly from participants, from other sources, or from existing records.

   When obtaining consent from research subjects, anticipate and request permission for secondary and indirect uses of the data, including linking to other datasets, and obtain consent appropriate for permitting independent replication of analyses and peer review.
4. Is aware of the legal restrictions on guarantees of privacy and confidentiality and refrains from making unwarranted claims or assuming there are protections for privacy and confidentiality where none exist.
5. Takes into account whether the necessary consents for research subjects were acquired before taking part in a study involving people or organizations, analyzing the data from such a study, and reviewing manuscripts for publication or internal use.

The statistician considers the treatment of research subjects (e.g., confidentiality agreements, expectations of privacy, notification, consent, etc.) when evaluating the appropriateness of the data source(s).

6. Declines to participate in the analysis if participating in an analysis of data from a particular source would be construed by the individuals who submitted the information as endorsing a violation of their rights.
7. Recognizes that statistical descriptions of groups may carry risks of stereotypes and stigmatization. Statisticians ought to think about and exercise caution when framing information to prevent undue harm to marginalized populations.

14.9.2 Responsibilities to Research Team

Since science and statistical practice are frequently carried out in groups of professionals with varying professional standards, the statistician needs to operate in an ethical manner in these kinds of settings.

The statistician with an ethical code:

1. Acknowledges the existence of standards and obligations in other professions, the variability of research practices and standards among disciplines, and the lack of obligation on the part of statisticians to adhere to standards that are at odds with these guidelines.
2. Verifies that all discussions and reports regarding statistical design and analysis adhere to these rules.
3. Prevents sacrificing scientific validity for the sake of convenience.
4. Seeks to encourage openness in the planning, carrying out, and presenting or reporting of every analysis.

14.9.3 Responsibilities to Other Statisticians

The entire spectrum of explanations for observed phenomena must be taken into account in the practice of statistics, and different observers, each drawing from a unique set of experiences, may reach different and possibly divergent conclusions regarding the plausibility of various explanations. When statisticians treat one another with mutual respect and concentrate on scientific principles, methodology, and the content of data interpretations, even in adversarial settings, discourse tends to be most successful. With consideration for other statisticians in the field, the ethical statistician:

1. Encourages the most appropriate and widespread sharing of techniques and data while maintaining confidentiality Makes the documentation appropriate for qualified researchers' use in replicate analyses, metadata studies, and other types of research.
2. Supports the improvement of others' work by conducting appropriate peer reviews, which respect differences of opinion and evaluate processes rather than people. Makes an effort to finish review assignments quickly, carefully, and completely.
3. Gives learners and non-statisticians a sense of the usefulness of the ideas and techniques they are studying or applying.
4. Bases decisions about hiring, firing, promoting, work assignments, publications and presentations, nominations for offices and awards, funding, or approval of research, and other professional matters for statistical practitioners on their professional qualifications and contributions.

14.9.4 Responsibilities on Allegations of Misconduct

An ethical statistician is able to distinguish between unethical statistical, scientific, or professional practices and unethical behavior. The moral statistician steers clear of all of the above and understands the appropriate way to handle each. The statistician with ethics:

1. Steers clear of endorsing or giving the impression of endorsing scientific, statistical, or professional misconduct.
2. Understands that disagreements and sincere mistakes do not equal wrongdoing; rather, they should be discussed rather than accused.
3. Is familiar with the terms and protocols associated with misconduct (e.g. falsification or actions against the law). If involved in an investigation into misconduct, adheres to the guidelines.
4. Upholds confidentiality throughout the investigation but, as soon as the results are available, honestly notifies the relevant parties and stakeholders.
5. Encourages everyone involved, including those who report potential scientific errors or misconduct, to make the necessary efforts to resume their careers as soon as possible after an investigation of misconduct.
6. Prevents and works to prevent reprisals against or harm to the employability of individuals who responsibly bring attention to potential scientific errors or other professional misconduct.

14.9.5 Responsibilities of Employers

Those employing any person to analyze data are implicitly relying on the profession's reputation for objectivity. Nonetheless, this places a duty on the employer to comprehend and honor statisticians' duty of objectivity.

Employers of statisticians are anticipated to:

1. Recognize that ethical guidelines exist and were instituted for the protection and support of the statistician and the consumer alike.
2. Ensure that there is no intimidation or discrimination in the workplace based on individual traits, compulsion, unwanted physical contact (including sexual contact), or other forms of harassment.

3. Recognize that valid findings result from competent work in a moral environment. For any data analysis, employers, funders, or those who commission statistical analysis have an obligation to rely on the knowledge and discretion of qualified statisticians.

   This duty could be particularly pertinent in analyses that are known or expected to have tangible physical, financial, or psychological effects.
4. Recognize that the results of valid statistical studies cannot be guaranteed to conform to the goals or expectations of the statistical practitioner(s) or the people who commissioned the study. Acknowledge that reporting or relying solely on results that meet expectations is against these guidelines and that decisions about which results to report, use, or cite must be made based on competing findings.
6. Acknowledge statistical practitioners as authors or contributors to projects or publications only with their express consent, as this implies support for the work.
7. Encourage rigorous statistical analysis and call out dishonest or corrupt statistical practices.
8. Strive to protect the professional freedom and responsibility of statistical practitioners who comply with these guidelines.

14.10 Future Line of Work

Education is an ongoing journey that necessitates a persistent desire for both faculty and students to engage with new concepts, deepen their understanding, and apply their knowledge effectively to current situations and environments. This book is proposed as part of the curriculum for undergraduate and postgraduate students pursuing studies in life sciences. Both educators and learners seek relevant and appropriate examples in the future to facilitate a meaningful grasp of the concepts presented. To enhance comprehension, we have included a chapter dedicated to the diagrammatic or pictorial representation of statistics within the context of life sciences. Moving forward, it is essential to develop and test a greater number of relevant examples. Topics such as measures of central tendency, variability, and probability will be examined through innovative illustrations tailored to the life sciences field. Additionally, advanced concepts like non-parametric regression, bootstrapping, and small-area statistics will be investigated. Statistical inference, encompassing hypothesis testing and estimation, can be simplified through the use of suitable examples and illustrations relevant to life sciences. With the rise of computer science and technology, the application of statistical software should be developed, tested, and implemented. Practical applications in genetics, particularly in molecular genetics and genetic engineering, will be introduced using appropriate software such as R, STATA, and SAS. Given the challenges posed by data scarcity, the importance of sampling and its size cannot be overstated for research. Furthermore, machine learning and artificial intelligence tools, accompanied by interdisciplinary research in life sciences, should be integrated and enhanced within the postgraduate curriculum. An increasing number of meta-analyses utilizing life sciences data will be incorporated into the teaching framework. Project work on these subjects may be proposed and presented as seminars, forming an integral part of the course curriculum. As some students today may pursue administrative roles in the future, it is crucial to equip them with the necessary skills and knowledge.

REFERENCES

Andrade, C. Sample size and its importance in research. *Indian Journal of Psychological Medicine* **42**, 102–103 (2020).

Brenner, D. R. et al. Methods sources of data statistical analysis ethics approval. *CMAJ* **192**, 1–7 (2020).

Correia, M. I. T. D. Ethics in research. *Clinical Nutrition Open Science* **47**, 121–130 (2023).

Drolet, M. J., Rose-Derouin, E., Leblanc, J. C., Ruest, M. & Williams-Jones, B. Ethical issues in research: Perceptions of researchers, research ethics board members and research ethics experts. *Journal of Academic Ethics* **21**, 269–292 (2023).

Georgiou, A. V. How to fund official statistics production to support statistical principles and ethics. *Statistical Journal of the IAOS* **38**, 425–434 (2022).

Hernández-Zambrano, D., Herrera, W., Moreno Barbosa, E., Guzmán Botero, A. & Baquero Quevedo, R. Building ethical guidelines to produce official statistics: the statistical ethics system (SETE) for the national administrative department of statistics (DANE) in Colombia. *Journal of Global Ethics* **19**, 410–425 (2023).

Heston, T. F. Statistical ethics in medical research: A narrative review. *Journal of Clinical Medical Research* 1–10 (2023). https://doi.org/10.46889/jcmr.2023.4308

Kambhampati, S. B. S., Menon, J. & Maini, L. Ethics in research and publications. *Indian Journal of Orthopaedics* **57**, 1722–1734 (2023).

Lesser, L. M. & Nordenhaug, E. Ethical statistics and statistical ethics: Making an interdisciplinary module. *Journal of Statistics Education* **12**, 345–351 (2004).

Miranda-Novales, M. G. & Villasís-Keever, M. Á. Research protocol VIII. The ethics of research on human subjects. *Revista Alergia Mexico* **66**, 115–122 (2019).

Razak, M. M., Othman, M. F., Kutty, R. M., Mohamed, N. N. N. & Rahim, N. A. Statistical methods for ICT ethics studies, in *3rd International Conference on Cybernetics and Intelligent Systems, ICORIS 2021* (Institute of Electrical and Electronics Engineers Inc., 2021). https://doi.org/10.1109/ICORIS52787.2021.9649488

Saltelli, A. Ethics of quantification or quantification of ethics? *Futures* **116**, 102509 (2020).

Utts, J. Enhancing data science ethics through statistical education and practice. *International Statistical Review* **89**, 1–17 (2021).

Wester, K. L., Deaton, J. D. & Jayamohan, H. in *Reimagining Research: Engaging Data, Research, and Program Evaluation in Social Justice Counseling*, 45–72 (Taylor and Francis, 2023). https://doi.org/10.4324/9781003196273-3

Williamson, P. et al. Statistical review by research ethics committees. *Journal of the Royal Statistical Society. Series A: Statistics in Society* **163**, 5–13 (2000).

Index

Note: **Bold** page numbers refer to tables and *italic* page numbers refer to figures.

For Product Safety Concerns and Information please contact our
EU representative GPSR@taylorandfrancis.com Taylor & Francis
Verlag GmbH, Kaufingerstraße 24, 80331 München, Germany